Mayayuana

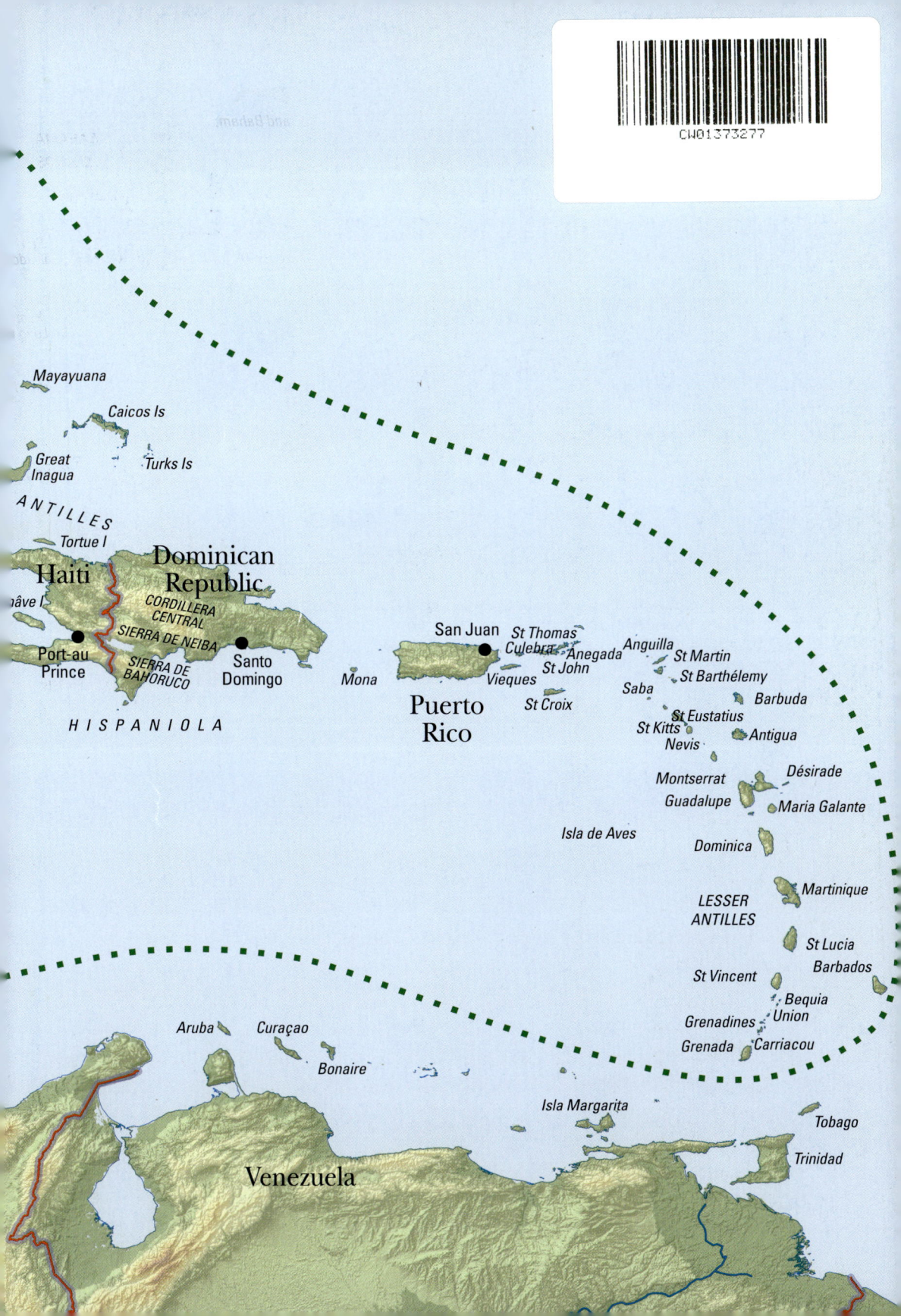

Caicos Is

Great Inagua

Turks Is

ANTILLES

Tortue I

Haiti

Dominican Republic

CORDILLERA CENTRAL
SIERRA DE NEIBA
SIERRA DE BAHORUCO

Port-au-Prince

Santo Domingo

HISPANIOLA

Mona

San Juan

Puerto Rico

St Thomas
Culebra
Vieques
St John
St Croix

Anegada
Anguilla
St Martin
St Barthélemy
Saba
St Eustatius
St Kitts
Nevis

Barbuda
Antigua

Montserrat
Guadalupe
Désirade
Maria Galante

Isla de Aves

Dominica

LESSER ANTILLES

Martinique

St Lucia
Barbados

St Vincent
Bequia
Grenadines Union
Grenada Carriacou

Aruba
Curaçao
Bonaire

Isla Margarita

Tobago
Trinidad

Venezuela

Birds of the West Indies

Guy M. Kirwan, Anthony Levesque, Mark Oberle & Christopher J. Sharpe

Lynx and BirdLife International Field Guides

Birds of the West Indies

Guy M. Kirwan, Anthony Levesque, Mark Oberle & Christopher J. Sharpe

with
Gustavo A. Rodríguez, Alex Berryman, Nárgila Moura, Juan F. Freile & Chris Bradshaw

Colour plates by

Richard Allen
Norman Arlott
Hilary Burn
Clive Byers
Martin Elliott
Al Gilbert
Alan Harris
Ren Hathway
Mark Hulme
Àngels Jutglar
Francesc Jutglar
Ian Lewington
Toni Llobet
Alex Mascarell
Dave Nurney

Douglas Pratt
David Quinn
Chris Rose
Lluís Sanz
Brian Small
Lluís Solé
Juan Varela
Ilian Velikov
Etel Vilaró
Lyn Wells
Jan Wilczur
Ian Willis
Martin Woodcock
Tim Worfolk

Lynx

BirdLife INTERNATIONAL

For Mike Flieg, who loved the West Indies and the islands' birds, and was taken from his family too soon.—GMK

For contributions by family, see CONTENTS:

Text preparation
- AB — Alex Berryman
- CB — Chris Bradshaw
- JFF — Juan F. Freile
- GMK — Guy M. Kirwan
- NM — Nárgila Moura
- GAR — Gustavo A. Rodríguez
- CJS — Christopher J. Sharpe

Illustrations
- RA — Richard Allen
- NA — Norman Arlott
- HB — Hilary Burn
- CB — Clive Byers
- ME — Martin Elliott
- AG — Al Gilbert
- AH — Alan Harris
- RH — Ren Hathway
- MH — Mark Hulme
- ÀJ — Àngels Jutglar
- FJ — Francesc Jutglar
- IL — Ian Lewington
- TL — Toni Llobet
- AM — Alex Mascarell
- DN — Dave Nurney
- DP — Douglas Pratt
- DQ — David Quinn
- CR — Chris Rose
- LS — Lluís Sanz
- BS — Brian Small
- LSo — Lluís Solé
- JV — Juan Varela
- IV — Ilian Velikov
- EV — Etel Vilaró
- LW — Lyn Wells
- JW — Jan Wilczur
- IW — Ian Willis
- MW — Martin Woodcock
- TW — Tim Worfolk

First Edition: June 2019

© Lynx Edicions – Montseny, 8, 08193 Bellaterra, Barcelona, www.lynxeds.com

Recommended citation:
Kirwan, G.M., Levesque, A., Oberle, M. & Sharpe, C.J. (2019). *Birds of the West Indies*. Lynx and BirdLife International Field Guides. Lynx Edicions, Barcelona.

Project co-ordinators: Arnau Bonan-Barfull, Amy Chernasky, Guy M. Kirwan
Map production: Christian Jofré, Anna Motis
Cover design: Susanna Silva
Interior book design: Xavier Ruiz

Cover illustration by Ilian Velikov
Blue-headed Quail-dove (*Starnoenas cyanocephala*)

Printed and bound in Barcelona by Ingoprint, S.A.
Legal Deposit flexi bound: B 15614-2019
Legal Deposit hardcover: B 15615-2019
ISBN flexi bound: 978-84-16728-17-6
ISBN hardcover: 978-84-16728-18-3

All rights reserved. No form of reproduction, distribution, public communication or transformation of this work may be carried out without the authorization of its copyrights holders, except that foreseen by the law. Those needing to photocopy or electronically scan any part of this work should contact Lynx Edicions.

CONTENTS

INTRODUCTION 7
USING THE FIELD GUIDE 19
ABOUT THE AUTHORS 23
ACKNOWLEDGEMENTS 23
BIRDLIFE INTERNATIONAL 24

SPECIES ACCOUNTS
    Chachalacas Cracidae – AB; *ÀJ* 26
    Guineafowl Numididae – AB; *FJ* 26
    New World Quails Odontophoridae – AB & GMK; *ÀJ* 26
    Pheasants Phasianidae – AB; *FJ, ÀJ, EV* 28
    Ducks, Geese and Swans Anatidae – CB & GMK; *FJ, ÀJ, AM* 28
    Grebes Podicipedidae – AB & GMK; *AM, LSo* 46
    Flamingos Phoenicopteridae – AB; *FJ* 46
    Tropicbirds Phaethontidae – GAR; *LS, JV* 48
    Pigeons and Doves Columbidae – AB & GMK; *HB, CB, ME, AM, JV, LW, JW* 48
    Potoos Nyctibiidae – GMK; *DP* 56
    Nightjars Caprimulgidae – GMK & CJS; *ÀJ, DN* 58
    Swifts Apodidae – GMK; *IL, AM* 62
    Hummingbirds Trochilidae – AB & GMK; *NA, HB, TL, DN, JW* 66
    Cuckoos Cuculidae – GAR & GMK; *JW, IW* 74
    Rails, Gallinules and Coots Rallidae – AB; *NA, HB* 80
    Limpkin Aramidae – AB; *LS* 84
    Cranes Gruidae – AB; *HB* 84
    Loons/Divers Gaviidae – AB; *ÀJ* 84
    Southern Storm-petrels Oceanitidae – GAR; *JV* 86
    Northern Storm-petrels Hydrobatidae – GAR; *JV* 86
    Albatrosses Diomedeidae – GAR & GMK; *FJ* 86
    Petrels and Shearwaters Procellariidae – GAR, GMK & CJS; *AM, JV* 88
    Storks Ciconiidae – AB; *FJ, AM* 92
    Ibises and Spoonbills Threskiornithidae – AB & GMK; *FJ, AM* 94
    Herons Ardeidae – AB & GMK; *FJ, AM, JV, IV* 96
    Pelicans Pelecanidae – GAR; *FJ* 104
    Frigatebirds Fregatidae – GAR; *FJ* 104
    Gannets and Boobies Sulidae – GAR; *FJ, JV* 106
    Cormorants Phalacrocoracidae – AB; *LS* 108
    Darters Anhingidae – AB; *LS* 108
    Thick-knees Burhinidae – CJS; *AM, IW* 108
    Oystercatchers Haematopodidae – CJS; *IW* 108
    Avocets and Stilts Recurvirostridae – CJS; *FJ, AM, IW* 110
    Plovers Charadriidae – CJS; *FJ, ÀJ, AM, JV, EV* 110
    Jacanas Jacanidae – CJS & GMK; *HB, AM* 116
    Sandpipers and allies Scolopacidae – CJS; *FJ, AM, JV, EV* 116
    Coursers and Pratincoles Glareolidae – CJS; *HB* 132
    Gulls and Terns Laridae – GAR & GMK; *FJ, IL, JV, IW* 132
    Skuas Stercorariidae – AB; *IL, JV* 148
    Auks Alcidae – AB; *CR* 152
    Barn-owls Tytonidae – AB; *MH* 152
    Typical Owls Strigidae – AB & GMK; *HB, CB, IL, IW* 154
    New World Vultures Cathartidae – GAR; *FJ, ÀJ* 158
    Osprey Pandionidae – GAR; *LS, JV* 158
    Hawks and Eagles Accipitridae – GAR & GMK; *NA, AH, FJ, IL, AM, JV, IW* 158
    Trogons Trogonidae – GMK; *RA* 168
    Bee-eaters Meropidae – AB; *CR* 168
    Todies Todidae – GMK; *DP* 170
    Kingfishers Alcedinidae – AB; *TW* 172
    Toucans Ramphastidae – GMK; *AG* 172
    Woodpeckers Picidae – GMK; *HB, MH, IW* 172
    Falcons and Caracaras Falconidae – GAR; *HB, FJ, JV, IW* 180

Cockatoos Cacatuidae – AB; *IL* 182
Parrots Psittacidae – AB, GMK & CJS; *NA, ÀJ, LS, EV, MW* 182
Tityras and allies Tityridae – GMK; *CR* 192
Tyrant-flycatchers Tyrannidae – GAR; *NA, HB, IL, IW* 192
Vireos Vireonidae – JFF & GMK; *BS* 208
Shrikes Laniidae – GMK; *TW* 218
Crows Corvidae – GMK; *DQ* 218
Larks Alaudidae – GMK; *TW* 220
Swallows and Martins Hirundinidae – JFF & GMK; *HB, AM* 220
Nuthatches Sittidae – GMK; *HB* 228
Gnatcatchers Polioptilidae – GMK; *HB* 228
Wrens Troglodytidae – GMK; *HB* 228
Starlings Sturnidae – GAR; *HB* 232
Mockingbirds and Thrashers Mimidae – GAR; *IL* 232
Thrushes Turdidae – NM; *NA, RH, IW, TW* 236
Old World Flycatchers and Chats Muscicapidae – GAR; *TW* 246
Kinglets and Firecrests Regulidae – GAR; *HB* 246
Palmchat Dulidae – GMK; *DP* 246
Waxwings Bombycillidae – GAR; *CR* 246
Weavers Ploceidae – GAR; *TW* 248
Waxbills Estrildidae – GAR; *HB* 248
Whydahs Viduidae – GAR; *HB* 254
Old World Sparrows Passeridae – GAR; *TW* 254
Pipits and Wagtails Motacillidae – GAR; *RH* 254
Finches Fringillidae – GAR & GMK; *NA, HB* 256
Longspurs Calcariidae – GMK; *DQ* 260
New World Sparrows Passerellidae – GAR & GMK; *DQ, BS* 260
Cuban Warblers Teretistridae – GMK; *DQ* 268
New World Blackbirds Icteridae – NM & GMK; *DQ, TW* 268
New World Warblers Parulidae – NM & GMK; *RH, DQ* 284
Hispaniolan Tanagers Phaenicophilidae – GMK; *HB, DQ* 310
Spindalises Spindalidae – GMK; *HB* 312
Puerto Rican Tanager Nesospingidae – GMK; *HB* 316
Chat-tanagers Calyptophilidae – GMK; *HB* 316
Cardinals Cardinalidae – JFF & GMK; *HB, BS* 316
Tanagers Thraupidae – JFF & GMK; *HB, BS* 322

APPENDIX 1. REGIONAL CHECKLIST 335
APPENDIX 2. SPECIES/TAXA NOT INCLUDED 375
REFERENCES AND FURTHER READING 377
ENGLISH AND SCIENTIFIC INDEX 385
QUICK INDEX 400

INTRODUCTION

GENERAL INTRODUCTION

In attempting a guide of this nature, we must accept both the debt and burden of those simultaneously blessed and unfortunate to 'stand on the shoulders of giants'. Catherine Levy (2008) in her recent paper on the history of ornithology in the region (to which interested readers should refer) traces the earliest contributions on West Indian birds to the early 1500s, although it almost goes without saying that most information prior to the advent of Linnean nomenclature is principally anecdotal, including multiple reports of endemic macaws, the majority of which almost certainly involve more myth than reality (e.g. Wiley & Kirwan 2013, Lenoble 2015). With the turn of the 19th century, however, Caribbean ornithology started to acquire a far more solid footing.

Given such a long and rich history, covering 27 different countries and territories, it would be rather invidious and potentially futile to attempt the only superficial résumé of the region's ornithological history possible in the space available, and impossible to detail every relatively significant contribution to the canon. In any case, as just mentioned, our colleague Levy has already rather done this job for us. Suffice to say that not only do we benefit from riding a wave of ornithological exploration that stretches back to the likes of Gosse and March (in Jamaica), Gundlach and Poey (Cuba, the former also in Puerto Rico), Alexander Wetmore (Hispaniola and Puerto Rico), Stuart Danforth (widespread contributions from across the region), Maurice and Hazel Hutt (Barbados), and a host of others who have attempted overviews of the avifaunas of one or more of the West Indies, but our efforts will also be judged against previous regional works.

Born in 1900 and following his first contribution on the region's birds in 1927 (the description of a new subspecies of Golden Warbler on St Lucia), in 1936 James Bond of the Academy of Natural Sciences of Philadelphia published his first field guide to the *Birds of the West Indies*, which was unique compared to subsequent editions in that he treated subspecies (as indeed we do too). Between then and his death, in 1989, Bond updated the book seven times, in addition to producing four editions of his *Check-list of Birds of the West Indies*, and augmented the information in the latter's fourth incarnation in 1956 via subsequent near-annual supplements. Bond's field guide revolutionized birdwatching in the Caribbean, while his checklist and their supplements remain invaluable to the region's ornithologists fully more than 30 years after the last was published.

Nevertheless, in 1998, Herb Raffaele, together with his wife Janis, the Cuban professional tennis player and outstanding all-round naturalist Orlando Garrido (arguably the father of Cuban ornithology, if Gundlach be its grandfather), and American ornithologists Allan Keith and the late Jim Wiley, produced a new handbook-sized work to West Indian birds. This was followed in 2003 by a true field guide by the same team using the same high-quality artwork. Raffaele had already published an earlier guide to Puerto Rico and the Virgin Islands (1983, revised 1989), while Keith produced two keynote works, to the birds of St Lucia (Keith 1987) and Hispaniola (Keith *et al.* 2003), the latter with Wiley, Steve Latta and Joe Wunderle. Wiley's own staggering contribution to Caribbean ornithology (see Kirkconnell 2019), in the modern age rivalling Bond himself, could be said to have pinnacled in his 800-page bibliography to the region (Wiley 2000), a work that has remained close at hand throughout the gestation of the present volume.

Viewed against such illustrious predecessors, we might be deemed nigh-on impudent for believing that we might produce something even comparable. And yet, here we stand. Our aim has been to produce a guide that builds on the best traditions of those that preceded us; consequently, we have aimed for a book that is at once attractive, portable and discusses many of the key identification issues, but is also scientifically accurate, in the best traditions of James Bond.

The present work covers just over 700 species, of which a quite remarkable c. 190 are endemic to the West Indies, many of them to single islands. Furthermore, six families are confined to the region, all restricted to the Greater Antilles, several of them recent taxonomic 'creations' as the result of detailed genetic investigations: Todidae (todies, five species, throughout the large islands), Dulidae (one species, Palmchat, confined to Hispaniola), Teretistridae (Cuban warblers, two species, as their name implies only in Cuba), Phaenicophilidae (Hispaniolan tanagers, four species, all on Hispaniola), Nesospingidae (one species, Puerto Rican Tanager) and Calyptophilidae (chat-tanagers, two species, both on Hispaniola). Another family is virtually confined to the West Indies, Spindalidae (spindalises, four species). In addition, many genera are endemic, among them *Starnoenas* (Blue-headed Quail-dove, recently suggested to represent the sole member of a subfamily grouping), *Siphonorhis* (poorwills, but just one definitely extant species), *Eulampis* (two species of hummingbirds), *Cyanophaia* (Blue-headed Hummingbird), *Cyanolimnas* (Zapata Rail), *Pseudoscops* (Jamaican Owl), *Margarobyas* (Bare-legged Screech-owl), *Priotelus* (Cuban Trogon, sometimes also Hispaniolan Trogon), *Nesoctites* (Antillean Piculet), *Xiphidiopicus* (Cuban Green Woodpecker), *Ferminia* (Zapata Wren), *Allenia* (Scaly-breasted Thrasher), *Torreornis* (Zapata Sparrow), *Nesopsar* (Jamaican Blackbird) and *Euneornis* (Orangequit).

Consequently, for birders of whatever 'stripe'—family collectors, world listers or those fascinated by taxonomic 'oddballs'— the West Indies couples a hotbed of avian endemism with an *al fresco* climate, generally safe travel opportunities and welcoming peoples.

GEOGRAPHICAL SCOPE

Covering approximately 115 islands, but not including rocky islets and small, often sandy cays (Cuba alone possesses more than 1000 of the latter!), the land area of the West Indies is c. 240,000 km^2, of which almost half (105,000 km^2) belongs to Cuba. Three major physiographic divisions constitute the West Indies: the Greater Antilles, comprising the islands of Cuba, Jamaica, Hispaniola (Haiti and the Dominican Republic) and Puerto Rico; the Lesser Antilles, including the Virgin Islands, Anguilla, St Martin/St Maarten, St Barthélemy, Saba, St Eustatius, St Kitts and Nevis, Antigua and Barbuda, Montserrat, Guadeloupe, Dominica, Martinique, St Lucia, St Vincent and the Grenadines, Barbados and Grenada; and the isolated island groups of the North American continental shelf—the Bahamas and the Turks and Caicos Is—and those of the South American shelf, including Trinidad and Tobago, Aruba, Curaçao and Bonaire. However, in terms of their fauna and flora, biogeographers consider all of those islands lying on the

South American shelf to be more closely related to the continent, and consequently exclude them from faunal treatments of the West Indies. This has been as true for birds as any other group, and we do nothing more than follow Bond and others in defining the southernmost extent of the region. This is true too of our approach to the Colombian archipelago of San Andrés, Providencia and Santa Catalina, in the south-west corner of the region, which we, like Bond and Raffaele et al., consider to be unequivocally Antillean given the obvious affinities of its endemic birds. All commentators, unsurprisingly, have treated the Cayman Is as part of the West Indies based on both geography and fauna, but the Swan Is (Honduras) were excluded from the region by Raffaele et al. Again in unison with Bond, we have no hesitation about including them, given that Vitelline Warbler is unique to the Swan and Cayman Is, and the Swan Is also formerly supported a population (and perhaps even separate race) of Red-legged Thrush, a species group unique to the West Indies. We also cover the Venezuelan islet of Isla de Aves, 185 km south-west of Montserrat, which was not explicitly treated in previous guides.

The shape and alignment of the Greater Antilles are determined by an ancient chain of folded and faulted mountains that originally extended west to east from Central America through the region. Nowadays, it is now mostly submerged, but remnants are visible in the Blue Mts of Jamaica, the Sierra de los Órganos and Sierra Maestra at the western and eastern extremities, respectively, of Cuba, and Mt Duarte, in the Dominican Republic, which at 3175 m is the highest peak in the Caribbean. Running north to south, another mostly submerged mountain chain forms the double arc comprising the Lesser Antilles. Stretching from St Kitts to Grenada, the mountainous inner arc consists of volcanic cones, some still active. The outer arc—between Anguilla and Barbados—comprises low, flat islands with limestone surfaces overlying older volcanic or crystalline rocks.

While the faunal division between the Greater and Lesser Antilles is typically considered to be straightforward, in fact it is subject to debate, some vague statements, and is potentially even dynamic; there is certainly a lack of unanimity on this issue. The Virgin Is have often been considered to be part of the Lesser Antilles. Nevertheless, it has been suggested, with a reasonable degree of authority, that the deep-water Anegada Passage, which was apparently formed by the fault between the islands of Anegada and Sombrero, and many of the northern Virgin Is west of Anegada are sited on the relatively shallow Puerto Rican Bank, and were connected to Puerto Rico in the geologically recent past. Allocation of St Croix, the southernmost of the Virgin Is, is more problematic, it being separated from the Puerto Rican Bank to the west and the Anguilla Bank to the east by deep channels.

Comparatively few Greater Antillean endemic taxa persist as far east as the Virgin Is, among them Puerto Rican Mango (declining), Puerto Rican Screech-owl (probably extinct there) and Puerto Rican Flycatcher (still widespread). On the other hand, this part of the region, which might almost be viewed as a transition zone, has been invaded as a result of the recent northward spread of Lesser Antillean birds, such as Lesser Antillean Bullfinch. Although not biogeographically significant, the Virgin Is also represent the south-easternmost point at which many Nearctic–Neotropical passerine migrant birds are still reasonably numerous in winter; further south, their numbers tend to decline noticeably.

The final issue to mention is that the Old Bahama Channel, despite being just 19–24 km wide, is generally considered to represent a boundary between Cuba and thus the Neotropics (to the south) and the Holarctic, represented by the Bahamas, to the north. We concur with this view, widely accepted since the publication of Voous (1973), although it is nonetheless important and interesting to note the degree of faunal interchange between the Bahamas and Cuba. So, for example there are relict (?) populations of Bahama Mockingbird in northern Cuba (and Jamaica), Bahamian populations of Bananaquit are perhaps invading the north-central Cuban cayerías, and two undeniably closely related taxa of lizard-cuckoos (usually treated as subspecies, but here as species) occur either side of the channel. Nevertheless, none of these facts impacts our recognition of this narrow strait as a significant biogeographic divide.

Note that in the main text, use of the term 'Bahamas' often refers to the Commonwealth of The Bahamas and the UK Overseas Territory of the Turks & Caicos Is, in recognition of their shared geological and biogeographic characteristics. This implies no political opinion.

CLIMATE

The West Indies have a tropical maritime climate. All of the islands, especially the larger ones, are arguably subject to as many, if not more, local variations in temperature than seasonal differences. Seasonal variation is most apparent in rainfall, rather than temperature shifts. Daily maximum temperatures (°C) over most of the region range from the upper 20s in December to April, to the low 30s in May to November. Nocturnal temperatures are c. 6°C cooler. In Cuba, for example, annual mean temperature ranges from 24°C to 27.6°C, with regional monthly mean differences of c. 5°C and c. 7°C, while means from various islands in the Bahamas range between 24.3°C and 27.8°C. Similar ranges are registered on the Cayman Is, Hispaniola and St Lucia. Nevertheless, in Cuba, temperatures can reach 40°C in summer (and even 43°C in the Dominican Republic) but have dipped as low as 1–2°C on Pico Turquino, Cuba's highest peak, in winter. Sub-zero temperatures have occasionally been recorded in the Dominican Republic and are perhaps regular in the country's highest mountains. Such conditions are unlikely to be experienced on any of the other islands in the region, even in the highest areas of Jamaica. For instance, the lowest recorded temperature on Grand Cayman is c. 11°C. The coolest months on Jamaica are likely to be December to February, but this is relative.

Most islands experience a wet and a dry season; annual rainfall totals typically range from 300 to 2000 mm (e.g. typically 740–1360 mm across the Bahamas, 1100–1800 mm in the Cayman Is and 1300–1800 mm in Barbados), but reach up to 7500 mm on the highest peaks, with some of the areas of highest precipitation being Jamaica's John Crow Mountains, the Moa / Baracoa region of eastern Cuba, and parts of the Massif de la Hotte in Haiti. Ironically, one of the driest, southern Guantánamo province, is also in eastern Cuba, and north-west Haiti also generally experiences very little rainfall, as do those parts of Hispaniola in the rain shadow of the higher mountains. Mean annual rainfall scarcely exceeds 1100 mm anywhere on the Cayman Is. In Cuba and Jamaica, there is also a trend for rainfall to increase away from the coast. The region's moisture-laden trade winds produce heavy rainfall on the windward sides of the higher islands. Relative humidity is high throughout the year. The dominant wind direction is north-east (85% of the time on St Lucia), whilst relative humidity is 75% on St Lucia, 80% in Cuba and can reach almost 87% in some parts of the Dominican Republic. Drier regions may experience relatively lower average humidity, exceptionally values below 40% are experienced, albeit rarely, on the Cayman Is.

It can rain at any season (so a small umbrella is always useful to carry), but downpours rarely occur in the dry season (October–April in Cuba, November–April on Hispaniola and February–April on Barbados). May and October are generally the wettest months on Jamaica, while June–November represents the wet season in the southern Lesser Antilles, e.g. in St Lucia and Barbados. Given that it can be cool in montane regions, such as La Güira on Cuba, the Blue Mountains of Jamaica, and in the south-west Dominican Republic on winter mornings, cool-weather clothing is advisable. A light jacket will also be useful if travelling to Zapata between late November and late February; evenings and nights can be quite surprisingly and deceptively cool in this region at this time of year. Cold fronts from North America can bring storms and lower temperatures to the Bahamas, Cuba and the Cayman Is, much more rarely the other islands, between December and March.

Hurricanes pass through the Caribbean region almost annually in June–October, occasionally as late as November. Winds in these storms regularly reach almost 120 km/h and up to 500 mm of rain can fall in just 24 hours. The low-lying Cayman Is can be extensively inundated by high seas during tropical storms, one of which, Hurricane Flora, brought winds of up to 320 km/hour to Hispaniola. Fortunately, the chances of witnessing one at any given location are very low, but most birders will want to avoid visiting the region at this season.

HABITATS

Tropical and subtropical humid broadleaved forests

These grow in areas of high annual rainfall (2000 mm), especially on windward mountain slopes, such as in eastern Cuba, northern Jamaica, eastern Hispaniola, northern Puerto Rico, and at higher elevations in the Lesser Antilles; also in some lowlands subject to prevailing north-easterly winds. These forests are evergreen, often with conspicuous presence of mosses, ferns and epiphytes. At higher elevations, they can receive as much moisture horizontally (from mist) as from rainfall, in which case they are known as cloud forests. These latter are the primary wintering habitat of the North American-breeding Bicknell's Thrush. Much of the original humid forest on the West Indian islands has been lost, and the remaining tracts persist largely in montane regions, where they are a refuge for threatened endemics such as Red-necked, St Lucia, Imperial and St Vincent Amazons, and Semper's, Whistling and Elfin Woods Warblers, although the first-named of these three warblers has not been definitely recorded for several decades.

Tropical and subtropical dry broadleaved forests

Found in areas of lower (1300 mm per year) and seasonal rainfall throughout, particularly on the leeward slopes of mountains. More open in structure than humid forests, tree species are semi-deciduous or deciduous, and there is often a well-developed understorey. Climate, fertile soils and adequate rainfall make these forests attractive for agriculture, and their area has been greatly reduced. Characteristic birds of dry forest include Caribbean Dove, Grenada Dove, White-breasted and St Lucia Thrashers, and Puerto Rican Nightjar.

Tropical and subtropical pine forests

Lowland pine forests occur on poor soils in the Bahamas and Turks & Caicos, with both lowland and montane pine forests in Cuba and Hispaniola. Characteristic birds of Bahamian pine barrens include Bahamas Hairy Woodpecker and the now perilously close to extinction Bahama Nuthatch, while montane pine forests in the Greater Antilles are favoured by Hispaniolan Pewee, Hispaniolan Palm Crow, Olive-capped Warbler, and Hispaniolan Crossbill. Formerly, they were also a final bastion of the potentially extinct Cuban Ivory-billed Woodpecker.

Savannas

Natural savannas occur in areas of seasonal drought in the Bahamas, Cuba and Hispaniola. Burning and clearance by humans has led to a great increase in their area, and they are now found on all islands where natural vegetation has been cleared. Savanna specialists include Double-striped Thick-knee, Burrowing Owl and Palmchat.

Karst forest

Restricted to limestone outcrops in the Greater Antilles, notably in the *mogotes* of Puerto Rico, the Cordillera de Guaniguanico in western Cuba and in Cockpit Country in Jamaica. Floral components tend to be highly endemic. Typical bird species in the second-named country include Ring-tailed Pigeon, Crested Quail-dove, Chestnut-bellied Cuckoo, Black-billed Amazon, Jamaican Crow and Rufous-throated Solitaire.

Shrubland

This type of vegetation is either a product of dry climatic, or impoverished soil, conditions or the result of forest removal. It is widespread, for example, in severely deforested Haiti, where it covers about one-third of the land area. The resultant avifauna tends to be a mix of dry forest birds and those of non-forest habitats.

Dry thorn and cactus scrub

Found in drier areas of Cuba, throughout the Lesser Antilles and on many small islands. Characteristic species include a trio of warblers, namely Vitelline Warbler (Cayman Is and Swan Is), Adelaide's Warbler (Puerto Rico and recently some of the Virgin Is) and the eponymous Barbuda Warbler. In eastern Cuba, species like Cuban Grassquit and Zapata Sparrow can be common in such habitat, while in Dominican Republic the endemic Least Poorwill and Flat-billed Vireo are just two of the birds found in these areas. The probably extinct Jamaican Poorwill might also have been found in very dry habitats, and Bahama Mockingbird still is on the same island.

Mangroves

Mangrove is a generic term applied to a suite of tree species that grow in tidal conditions around tropical coasts. Four species of mangrove occur in the West Indies, displaying different tolerances to salinity and submersion. The region is home to some of the most extensive and important mangrove forests in the world, with Cuba alone harbouring at least 5300 km². Mangroves provide nesting sites for boobies and Magnificent Frigatebirds, as well as vital stopover and wintering habitat for migrants, especially waders. Typical species found in mangrove include West Indian Whistling-duck, White-crowned Pigeon, Puerto Rican Emerald, Mangrove Cuckoo, Clapper Rail, Cuban Black Hawk, Thick-billed Vireo, Martinique Oriole and Golden Warbler.

Beaches

The large number of Caribbean beaches offer key sites for nesting waders, particularly Wilson's Plover, as well as wintering habitat for several North American breeders, the most important of which is Piping Plover.

Coastal and freshwater wetlands

The 5000 km² Zapata Swamp, home to the eponymous rail, wren and sparrow, and much else besides, is the region's largest wetland of international importance, and designated a Ramsar Site. It comprises freshwater, brackish and saline environments. Inland lakes, reservoirs and pools are a feature of all of the islands, providing habitat for a diversity of resident waterbirds as well as migrant waterfowl and waders. Artificial reservoirs, desalination and sewerage treatment plants have increased the availability of freshwater habitats, providing birders with welcome opportunities to find vagrant birds close to human habitation.

Marine habitat

Marine environments account for the lion's share of the surface area of the West Indies. As a bird habitat, the Caribbean Sea remains little known due to its vast area, relative inaccessibility of pelagic regions, low average densities of birds, their often exclusively seasonal presence and difficulties in identification. Some exciting records are emerging from seawatches in, for example, Guadeloupe. To complement direct observations, the use of modern tracking technologies is just starting to reveal the use that seabirds make of the marine environment. This is particularly true of some jaegers and skuas, Arctic Tern, storm-petrels and the globally threatened Black-capped Petrel, which is endemic as a breeder to the region.

Agricultural land

Makes up the majority of land coverage in the Greater Antilles and is home to a range of widely distributed generalist species. Sugarcane—widely planted in Cuba (especially since the Revolution) and Hispaniola—is the least attractive habitat for birds, and pastureland scarcely better, whereas mixed silvicultural systems and plantations of shade coffee and cocoa can provide relatively good analogues of natural forest cover, especially if adjoining blocks of original vegetation, and can support more complex communities and appreciable numbers of migrants.

Urban environments

Towns, cities and tourist infrastructures are home to a variety of birds, some of which are endemic to the West Indies. Cuban Martin is perhaps the most obvious example, nesting under the roof tiles in Havana's historic centre, but one of Hispaniola's endemic families, Palmchat, has adapted well to urban areas, while Cuban Blackbird and Antillean Palm-swift are two other species unique to parts of the West Indies that are regularly found in close proximity to man. One or other species of attractive red-legged thrush can usually be found in parks and gardens in the Greater Antilles and the Bahamas, while Barbados Bullfinch is common in Bridgetown. During the boreal winter, any plot of land with trees can attract migrant Nearctic warblers. Even enigmatic rarities like Stygian Owl can be seen taking advantage of streetlights in Cuba. Apart from native birds, built-up areas are often now habitat to a growing number of introduced species. Eurasian Collared-dove is as common in some towns and cities in the region as it is in its native range.

BIRD CONSERVATION IN THE WEST INDIES

Of the 550 bird species that regularly occur in the West Indies 70 (12.7%) are threatened, and another 47 are considered Near Threatened. At least six bird species—at least four parrots (there is quite some dispute over the number of macaws that may have formerly existed in the region), a hummingbird and a thrush—have become extinct since European colonization of the region, due to hunting and habitat loss.

The major contemporary threats to birds are loss of habitats, introduced predators, hunting and the cagebird trade. Driven by the demand for land, the forests that once covered most of the West Indies were cut down in many areas by sugar plantation owners for firewood to heat their refining vats. This practice resulted in soil impoverishment and erosion. The region's forests have been lost at alarming rates and now cover only a fraction of their original extent. Destruction of primary forest has also occurred due to the advance of slash-and-burn agriculture, and (in the case of Haiti) for charcoal production. Other habitats have also been affected by humans, notably wetlands which have been drained for agriculture and coastal ecosystems that have been modified by development of infrastructure for tourism. Non-native species such as mice, rats, cats and, especially, mongooses are serious predators of eggs, young and adult birds on many islands. Mongooses have been implicated in the apparent disappearance of an endemic tubenose, Jamaican Petrel, not seen since the late 19th century. Hunting for food targets mainly pigeons, wildfowl and waders, and with the precipitous declines in populations of some of the latter, has now become a factor in declines. Capture of wild birds for the cagebird trade affects parrots and a variety of songbirds.

Added to those anthropogenic threats, birds with restricted ranges, particularly single-island endemics, are vulnerable to the stochastic effects of natural phenomena like hurricanes and volcanic activity. Following a series of volcanic eruptions during the late 1990s, the population of Montserrat Oriole declined dramatically after two-thirds of its forest habitat were destroyed. Recent hurricanes have triggered fears for the survival of several of the region's endemic parrots, but many other species too, like the globally threatened Fernandina's Flicker in Cuba, are also negatively impacted by such once-freak weather events. Given that most species recover their populations after these storms within a few years, and have lived with them for millennia, it is their increasing frequency and ferocity, apparently at least in part due to our rapidly changing climate, that pose the real risk, especially to large-bodied birds that require longer periods to recoup and are least able to withstand more closely spaced hurricanes. The effects on birds of climate breakdown are increasingly well documented, but the extent to which it will exacerbate existing threats depends on securing concerted global action to address its causes.

Mindful of the need to conserve the natural systems that sustain both life and a thriving tourism industry, many West Indian countries have committed to the creation and management of national parks and entire protected areas systems, something that perhaps reaches its peak expression in Cuba, which has created a complex of more than 200 conservation units protecting more than 12% of the total land surface, several of which are internationally recognised: six Ramsar sites, six UNESCO-MAB Biosphere Reserves and two UNESCO Natural World Heritage Sites. In Haiti, the Pic Macaya and Grande Colline National Parks protect key parts of the Massif de la Hotte, a mountain range harbouring one of the world's highest densities of threatened species. Elsewhere protected areas conserve bird habitats and safeguard natural ecosystems and the services that they provide.

To guide the process, BirdLife International has identified 12,000 Important Bird and Biodiversity Areas (IBAs) throughout the world. In the West Indies, of the 268 identified (including sea area covering almost 47,500 km^2), many have now been incorporated into conservation and development planning, and national legislation. For example, IBAs have been included in the National Master Plan for Protected Areas in the Bahamas, the 2009–2013 Strategic Plan for the National Protected Areas System in Cuba, Montserrat's National Physical Development Plan, and Grenada's System Plan for Protected Areas. They are also identified as high-priority eco-zones within the Nature Policy Plan for St Eustatius, Saba and (just outside our region) Bonaire.

Single-species conservation programmes have helped increase our knowledge of, for example, Black-capped Petrel. Targeted awareness-raising campaigns to protect flagship species have been highly successful for Imperial, St Lucia and St Vincent Amazons on Dominica, St Lucia and St Vincent and the Grenadines, respectively.

The BirdLife Americas Partnership is a growing network of national conservation NGOs working for the joined-up conservation of birds from the tundra to Tierra del Fuego. In the West Indies specifically, there are currently four BirdLife Partner programmes:

Bahamas National Trust (BNT)

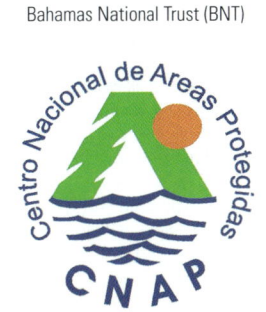

Cuba—Centro Nacional de Áreas Protegidas (CNAP)

Dominican Republic—Grupo Jaragua (GJI)

Puerto Rico (to USA)—Sociedad Ornitológica Puertorriqueña, Inc. (SOPI)

The official global conservation status, as defined by BirdLife International, and the main threats facing the most threatened West Indian birds (from a global perspective) are as follows.

Critically Endangered (Possibly Extinct)

Jamaican Poorwill—Habitat destruction, introduced predators (not recorded since 1860; endemic)
Jamaican Petrel—Introduced predators (last recorded 1879; endemic)
Eskimo Curlew—Hunting, habitat loss (last record 1963, on Barbados, where was probably regular on migration)
Bachman's Warbler—Habitat destruction, both on breeding and wintering grounds (last report 1988, last well-documented record 1964)

Critically Endangered

Grenada Dove—Habitat loss and fragmentation caused by hurricanes, clearance for infrastructural projects, grazing and predation by invasive species (endemic)
Zapata Rail—Invasive species, perhaps fires (endemic)
Cuban Kite—Habitat loss and fragmentation (endemic)
Ridgway's Hawk—Habitat loss and fragmentation, direct persecution (endemic)
Cuban Ivory-billed Woodpecker—Logging and clearance for agriculture, perhaps hunting (last record 1987; endemic subspecies)
Puerto Rican Amazon—Habitat loss (exacerbated by hurricanes), hunting for food and cagebird trade, parasitic botflies, predation (endemic)
Bahama Nuthatch—Forest loss, invasive species and hurricanes (endemic)
Bahama Oriole—Introduced species, infrastructural and agricultural development (endemic)
Semper's Warbler—Introduced mongooses, perhaps compounded by habitat loss (last definite record 1961; endemic)

Endangered

Blue-headed Quail-dove—Hunting, habitat loss (endemic)
White-fronted Quail-dove— Habitat loss, hunting, introduced predators (endemic)
Puerto Rican Nightjar—Habitat loss and fragmentation, introduced mongooses and feral cats, fire (endemic)
Bay-breasted Cuckoo—Habitat loss, hunting (endemic)
Atlantic Yellow-nosed Albatross—Longline fishery bycatch (vagrant)
Black-capped Petrel—Habitat loss and degradation, hunting, invasive predators (endemic breeder)
Gundlach's Hawk—Hunting, habitat fragmentation and loss, mining, fires (endemic)
White Cockatoo—Forest loss, cagebird trade (introduced)
Red-crowned Amazon—Cagebird trade, long-term habitat loss (introduced)
Imperial Amazon—Habitat loss (including hurricane-related damage), hunting for food, trapping for cagebird trade (endemic)
Giant Kingbird—Habitat loss, especially loss of tall trees used for nesting (endemic)
Bahama Swallow—Logging, planned housing developments (endemic)
Zapata Wren—Dry-season burning, wetland drainage, agricultural expansion, predation by introduced mongooses and rats, introduced plants (endemic)
White-breasted Thrasher / St Lucia Thrasher—Habitat clearance for agriculture, housing and tourism infrastructure (endemic)
Java Sparrow—Cagebird trade (introduced)

Hispaniolan Crossbill—Habitat loss due to logging, small-scale agriculture and uncontrolled fires (endemic)
Red Siskin—Cagebird trade (introduced)
Jamaican Blackbird—Habitat loss and fragmentation due to bauxite mining, agricultural expansion and the charcoal industry (endemic)
Yellow-shouldered Blackbird—Habitat loss due to residential and tourist development and agriculture, cowbird parasitism (endemic)
Whistling Warbler—Habitat loss and fragmentation (endemic)
Elfin Woods Warbler—Habitat loss and degradation (endemic)
Golden-cheeked Warbler—Habitat loss in both breeding and wintering grounds (vagrant)
St Lucia Black Finch—Agricultural clearance, introduced predators (endemic)

Vulnerable

West Indian Whistling-duck—Hunting, habitat loss and degradation (endemic)
Long-tailed Duck—Oil spills, gillnet fishery bycatch, hunting, avian cholera and influenza (vagrant)
Common Pochard—Loss of breeding habitat, hyper-eutrophication, hunting, human recreation activities, urban development, habitat destruction (vagrant)
Ring-tailed Pigeon—Illegal hunting, logging, clearance for plantation agriculture (endemic)
Grey-headed Quail-dove—Habitat loss, hunting pressure (endemic)
Black Swift—Climate change and pesticide use?
Chimney Swift—Loss of nesting habitat (migrant)
Leach's Storm-petrel—Cause(s) of declines unknown, but probably multi-faceted (migrant)
Trindade Petrel—Very small breeding range and population susceptible to human impacts and stochastic events (vagrant)
Atlantic Kittiwake—Oil pollution, climate change, longline fishery bycatch (vagrant)
Channel-billed Toucan—Deforestation, hunting (introduced)
Fernandina's Flicker—Logging and clearance for agriculture, hurricanes, loss of nest trees, competition with West Indian Woodpecker (endemic)
Black-billed Amazon—Habitat loss, fragmentation and degradation, trapping and predation (endemic)
Yellow-billed Amazon—Habitat loss, fragmentation and degradation, trapping (endemic)
Hispaniolan Amazon—Agricultural conversion, charcoal production, persecution, hunting, at least formerly international cagebird trade (endemic)
Red-necked Amazon—Habitat loss (including hurricane-related damage), hunting, illegal trade (endemic)
St Lucia Amazon—Small population and tiny range susceptible to stochastic events (endemic)
St Vincent Amazon—Small population and tiny range susceptible to stochastic events (endemic)
Cuban Parakeet—Habitat loss, trapping for cagebird trade (endemic)
Hispaniolan Parakeet—Habitat loss, persecution (endemic)
San Andres Vireo—Tiny range susceptible to stochastic events (endemic)
White-necked Crow—Habitat loss for timber and agricultural conversion, hunting for food, direct persecution (endemic)
Golden Swallow—Forest loss and fragmentation, predation by introduced mammalian predators, competition for nest sites (endemic)

Bicknell's Thrush—Habitat loss and degradation in both breeding and wintering ranges (winter visitor)
Forest Thrush—Human-induced deforestation, introduced predators, severe habitat loss due to volcanic eruptions (endemic)
La Selle Thrush—Habitat loss due to agriculture and timber extraction (endemic)
Sprague's Pipit—Conversion of prairie to seeded pasture, hayfields and cropland, inappropriate grazing causes habitat loss, degradation and increased juvenile mortality (hypothetical vagrant)
Zapata Sparrow—Habitat loss and degradation (endemic)
Martinique Oriole—Brood-parasitism by Shiny Cowbird, whose spread has been facilitated by deforestation (endemic)
Montserrat Oriole—Volcanic eruptions, hurricane-related habitat loss and degradation (endemic)
Rusty Blackbird—Destruction and conversion of boreal wetlands, mining, climate change, disease (hypothetical vagrant)
Cerulean Warbler—Habitat loss and fragmentation on both breeding and wintering grounds (migrant)
White-winged Warbler—Habitat loss, fragmentation and degradation (endemic)
Western Chat-tanager—Habitat loss, fragmentation and degradation due to logging and conversion to agriculture (endemic)

BIRDING IN THE WEST INDIES

The primary reason for most birders to visit any island in the Greater Antilles and many of those in the Lesser Antilles will be to find the endemic landbird species, and this is generally feasible at virtually any time of the year. However, most of the specialties are more easily found during the breeding season (generally March–July in the Bahamas and Greater Antilles, but nesting activity may well start considerably earlier and can be more prolonged on at least some of the Lesser Antilles).

For those birders interested in migration, and perhaps the chance of a vagrant, the best months are August–November, and March or April, but any time between August and May can be productive. Arguably the best areas to search for rarities are the Colombian-administered islands of San Andrés and Providencia (several boreal migrants are potential accidental visitors, and austral migrants from South America are also possible, as they are in the Lesser Antilles), and the southernmost Lesser Antilles, which are most influenced by the trade winds (for the chance of trans-atlantic vagrants, principally but not exclusively waterfowl and shorebirds). Most of these islands are severely under-watched. One of the most extraordinary products of the trade winds was the arrival in October 1988 of an estimated 100 million African desert locusts *Schistocerca gregaria* between Martinique and northern South America. Returning to migrant birds, do not ignore the Bahamas (particularly the larger, northernmost islands, mainly for unusual North American visitors, some from as far afield as the western USA) and westernmost Cuba (the Guanahacabibes Peninsula, where a previously unsuspected raptor passage, presumably taking the most direct route between Florida and the Yucatán Peninsula, has been detected in recent years).

For seawatching, Guadeloupe is the premier location, with most variety in the boreal spring. Dedicated pelagic trips are not yet available, but would surely yield additions to the regional list, and help revise the status of some of those species already known to occur. In addition to Guadeloupe, the waters around Barbados might prove to be particularly productive in this respect.

The best time to see a good selection of winter visitors is between October and March, when the overall numbers of species present on the islands, especially those in the north of the region, are greatest. The small number of summer residents generally start to arrive in March and April, e.g. Antillean Nighthawk, but note that Cuban Martin arrives on Cuba from the end of January, although it is not generally very widespread until March, and Caribbean Martin is mainly present in the region between February and October. The wintering grounds of all three of the just-mentioned species are fundamentally unknown, although usually speculated to be in South America.

Given the hot, often humid (and sometimes very wet) conditions that prevail in summer, most birders will find that a visit anytime between late November and late April is liable to be most productive. If observing a wide range and the largest numbers of North American migrants, especially wood-warblers, is a priority then it is best to visit the islands before the end of February, otherwise seeing large numbers of individuals will be to some extent dependent on experiencing a 'fall-out'.

As in many areas, the Caribbean is a popular part of the world to visit at either Christmas or Easter, thus airfares and hotel prices are frequently inflated at these times. If possible, they are best avoided or, if you must visit during one of these periods, then be sure to book both your flights and accommodation as far in advance as possible. Hurricanes pass through the Caribbean region almost annually in July–October, and this season of the year is generally best avoided. Tropical storms can also be bad news for birds and other wildlife, especially when they strike forested areas, leaving large areas of trees dead and dying. Their effects have been most closely monitored in our region on Puerto Rico, especially on the endemic and globally threatened parrot, but may have a visible impact on other species too. Populations of at least some species may be significantly depleted and be unusually difficult to find for months or even a year or two after a hurricane has hit an island. This can make local knowledge very useful when planning your trip. Nevertheless, should work or other circumstances leave you few options for visiting except during the tropical cyclone season, it should be noted that Cuba has a better than average weather report system in advance of such events, and has good experience in organising evacuations, so that casualties are minimal or non-existent.

Sound recording and playback

Like many countries in the tropics, birding in the region's forests is often heavily dependent on hearing birds first, with many of the most sought-after species being shy and skulking. Until the advent of the free-to-use xeno-canto and Internet Bird Collection (IBC) websites, as well as the Macaulay Library, birders were reliant for knowledge of avian vocalizations on audio compilations such as Reynard & Garrido's *Cantos de las Aves de Cuba*, Reynard's *Cantos de las Aves de la República Dominicana* (both of which were originally issued on vinyl LPs), Roché *et al.*'s *Oiseaux des Antilles*, Reynard & Sutton's *Bird Songs in Jamaica* and Oberle's *Cantos de Aves del Caribe* (which covers Puerto Rico and the Lesser Antilles, as well as the Bahamas, the Cayman Is and San Andrés); the latter three were published on CDs. Despite the availability of all of the free resources just mentioned, these commercial publications are still valuable, especially in recognition of the pioneering work of the late George Reynard in recording the region's bird sounds. Caution should, however, be observed with respect to using playback in the field and it should be kept to the absolute minimum. Prolonged use of playback and playing of calls at unnaturally high volume can cause distress to birds and should be avoided at all costs.

Health

No specific vaccinations are required for entry to any of the Antilles, although those arriving from or continuing to South America may be asked to present a yellow fever vaccination certificate. Proof of a cholera vaccination may also be demanded if travelling from an area that has recently suffered an outbreak. It is advisable to ensure that your vaccinations against typhoid, tetanus, hepatitis C and polio are up to date. Malaria has been eradicated across most of the region, but Hispaniola is one notable exception. However, dengue and the Zika virus infection is still a risk throughout. Furthermore, remember that some medicines can be difficult to obtain locally, especially in countries like Cuba, where apparently mundane items as painkillers for headaches and the like could be problematic to obtain locally, so prepare in advance.

Medical facilities on most islands in the region are generally good, in some cases excellent, but particularly expensive in the Bahamas, Puerto Rico and the Caymans, so ensure that your travel insurance is up to date. Most people are unlikely to suffer anything more severe than an upset stomach anywhere in the Antilles, and the risk of this will be minimized by drinking bottled water. Sunstroke is the other principal risk; drink plenty of water and wear a hat, especially during the middle of the day, as well as applying sufficient sunscreen.

There are several mildly poisonous plants in the Antilles, but you are unlikely to have any problems, especially if you keep to trails. The commonest belongs to the genus *Comocladia*. The plant is easily recognisable by its compound, bright green, saw-like leaves. No dangerous, poisonous animals inhabit most of the West Indies—the black widow spider and several scorpions are rather rare and spend daylight hours out of sight—but on Martinique and St Lucia be aware of the presence of *Bothrops lanceolatus* and *B. caribbaeus*, respectively, both which are extremely venomous snakes, although the latter is quite rare and of restricted range on the island. Sleeping outdoors anywhere should be generally avoided. Try to avoid direct contact with centipedes and wasps, all of which can have a painful sting.

Anywhere in the region during the summer and part of the rainy season (May–September), mosquitoes and sand flies can be very unpleasant, for example in Cuba's Zapata Swamp and on the country's cays. In November–April, their numbers are tolerable, and in some places they are absent. A good repellent is virtually essential, and wearing long-sleeved clothing is advisable at any season. It might be advisable to spray your clothing with a repellent before entering the bush, as seed ticks can be numerous and relentless. The bites are reminiscent of chiggers and will irritate for many days.

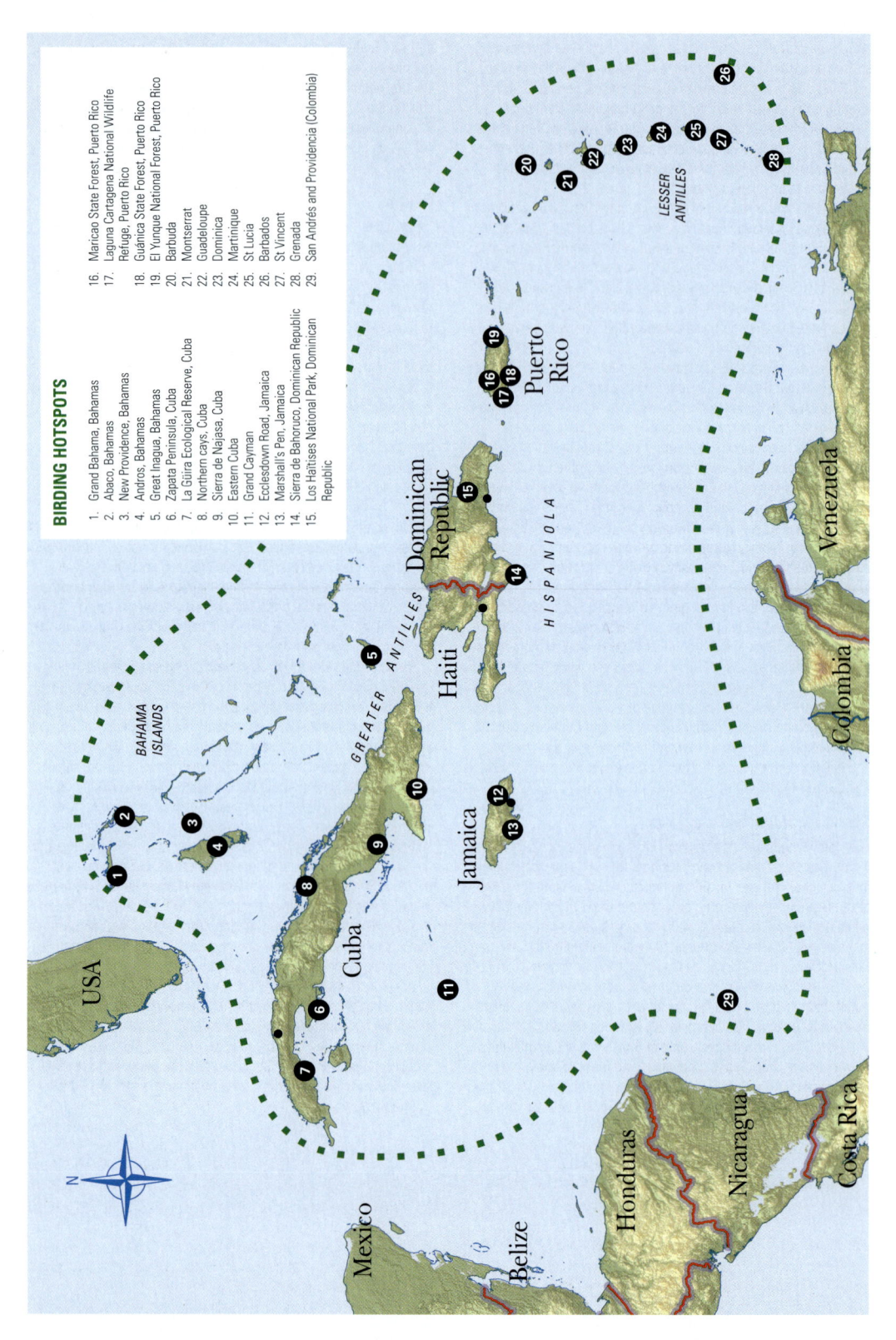

BIRDING HOTSPOTS

1. Grand Bahama, Bahamas
2. Abaco, Bahamas
3. New Providence, Bahamas
4. Andros, Bahamas
5. Great Inagua, Bahamas
6. Zapata Peninsula, Cuba
7. La Güira Ecological Reserve, Cuba
8. Northern cays, Cuba
9. Sierra de Najasa, Cuba
10. Eastern Cuba
11. Grand Cayman
12. Ecclesdown Road, Jamaica
13. Marshall's Pen, Jamaica
14. Sierra de Bahoruco, Dominican Republic
15. Los Haitises National Park, Dominican Republic
16. Maricao State Forest, Puerto Rico
17. Laguna Cartagena National Wildlife Refuge, Puerto Rico
18. Guánica State Forest, Puerto Rico
19. El Yunque National Forest, Puerto Rico
20. Barbuda
21. Montserrat
22. Guadeloupe
23. Dominica
24. Martinique
25. St Lucia
26. Barbados
27. St Vincent
28. Grenada
29. San Andrés and Providencia (Colombia)

BIRDING HOTSPOTS

1. **Grand Bahama, Bahamas**: the closest large island of the Bahamas archipelago to Florida plays home to Bahama Hummingbird, Bahama Swallow, Bahama Yellowthroat and Olive-capped Warbler, as well as a range of more widespread regional endemics. Considered a separate species herein, Bahama Nuthatch is entirely restricted to this island, but is now extremely rare and quite possibly on the verge of extinction. The other specialties can mostly be found in Lucayan National Park, and the long beach area west of the protected area is important for wintering Piping Plovers. The Garden of the Groves is a good site for the hummingbird, while Cuban Emerald is also present there. The west end of Grand Bahama is a regular haunt of vagrants, and the northern Bahamas in general are, with the exception of the southern Lesser Antilles, perhaps the most likely part of the West Indies to produce new birds for the region. Like all of the islands in the Bahamas, accommodation is relatively expensive.

2. **Abaco, Bahamas**: two islands in one, Great Abaco (180 km long) and Little Abaco (32 km) support a wide range of habitats and an excellent range of northern West Indian specialties, including all those present on Grand Bahama, with the exception (of course) of the nuthatch. In addition, one can find the recently split Bahama Warbler and an endemic race of Cuban Amazon. Little Abaco supports what is believed to be the oldest, and perhaps only remnant virgin, stand of pines in the Bahamas. One of the traditional localities to search for wintering Kirtland's Warblers is the quirkily named Hole-in-the-Wall, at the far southern tip of Great Abaco. Adjacent Abaco National Park offers an important stronghold for many of the endemics and near-endemics. Offshore Elbow Cay, which is well served by ferries, is one of the best sites in the northern Bahamas to find Key West Quail-dove.

3. **New Providence, Bahamas**: two-thirds of the human population of the Commonwealth of The Bahamas lives on this island at the hub of the archipelago, most of them in the capital, Nassau. Although there are no native species unique to the island (in the region the introduced Pied Imperial-pigeon is, however, only found here), its geographical position with ferry and airplane links to all of the other major islands, and its being a regular port of call for cruise ships, means that it receives regular coverage by visiting birders. Bahama Hummingbird is present as is Bahama Swallow, the latter albeit mainly in the non-breeding season, as well as Bahama Lizard-cuckoo, which is now rather rare here (but otherwise occurs only on neighbouring Andros and Eleuthera). New Providence is also the stronghold of the invading Boat-tailed Grackle. Harrold & Wilson Ponds National Park is one worthwhile birding area close to Nassau, and the island's south-east coastline hosts Piping Plovers in winter. Some of the surrounding small cays provide breeding sites for a variety of seabirds, mainly terns and gulls, in summer.

4. **Andros, Bahamas**: comprising several main islands and a maze of bays and creeks, Andros is the largest 'single' landmass in the Bahamas and supports remnant secondary and planted pine woodland, as well as mangroves and dense scrub. Andros is the last bastion of the globally threatened Bahama Oriole, which is regularly found around Andros Town, for example around Small Hope Lodge. Also present in the same general area are Bahama Hummingbird, Bahama Lizard-cuckoo, Bahamas Hairy Woodpecker, Bahama Swallow, Thick-billed Vireo, Bahama Mockingbird and Bahama Yellowthroat. Several of these, and West Indian Whistling-duck, can be found in the vicinity of San Andros Pond. It is possible to see all the key birds in a 24-hour trip from New Providence.

5. **Great Inagua, Bahamas**: the largest colony of American Flamingos outside of Cuba breeds here, on Lake Rosa, and this area also supports West Indian Whistling-duck and Cuban Amazon, but the principal target and reason for visiting the island is the recently split Lyre-tailed Hummingbird. Fortunately, it is common there and even visits feeders in the capital, Matthew Town, at least during certain months of the year. There are regular (but not daily) flights from New Providence to the island.

6. **Zapata Peninsula, Cuba**: located in southern Matanzas, just over 150 km from La Habana, the peninsula occupies an area of 4700 km^2 and supports a variety of habitats including semi-deciduous forest, swamp forest, mangrove, coastal thickets, mudflats, and extensive wetlands, including saltpans and fresh and brackish marshes. It is also the site of one of the greatest debacles in US foreign policy, the ill-fated 'Bay of Pigs' invasion. For birders, Zapata is arguably *the* keynote site in the entire region, but at least in the Greater Antilles. Almost 260 species have been reported in the area, among them all but four of Cuba's avian endemics, including the near-mythical Zapata Rail, but also Zapata Wren and Zapata Sparrow, as well as Red-shouldered Blackbird, Blue-headed and Grey-headed Quail-doves, Gundlach's Hawk, Fernandina's Flicker and Bee Hummingbird. In addition, more widespread but still much-desired species, such as West Indian Whistling-duck and Stygian Owl, will be targeted by many visiting birders. Furthermore, the forested areas afford prime wintering grounds for a wide variety of North American-breeding wood-warblers. There are a number of accommodation of options in the region, but a car is necessary to get around, and a local guide is a requirement for visiting areas inside the national park.

7. **La Güira Ecological Reserve, Cuba**: in the Sierra de los Órganos, 150 km west of La Habana, the area is easily visited as a day trip from the capital. However, the closest town, San Diego de los Baños, has several accommodations. Main habitats are semi-deciduous woodland, gallery forest, tropical karstic forest and pine forest, and the principal species of interest for visiting birdwatchers are Cuban Solitaire and Olive-capped Warbler. However, other targets include Cuban Grassquit (at one of its few strongholds in western Cuba) and Giant Kingbird, as well as a wide range of common endemics such as Cuban Trogon and Yellow-headed Warbler. For those with a little more sense of adventure, a visit to the Guanahacabibes Peninsula, at the westernmost extremity of Cuba, could be worthwhile, especially during migration seasons, particularly in autumn. Recent studies have detected a considerable raptor passage, presumably between Florida and the Yucatán Peninsula, and reported several new records of passerines for Cuba, and even the West Indies as a whole.

8. **Northern cays, Cuba**: most birders visit Cayo Coco, the second-largest cay in Cuba, which is connected to the mainland by a causeway and is served by many hotels. The island is mostly covered by woodland, including mangrove, as well as coastal scrub and lagoons. More than 200 species have been registered, including many rarities (e.g. the only Black-throated Grey Warbler to have been spotted in the Antilles), although the most interesting species are Ruddy and Key West Quail-Doves, a local race of Cuban Lizard-cuckoo, Cuban Tody, West Indian and Cuban Green Woodpeckers, Cuban Flicker, Cuban Gnatcatcher, a race of Zapata Sparrow, Western Spindalis and Cuban Bullfinch. Many waders can be seen in winter and during passage periods, and a regular feature is the largest aggregation of American Flamingos in the West Indies.

9. **Sierra de Najasa, Cuba**: located c. 70 km south-east of Camagüey (an old city with numerous hotel and other accommodation options), this is a protected area of open country with many palm groves, semi-deciduous woodland, and tropical karstic forest in the hills. Among the 120 species reported are now relatively rare and local birds such as Plain Pigeon, Cuban Parakeet, Cuban Amazon, Fernandina's Flicker, Giant Kingbird and, especially, Cuban Palm Crow.

10. **Eastern Cuba**: comparatively few of the now many birders that visit Cuba venture east of Camagüey, yet the region around the country's second city, Santiago de Cuba, and the almost equally famous Guantánamo offers some potentially exciting birdwatching in relatively untouched habitats. Most visitors will choose to base themselves in one of the principal conurbations. Two easily visited areas are the Pico Turquino, Cuba's highest mountain at fractionally below 2000 m, whose highest slopes (a five-hour walk) support cloud forest and wintering Bicknell's Thrushes. Both Blue-headed Quail-dove and Cuban Solitaire can be seen along the trail to the peak. Black-capped Petrel perhaps breeds in these mountains and is sometimes seen from shore in the evenings. Further east, in the vicinity of Baitiquirí, the xerophytic scrub supports a third race of the endemic Zapata Sparrow, as well as abundant Cuban Gnatcatcher, Oriente Warbler and Cuban Grassquit, while after-dark possibilities include Cuban Nightjar and Northern Potoo. There is also a small colony of White-tailed Tropicbirds in this region.

11. **Grand Cayman**: this low-lying island, the largest of the Cayman group, boasts several good birding sites, most notably the Mastic Reserve, Botanic Park and Governor Gore's Pond. The near-endemic Vitelline Warbler (shared only with the Swan Is) is common at the first two, and the recently split Grand Cayman (Cuban) Bullfinch reasonably so. Other species that can be expected include Cuban Amazon, La Sagra's Flycatcher, Loggerhead Kingbird (summer), Chinchorro Elaenia, Zenaida Dove, Greater Antillean Grackle (an endemic subspecies), Yucatan and Thick-billed Vireos (both of them represented by endemic races), Western Spindalis and Cayman Woodpecker (another endemic subspecies, of West Indian Woodpecker). West Indian Whistling-duck can be found at several sites across the island. Most visitors probably arrive by cruise ship, but it is possible to fly from Cuba, Jamaica or Miami. Accommodation options are reasonably numerous, but most are expensive.

12. **Ecclesdown Road, Jamaica**: this site at the base of the John Crow Mts is less than an hour from the town of Port Antonio. Along the road is productive habitat for a rich variety of endemics, including the globally threatened Ring-tailed Pigeon, as well as Crested Quail-Dove (which is most easily found soon after dawn), both parrots (Black-billed seems particularly common, but both are numerous), both endemic cuckoos, Jamaican Mango, Black-billed Streamertail, Vervain Hummingbird, Small Jamaican Elaenia, Rufous-tailed Flycatcher, Jamaican Becard, Blue Mountain, Black-whiskered and Jamaican Vireos, Jamaican Crow, Rufous-throated Solitaire, both endemic thrushes, Arrowhead Warbler, Yellow-shouldered Grassquit, Orangequit and Jamaican Blackbird, which seems unusually easy to find in this area.

13. **Marshall's Pen, Jamaica**: a 300 ha-property on the outskirts of Mandeville is the home of Ann Sutton, whose husband Robert, the island's pre-eminent ornithologist, was tragically murdered in July 2002. Ann, an ornithologist in her own right, has carried on the tradition of providing a fine base for birders visiting Jamaica. This is arguably the best site for Jamaican Owl and White-eyed Thrush. Other single-island endemics include: Crested Quail-Dove, Chestnut-bellied Cuckoo, Jamaican Lizard-cuckoo, Jamaican Mango, Red-billed Streamertail, Jamaican Tody, Jamaican Woodpecker, Jamaican Pewee, Jamaican Becard, Jamaican Crow, Jamaican Euphonia, Jamaican Spindalis, Jamaican Oriole, and Yellow-shouldered Grassquit. Other species of interest, some of them multi-island endemics, include: Caribbean Dove, Vervain Hummingbird, Black Swift, Northern Potoo, Antillean Nighthawk (summer), Rufous-throated Solitaire, Cave Swallow and Greater Antillean Bullfinch.

14. **Sierra de Bahoruco, Dominican Republic**: for devotees of Caribbean birding, this region, in the south-west of the country on the border with Haiti, is surely one of those must-visit areas, despite ongoing and disheartening levels of deforestation. There are two main routes into the sierra, one from the east, via Duvergé and Puerto Escondido, and the other from the south, via Pedernales. The first-named, especially, permits the observer to drive a transect from lowland xerophytic scrub through drier to more humid forest and finally pines with the concomitant changes in avifauna. The following are all possible, although some species are distinctly difficult: Black-capped Petrel (winter breeder; it would probably be necessary to camp in the mountains to hear birds visiting the colony, and in any case most breed in adjacent Haiti), Key West Quail-dove (uncommon; best looked for in drier areas), White-fronted Quail-dove, Hispaniolan Parakeet, Hispaniolan Parrot, Hispaniolan Lizard and Bay-breasted Cuckoos (the latter is one of the most difficult endemics), Ashy-faced Owl (rare), Least Poorwill (uncommon), Hispaniolan Nightjar (common), Hispaniolan Trogon, Broad-billed and Narrow-billed Todies, Hispaniolan Woodpecker, Hispaniolan Pewee, Golden Swallow (rare but regularly seen), White-necked Crow, Hispaniolan Palm Crow (both crows rather uncommon, but can also be found around the nearby Lago Enriquillo), La Selle Thrush and (in winter) Bicknell's Thrush, Rufous-throated Solitaire, Flat-billed Vireo, Pine Warbler, Antillean Siskin, Hispaniolan Crossbill, Black-crowned Palm-tanager, Green-tailed Warbler, White-winged Warbler (usually rather uncommon), Western Chat-tanager, Hispaniolan Spindalis and Hispaniolan Oriole.

15. **Los Haïtises National Park, Dominican Republic**: currently, this is the only regular site for Ridgway's Hawk. The 200 km² protected area is one of the wettest parts of Hispaniola and comprises dense lowland broadleaf forest growing on karstic limestone, as well as mangrove. Forested hills of limestone, called mogotes, dot the landscape. Accessing the park is comparatively difficult, which is three hours from Santo Domingo on the south side of Samaná Bay. The best place to stay is probably Caño Hondo. Besides Ridgway's Hawk, other species include the globally threatened Plain Pigeon, Hispaniolan Parrot, Mangrove Cuckoo, Least Poorwill, Antillean Piculet, White-necked Crow (the park is one of the species' strongholds and it is easily seen around Caño Hondo) and a variety of common endemics.

16. **Maricao State Forest, Puerto Rico**: accessed via the PR 120 / Monte del Estado, the best birding is at higher elevations, where Puerto Rican Woodpecker and Puerto Rican Vireo can be found. Ruddy Quail-dove may fly across the road and can be heard early morning. Trails from the picnic area at km 16.2 are reliable for Elfin-woods Warbler. Listen for its rattling trill of a song, or its odd, buzzy call. Also here are hummingbirds, Puerto Rican Pewee (a race of Lesser Antillean Pewee), Puerto Rican Bullfinch, Puerto Rican Spindalis, Puerto Rican Tody, Puerto Rican Tanager and migrant warblers. The concrete observation tower at km 14 offers views of the canopy and an open area to look for swifts, swallows and soaring hawks.

17. **Laguna Cartagena National Wildlife Refuge, Puerto Rico**: lowland marsh, fields and ponds. Follow PR 101, and at km 12.2 turn south onto PR 306 to Laguna Cartagena (www.fws.gov/refuge/laguna_cartagena). US Fish & Wildlife Service is restoring the wetlands, which is a regular site for ducks and rails. At the lake edge look for Northern Red Bishop, Venezuelan Troupial, Loggerhead Kingbird and Puerto Rican Vireo. An observation platform offers views of American Coot, gallinules, migrant ducks and West Indian Whistling-duck. An hour's drive towards Cabo Rojo (www.fws.gov/refuge/cabo_rojo), PR 301 leads to the south-west tip of Puerto Rico and passes salt flats with Snowy Plover, Stilt Sandpiper and other migrant shorebirds, plus views of seabirds from the cliffs at the lighthouse.

18. **Guánica State Forest, Puerto Rico**: dry forest accessed from the PR 116. Turn south-west onto the PR 333 one block east of the Guánica Burger King. After c. 3 km, walk the dirt road to Playa Jaboncillo (marked by a gate and sign). This area is productive for Adelaide's Warbler, Puerto Rican Vireo, Puerto Rican Flycatcher, Puerto Rican Tody, Key-West Quail-dove, Puerto Rican Lizard-cuckoo and Mangrove Cuckoo. At night, the endemic Puerto Rican Nightjar can be heard along PR 333. At the end of the pavement on PR 333 is a large sandy parking lot. Walk north on the trail that heads inland through a wire gate. Puerto Rican Bullfinch, Puerto Rican Flycatcher, Puerto Rican Nightjar, Venezuelan Troupial, Caribbean Elaenia and other dry-forest species are regular here. An elaborate trail system in the forest can be accessed by returning to the PR 116: just north of the intersection with PR 333 turn east onto PR 334, then bear left at two forks in a residential area, to the entrance gate. Puerto Rican Screech-owl and Puerto Rican Nightjar have been regular along the entrance road. The road continues to the headquarters, where a map of the trail system can be acquired. In the nearby tourist town of La Parguera—famous for its Phosphorescent Bay—Yellow-shouldered Blackbirds feed on table scraps at Villa Parguera hotel. The blackbirds can be seen in the non-breeding season flying at dawn and dusk to and from their roosts on nearby mangrove islands.

19. **El Yunque National Forest, Puerto Rico**: the best-known montane forest area on the island, being close to, and visible from, San Juan, but can be crowded with tourists. PR 191 from the north or an unconnected, separate segment of PR 191 from the south are the major access points. Trails provide access to forest where hummingbirds, Puerto Rican Tody, Puerto Rican Screech-owl and Puerto Rican Lizard-cuckoo occur (the last was regular at the El Portal Visitor Center parking lot). The Puerto Rican Parrot population in this forest was wiped out by the 2017 hurricanes, but the species will be reintroduced from government aviaries. Visitor infrastructure was badly damaged by the storms. Many roads, trails and the main visitor centre were severely damaged. Check conditions and opening hours at www.fs.usda.gov/elyunque. Also plan ahead for the famous traffic jams leaving San Juan.

20. **Barbuda**: this tiny, low-lying, limestone island (161 km^2) supports a single endemic species, Barbuda Warbler, and forms a nation state together with the neighbouring islands of Antigua and Redonda. The warbler can be found in the dry scrub around the only Important Bird Area on Barbuda, Codrington Lagoon, an almost fully enclosed body of salt water bordered by mangroves and sand ridges. Other avian attractions at the site include one of the Caribbean's largest breeding colonies of Magnificent Frigatebirds and important numbers of West Indian Whistling-duck.

21. **Montserrat**: this UK Overseas Territory is most likely to impinge on the public's consciousness for the devastating volcanic eruption of Chances Peak, in June 1997, which destroyed the capital, Plymouth, and forced a large part of the island's human population to evacuate. However, the event also impacted Montserrat's only endemic bird, Montserrat Oriole, which lost approximately 60% of its habitat in the Soufriere Hills as a result. Two trails in the Centre Hills, the eponymous Oriole Trail and the Blackwood Allen Nature Trail, give visitors the chance to still find Montserrat Oriole, as well as other regional endemics including Bridled Quail-dove, Forest Thrush, Pearly-eyed Thrasher, Brown Trembler and Lesser Antillean Bullfinch.

22. **Guadeloupe**: the Route des Mamelles (D23) crosses the Basse-Terre Mts from west to east, between Mahaut and Vernou, along which a number of picnic areas (e.g. Cascade aux Ecrevisses, Petit Bras David, Maison de la Forêt, Corossol) provide easy opportunities to find many regional endemics, such as Guadeloupe Woodpecker, Forest Thrush, Bridled Quail-dove, Plumbeous Warbler, hummingbirds and the less numerous Lesser Antillean Pewee and Lesser Antillean Flycatcher. Also check the sky for Lesser Antillean and Black Swifts. The Pointe des Châteaux, a peninsula at the extreme east end of Grande-Terre, is the premier tourist destination in Guadeloupe, and where the D118 road ends. It is often crowded with people between December and April, but they tend to stay on the beach or climb to the cross for great views of the nearby islands. This is the best spot for shorebirds in autumn, and the best seawatching spot anywhere in the Caribbean basin. Shearwaters, storm-petrels, jaegers and terns are all possible during peak migration (March–June). Don't forget your telescope and sunscreen.

23. **Dominica**: one of the best places on this relatively small and comparatively mountainous island is the Syndicate Nature Trail within the Northern Forest Reserve. To access the area, after 'Dublanc' (in the municipality of St-Peter) head north and then turn right in the direction of Morne Diablotin, the island's highest peak at just below 1450 m and named after the Black-capped Petrel. At the end is the nature trail with suitable lookouts over the forested Picard Valley. Here it is possible to find both the Endangered Imperial Amazon and the Vulnerable Red-necked Amazon. Furthermore, this is also home for all four hummingbird species that occur on the island, Plumbeous Warbler, Forest Thrush, Lesser Antillean Saltator, Lesser Antillean Pewee, Lesser Antillean Flycatcher, Antillean House Wren, Scaly-breasted Thrasher and others. Dominica is one of the poorest and least touristic islands in the Caribbean, although it is served by direct international flights. It was hard hit by Hurricane Maria in autumn 2017. Most accommodation is on the coast.

24. **Martinique**: Plateau Boucher lies c. 1 km south of the crossroads where the N3 and D1 meet in the montane forest in the north of the island. From the parking area, there is a trail to Piton Boucher where you can find most of the interesting species on the island, especially Blue-headed Hummingbird, Grey Trembler and Martinique Oriole, but also Rufous-throated Solitaire, Purple-throated Carib and Lesser Antillean Saltator. Chestnut-bellied Seed-finch is also found here, an introduced species from South America that in our region occurs only on Martinique. To visit La Caravelle, a peninsula on the north-east side of the island, leave road N1 at La Trinité, and take the D2 to 'Presqu'île de la Caravelle'. This area supports dry-forest habitat and is *the* place to find the Endangered White-breasted Thrasher, which is otherwise found only on St Lucia (where represented by a different subspecies group). On Martinique the thrasher occurs only here. Other species in the same area include Martinique Oriole, Lesser Antillean Saltator and Scaly-breasted Thrasher, and this is the best place to see

White-tailed Nightjar, which in the Caribbean is found only on Martinique.

25. **St Lucia**: a major tourist destination, with plentiful albeit somewhat expensive accommodation, St Lucia is one of the most popular islands in the Lesser Antilles with birders. A 4WD-vehicle is virtually essential to reach some of the best birding sites, but not to access the Millet Bird Sanctuary Reserve more or less in the centre of the island. South of Castries airport along the main road following the west coast, pass Marigot Bay and then over the Roseau River, after which turn left to Morne d'Or (with the Millet Reserve also signed). The reserve (small entrance fee) is reached after c. 5 km. Possible birds include St Lucia Parrot, St Lucia Warbler, St Lucia Black Finch, St Lucia Oriole, St Lucia Pewee (a race of Lesser Antillean Pewee), Grey Trembler and Rufous-throated Solitaire. In contrast, getting to Gran Anse and Ravine La Chaloupe, on the east coast near Desbarra, where possibilities include St Lucia (Rufous) Nightjar, St Lucia Thrasher (currently regarded as a race of White-breasted Thrasher, although a recent genetic analysis has advocated their separation at species level) and Antillean House Wren, will necessitate a 4WD and quite probably also a local guide.

26. **Barbados**: the easternmost of the Antilles, 165 km beyond St Vincent, isolated Barbados (432 km^2) is the best-placed of the West Indian islands to receive vagrants from the Old World, and thanks to a very small but dedicated team of resident observers it has a long track record in producing rarities, among them many regional 'firsts'. Fewer than 30 species breed, but these include the highly ubiquitous Barbados Bullfinch, the country's sole endemic. Most birders will want to focus on the island's wetlands, among them the various shooting swamps, which are all privately owned and strictly managed for sport hunting. Two protected areas, Graeme Hall Swamp and Chancery Lane Swamp, situated relatively close to one another in the south of Barbados, offer safer refuge for migrant birds, and the former is also the site of the New World's first breeding colony of Little Egrets. In addition to migrants and the chance of a rarity in season, Green-throated Carib and Antillean Crested Hummingbird can regularly be found around the Chancery Lane site.

27. **St Vincent**: situated almost at the southern tip of the Lesser Antillean chain, St Vincent is another rather small and mountainous island that reaches almost 1300 m above sea level. Tourism tends to be focused on the more exclusive end of the market. Unsurprisingly, given the island's topography, most accommodation options are sited around St Vincent's coastline. The Vermont Forest Reserve (St Andrew) in the centre-south of the island supports an extensive area of montane forest and is accessed via a side road heading inland from the road between Kingston and Layou. Follow the Loop Trail for chances to find St Vincent Parrot, Whistling Warbler, Grenada Flycatcher and St Vincent (Lesser Antillean) Tanager, as well as a cross-section of other Lesser Antillean regional endemics.

28. **Grenada**: most parts of this, yet another rugged, island are accessible using taxis or minibuses, although most visitors will probably nevertheless prefer to hire their own vehicle. Most accommodation and infrastructure for tourists is concentrated in the south-west of Grenada. Here too is the Mount Hartman Estate (Saint George), part of which is now a national park supporting an area of dry forest and almost 40% of the total population of what was until recently Grenada's sole endemic, the Critically Endangered Grenada Dove. Also possible here are Grenada Flycatcher and the newly split Grenada (Lesser Antillean) Tanager, as well as Hook-billed Kite (a rare endemic subspecies), Scaly-naped Pigeon, Brown Trembler, Antillean House Wren and Lesser Antillean Bullfinch.

29. **San Andrés and Providencia (Colombia)**: despite lying closer to Nicaragua in a relatively remote corner of our region, these islands are easily accessible only via flights from Panama, and Colombia, which owns both. However, the culture is definitely Caribbean in flavour and the main language is Creole, rather than Spanish. San Andrés (St Andrew), especially, is a very popular tourist destination, with plenty of accommodation options. Two interesting sites to visit within walking distance of some hotels are the Big Pond (La Laguna) and the Botanic Garden. The endemic San Andres Vireo is easily found at both, with a supporting cast potentially including White-crowned Pigeon, Caribbean Dove, Green-breasted Mango, Chinchorro Elaenia, Tropical Mockingbird, etc. It is possible to take a boat between San Andrés and Providencia (Old Providence), although most people fly to the latter from Bogotá, Colombia's capital. Arguably the best site on the island is the trail between Casa Baja and El Pico, the highest peak on Providencia (at 350 m). Here the main prize will be Providence Vireo, treated here as a subspecies group in the Mangrove Vireo complex, but potentially a species in its own right, although it has also been considered a race of Thick-billed Vireo. Whatever the most appropriate classification, it represents a very interesting taxonomic case and one that certainly merits additional study. Freshwater Dam and McBean Lagoon are both worth checking for waterbirds, potentially including rarities (at least for Colombia!), and the vireo is also common at the latter. Although there is relatively little public transport on Providencia (compared to its neighbour), it is quite possible to hitch rides (although be prepared to pay).

USING THE FIELD GUIDE

SYSTEMATIC AND TAXONOMIC TREATMENT

The systematics and taxonomy used in this field guide follow the two volumes of the HBW and BirdLife International *Illustrated Checklist of the Birds of the World* (del Hoyo & Collar 2014, 2016) with some updates based on subsequent research that have already been adopted by BirdLife International and HBW Alive. This Checklist is the product of a truly integrative system of taxonomic evaluation. To start, the results and insights of recent molecular studies have been constantly accounted for and incorporated, as well as additional data from the rest of the literature. For differences in plumage pattern and structure, bare-part colours and formations, morphometrics, vocalizations, ecological factors and behavioural traits, the scoring system based on the Tobias criteria (del Hoyo & Collar 2014: 30–41) also has been proactively used to increase the consistency of the taxonomic decisions and, particularly, to evaluate the many hundreds of taxa for which no recent studies have been published.

EXPLANATION OF SPECIES ACCOUNTS

Each family section starts with the name and a breakdown of the number of species in each that are recognized worldwide (according to current HBW/BirdLife taxonomy), and those that have occurred in the region, with the number considered to be vagrants, introduced or extinct listed separately.

Subspecies groups: This field guide is novel, compared to most others that readers will have used, in providing separate accounts and illustrations for each subspecies group recognized under current HBW/BirdLife taxonomy. Subspecies groups are informal taxonomic units used in several recent world checklists to highlight seemingly monophyletic groups of taxa (sometimes single subspecies) that at present appear to sit between the species and subspecies levels (although in *some* cases it is plausible that fuller scrutiny and better evidence will result in their being awarded species rank). Such groups are identified by their possession of one or several reasonably distinct characters and which therefore seem worthy of attention. In some cases they may already have been recognized as species in other lists or accorded a taxonomic status such as 'megasubspecies' or 'allospecies'.

Accordingly, where two or more such subspecies groups belonging to a single species occur in the region, this necessitates inclusion of more than one account. However, the focus is always on that subspecies group most widespread or likely to be encountered by birdwatchers in the region, with more comparative details alone being provided for other subspecies groups of the same species.

Because many subspecies groups comprise multiple races and the name of the group is always determined by the first of the relevant subspecies to have been named, readers should be aware that the subspecies group name and the race/s belonging to that group recorded in the West Indies are often different. For example, a subspecies present in West Indies may form part of a subspecies group bearing the name of a subspecies that is not present in the region but whose name is older, which follows the principle of priority of the International Code of Zoological Nomenclature (ICZN) for naming the group. This may initially prove confusing to some users, but we decided that the benefits of giving separate coverage to reasonably well-defined groups of taxa, currently treated below species level, outweighed the disadvantages.

QR codes: each species has a QR code that links to a relevant website page with complementary audiovisual material, so that you can view photos and videos, and listen to sounds of the species in question. Often there are hundreds of audiovisual records for a single species and in many cases they illustrate behaviours and aspects of biology. To access this information directly from the book you must first download any one of the free smartphone apps that scans QR codes. Then open the QR code reader, and hold your device over the QR code that you wish to scan so that it is clearly visible within the phone's screen; depending on the type of phone, it will either automatically scan the code, or you may need to click a button, in a manner not unlike taking a photo using your phone. Within a few seconds your phone should have navigated to the relevant page when you have an internet connection.

Each species account is given in the following format:

Name, conservation status and presence: common English names are followed by the scientific name in italics. Unlike other guides in this series, we do not include any vernacular names in exclusively local use because of their great abundance (e.g. Spanish names in Cuba, the Dominican Republic and Puerto Rico, French names in Guadeloupe and Martinique, etc.), fully reflecting that the Caribbean is a multi-lingual and multi-cultural conglomerate. Capitalization and hyphenation of species names used here sometimes differ from other widely used taxonomies, e.g. by North American authorities. The most striking differences are noted, but simple issues of variance between UK and US English names are not noted, e.g. 'Grey' instead of 'Gray', as in the catbird, or 'Tricoloured' vs. 'Tricolored' as in the munia. The global conservation status is denoted by the standard two-letter codes summarizing the IUCN Red List category of extinction risk, as follows: **LC** (Least Concern), **NT** (Near Threatened), **VU** (Vulnerable), **EN** (Endangered), **CR** (Critically Endangered), **CR(PE)** (Critically Endangered [Possibly Extinct]) and **DD** (Data Deficient). Immediately below this the reader will find a general statement of relative abundance and temporal status (whether migrant or resident) in the West Indies. In this section alone, for sake of brevity we use the following abbreviations: **GA** (= Greater Antilles), **LA** (= Lesser Antilles) and **O** (= Other, in reference to more than one of the following, the Bahamas, Cayman Is, Swan Is and the Colombian archipelago of San Andrés and Providencia). It also bears mention that it is usually the case, again due to limitations of space, that we use the term 'Bahamas' to also include the Turks & Caicos Is.

Size: 'L' denotes length in centimetres. Where available, 'WS' denotes wingspan in centimetres. WS measurements should however be treated with caution, as this is not a measurement that is typically recorded by field researchers, and cannot accurately be measured from museum specimens. As a consequence, the sizes given may in some cases be based on small sample or unrepresentative or inaccurate data. On the plates, note that most figures of birds in flight are not to the same scale as those at rest.

19

Subspecies: each account commences with a list of recognized subspecies that occur in the region and their general ranges in the country; use of superscript letters is keyed to the relevant map, wherein the approximate boundaries between different races of resident or breeding species with contiguous or largely continuous ranges are delimited, as best as possible. If a species is monotypic, this is clearly stated.

Habitat and habits: a brief description is provided of the species in the region, its preferred habitat and elevational range. Where relevant, notes are also provided on general habits, especially where these can assist identification. For vagrants, or for some especially rare and localized taxa, specific details are often provided concerning occurrence. Any species or subspecies group that is Endemic (E) or almost so (Near-endemic; NE) to the region is clearly indicated, and the appropriate codes are also used next to the maps.

Identification (ID): unless otherwise stated, the initial description involves the nominate race. Adult plumages are generally listed before immature or juvenile plumages. Additional notes are provided on seasonal differences as appropriate.

Geographical variation (GV): differences from the main description are given for all other races known to have occurred in the region and, where deemed relevant, for races thought likely to occur.

Voice: vocalizations are typically based on a description of the song, followed by any regularly heard calls. For some non-breeding migrants that are very unlikely to be heard singing, only call is given. For some species, such as seabirds, vocalizations are not transcribed as they are unlikely to aid field identification.

Similar species (SS): additional notes are given in cases where another species in the region (or occasionally potential vagrants) may cause confusion in identification.

Taxonomic notes (TN): these are focused on recent, species-level changes in taxonomy, with special reference to those arrangements introduced in either volume of the HBW and BirdLife International *Illustrated Checklist of the Birds of the World* on which sequence this book is based. Notable deviations from the latter, especially in relation to the list maintained by the American Ornithological Society (formerly American Ornithologists' Union), are explained. In addition, notable differences in coverage of subspecies compared to other recent works to the same region are explained. In contrast, genera-level differences are rarely covered.

Alternative names (AN): where other common English names are, or occasionally have been, in use, especially in other works likely to be consulted by visiting or resident birdwatchers, such alternative names are stated. However, we generally do not provide spelling variants or alternative names that differ only slightly from the name employed.

ABBREVIATIONS USED IN THE FIELD GUIDE

N	north
S	south
E	east
W	west
C	central
NE	north-east
NW	north-west
SE	south-east
SW	south-west
NC	north-central
SC	south-central
EC	east-central
WC	west-central
Ad/ad	adult
br	breeding plumage
non-br	non-breeding plumage
Juv/juv	juvenile
Imm/imm	immature
1st-w/2nd-y	first-winter/second-year (plumages), etc.
km	kilometres
m	metres
cm	centimetres
I	Island
Is	Islands
L	Lake
R	River
Mt	Mount
Mts	Mountains
GA	Greater Antilles
LA	Lesser Antilles
O	Other (refers to some or all of the Bahamas, Cayman Is, Swan Is, and San Andrés and Providencia)
AMNH	American Museum of Natural History (New York)
ANSP	Academy of Natural Sciences (Philadelphia)
COP	Colección Ornitológica Phelps / Phelps Ornithological Collection (Caracas)
FMNH	Field Museum of Natural History (Chicago)
LSUMZ	Louisiana State University, Museum of Zoology (Baton Rouge, LA)
MCZ	Museum of Comparative Zoology (Cambridge, MA)
NHMUK	Natural History Museum (Tring)
USNM	US National Museum of Natural History (Washington DC)
AB	*American Birds*
FN	*National Audubon Society Field Notes*
NAB	*North American Birds*

MAP KEY

■ Resident/present all-year round.

■ Breeding visitor.

■ Winter/Non-breeding visitor: indicates zones occupied by the species outside of the breeding season. This covers the ranges of a wide variety of species present at different times of year, including a variety of seabirds that breed in the Southern Ocean but visit the Caribbean mainly during the boreal summer, and many shorebirds and passerines that breed in North America and spend the boreal winter in the West Indies.

■ Passage migrant: indicates areas used by a species only when moving between breeding and non-breeding ranges; but only when such regions are particularly well known are these mapped. Large areas where the species might be seen or occasionally appears on migration are not mapped, but may be highlighted in the text. Note that (usually) very small numbers of a handful of species (e.g. some hirundines) that migrate across our region between their breeding and wintering grounds in North and South America, respectively, winter locally on some islands. However, because the species concerned are principally passage migrants, such instances of wintering are only mentioned in the text.

■ Introduced: populations either wholly or partially the result of actions by humans, both deliberately introduced or the product of escapes from captivity. We have attempted to conservatively map only those populations that are, or appear to be, self-sustaining, although the situation is often rather fluid.

■ Best-guess ranges of recently extinct (or probably extinct) species in the region; usually the last known areas, rather than the entire historical distribution.

A B Resident or breeding subspecies; letters correspond to those in superscript in the relevant text.

∼ Approximate boundaries between different resident or breeding subspecies.

↙ Highlighting presence (resident or breeding populations) within certain small, well-circumscribed areas (usually small islands). Keep in mind that very small pockets of distribution are often deliberately amplified on the maps, to enable readers to see them more easily.

? Highlighting historical collection but current status unknown, or species whose current status is uncertain.

E Endemic.

NE Near-endemic.

BIRD TOPOGRAPHY

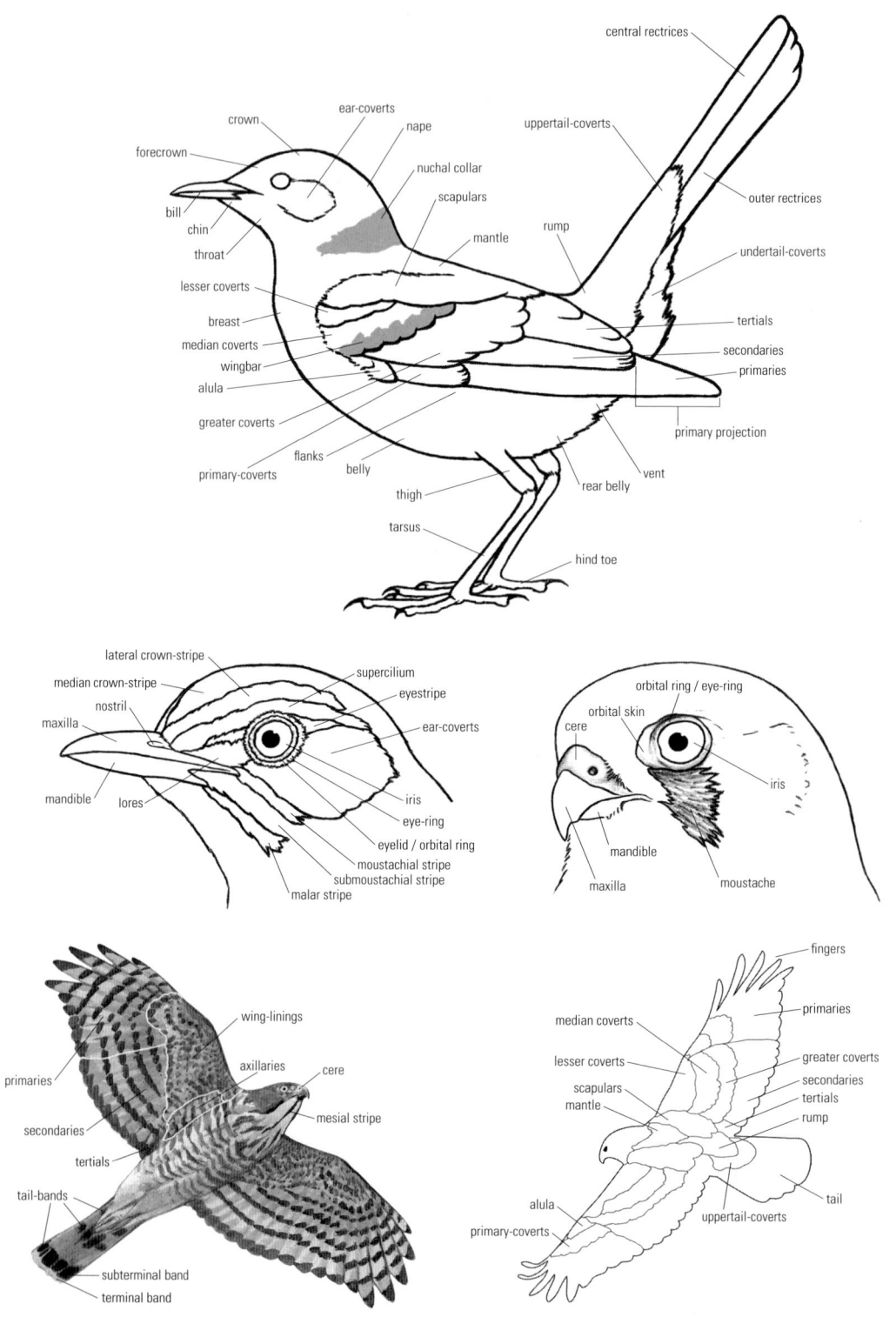

ABOUT THE AUTHORS

Guy Kirwan

Guy is a freelance editor and ornithologist currently based in the UK but working for Lynx Edicions, following a lengthy sojourn in Brazil. A regular visitor to the West Indies since the early 1990s, he is currently finalizing a detailed checklist to the birds of Cuba. Author of more than 150 communications in the technical literature, and several previous books, Guy is a Research Associate of the Field Museum of Natural History, Chicago, and the Museu Nacional, Rio de Janeiro. His main avian interests are in taxonomy and the breeding biology of passerines in the New World tropics.

Mark Oberle

Professor Emeritus at the University of Washington, Mark trained in biology and medicine at Harvard and Johns Hopkins Universities. He has worked in biology and ornithology in the American tropics for four decades and is a founding member of the Washington Ornithological Society. Mark has published several books, audio CDs, and a smartphone app for birds of Puerto Rico and the Virgin Islands.

Anthony Levesque

Anthony arrived in Guadeloupe from France in 1998 after studying Wildlife Management and Nature Protection. Following work at a nature reserve and as a wildlife consultant for the National Hunting and Wildlife Agency, he now owns his own environmental consultancy. An active birdwatcher, he has found more than 50 species new to Guadeloupe, the Lesser Antilles and, in some cases, the Caribbean as a whole, and is an eBird reviewer for the Lesser Antilles. Anthony is the founder of AMAZONA—the bird conservation NGO in Guadeloupe.

Chris Sharpe

Chris is a biologist who has worked on the conservation of Neotropical birds for more than 30 years, having been based for most of that time in Venezuela, where he is a Research Associate of the Phelps Ornithological Collection (COP) and the NGO Provita, and a Founder Member of the Venezuelan Ornithologists' Union. Chris is an editor of *HBW Alive* and of the *Lynx Edicions and BirdLife International Field Guides* series.

ACKNOWLEDGEMENTS

Josep del Hoyo has our long-lasting gratitude for agreeing to this project. Among many valued colleagues at Lynx Edicions, our consistently upbeat coordinators and cheerleaders, Amy Chernasky and Arnau Bonan-Barfull, have proffered liberal and equal doses of patience and expertise throughout the last year. Daniel Roca helped check the page proofs and saved us from many errors of omission and commission. Oriol Cabrero stoically coped with the many requests we made of him with the book already in proof. We also thank all of the artists who have contributed their superb artwork, especially Alex Mascarell Llosa and Francesc Jutglar for preparing the additional figures needed here, and of course Ilian Velikov for the superb Blue-headed Quail-dove 'cover star'. Anna Motis and Christian Jofré are thanked profoundly for their assistance in making the range maps 'shipshape', also for quietly accommodating changes in proof. Finally, we thank Chris Bradshaw, Alex Berryman, Juan Freile, Nárgila Moura and Gustavo Rodríguez for their initial work on some of the texts; their contributions are specifically credited elsewhere.

The following have offered much valuable support, either by supplying details of, or discussing, records, permitting access to specimens and library materials, or reviewing various parts of this work: Mark Adams (NHMUK), Mike Akresh, Bertrand Jno Baptiste, Bill Boyle, Elwood Bracey, Matt Brady, Adam C. Brown (Environmental Protection in the Caribbean), Thibaut Chansac, Sean Christensen, Mat Cottam, Jack Eitniear (Center for the Study of Tropical Birds), Binkie van Es, Mark Garland, Jeff Gerbracht (Cornell Lab of Ornithology), Hein van Grouw (NHMUK), Percival Hanley, Bruce Hallett, Karen Halliday, Alison Harding (NHMUK), Claudia Hermes (BirdLife

International), Julian Hough, Lenn Isidore, Alvaro Jaramillo, Anthony Jeremiah, Lyndon John (Royal Society for the Protection of Birds), Tom Johnson, Allan Keith, Alexander C. Lees, Tony Leukering, Catherine Levy, Kevel Lindsay, James Lowen, Tim Marlow, Eddie Massiah, Nick Moran (British Trust for Ornithology), Robert L. Norton, Michael O'Brien (Victor Emanuel Nature Tours), Ryan Porteous, Robert Prŷs-Jones (NHMUK), Willy Raitière, Luis Ramos (SOPI), Martin Reid, Craig Robson, Forrest Rowland, Yvan Satgé and Denny Swaby.

Guy is very grateful to some long-time friends, Mark Elwonger, the late Mike Flieg, Arturo Kirkconnell (and his family), Andy Mitchell, William Suárez, and the late Jim Wiley, for their comradeship in both the field and museum over many years. His work on the present volume was conducted in tandem with a forthcoming checklist to Cuba, in partnership with Kirkconnell, Mitchell and Wiley, as well as Orlando Garrido, to be published by the British Ornithologists' Club in its Checklist series. He is especially grateful for the interest and support shown by Allan Keith in the latter project. Many years ago, Travis Rosenberry, formerly of the Peregrine Fund library, kindly made PDF files for Guy of almost all of James Bond's supplements to his West Indian checklist; they have been invaluable companions on the journey. Nigel Collar generously enabled several of Guy's museum visits, plied him with G&Ts, and provided lively discussion of all manner of subjects from Brexit to Shostakovich. Chris Bradshaw and Rob Williams were first-rate companions on Guy's 'formal' introduction to the region. Many other friends have joined him in the field in the West Indies, too many to mention by name, but he thanks them all for sharing some great times. Another longstanding comrade Hadoram Shirihai entirely financed one of Guy's trips, and Veronica Nogueira Gama, who even joined him on one visit to Cuba, kept the 'home fires' burning during the many periods when he was in the field.

Anthony personally thanks his best friend, Frantz Delcroix, for the many hours they have spent in the field or on the computer to protect the birds of Guadeloupe or discover the birds of the Caribbean. He also thanks passionate Caribbean birders Eric Delcroix, Antoine Chabrolle, Laurent Malglaive (all from Guadeloupe), Eddie Massiah (Barbados), Jorge Brocca (Dominican Republic), the Bahamas team (Erika Gates, Carolyn Wardle, Martha Cartwright, Woody Bracey), Lisa Sorenson (Birds Caribbean), Jeff Gerbracht (Cornell Lab of Ornithology, eBird) and many others from all over the Caribbean. Also the local partners—Office national de la chasse et de la faune sauvage (Blandine Guillemot, David Laffite, André Lartiges & David Rozet), Réserve Naturelle de Petite Terre (René Dumont, Alain Saint-Auret) and Parc national de la Guadeloupe (Maurice Anselme, Arnaud Larade)—who have enabled his studies of birds and from whom he has consequently learned so much. Last but certainly not least, his family, in particular his wife Marie-Josée and kids (Yasmine, Marine, Noah) who sometimes have suffered to live with a passionate birder.

Mark would like to thank the hundreds of local partners who have helped along the way, especially Sergio A. Colón López, Rafael Rodríguez Mojica, Giff Beaton, Leopoldo Miranda Castro, José Ramos Gómez, José F. Cordero, Manuel Cruz, Raul Pérez-Rivera and Joseph Wunderle, Jr. And of course, special thanks for the eternal support from my wife Mardie Lynne and family, Niña T. Oberle, William P. Oberle, William S. Oberle and Annie Farman.

Chris particularly thanks Chris Batty, Peter Boesman, William S. (Bill) Clark, Daniel López-Velasco, Christopher M. Heckscher (Delaware State University), Tom Hince, Curtis Marantz, John Murphy and Sébastien Reeber for generously sharing their expertise in confirming identification of some critical groups. Juliana Coffey, Sergio A. Colón López, Thomas Donegan, Peter Davey (National Trust for the Cayman Islands), Steven L. Hilty, Niels Larsen, Catherine Levy, Larry Manfredi, Robert L. Norton, Bruce Purdy, Michael H. Ryan and Lisa Yntema patiently answered multiple queries on national records. Chris is most grateful to colleagues Miguel Lentino, Margarita Martínez and Robin Restall at the Phelps Ornithological Collection, Caracas, and all of the staff at BirdLife International—especially David C. Wege, Ian Burfield and Stu Butchart—for decades of assistance. For their constant support Chris thanks his wife, Iokiñe Rodríguez, and family.

BIRDLIFE INTERNATIONAL

BirdLife International is a global partnership of conservation organizations (NGOs) that strives to conserve birds, their habitats and global biodiversity, working with people towards sustainability in the use of natural resources. Together we are over 100 BirdLife Partners worldwide—one per country or territory—and growing.

We are driven by our belief that local people, working for nature in their own places but connected nationally and internationally via our global Partnership, are the key to sustaining all life on this planet. This unique local-to-global approach delivers high-impact and long-term conservation for the benefit of nature and people.

BirdLife is widely recognized as the world leader in bird conservation. Rigorous science informed by practical feedback from projects on the ground in important sites and habitats enables us to implement successful conservation programmes for birds and all nature. Our actions are providing both practical and sustainable solutions significantly benefiting nature and people.

The Lynx and BirdLife International Field Guides project came to life after years of conversations motivated by a shared idea between Lynx and BirdLife that the existence of country field guides is a basic element for the 'emergence' and education of birdwatchers, ornithologists, bird guides and naturalists in any country, which, in turn, has important repercussions for the conservation of nature and biodiversity, both locally and globally.

In addition to the principal goal of the collection to produce modern, standardized field guides, especially for countries without any recent or country-level guide, we have a secondary goal of publishing a number of the titles in their local languages, to enhance the impacts of the work.

Apart from sharing and promoting the important vision of the collection, BirdLife also collaborates on the field guides via application of the systematics and taxonomy of the HBW and BirdLife International *Illustrated Checklist of the Birds of the World*, as well as with the distribution maps, which are based on distribution data compiled by BirdLife International and the *Handbook of the Birds of the World* for all the world's birds. These are then modified and improved by the authors of the field guides. Of course, the global conservation status for each species in the collection reflects the current version of the IUCN Red List, for which BirdLife International acts as the authority for birds.

SPECIES ACCOUNTS

CRACIDAE
Chachalacas
56 extant species, 1 introduced in region

Rufous-vented Chachalaca *Ortalis ruficauda* **LC**
Introduced; uncommon to rare resident (**Grenadines**).
L 53–61 cm. Two subspecies, of which nominate occurs in region. Tall scrub and disturbed woodland on Union I and Bequia; probably introduced from Tobago. Forages in small groups. **ID** Large, long-tailed, primarily terrestrial gamebird. **Ad** has slate-grey head, bare red throat, brown upperparts, and buffish underparts becoming rich rufous on undertail-coverts. Tail black with chestnut tip. **Juv** is indistinguishable from ad. **Voice** Loud, noisy calls, mainly heard early morning, apparently given in duets within groups: one bird giving low-pitched grating "OTRA MAS", answered by much higher-pitched "WATCH-a-lak". **SS** None. **AN** Rufous-tipped Chachalaca.

NUMIDIDAE
Guineafowl
8 extant species, 1 introduced in region

Helmeted Guineafowl *Numida meleagris* **LC**
Nine subspecies divided into four subspecies groups, of which one occurs in region.

West African Guineafowl *Numida (meleagris) galeatus*
Introduced, locally fairly common to rare resident (**GA**, **LA**).
L 53–63 cm. Two subspecies in group, of which both (*sabyi*, *galeatus*) occur in region. Long-established and locally common in Cuba (including I of Pines and some cays) and on Hispaniola (but declining), rare Puerto Rico, Virgin Is, St Martin and Barbuda (latter attributed to *sabyi*). Dry scrub and nearby farmland. Status confused by domestic birds in farmyards. Usually in groups. **ID** Large, rotund gamebird with horn-coloured bony casque. **Ad** (*galeatus*) is mainly blackish grey with conspicuous white spots and vermiculations, very pale bluish face skin, red facial appendages and black filoplumes on hindneck. **Juv** is dull brown with feathers tipped reddish cream. Casque, wattles, etc. smaller; retains natal down on head almost until ad plumage. **GV** Race *sabyi* larger and slightly blacker, with taller casque and longer filoplumes. **Voice** Raucous cackling that varies in intensity and length, e.g. nasal, staccato "kek-kek-kek-kek-kek-kek-kek...". Flocks maintain contact with metallic "chenk", and more plaintive, far-carrying "CHER-cheeng, CHER-cheeng...". **SS** None.

ODONTOPHORIDAE
New World Quails
35 extant species, 2 introduced in region

Northern Bobwhite *Colinus virginianus* **NT**
Twenty-two subspecies divided into four subspecies groups, of which two occur in region.

Eastern Bobwhite *Colinus (virginianus) virginianus*
Introduced, common to rare resident (**Bahamas**, **Haiti**).
L 20–28 cm. Eight subspecies in group, of which two (nominate[A] S Haiti, *floridanus*[B] N Bahamas, on New Providence, Abaco and Andros) occur in region. Increasingly local in Haiti, where recorded to 1640 m. Prefers similar habitats to, and behaviour much as, Black-breasted Bobwhite, e.g. pine barrens, wasteland, rough pasture and arable land in Bahamas. Reportedly introduced to latter in 19th century, and Haiti probably in 18th century (Lever 2005). **ID** Structure as Black-breasted Bobwhite. **Ad** ♂ has much less deeply rufous-saturated plumage than same sex of Black-breasted Bobwhite, with much more heavily white-spotted underparts. Lacks black on breast. **Ad** ♀ lacks black barring on underparts. **Juv** is similar to ♀. **GV** Race *floridanus* has narrower white supercilium and fewer white markings on underparts. **Voice** As Black-breasted Bobwhite. **SS** See remarks under Black-breasted Bobwhite.

Black-breasted Bobwhite *Colinus (virginianus) pectoralis*
Introduced, common to rare resident (**GA**, **O**).
L 24–28 cm. Six subspecies in group, of which one (*cubanensis*) definitely occurs in region. Common Cuba (including I of Pines), scarce and declining Dominican Republic and rare Puerto Rico. Formerly St Croix. Unsuccessful introductions St Kitts, Guadeloupe, Martinique and Barbados. Favours scrub and pasture; to c. 1500 m on Hispaniola, once 1930 m. Usually well hidden in or close to grassy cover; when flushed, coveys fly low and scatter widely. Flocks of up to 30. **ID** Plump, short-tailed, sexually dimorphic quail. **Ad** ♂ has blackish-chestnut head with bold white supercilium and throat. Upperparts and underparts reddish brown with white-spotted flanks. **Ad** ♀ has warm buff supercilium and throat, and less blackish crown and head-sides. Often more white on underparts. **Juv** is like ad ♀, but primary-coverts tipped buff and often has indistinct brown throat. **Voice** Loud whistled upslurred "bob-wheet!" given by both sexes. Coveys keep contact with sharp whistles and soft calls. **SS** Contrasting supercilium and throat should make species readily identifiable. No overlap with Crested Bobwhite. **TN** Race *cubanensis* (described for Cuban populations) unstable, and may be based on introductions. Considerable debate exists, but no evidence, e.g. from fossil record, that *Colinus* is native to West Indies. Furthermore, strong possibility that birds similar to races *floridanus* and *texanus* (of Eastern Bobwhite) occur in mainland Cuba, and that those on I of Pines recall one or more of Mexican races. Those in Dominican Republic and Puerto Rico emanate from translocations from Cuba. Consequently, we treat this species as a long-established non-native.

Crested Bobwhite *Colinus cristatus* **LC**
Introduced resident (**Grenadines**).
L 18–23 cm. Thirteen subspecies, of which one (*sonnini*) occurs in region. Current status largely unknown, though only on Mustique, where introduced 1885 but no recent records (Paice & Speirs 2010). Introduced in mid 1800s to Virgin Is (St Thomas), ultimately unsuccessfully. One reported on Andros (Bahamas) Jul 2006 (*NAB* 60: 589). Scrub, pasture and cultivation. Shy and typically keeps to cover, but can be confiding. Usually in pairs or small coveys. **ID** Prominent crest distinctive. **Ad** ♂ has buffish head and crest, black-and-white ring around neck, and relatively uniform reddish-brown underparts with black-and-white scaling on flanks. Upperparts dull brown with fine grey, black and white vermiculations. **Ad** ♀ has shorter crest, and is duller and less strongly marked, with buffier head and underparts. **Juv** is like ♀ but has buff-tipped primary-coverts and less distinct markings on underparts. **Voice** Call, given year-round, a loud whistled upslurred "bob-wheet!", less rich and mellow than Northern Bobwhite and sequence faster. Coveys maintain contact with chirps and cheeps. **SS** None in range.

Rufous-vented Chachalaca
ruficauda

West African Guineafowl
(Helmeted Guineafowl)
galeatus

Eastern Bobwhite
(Northern Bobwhite)

virginianus

floridanus

♂ ♀ ♂

Black-breasted Bobwhite
(Northern Bobwhite)
cubanensis

♂ ♀

Crested Bobwhite
sonnini

♂ ♀

27

PHASIANIDAE
Pheasants
187 extant species, 3 introduced in region

Indian Peafowl *Pavo cristatus* `LC`
Introduced, uncommon and local resident (**Bahamas**).

L ♂ 180–230 cm (tail 40–45 cm, train 140–160 cm); ♀ 90–100 cm (tail 32.5–37.5 cm). Monotypic. Present Little Exuma (since 1950s), but not frequently seen. Widely introduced elsewhere (e.g. Virgin Is), to farms, parks and large gardens, but no evidence of self-sustaining feral populations. **ID** Very large; ♂ with very long bushy train. **Ad** ♂ has velvet-blue head to breast, wire-like crest on rear crown, green back with narrowly barred black-and-white wing-coverts and tertials, and long glossy green train with large ocelli. **Ad** ♀ has white belly, brown upperparts, wings and tail, long uppertail-coverts and glossy green neck. **Juv** resembles ♀ but has less developed crest, paler head and inconspicuously barred upperparts. **2nd-y** ♂ resembles ad ♂, but train shorter without ocelli; length of train increases with age, markedly so until 4th-y. **Voice** Territorial call a far-carrying, penetrating "keeaaah" or "nyow" with nasal wailing quality, first note longest, rest shorter, decelerating at end. Also a loud, urgent "ka-an..ka-an..ka-an…", likened to braying donkey. **SS** None. **AN** Common Peafowl.

Red Junglefowl *Gallus gallus* `LC`
Introduced, local and uncommon resident (**GA, LA, O**).

L ♂ 65–78 cm; ♀ 41–46 cm. Five subspecies, of which one (perhaps nominate) occurs in region. Woodland and thickets; Little San Salvador, Inagua? (Bahamas), Cayman Is, parts of Dominican Republic, Puerto Rico (mainly Mona and possibly Culebra), Saba (Boeken 2018), Guadeloupe, Martinique and Grenadines. Domestic birds common throughout region. **ID** Resembles domestic chicken. **Ad br** ♂ has head and neck golden, hackles golden-yellow, back deep russet, rump golden, wings with green sheen and large yellowish panel in primaries, underparts dark. Tail has long, decurved central feathers. Fleshy red comb, red wattle and skin around eye. **Ad non-br** ♂ has reduced comb and wattles, lacks hackles, with crown and neck all dark. **Ad** ♀ is smaller and lacks bright colours of ♂. Short golden hackles on hindneck, upperparts brown vermiculated dark and streaked white, underparts paler. Tail rather short and blunt. Bare skin around eye pale red/pink. **Juv** is much like ad ♀ but has darker underparts; young ♂ soon develops reddish-yellow hackles and red on back, with blacker tail. **Voice** Similar to but shriller than well-known "cock-a-doodle-do" of village chicken. Also varied clucks and cackles. **SS** None.

Common Pheasant / Ring-necked Pheasant `LC`
Phasianus colchicus

Thirty subspecies divided into five subspecies groups, of which one occurs in region.

Grey-rumped Pheasant *Phasianus (colchicus) torquatus*
Introduced, local and uncommon resident (**GA, O**).

L ♂ 75–89 cm (tail 42.5–59 cm); ♀ 53–62 cm (tail 29–31 cm). Seventeen subspecies in group, of which one (*torquatus*) occurs in region. Wooded edges, adjacent grassland and cultivation, in Bahamas on Eleuthera (still?) and New Providence, locally in far W, SC & extreme E Cuba (and I of Pines) and SE Dominican Republic, at least. **ID** Large and long-tailed. **Ad** ♂ has dark green head and neck, bare red skin around eye, prominent white collar broken at front, purplish-red central breast, bluish belly, golden-yellow flanks, greenish-grey wing-coverts back and rump. **Ad** ♀ is dull sandy brown mottled darker on neck and breast-sides, otherwise almost unmarked below. Buffy-brown tail barred dark brown and buff. **Voice** Good indicator of presence. Two harsh crows "corrr … corrr", a series of grating doubled or tripled notes "kutuk … kutuk … kutuk" in alarm and, when flushed, a rapid set of "guduk-guduk-guduk…" notes. **SS** Only pheasant in region.

ANATIDAE
Ducks, Geese and Swans
165 extant species, 18 in region + 22 vagrants + 1 introduced

White-faced Whistling-duck *Dendrocygna viduata* `LC`
Vagrant (**GA, LA**).

L 38–48 cm. Monotypic. Freshwater and brackish wetlands. Records Cuba (last in 1947), Dominican Republic (once, May 1926), Puerto Rico (Jul 2005, Jul 2007), Guadeloupe (several shot by hunter, Sept 2011) and Barbados (large flock Sept 1887). **ID** Long-legged duck. **Ad** has white face with black neck and line bisecting white of chin and rest of head. Lower neck and upper breast rich chestnut, rest of underparts to tail black, except ochre-white breast-sides and flanks barred black. Back olive-brown with buff-brown fringes, scapulars grey-brown, lesser coverts dark chestnut and rest of wing bluish grey to black. **Juv** has greyish buffy-white or ash-grey face, throat and underparts, and less extensively and vividly chestnut breast. **Voice** A sibilant trisyllabic whistle "swee-swee-sweeoo", given in flight and when feeding, repeated after brief pause, often in chorus; a single "wheee" in alarm. **SS** Other *Dendrocygna* lack white on head.

Black-bellied Whistling-duck `LC`
Dendrocygna autumnalis

Two subspecies placed in separate subspecies groups, of which both occur in region.

Northern Black-bellied Whistling-duck
Dendrocygna (autumnalis) fulgens
Rare visitor or resident (**Bahamas, Cuba**).

L 43–53 cm. One subspecies in group. Freshwater marshes, brackish lagoons and rice fields. Very rare resident W & C Cuba and perhaps wanders to Bahamas, but has been released on Abaco, Eleuthera and New Providence (*NAB* 69: 308). **ID** Distinctive. **Ad** differs from conspecific Southern Black-bellied Whistling-duck (has also occurred Bahamas) principally in lack of grey lower breast to hindcollar. **Juv** is duller with paler bill. **Voice** As Southern Black-bellied. **SS** Like Southern Black-bellied, large white wing patch, black belly and reddish bill distinguish from other *Dendrocygna*.

Southern Black-bellied Whistling-duck
Dendrocygna (autumnalis) autumnalis
Rare visitor or resident (**Guadeloupe, Martinique, Barbados**), increasing vagrant elsewhere.

L 43–53 cm. One subspecies in group. Freshwater marshes and brackish lagoons. Rare Barbados, where probably only a visitor, mainly summer/autumn (and note Northern Black-bellied Whistling-duck occurs in captivity), annual Guadeloupe and Martinique. Elsewhere an apparently increasing vagrant, with widely scattered records from, e.g., Bahamas (Long I), Jamaica, Haiti, Puerto Rico, Virgin and Cayman Is, and virtually throughout Lesser Antilles (more than annual on some). Proof that all pertain to present taxon, rather than Northern Black-bellied, lacking but specimens available from Puerto Rico and St Vincent (Hellmayr & Conover 1948, Bond

Indian Peafowl

Red Junglefowl
gallus

Grey-rumped Pheasant
(Common Pheasant / Ring-necked Pheasant)
torquatus

White-faced Whistling-duck

Northern Black-bellied Whistling-duck
(Black-bellied Whistling-duck)

Southern Black-bellied Whistling-duck
(Black-bellied Whistling-duck)

1952, 1980, Banks 1978) and photographic evidence increasingly is clarifying their relative status. **ID** Distinctive duck with pinkish-red bill. **Ad** has chestnut-brown crown, nape and neck, grey upper breast and collar, solid black lower breast and belly, mottled white on ventral area. Buffish-grey head-sides and upper neck. Upperparts chestnut-brown with large white wing patches. **Juv** is duller with paler bill. **Voice** Whistling "pee-chee-chee", uttered particularly vociferously in flight, becomes an almost constant chatter from resting flocks. **SS** Large white wing patch, black belly and reddish bill distinguish from other *Dendrocygna*.

West Indian Whistling-duck *Dendrocygna arborea* VU
Locally common to rare resident (mainly **GA**, **O**), vagrant (most of **LA**).

L 48–58 cm. Monotypic. Endemic. Largely restricted to Greater Antilles, Bahamas and Cayman Is, but also breeds Antigua, Barbuda and Guadeloupe. Swamps and marshes surrounded by trees, especially mangroves, also other shallow, freshwater, brackish or saline ponds, dams and rice fields. Forages mainly at night. **ID** Large *Dendrocygna*. **Ad** is generally rather dark with distinctively spotted flanks, paler greyish-buff face and foreneck, and fine blackish streaks on neck forming necklace. **Juv** is duller with less distinct spotting on flanks and lacks necklace. **Voice** A five-syllable whistle given in flight and when foraging, and harsh, high-pitched "peep" notes, given continuously when nervous or alarmed. Groups give scarcely audible low-intensity chittering. **SS** Fulvous Whistling-duck has tawny-coloured underparts, distinct white flank streaks, unspotted white tail-coverts and slightly greater contrast between coverts and remiges on upperwing in flight. Compare Black-bellied and White-faced Whistling-ducks.

Fulvous Whistling-duck *Dendrocygna bicolor* LC
Uncommon to common resident (**GA**), rare visitor (**O**) or vagrant (**LA**).

L 45–53 cm. Monotypic. Rare visitor Bahamas, Cayman Is, Jamaica, Antigua, St Barthélemy and Barbados, vagrant Virgin Is and most of Lesser Antilles. Some evidence of winter immigration from North America, e.g., to Cuba. Status confused in some areas by introductions, e.g. Barbados. Rice fields and marshes with abundant vegetation. Usually in flocks. **ID** Comparatively large and fairly long-necked duck. **Ad** has plain buff-brown head and neck, latter with dark line at rear, extensive white patch streaked black on throat, buff-brown underparts with long white flank stripes and dark brown upperparts with rufous fringes. **Juv** is duller with uppertail-coverts tipped brown, paler underparts and little chestnut on upperwing. **Voice** A squealing whistle on take-off, high-pitched bisyllabic "pit-tu", "kit-dee" or "peee-chee" from ground or in flight, and conversational "cup-cup-cup-cup" of variable length. **SS** See West Indian Whistling-duck.

Masked Duck *Nomonyx dominicus* LC
Uncommon to locally fairly common resident (**GA**), rare to very rare elsewhere.

L 30–36 cm. Monotypic. Resident Cuba (including I of Pines), Hispaniola and Jamaica. Rare Puerto Rico (though one record of large flock), Guadeloupe, Martinique and Barbados (perhaps increasing); very rare or vagrant elsewhere S to St Lucia and Grenada. Intra-regional and off-island movements probably occur (Eitniear & Colón-López 2005, Buckley *et al.* 2009). Freshwater lakes, pools, swamps, marshes and slow-flowing rivers with abundant emergent and floating vegetation, often with tree cover. **ID** Distinctive duck with large white patch on upperwing conspicuous in flight. **Ad** ♂ has black mask reaching rear crown and neck, chestnut nape, strongly black-spotted flanks and upperparts, whitish breast and belly, blackish wings and tail. Bill pale blue with black tip. **Ad** ♀ has two dark bands on face, buff throat and upper neck. Rufous-brown body heavily streaked dark brown, especially on upperparts. Tail and wings blackish. Bill slate-grey with blackish nail. **Eclipse** ♂ is similar to ♀ but has larger wing patch, less contrasting head pattern, broader lower cheek-stripe and paler more contrasting flanks. **Juv** resembles ♀ but has more uniform underparts, darker crown and more extensively pale-barred upperparts. **Voice** In display, ♂ makes distinctive "kirri-kirroo-kirri-kirroo kirroo kirroo kirroo" and dull-sounding, throaty "oo-oo-oo", somewhat reminiscent of domestic pigeon. **SS** More robust ♀ Ruddy Duck has single dark facial stripe and lacks white wing patch.

Ruddy Duck *Oxyura jamaicensis* LC
Locally uncommon resident and fairly common winter visitor (mainly **GA**), uncommon and local resident, visitor or vagrant (**LA**, **O**).

L 35–43 cm. Monotypic. At least Greater Antilles receive winter visitors from North America. Locally common resident New Providence, uncommon and local in rest of Bahamas. Rare Virgin Is and Barbados, mainly vagrant Cayman Is and rest of Lesser Antilles, but breeds S to Grenada. Freshwater marshes, lakes and pools with emergent vegetation and open water. **ID** Small and dumpy, with short neck and broad, concave-ridged bill and prominent head. **Ad** ♂ has striking white cheeks, black crown and nape, mostly chestnut-brown body. Bill bright blue. **Ad** ♀ has dark rufous-brown crown, buff cheeks and neck mottled darker, with dark brown stripe from bill base to rear of ear-coverts. Upperparts dark brown vermiculated paler. Bill slate-coloured, sometimes patchily blue. **Juv** is similar to ♀ but has paler cheeks and bolder barring above, on flanks and rump. **Voice** Generally silent outside breeding season. Bubbling display of ♂ commences with popping sounds as bill is slapped against breast and terminates in bisyllabic "burp". ♀ produces varied hissing and squeaking noises during agonistic encounters. **SS** ♂ unmistakable. See Masked Duck.

Tundra Swan *Cygnus columbianus* LC
Two subspecies placed in separate subspecies groups, of which one occurs in region.

Tundra Swan *Cygnus (columbianus) columbianus*
Vagrant (**GA**, **LA**).

L 120–150 cm; **WS** 167–225 cm. One subspecies in group. Larger wetlands, estuaries, agricultural fields. All records (from Cuba, Puerto Rico, St Thomas, Antigua and St Kitts) in Nov–Feb, perhaps displaced by severe weather in E North America. **ID** Small swan with relatively short neck and rather peaked crown. **Ad** has variable yellow in front of eye, sometimes extending to base of maxilla, black over rest of bill with narrow yellow stripe on cutting edge at base of mandible; all-white plumage. **Juv** has pale dusky-brown plumage and mostly pale pinkish bill. White feathering acquired with age. **Voice** Yapping or honking notes typically delivered singly or in pairs. **SS** Extralimital **Mute Swan** *C. olor* (mainly Palearctic) has orange bill with black knob and is larger and longer-necked; recorded at least five times in Bahamas (mainly Abaco) and once on St Kitts, presumably from introduced population in USA (*NAB* 63: 518). Compare vagrant Snow Goose. **AN** Whistling Swan.

West Indian Whistling-duck

Fulvous Whistling-duck

Masked Duck

Ruddy Duck

juvenile
Tundra Swan
adult

Mute Swan

31

Brent Goose / Brant *Branta bernicla* `LC`
Four subspecies divided into three subspecies groups, of which two occur in region.

Pale-bellied Brent Goose *Branta (bernicla) hrota*
Vagrant (**Barbados**).
L 55–66 cm; **WS** 110–120 cm. One subspecies in group. Only record Nov 1876 (Feilden 1889) generally ascribed to *hrota*, but *B. b. nigricans* has recently occurred in Puerto Rico. In winter frequents estuaries, sandy shores, saltmarshes and nearby arable fields. **ID** Small, relatively short-necked goose. **Ad** has pale brownish-grey upperparts, brownish-white sides contrasting with blackish neck and upper breast; white necklace sometimes visible only close to. **Juv** has conspicuous white fringes on upperparts, which are generally duller and browner; acquires white marks on neck gradually. **Voice** Often silent. Most characteristic a hard, rolling "raunk, raunk" or softer "ronk"; in alarm a higher-pitched "wauk". **SS** See Black-bellied Brent Goose. **AN** Brent, Light-bellied Brent.

Black-bellied Brent Goose *Branta (bernicla) nigricans*
Vagrant (**Puerto Rico**).
L 55–66 cm; **WS** 110–120 cm. Two subspecies in group, of which one (*nigricans*) occurs in region. Records Dec 1977, Mar 2009 (Pérez-Rivera 1987, eBird). Two undocumented Bahamas reports (Andros, c. 1970s, Eleuthera, Nov 2004) most likely to have involved this race. Coastal bays and wetlands. **ID** Small, compact, short-necked goose with stubby bill. **Ad** has black head and neck with inverted V of white feathers on throat, dark brown underparts with broad paler fringes on flanks, white rear belly and vent, largely white uppertail-coverts (longest black) and black tail. **Juv** has conspicuous white fringes to upperparts, which are typically duller brown; acquires white on throat gradually, and underparts overall paler, while tail feathers may be tipped white, sometimes also primary- and median wing-coverts. **Voice** As Pale-bellied Brent Goose. **SS** Most similar to Pale-bellied Brent, but has broader white patches on neck and striking white triangular-shaped flank patch barred darker, as well as much darker rear belly contrasting with snowy-white undertail-coverts, whereas contrast between lower neck/breast vs. mantle is hardly visible. However, considerable variation exists in coloration of mantle and scapulars, prominence of white flank patch, and shape and extent of neck collar. Compare larger Canada Goose. **AN** Black Brant.

Canada Goose *Branta canadensis* `LC`
Vagrant (**GA**, **O**).
L 55–110 cm; **WS** 122–183 cm. Seven subspecies, of which at least nominate occurs in region. Records (mainly Oct–Apr), perhaps not all of wild provenance (AOU 1998), in Bahamas (Abaco, Andros, Cat I, Eleuthera, Grand Bahama, Great Exuma, New Providence), Cuba, Dominican Republic and Puerto Rico; older reports from Cayman Is and Barbados rejected (Bradley 2000, Buckley *et al.* 2009), but since reported again (Dec 2013) on former group (eBird). Also said to have been introduced in Caicos Is, but only records in winter (Buden 1987). Saline lagoons to freshwater swamps. **ID** Large, long-necked and long-billed goose. **Ad** has black head and neck with white 'chinstrap' patch, underparts brown with paler fringes, often whitish on upper breast and belly, and rear belly and undertail-coverts white. Upperparts darker with white uppertail-coverts and black tail. **Juv** is duller, especially head pattern, with white of cheeks tinged brown and less sharply separated from black; breast and flanks mottled not barred. **Voice** Typically a loud disyllabic "h-ronk". **SS** None, but compare Pale-bellied Brent Goose.

Snow Goose *Anser caerulescens* `LC`
Vagrant (**GA**, **LA**, **O**).
L 66–84 cm; **WS** 132–165 cm. Two subspecies, of which nominate occurs in region. Agricultural areas, freshwater ponds and swamps; Sept–Mar, rarely until May. Records (few recent outside first-named archipelago) from Bahamas (reasonably regular in N, rare in S), Cuba, Cayman Is, Puerto Rico, Virgin Is (St Croix, St Thomas) and Guadeloupe (Nov 2007–Feb 2008, Jan–Apr 2011). Barbados report not accepted (Buckley *et al.* 2009). **ID** Fairly small, stocky and short-necked goose. **Ad** has three colour morphs, at least two of which have occurred. **White morph** is all white except black primaries; bare parts pink. **Blue morph** is dark blue-grey with white head, variable white on neck, pale wing-coverts and white undertail-coverts and tail. **Intermediate morph** shows complete range of variation in amount of dark and white feathering. **Juv white morph** has mottled greyish head and neck, ash-brown back and scapulars with white edges; bill and legs dusky. **Juv blue morph** is typically dark grey-brown, lacking white on head and neck, with pale grey greater coverts. **Voice** Hard cackling, rather nasal and high-pitched "la-luk", likened to bark of small dog; family groups keep contact with low guttural grunts or long series of "uh-uh-uh" notes in rapid succession, and flight chorus a mix of honks, higher-pitched quacks or squawks. **SS** See Ross's Goose.

Ross's Goose *Anser rossii* `LC`
Vagrant (**Bahamas**).
L 53–66 cm. Monotypic. White morph photographed, Grand Bahama, Nov 2006 (*NAB* 61: 159), Abaco, Feb 2011 (*NAB* 65: 356). Open country, including agriculture and even golf courses. **ID** Relatively small-bodied, short-necked and round-headed goose. **White morph** has black wingtips and greyish primary-coverts; pinkish bill with paler nail, legs pinkish and eyes dark. ♀ averages smaller with shorter neck, slightly flatter forehead, reduced or no warty growths (caruncles) at base of bill and less obvious furrows on neck. **Blue morph** (very rare) is similar to that of Snow Goose, but apparently shows more white below and less on head and neck. **Juv** (white morph) has brownish-grey head, back, scapulars and upper breast-sides, with greyer or browner flight feathers. **Voice** Sad, murmuring "mmmmm" in contact, or repeated "uuggh uuggh uuggh" (higher-pitched in ♀) largely similar to Snow Goose but less piercing and overall higher-pitched, with a high-pitched "keek keek" and occasional harsh "kork" or "kwok" in flight. **SS** Smaller than Snow Goose, always lacks rusty stains on head common in latter, while shorter bill has different pattern and colour (lores meet base of maxilla in a straight line instead of a forward-curved arc), and lacks 'grinning patch' and caruncles. In addition to size differences, juv has less and paler grey than same-age Snow Goose.

Greater White-fronted Goose *Anser albifrons* `LC`
Vagrant (**Bahamas**, **Cuba**).
L 65–86 cm; **WS** 130–165 cm. Six subspecies, of which one (*gambelli*) occurs in region. Abaco, Feb–Mar 2010 (*NAB* 64: 337–338), and Grand Bahama and Grand Exuma, Bahamas, Jan–Mar 2018 (eBird); Zapata Peninsula, Cuba, late Oct–late Mar in 19th century, then Mar 1916 and Feb 2012 (Kirkconnell *et al.* in press). Freshwater ponds or lagoons, open farmland. **ID** Comparatively small, slender-necked grey goose. **Ad** has white blaze at base of bill, uniform grey-brown plumage with pale-fringed upperparts, irregular black bars on belly and darker rear flanks, white undertail-coverts and lower uppertail-coverts, and grey-brown tail with white terminal band. **Juv** lacks white blaze at bill base and dark markings on belly.

Pale-bellied Brent Goose
(Brent Goose / Brant)

Black-bellied Brent Goose
(Brent Goose / Brant)
nigricans

Canada Goose
canadensis

Snow Goose
caerulescens

white morph

blue morph

Ross's Goose

white morph

blue morph

Greater White-fronted Goose
gambelli

33

Voice In flight a ringing "kyow-lyow" and high-pitched "widawink widawink...". **SS** The only grey goose recorded in region.

Long-tailed Duck *Clangula hyemalis* VU
Vagrant (**Anguilla**).

L ♂ 51–60 cm (including tail up to 15 cm); ♀ 37–47 cm; **WS** 73–79 cm. Monotypic. ♂, Oct 2008 (Mukhida *et al.* 2011). Winters mostly at sea, but also on large, deep freshwater lakes or brackish lagoons. **ID** Small, compact, round-headed duck with fairly short bill. **Ad non-br** ♂ has white head and neck, dusky face and dark brown patch below, dark brown chest-band, dark brown upperparts with long pale grey/white scapulars, narrow dark brown band on rump and white tail with long dark brown central rectrices. Bill dark with broad pink band. **Ad br** ♂ has head and neck dark, and bright fringes to scapulars (lower elongated, though less so than in non-br plumage); bill can be all dark. **Ad** ♀ has dark brown central crown, white head and neck, brown face, pale brown breast-band, white underparts and brown upperparts. Bill grey. **Eclipse** ♂ resembles br ♂ with shorter lower scapulars, the fringes of which are less bright. **Juv** resembles ♀, face dusky but lacks dark brown patch on face-sides, blue-grey bill acquires pink in ♂ (from Oct). **Voice** Unlikely to be heard in region. Yodelling "ow-ow-owlee" given by ♂ during courtship and on migration, when both sexes also give nasal "gak". **SS** Shape distinctive and bold face patch, though variable, is unique among diving ducks in region as is pink band on bill of ♂ in non-br plumage. **AN** Oldsquaw.

Surf Scoter *Melanitta perspicillata* LC
Vagrant (**Cuba**).

L 43–56 cm; **WS** 76–92 cm. Monotypic. 1st-w ♂, Canasí Reservoir, Jan 2015 (Garrido *et al.* 2016). Winters mostly at sea or in shallow bays, estuaries and river mouths. **ID** Medium-large sea duck with angular head and large bill. **Ad** ♂ has patches of orange-yellow, scarlet, white and black on swollen bill. Black with white forehead and large white patch on nape. **Ad** ♀ is brown with dark crown and paler face with whitish patches at bill base, behind eye and sometimes on nape. **Juv** resembles ♀ but has paler breast, sometimes lacks whitish nape, but has more conspicuous whitish markings on head. Bill pattern of ♂ develops in 1st-w. **Voice** Largely silent in winter; ♂ gives low whistles or sharp "puk-puk". **SS** Bill pattern and white patches of ♂ diagnostic. Large, swollen bill of ♀ (with straight or convex culmen) distinctive; when present, pale patch on nape diagnostic. Compare White-winged Scoter (not yet definitely recorded in region).

White-winged Scoter *Melanitta deglandi* LC
Hypothetical (**Bahamas**, **Cuba**).

L 50–58 cm; **WS** 86–99 cm. Monotypic. Two undocumented reports: Abaco (Bahamas), Nov 2010 (*NAB* 65: 181); Cayo Coco (Cuba), Jan 2016 (Kirkconnell *et al.* in press). **ID** Large, long-bodied and thick-necked sea duck with long uniquely shaped bill. **Ad** ♂ has largely orange-yellow maxilla, starting just behind nostril, with black raised knob at base and largely black-based mandible. Black, except brown flanks, with long white eye patch and white secondaries visible on folded wing. **Ad** ♀ is brown with large whitish patches at bill base and behind eye, sometimes joined in single band across face.

Juv recalls ♀ but has paler breast and more conspicuous whitish markings. **Voice** Largely silent in winter. **SS** Most liable to confusion with Surf Scoter, but note head and wing patterns. **TN** Usually treated as race of extralimital Velvet Scoter *M. fusca* (N Europe to C Asia) or conspecific with extralimital Siberian Scoter *M. stejnegeri* (E Asia) when latter separated from Velvet Scoter.

Bufflehead *Bucephala albeola* LC
Vagrant (**GA**, **LA**, **O**).

L 32–40 cm; **WS** 53–61 cm. Monotypic. Records Bahamas (Abaco, Eleuthera, Grand Bahama), Cuba, Jamaica, Puerto Rico and St Martin span Oct–Mar. Shallow coastal bays, reservoirs and lagoons. **ID** Small duck with large head and tiny bill. **Ad br** ♂ has black head with striking iridescent gloss on sides (rarely visible), white V-shaped wedge on nape, dark back and tail, white breast, belly and flanks, bluish-grey bill, and bright pink legs and feet. In flight, black wings with broad white band over most secondaries and many coverts. **Ad** ♀ has dark brownish head, white oval-shaped patch on ear-coverts, dark brownish-black back, brownish-grey flanks, and dull white breast and belly; legs and feet dark bluish grey. In flight dark brownish-grey wings with white patch in secondaries. **Eclipse** ♂ has dark head and back, large pale patch on ear-coverts, greyish underparts and white wing-coverts. **Juv** is browner and duller, with smaller white head patch, although 1st-w ♂ acquires blacker head with larger pale patch on ear-coverts, more white on flanks and noticeably paler rump. **Voice** Largely silent. **SS** Black head with large white wedge across nape of ♂ unique; ♀ easily distinguished from extralimital **Common Goldeneye** *B. clangula* (Holarctic) by dark head with white oval patch behind eye. Based on 19th-century claim, Common Goldeneye sometimes admitted as vagrant to Bahamas (Raffaele *et al.* 1998), and even less frequently Cuba and Barbuda, but we follow Bond (1956) in considering all of these reports unsatisfactory.

Hooded Merganser *Lophodytes cucullatus* LC
Vagrant (**GA**, **LA**, **O**).

L 42–50 cm; **WS** 56–70 cm. Monotypic. Records (mid Nov–May, mainly Dec–Feb) in Bahamas, Cayman Is, Cuba, Jamaica, Hispaniola, Puerto Rico, Virgin Is (St Croix, St John), Anguilla, Antigua, Guadeloupe, Martinique and Barbados, at inland ponds and lagoons. **ID** Medium-sized, long-bodied duck with long bill and steep forehead, and long flat crown with crest at rear, creating distinctive profile. **Ad br** ♂ has black head with large white patch, elongated when flat and broad when crest raised. White-striped tertials, white breast with two black vertical stripes at rear, rusty-buff flanks, black bill, yellow eye, and blue-grey patch on upperwing-coverts. **Ad** ♀ has bushy, dull rufous crest, pale chin, largely grey body, with white belly and white-striped tertials, pale orange mandible and cutting edges, and dark eye. **Eclipse** ♂ recalls ♀ but crest small and dusky; eye colour and wing pattern as ad br. **Juv** resembles ♀ with shorter (sleeker) crest and paler upperparts, less white in wings and white chin until 2nd-w; 1st-y ♂ may show evidence of white crest. **Voice** Unlikely to be heard. **SS** Head shape unique; black, white and rusty-buff plumage of ad ♂ distinctive; bushy rufous crest of ♀ contrasting with paler face separates it from Red-breasted Merganser and Goosander.

Long-tailed Duck

♂ breeding

♂ non-breeding

♀

Surf Scoter

♂

♀

White-winged Scoter

♂

♀

Bufflehead

♂

♀

Common Goldeneye

♂

♀

Hooded Merganser

♂

♀

Goosander / Common Merganser `LC`
Mergus merganser
Hypothetical (**Cuba**).

L 58–72 cm; **WS** 78–102 cm. Three subspecies, of which one (*americanus*) is most likely to occur in region. ♀, Cayo Guillermo, Dec 2013 (Parada Isada *et al.* 2014) is undocumented, but was seen by at least one observer with ample experience of the species. **ID** Large, long-bodied and long-billed duck with fairly thick neck. **Ad br** ♂ has green head, white breast, belly and flanks, black mantle and upper scapulars, white lower scapulars and wing-coverts, dusky-grey rump and tail, and red bill. **Ad** ♀ has reddish-brown head with shaggy crest, white chin and pale grey body. **Eclipse** ♂ recalls ♀ but has shorter crest and large white patch on wing-coverts. **Juv** resembles ♀ but paler-headed with pale stripe on lores, larger, more diffuse whitish chin, and shorter crest. **Voice** Unlikely to be heard.
SS Compare Red-breasted Merganser.

Red-breasted Merganser *Mergus serrator* `LC`
Locally common (**Cuba**) to rare winter visitor (rest of **GA**, **O**).

L 52–58 cm; **WS** 67–86 cm. Monotypic. Records Bahamas S to Cayman Is and San Andrés, E to Puerto Rico and St Martin, mid Oct to early May, very rarely from late Aug. Nearshore and offshore waters, sandy estuaries, sheltered bays and brackish lagoons. **ID** Long body, long narrow neck with steep forehead, shaggy double crest and long, narrow-based, bill. **Ad br** ♂ has green head, white neck-ring, rufous-brown breast streaked dark, bordered by black area with conspicuous white spots; vermiculated upperparts with white wing-coverts. **Ad** ♀ has light reddish-brown head with pale loral-stripe, pale chin and greyish body. **Eclipse** ♂ resembles ♀ but crown darker, dark upperparts have broad grey fringes, and mostly grey with pale fringes on breast and flanks. **Juv** recalls ♀ but has black upperparts with large white patch on wing-coverts and much shorter crest. **Voice** Largely silent; ♂ utters complex, wheezy, cat-like "yeow-yeow", single "yeow" or loud, rough-sounding purr; ♀ gives deep gruff-sounding "gra gro garr" or "grack" in alarm.
SS See Hooded Merganser and Goosander.

Orinoco Goose *Neochen jubata* `NT`
Vagrant (**Jamaica**, **Barbados**).

L 61–66 cm. Monotypic. ♀ specimen, Jamaica, Oct 1865 (Banks & Hole 1992; USNM 42013); four seen, Barbados, Mar 1995 (Buckley *et al.* 2009). In native range, along forested rivers, with rocky or sandy beaches, and wooded islands; also more open country, in wet savanna and around large freshwater wetlands. **ID** Unmistakable goose. **Ad** has pale greyish-buff head and broad, often ruffled-looking, neck, rufous upper mantle and scapulars, dark green wings with broad white band on secondaries, glossy green speculum, chestnut flanks and belly, and white undertail-coverts. **Ad** ♀ is slightly smaller with shorter, paler neck feathers. **Juv** is a duller more washed-out version of ad, lacking gloss on wings and tail. **Voice** ♂ gives loud "shewit shewit" and ♀ "a-ohk" in flight.
SS None. **TN** Some authorities place in genus *Oressochen*.

Common Shelduck *Tadorna tadorna* `LC`
Vagrant (**Barbados**, **Martinique**).

L 56–67 cm; **WS** 100–133 cm. Monotypic. 1st-y ♀, Barbados, Nov 2013 (*NAB* 68: 162); one shot, Martinique, Nov 2015 (Belfan & Conde 2016). Status in New World compromised by likelihood that many reports involve escapees, but natural vagrancy plausible (Howell *et al.* 2014) and recently accepted to occur by AOU (Chesser *et al.* 2018). Most likely at coastal wetlands. **ID** Large, bulky duck with long neck and goose-like proportions. **Ad br** ♂ has dark green head and white body, with broad rusty-brown breast-band forming collar, black scapulars, iridescent green secondaries and dark primaries. Bill bright red with swollen basal knob, legs pink. **Ad** ♀ is smaller with white face mottling, narrower breast-band and smaller bill knob than br ♂. In flight, white wing-coverts and black flight feathers with green-glossed secondaries. **Eclipse** ♂ has white face mottling, less distinct breast-band and smaller bill knob. Eclipse ♀ is duller with more extensive white face mottling than eclipse ♂ and an even less distinct breast-band. **Juv** is much duller than ad, lacks breast-band and has rather ill-defined brownish head and body markings. **1st-y** is characterized by white trailing edge to secondaries and inner primaries. **Voice** Perhaps unlikely to be heard. Vocalizations sexually dimorphic, with ♂ uttering whistles, especially a soft, clear "whee-chew", while ♀ is typically louder and lower-pitched with a cackling quality. **SS** Rusty breast-band on predominantly white body, combined with green head, bright red bill and pink legs of ad, unique among wildfowl in region. Compare **Egyptian Goose** *Alopochen aegyptiaca*, introduced into Florida (USA) and recorded several times in Bahamas (*NAB* 69: 508, eBird).

Muscovy Duck *Cairina moschata* `LC`
Introduced (**Cuba**, **St Martin**, **Martinique**).

L 66–84 cm. Monotypic. Present in Cuba since at least mid 1800s, but extent to which pure phenotype survives is very poorly known; widespread on ponds, slow-flowing rivers, and lakes. Regularly observed and breeds on St Martin (since 2001). Domestic and semi-feral throughout region. **ID** A distinctive, large bulky duck. **Ad** ♂ has mostly blackish plumage with short mane and striking white upperwing-coverts; bill black with bluish-white band, hooked tip and enlarged basal knob. **Ad** ♀ is similar to ♂ but smaller and lacks knob at base of bill. **Juv** is browner with little or no white on upperwing. **Voice** Considered almost mute: ♂ gives asthmatic wheeze and ♀ single quack when alarmed. **SS** None. Domesticated Muscovy Ducks look smaller, with variable white on body.

Wood Duck *Aix sponsa* `LC`
Uncommon resident and winter visitor (**Cuba**), rare passage migrant and winter visitor (**Bahamas**), vagrant (**LA**, **O**).

L 43–54 cm; **WS** 68–74 cm. Monotypic. Mainly Sept–Apr. Vagrant: Cayman Is E to St Kitts and perhaps Guadeloupe; at least some perhaps from Cuba, rather than North America. Freshwater swamps, marshes, pools and lakes, usually surrounded by trees. In winter, on more open waters, occasionally brackish. Introduced Cayman Is (mid 1980s, but failed) and Barbados (1999/2000). **ID** Unmistakable by crest and striking facial pattern. **Ad** ♂ has metallic-green head with white lines and patches, white chin and throat, dark greenish-brown mantle, metallic-blue, green and black wings with white tips to secondaries and tertials, and long, curved black coverts streaked golden-buff. Upper breast vinous spotted white; bill multicoloured. **Ad** ♀ is dowdy in comparison, olive-brown above and white below, with white chin and throat, white line around eyes and bill base, mottled breast and flanks; crest and mantle glossed purple and green. **Juv** resembles ♀ but has streaked and mottled brown belly. **Voice** Rather quiet, ♂ gives soft, high, rising whistle in contact and burping "pfits" in courtship, and ♀ various calls including loud squealing "we-eek, we-eek". **SS** None. **AN** Carolina Duck.

Goosander / Common Merganser
americanus

Red-breasted Merganser

Orinoco Goose

Common Shelduck

Egyptian Goose

Muscovy Duck

Wood Duck

37

Common Pochard *Aythya ferina* VU
Vagrant (**Barbados**).

L 42–49 cm; **WS** 67–82 cm. Monotypic. One ♂ and three ♀, Feb 2011 (*NAB* 65: 356) during unusually cold weather in W Europe (Howell *et al.* 2014). In winter, prefers larger lakes, reservoirs, brackish coastal lagoons and tidal estuaries. **ID** Long bill with concave culmen, long, gently sloping forehead and peaked rear crown give distinctive profile. **Ad** ♂ has chestnut head, black breast and grey body with black tail-coverts. Bill dull grey to blackish, with black tip and pale grey subterminal band. In flight has grey bar on upperwing. **Ad** ♀ has dull brown head with pale grey spectacles and eyestripe, throat, lores and cheeks (pattern varies), greyish-brown body, darker above. In flight upperwing duller than ♂. **Eclipse** ♂ is dull with brown-tinged plumage. **Juv** recalls ad ♀ but has duller head lacking spectacles and eyestripe. **Voice** Unlikely to be heard, with ♂ usually silent (except on breeding grounds); ♀ gives soft growl when flushed. **SS** Redhead has steeper forehead, more rounded crown and thicker, less tapering bill that is much paler grey basally, overall body colour darker grey in ad ♂ and darker brown in ♀; eye of ♂ yellow to yellowish orange, vs. red in Common Pochard. Canvasback is larger, paler-bodied, with all-dark bill (can be shown by ♀-type Common Pochard) and in flight has less contrasting upperwing. Both principal confusion species generally very rare in region, and neither has been recorded in Barbados (where the only Common Pochards recently seen), thus any of these species should be carefully scrutinized.

Redhead *Aythya americana* LC
Very rare winter visitor (**Bahamas**, **Cuba**), vagrant (**Cayman Is**, **Hispaniola**).

L 41.5–54 cm; **WS** 74–85 cm. Monotypic. Brackish and freshwater lagoons, also reservoirs; late Nov–early Mar. Reported once in large numbers (Keith *et al.* 2003). **ID** Mid-sized duck with steep forehead and rounded crown. **Ad** ♂ has very bright chestnut head, black breast and grey body with black tail-coverts. Bill pale blue-grey with broad black terminal band. **Ad** ♀ has yellowish-brown head and neck, darker on crown, paler at bill base and behind eye, with pale eye-ring and scattered white feathers on nape, brown breast, greyish-brown back and sides sometimes flecked white, white belly and greyish-brown tail-coverts (whiter terminally) and tail. In flight darker grey wing-coverts contrast with paler wingbar. **Eclipse** ♂ is duller with less sharply defined blackish breast. **Juv** recalls ad ♀ but is more heavily mottled; **1st-w** ♂ resembles eclipse ad. **Voice** In winter ♂ gives a faint "zoom-zoom", especially at night. ♀ utters a loud, clear-sounding "squak" on take-off. **SS** Canvasback has longer, more tapering, uniformly dark bill, longer, more gently sloping forehead, flatter crown, paler body and in flight a less contrasting upperwing. See Common Pochard.

Canvasback *Aythya valisineria* LC
Vagrant to very rare winter visitor (**GA**, **LA**, **O**).

L 48–61 cm; **WS** 74–90 cm. Monotypic. Recorded Bahamas, Cuba (where probably very rare rather than vagrant), Hispaniola, Puerto Rico and Antigua, Oct–Apr, on larger lakes, reservoirs and coastal wetlands. **ID** Large *Aythya* with very long bill, long sloping forehead and long neck creating distinctive profile. **Ad** ♂ has chestnut head, darker forecrown, black breast and uniform whitish-grey body with black tail-coverts. **Ad** ♀ has greyish-brown head and neck, darkest on crown and nape, palest on chin, throat and eyestripe, brown breast, whitish-grey flanks, white belly, whitish-grey back and scapulars often mottled brown, dusky-brown rump, and whitish vent. In flight upperwing lacks strong contrast between grey lesser and median coverts, and flight feathers. **Eclipse** ♂ is duller and has dark grey breast and tail-coverts. **Juv** recalls ad ♀ but plain greyish brown overall with brownish belly; ad characters gradually acquired during late autumn/winter. **Voice** Vocalizations include "ick ick cu-oo" of ♂ ('Kink-neck' call), an inconspicuous coughing "hiff" and "rrr-rrr-rrr" in aggression. **SS** See Redhead and Common Pochard.

Ring-necked Duck *Aythya collaris* LC
Rare to common winter visitor (**GA**, **O**), rare to vagrant (mainly **LA**).

L 37–46 cm; **WS** 61–75 cm. Monotypic. Uncommon to common Bahamas, Cuba, Jamaica, Hispaniola, Puerto Rico, Virgin Is and Guadeloupe, rare Barbados, San Andrés, vagrant Cayman Is and Lesser Antilles. Open freshwater wetlands, including reservoirs, mainly Sept–Apr, but probably occasionally oversummers. **ID** Fairly small, short-bodied *Aythya* with long neck and high crown, peaked at rear creating distinctive profile. **Ad** ♂ has black head, neck, breast, scapulars and back, with dark purple sheen to most feathers, inconspicuous vinous collar, white belly, grey-vermiculated flanks (with vertical white wedge at front edge), black tail and coverts. Grey bill with narrow white band at base, broad white subterminal and black terminal bands. In flight, grey wingbar on secondaries and primaries. **Ad** ♀ has dusky to brown head and neck, darkest on crown and nape, palest near bill base, on throat, foreneck and around eyes. Brown breast and flanks, white belly, and dark brown back, tail and scapulars. **Eclipse** ♂ recalls ♀ but duller with reduced whitish blaze at bill base and brown-mottled flanks. **Juv** resembles ♀ but has pale-mottled underparts. **Voice** Usually silent in region. **SS** ♂ Lesser Scaup has greyer, vermiculated upperparts, lacks subterminal bill-band, has less angular head, whiter in inner flight feathers, but broader black trailing edge to secondaries and tertials. ♀ Lesser Scaup darker brown, with less prominent face markings, no band on bill, less angular head. See Tufted Duck.

Tufted Duck *Aythya fuligula* LC
Vagrant (**Puerto Rico**, **Guadeloupe**, **Barbados**).

L 40–47 cm; **WS** 65–72 cm. Monotypic. Old World vagrant: ♀, Puerto Rico, Nov–Dec 2012 (*NAB* 67: 356), ♀, Barbados, Mar 2017 (eBird), ♂, Guadeloupe, Mar 2019 (eBird). In winter, on brackish lagoons, tidal bays and freshwater wetlands. **ID** Mid-sized *Aythya* with crest or tuft on rear crown. **Ad** ♂ has black head glossed dark purple, with long drooping crest, black breast and upperparts, white flanks and black tail-coverts. Grey bill has indistinct pale subterminal band and broad black tip. **Ad** ♀ is brown with slightly darker back and paler flanks, often with white blaze at bill base; tuft on rear crown. **Eclipse** ♂ has reduced crest and brownish cast. **Juv** recalls ♀, but ♂ has darker head and foreneck. **Voice** Rarely vocalizes in winter. Occasionally a soft "kack". **SS** Vs. Ring-necked Duck in non-ad ♂ plumage, may show whitish fore-flanks, but these are distinctly S-shaped in Ring-necked, which is also shorter-tailed and longer-bodied. Also separable from ♂ Ring-necked by bill pattern and lack of collar; in ♀-like plumage, has darker head, breast and upperparts than Ring-necked, pale brown breast-sides and flanks, no white at bill base and whiter underparts. Vs. ♀ Lesser Scaup, ♀ of present species always has short crest on nape, squarer head, more black on bill tip, strong contrast between upperparts and flanks, and longer white wingbar. Greater Scaup lacks crest or tuft, is larger with more rounded head and much less black on bill tip: ♂ has whitish scapulars with grey vermiculations, ♀ has broader white blaze at bill base and usually a pale patch on face.

Common Pochard

Redhead

Canvasback

Ring-necked Duck

Tufted Duck

variant

39

Greater Scaup *Aythya marila* `LC`
Vagrant (**GA**, **LA**, **O**).

L 40–51 cm; **WS** 72–84 cm. Two subspecies, of which one (*nearctica*) occurs in region. Bahamas (Cat I, Jan 1960; New Providence, Jan 1971, Dec 2011–Feb 2012, Jan 2013, Nov 2013, Dec 2014; Eleuthera, Jan 1972; Andros, Jan 1996; Great Inagua, Dec 2012; Turks & Caicos, Mar 2013, Apr 2014, Apr 2015), Jamaica (winter 1994), Virgin Is (St Croix, Dec 1984–Feb 1985, Nov 1985) and Barbados (Nov 1979, Feb 2003, Nov–Dec 2008). Claimed in Cuba and Dominican Republic, but inadequately documented. Prefers deep saline pools and brackish ponds. **ID** Large *Aythya* with very rounded head. **Ad** ♂ has black head glossed green or purple, black breast, white belly and flanks, whitish scapulars with fine grey vermiculations and black tail-coverts and tail. Bill pale grey with black restricted to small area around nail. In flight upperwing has white bar, dusky on outer primaries. **Ad** ♀ is brown with scapulars variably vermiculated grey, broad white blaze at bill base and sometimes a large pale cheek patch. **Eclipse** ♂ is duller with brown cast to much of plumage. **Juv** recalls ♀ but has less white at base of bill, grey often appears in scapulars by late autumn. **Voice** Generally rather silent in winter. **SS** See Lesser Scaup. Hybrid *Aythya* might also be confused with pure Greater Scaup but generally show more extensive black on bill tip; consult specialist literature.

Lesser Scaup *Aythya affinis* `LC`
Locally fairly common to uncommon winter visitor (**GA**, **O**), rare to very rare (**LA**).

L 38–48 cm; **WS** 64–77 cm. Monotypic. Locally fairly common Bahamas and Cuba, uncommon Jamaica, Hispaniola, Puerto Rico, Cayman Is and Guadeloupe, and rare to very rare Virgin Is, Lesser Antilles and San Andrés. Freshwater and brackish lakes, ponds and reservoirs; mid Aug to late Apr. **ID** Fairly small *Aythya*; steep forehead and rounded crown with slight peak at rear give distinctive profile; bill rather narrow with concave culmen. **Ad** ♂ has black head glossed purple or green, black breast and upper mantle, grey scapulars coarsely vermiculated dusky grey, white flanks finely vermiculated grey, and black undertail-coverts. Black on bill restricted to nail. In flight two-toned wingbar, white on secondaries and grey-brown on primaries. **Ad** ♀ is brown with broad grey-fringed scapulars and flanks vermiculated darker, white blaze at bill base and sometimes has (especially in summer) pale patch on face. **Eclipse** ♂ is dull with brownish flanks. **Juv** recalls ♀ but is duller, with darker eye, plain scapulars, paler and browner underparts, and white on face less clearly defined. **Voice** Generally silent in winter; ♂ gives frequent whistled "whew" during group displays, and faint "whee-ooo". **SS** Greater Scaup is larger with broader body, broader bill with straighter culmen, more rounded head peaking on forecrown, slightly more narrowly and less coarsely vermiculated body, and more extensive wingbar extending to inner primaries; ♀ typically has more extensive white at bill base and larger pale cheeks. Hybrids between various *Aythya*, perhaps unlikely to be observed in region, could be confused with pure Lesser Scaup.

Garganey *Spatula querquedula* `LC`
Vagrant (**GA**, **LA**).

L 37–41 cm; **WS** 58–69 cm. Monotypic. Old World vagrant: Puerto Rico (Jan–Mar 1978), Guadeloupe (♂ Jan–Mar 2006, possibly two ♂♂, Mar 2007), Martinique (bird shot, c. Jan 2000) and Barbados (Aug 1960, Dec 2000–Jan 2001, Nov 2006–Mar 2007, Dec 2007). First North American record in 1957 (Howell *et al.* 2014). Could appear at freshwater marshes, pools and small lakes with emergent vegetation, coastal marshes or lagoons, provided there is some partially submerged fringe vegetation. **ID** Small; ♂ with broad white eyebrow. **Ad** ♂ has brownish face and breast, greyish vermiculated flanks, white central belly, and brown upperparts with long scapulars. In flight pale grey upper forewing and green speculum bordered by white bars. **Eclipse** ♂ resembles ♀ except wing pattern. **Ad** ♀ has pale supercilium, dark eyestripe, pale cheeks with dark horizontal stripe and obvious pale spot near bill base; darker forewing in flight. **Juv** resembles ♀ but belly less clean. **Voice** Readily vocalizes only when breeding, ♂ giving harsh wooden rattling note, "rrar ... rrar ... rrar". **SS** ♂ Green-winged and Eurasian Teals have green in head, while ♀ lacks spot at bill base. See Blue-winged and Cinnamon Teals.

Northern Shoveler *Spatula clypeata* `LC`
Uncommon to common winter visitor (**GA**, **O**), uncommon to vagrant (**LA**).

L 43–56 cm; **WS** 70–85 cm. Monotypic. Uncommon to common winter visitor on most of Greater Antilles, Bahamas and Cayman Is, rare Jamaica and Virgin Is; uncommon to rare Antigua, Guadeloupe and Barbados, vagrant elsewhere in Lesser Antilles. Freshwater and brackish ponds and lagoons, mainly Oct–May, but from Jul. **ID** Mid-sized with long, broad, spatulate-tipped bill. **Ad** ♂ has green head, white underparts with rufous flanks and belly, and largely black-and-white upperparts. In flight bright turquoise-blue forewing with white rear border and green speculum. **Ad** ♀ is brown with pale feather fringes and slightly paler face; upper forewing pale grey without white border and speculum duller. **Eclipse** ♂ is similar to ♀ but typically darker overall, with yellow iris, bluer forewing and often blacker tertials. **Juv** recalls ♀ but has densely speckled breast to belly, and sharply pointed upper flanks feathers. **Voice** Unlikely to be heard. **SS** Bill shape unique in region.

Cinnamon Teal *Spatula cyanoptera* `LC`
Vagrant (**GA**, **LA**, **O**).

L 35–48 cm. Five subspecies, of which one (*septentrionalium*) occurs in region. Records Bahamas, Cuba (Nov–Dec), Jamaica, Puerto Rico, Virgin Is (Sept–Nov), Antigua and Grenada; old reports on Barbados not accepted (Buckley *et al.* 2009). Hybrid with Blue-winged Teal seen Cayman Is. Shallow freshwater or brackish wetlands, with abundant emergent and fringe vegetation. **ID** Bright rusty dabbling duck with blue forewing and bold white underwing. **Ad** ♂ has head, neck, flanks and belly reddish brown to rich chestnut, with blacker crown, chin, throat and nape. Black undertail-coverts. Back, rump, uppertail-coverts and tail brownish black. Bill black and slightly spatulate; legs and feet yellow to orange; iris reddish. **Ad** ♀ is like ♀ Blue-winged Teal but warmer-toned with plainer head pattern. **Juv** lacks warm brown tones, has streaked rather than spotted breast, and greyish legs. **Voice** Not very vocal. **SS** ♂ unmistakable. ♀ Blue-winged Teal has distinct whitish loral spot, greyer plumage and brown iris. ♀ lacks dark cheek-stripe of same-sex Garganey.

Greater Scaup
nearctica

Lesser Scaup

Garganey

Northern Shoveler

Cinnamon Teal
septentrionalium

41

Blue-winged Teal *Spatula discors* `LC`
Fairly common to locally common winter visitor and passage migrant (**GA**), less common elsewhere.

L 35–41.5 cm; **WS** 60–64 cm. Monotypic. Less frequently recorded, and in smaller numbers, in Lesser Antilles. Can occur any month, but most frequent Oct–Apr (Aug–Jun), on shallow fresh, brackish or saline pools. **ID** Relatively small duck. **Ad** ♂ has steel-blue head and neck, with white crescent between bill and eye, black crown, forehead, chin and border of crescent, brownish body, flanks and belly with darker brown spots, very dark tail-coverts and white rump-sides. In flight bright turquoise-blue forewing. **Ad** ♀ has very distinct pale loral spot. **Eclipse** ♂ resembles ♀, but can show hint of pale facial crescent. **Juv** recalls ♀ but has more heavily spotted underparts. **Voice** Unlikely to be heard. **SS** Blue-grey head of ♂ diagnostic; ♀ lacks dark cheek-stripe of same-sex Garganey, and ♀ Green-winged and Eurasian Teals lack or have much smaller pale loral spot. Hybrids with Northern Shoveler typically have rather long, somewhat spatulate bill, but can resemble pure examples of present species. See Cinnamon Teal.

Gadwall *Mareca strepera* `LC`
Rare to very rare winter visitor (**Bahamas, Cuba**), vagrant elsewhere.

L 45–58 cm; **WS** 84–95 cm. Two subspecies, of which nominate occurs in region. Mainly Oct–Mar (Sept–May), on variety of freshwater ponds and lagoons. Vagrants Jamaica, Hispaniola (Apr 2005), Guadeloupe (unconfirmed), St Lucia (Nov 1992) and San Andrés (Oct 2005). **ID** Fairly large dabbling duck with prominent white speculum. **Ad** ♂ has greyish-brown head, sometimes suffused cinnamon on forehead and chin. Grey mantle, back and flanks, vermiculated black and white. White speculum, black-scaled breast, white belly freckled dark, and grey-brown tail with black coverts. In flight black-and-chestnut inner greater coverts. **Eclipse** ♂ resembles ♀ but retains wing pattern. **Ad** ♀ has head and neck buffish brown, streaked darker. Paler chin and upper breast. Darker crown, nape and diffuse eyestripe. Body mainly warm brown, heavily scalloped darker, especially flanks. Whitish belly. Lacks or has less chestnut in greater coverts. **Juv** resembles ♀ but greyer on head and neck, and more heavily streaked (less scalloped) below. **Voice** ♂ calls include loud, deep croaks, sometimes combined with whistles, ♀ mainly gives quacking or chattering calls. **SS** Lack of bold pattern and bright colours on head separate ♂ from other mainly grey-bodied ducks in region. Compare American Wigeon, Eurasian Wigeon, Mallard and Northern Pintail.

Eurasian Wigeon *Mareca penelope* `LC`
Vagrant (**GA, LA**).

L 42–51 cm; **WS** 71–86 cm. Monotypic. Grand Bahama, Mar 2003 (*Cotinga* 21: 82); Cuba, Mar 2014 (Stott 2015); Hispaniola, early 1950s, Dec 1972, Dec 1997, Feb 2014, Mar 2015 (Keith *et al.* 2003, eBird); Puerto Rico, Feb 1958, Jan 2015, Jan 2016 (Bond 1959, *NAB* 69: 306, 70: 241); St Croix, Nov 2003 (McNair *et al.* 2006), Anguilla, Dec 2014–Jan 2015 (*NAB* 69: 306), Barbuda, Oct 1937, bird ringed in Iceland (Cooke 1945), Martinique, c. Nov 2014 (eBird), Barbados, records Oct–Mar (Buckley *et al.* 2009) and Grenada (Jan–Feb 2001). Usually on freshwater ponds and lagoons, often with American Wigeon. **ID** Fairly large with rounded head, short, stubby bill and pointed tail. **Ad** ♂ has chestnut head and neck, yellowish crown, pinkish-grey breast, grey-vermiculated upperparts and flanks, white lower breast and belly, black surround to white-grey tail, and long grey scapulars. In flight large white patch on forewing, dark green speculum edged black and axillaries grey, but can appear largely white. Pale grey bill with black tip. **Ad** ♀ has grey-brown head and mantle, brown back, breast and flanks often with chestnut tone, white belly; in flight, lacks white patch in forewing and has duller speculum. **Eclipse** ♂ is like ♀ but often more richly coloured, particularly flanks, and has white in forewing. **Juv** resembles ♀ but has mottled belly and even duller speculum. **Voice** Whistled "whee-OOO". **SS** Chestnut head combined with yellowish crown of ♂ unique. ♀ American Wigeon has on average greyer head, contrasting more strongly with breast and often dark mottling around eye and at bill base; also white axillaries (sometimes tipped grey) and broad whitish bar on greater coverts. Grey bill separates ♀ from Gadwall and Mallard. See Northern Pintail.

American Wigeon *Mareca americana* `LC`
Fairly common to rare passage migrant and winter visitor (**throughout**).

L 45–56 cm; **WS** 76–89 cm. Monotypic. Commonest Cuba and Hispaniola, uncommon Puerto Rico, Bahamas, Virgin and Cayman Is, rare Jamaica and Lesser Antilles. Mostly Oct–Apr (Aug to mid May), at wetlands including freshwater, brackish and saline pools. **ID** Fairly large dabbling duck with rounded head, steep forehead and full nape, short, stubby bill and pointed tail. **Ad** ♂ has whitish forehead and central crown, dark green band from eye to nape, greyish head- and neck-sides, pinkish-brown breast and flanks, with white band separating them from black undertail-coverts, greyish-brown upperparts and white wing-coverts. In flight axillaries and median underwing-coverts white. **Ad** ♀ has rather grey or grey-brown head, brown back with paler, often rufous-tinged fringes, quite bright rufous breast and flanks, and white belly. In flight, lacks white in forewing but has broad whitish bar on greater coverts and duller speculum. **Eclipse** ♂ recalls ♀ but has rather rufous head and body, and more variegated upperparts. **Juv** resembles ♀ but has plainer upperparts. **Voice** Generally weaker and more trisyllabic (less disyllabic) than Eurasian Wigeon. In courtship ♂ produces three distinct whistles, the second obviously higher-pitched (louder "wheoo" of Eurasian Wigeon is slurred, not interrupted); ♀ alarm a low growling "krr". **SS** See Eurasian Wigeon. Grey bill separates ♀ from Gadwall and Mallard. See Northern Pintail.

Mallard *Anas platyrhynchos* `LC`
Very rare winter visitor (**N Bahamas, Cuba**), vagrant elsewhere.

L ♂ 55–70 cm; ♀ 50–60 cm; **WS** 75–100 cm. Two subspecies, of which nominate occurs in region. Mainly Oct–Apr in N Bahamas and Cuba. Vagrant Jamaica, Hispaniola, Puerto Rico, Virgin and Cayman Is, and many of Lesser Antilles (but some records > 100 years old). Could be recorded at almost any pond, lake or marsh with shallow, open water. Introduced on many islands and proof of genuine vagrancy rarely available. **ID** Large, stocky and short-tailed. **Ad** ♂ has green head, white neck-ring, dark brown breast, grey body, black tail-coverts and pale tail. Bill yellow, legs orange. In flight blue speculum bordered white. **Eclipse** ♂ resembles ♀ but note wing pattern and bill colour. **Ad** ♀ is mottled pale brown to buff, with dark-streaked crown, paler supercilium, dark eyestripe, brown body with paler feather fringes. Bill dark with orange sides. **Juv** resembles ♀ but generally darker with more streaked underparts, black crown and eyestripe, and bill more diffusely orange without spots. **Voice** ♂ gives nasal, weak-sounding, repeated "raeb", commonest ♀ vocalization a deep, harsh-sounding "quack". **SS** ♀ Gadwall has whitish belly, white speculum, steeper forehead, smaller and more evenly orange-sided bill. See American Black and Mottled Ducks, American Wigeon and Northern Pintail.

Blue-winged Teal

Gadwall
strepera

Eurasian Wigeon

American Wigeon

Mallard
platyrhynchos

43

American Black Duck *Anas rubripes* **LC**
Vagrant (**GA**, **LA**, **O**).

L 53–61 cm; **WS** 85–96 cm. Monotypic. Unknown island in Bahamas, ring recovery, Nov 1972 (Buden 1991); Grand Turk, Dec 1985 (*AB* 40: 1029); Puerto Rico, Dec 1935 (?), Jan 1980 (Barnés 1936, Colón 1982); St Thomas, Dec 2002 (*NAB* 57: 272); St Martin, Oct 2014 (*NAB* 69: 169); Guadeloupe, Jan–Feb 2006, Sept 2011 (Levesque *et al.* 2012). Brackish or fresh, shallow waterbodies. **ID** Large, dark, stocky and short-tailed dabbling duck. **Ad** ♂ has dark brown body with pale fringes, blackish on tail and its coverts, and paler face with darker crown, nape and eyestripe. Yellow bill. **Ad** ♀ is paler, with more heavily marked face and head, better-defined eyestripe, smaller speculum and more olive bill. **Eclipse** ♂ feathering is shorter and broader, and bare parts darker, with dark bluish or greenish bill. **Juv** has more heavily variegated breast and underparts with buff fringes. **Voice** Very similar to Mallard, ♀ utters noisy lower-pitched quacks and ♂ reedy grunts and clear whistles. **SS** ♀ Mallard has orange bill with dark centre, paler, more patterned body, tail with whitish sides and, in flight, has broad white borders to less deeply violet speculum. See Mottled Duck.

Mottled Duck *Anas fulvigula* **LC**
Vagrant (**Bahamas**, **Cuba**).

L 53–58 cm. Three subspecies, of which one (*fulvigula*) occurs in region. Two, Grand Bahama, Feb 2019 (AL); one ringed in S Florida, USA, shot in E Cuba during 2009 (Kirkconnell *et al.* in press). **ID** Dull brown duck with characteristics intermediate between American Black Duck and ♀ Mallard. **Ad** has rather plain head that contrasts with darker brown body and lacks white in tail. Blue-green speculum lacks white margin to greater coverts. Bill uniformly yellow. **Voice** Similar to Mallard, a low raspy "raeb", single "raeb" in alarm and double-noted "raeb-raeb" in courtship or 'conversation'. **SS** American Black Duck has much darker body, with more contrastingly paler head and violet speculum. ♀ Mallard has less contrast between head/neck vs. paler brown body and white tail-sides. ♀ Mallard × American Black Duck hybrids closely recall Mottled Duck, but show narrow white band at base of speculum. **TN** Previously considered conspecific with Mallard.

White-cheeked Pintail *Anas bahamensis* **LC**
Locally common resident (**GA**, **LA**, **O**).

L 38–51 cm. Three subspecies, of which nominate occurs in region. Locally common resident Bahamas, Greater Antilles, Virgin Is and N Lesser Antilles from Anguilla S to Guadeloupe; vagrant further S, e.g. Barbados. Mangroves, small pools and saline or brackish lagoons; in Bahamas regularly on waters that are distinctly more salty than sea. Also reservoirs and sewage ponds, and at least locally, e.g. in Cuba, rice fields. **ID** Unmistakable duck with contrasting white cheek and throat patches. **Ad** has dark brown crown and nape, with breast and underparts pale brown spotted black, upperparts medium brown. Long dark brown tertials with buff fringes. Iridescent green speculum with narrow black subterminal band and broader buff terminal edge, and pale buff tail and tail-coverts. Red spot at base of otherwise of dark bluish-grey bill; ♂ lacks eclipse plumage. ♀ is slightly duller, especially on crown and back, with shorter tail and smaller and more orange-red spot on bill. **Juv** resembles ♀ but has less iridescent speculum, paler brown crown and back washed greyish, and bill spot duller. **Voice** In courtship, alarm or contact between pair, ♂ gives rising "bzzzzzz", while in 'down-up' display utters loud two-noted whistle; ♀ gives typical quacking notes. **SS** Extremely unlikely to be confused. **AN** Bahama Pintail.

Northern Pintail *Anas acuta* **LC**
Uncommon passage migrant and winter visitor (**GA**), largely rare visitor or vagrant elsewhere.

L 50–76 cm (including up to 10 cm-long central rectrices); **WS** 80–95 cm. Monotypic. Uncommon Cuba, Hispaniola and Puerto Rico, rare Bahamas, Virgin Is, St Barthélemy, Guadeloupe and Barbados (where some might be transatlantic waifs), and vagrant Jamaica, Cayman Is and rest of Lesser Antilles. Freshwater, brackish and saline pools. Mainly late Aug–Apr. **ID** Small-headed, slender-bodied and slender-necked with pointed tail. **Ad** ♂ has chocolate-brown nape, throat and head, white stripe on neck-sides, largely grey-vermiculated upperparts and flanks, white foreneck, breast and central underparts, becoming yellow-buff ventrally, black tail-coverts and tail with long black central rectrices, pale at sides. In flight black-green secondaries bordered buff at front and white at rear. **Eclipse** ♂ resembles ♀ but retains wing pattern and lacks all-dark bill. **Ad** ♀ has greyish-brown head and neck, finely flecked and mottled darker, paler breast, brown flanks with darker crescentic markings, brownish-grey upperparts with intricate markings. In flight brown secondaries with broad white border. **Juv** has darker, browner crown, plainer, less buffy upperparts and flanks, more spotted below, and sepia-brown tail feathers fringed buff. **Voice** ♂ gives monosyllabic "geeeee", while ♀ utters series of low quacks, similar to Mallard. **SS** Head pattern of ♂ unique. Small head and slim body, long slim neck and attenuated rear best features to separate pale-bodied ♀ from other ducks, including ♀ Gadwall, Eurasian and American Wigeons.

Common Teal *Anas crecca* **LC**
Two subspecies placed in separate subspecies groups, both occur in region.

Eurasian Teal *Anas (crecca) crecca*
Vagrant (**GA**, **LA**, **O**).

L 34–38 cm; **WS** 58–64 cm. One subspecies in group. All records of ad ♂ (other plumages easily overlooked vs. Green-winged Teal): Bahamas, Feb 2017 (eBird); Puerto Rico, Mar 2016 (eBird); Guadeloupe, Jan–Feb 2014 (*NAB* 68: 290); Barbados, Jan–Mar 1996, Dec 1997–Mar 1998, Jan 1999, Jan 2004 (Buckley *et al.* 2009). **ID** Small, short-billed duck. **Ad** ♂ has chestnut head with green patch bordered yellow, extending to nape, pale breast spotted dark, grey body heavily vermiculated, with white scapulars forming horizontal line, and black tail-coverts with buff patch at sides. In flight glossy green speculum bordered by white bars that broaden distally on upper edge. **Eclipse** ♂ resembles ♀, but tends to have greyer upperwing-coverts, darker and less contrasting supercilium, pale markings on scapulars more barred. **Ad** ♀ has pale brown head with dark eyestripe, darker brown body with intricate markings, and narrow white patch on tail-sides. **Juv** recalls ♀ but has speckled breast to belly, more pointed flanks feathers with rather triangular markings, while young ♀ has rounder flanks feathers with crescentic markings. **Voice** As Green-winged Teal. **SS** See Green-winged Teal for confusion species, but greatest risk of confusion is with latter. ♂ differs from ♂ Green-winged in lack of vertical white breast-side line and white horizontal scapular stripe, and lack of narrow buff supercilium (above broad green 'eyestripe'). ♀-type plumages extremely difficult to separate, but focus on upperwing pattern; present species has wingbar whiter throughout its length (reddish inwards in Green-winged) and broader, while white trailing edge is slightly narrower (even-width in Green-winged). In addition, head pattern is less contrasting, bill usually has much more obvious pale base to mandible and breast is warmer in Eurasian Teal (Reeber 2015). In all but br ♀, outermost tertial has

American Black Duck

Mottled Duck
fulvigula

White-cheeked Pintail
bahamensis

Northern Pintail

Eurasian Teal
(Common Teal)

Green-winged Teal
(Common Teal)

45

shorter dark submarginal stripe with broadening pale fringe towards feather tip in Green-winged (pale fringe much narrower and tapering pale fringe with broader and longer submarginal stripe in Eurasian). Beware possibility of hybrids. **TN** Usually considered conspecific with Green-winged Teal, but some evidence to suggest that two-species option more appropriate.

Green-winged Teal *Anas (crecca) carolinensis*
Rare to very rare winter visitor or vagrant (**throughout**).
L 34–38 cm; **WS** 58–64 cm. One subspecies in group. Regular but rare Bahamas, Cuba, Virgin and Cayman Is, Guadeloupe and Barbados, Aug–Apr; very rare elsewhere. Shallow, freshwater pools and marshes with emergent vegetation. **ID** Small, short-billed dabbling duck. **Ad** ♂ has chestnut head with green patch behind eye to nape, pale breast spotted dark and often washed salmon-pink, vertical white line on breast-sides, heavily vermiculated grey body, and black tail-coverts with buff patch at sides. **Ad** ♀ has dark eyestripe, darker brown body with intricate markings, and narrow white patch on tail-sides. **Eclipse** ♂ resembles ♀. **Juv** recalls ♀ but is slightly darker. **Voice** ♂ gives melodious, disyllabic and far-carrying whistle "prip-prip" during winter/spring, and ♀ utters "quack" notes. **SS** Head pattern of ♂ unique (but see vagrant Eurasian Teal for detailed discussion of these conspecifics). See Blue-winged and Cinnamon Teals, and Garganey.

PODICIPEDIDAE
Grebes
20 extant species, 2 in region + 1 vagrant

Least Grebe *Tachybaptus dominicus* `LC`
Common to uncommon resident (**GA**, **O**), rare or vagrant (**LA**).
L 21–26 cm. Five subspecies, of which nominate occurs in region. Common Bahamas, Cuba and Jamaica, uncommon Hispaniola and Puerto Rico, rare Virgin Is, and vagrant elsewhere. Freshwater cattail swamps and ponds with abundant plant cover. **ID** Small grebe with diagnostic yellow eyes. **Ad br** is uniform greyish, with browner upperparts and flanks. In flight (prefers to dive in response to danger), diagnostic white wing patch. **Ad non-br** is browner, with whitish chin and throat, cleaner abdomen, and paler mandible. **Juv** resembles ad non-br, but has irregular stripes on head-sides. **Voice** Commonest call a trill, often protracted, given at varying speed. Rather high-pitched loud "gump" in advertisement, and "beep" in alarm. **SS** All-grey plumage, neater profile, thinner, unicoloured bill, yellow eye, and white wing patch distinguish from slightly larger Pied-billed Grebe.

Pied-billed Grebe *Podilymbus podiceps* `LC`
Common resident and winter visitor (**throughout** but scarcer on smaller islands).
L 30–38 cm; **WS** 56–64 cm. Three subspecies, of which at least two (nominate probably winters throughout, *antillarum* resident Greater and Lesser Antilles) occur in region. Specimen from Grenada has characters close to mainly South American *antarcticus* (Bond 1956). Generally less common S Lesser Antilles. Nests on freshwater lakes and ponds; in winter, wider range of waterbodies including brackish lagoons. **ID** Thickset grebe with robust bill. **Ad br** (*antillarum*) is dark brown with white vent; black throat contrasts with paler, slightly warmer neck. Bill pale grey crossed by black band; iris brown. **Ad non-br** has dark crown and hindneck, buffish face and foreneck. Bill dull yellow-brown, dark band usually obscured. **Juv** is like ad non-br, but head and neck with blackish-brown and white stripes. **GV** Nominate larger, with even thicker bill and more black on throat in breeder; *antarcticus* (not definitely recorded) even larger, with greyer sides to head, foreneck and breast, darker nape and hindneck, and larger black bib. **Voice** Song a series of barking "caow" notes, steadily increasing in volume and speed. Also short "kek" in alarm. **SS** See Least Grebe.

Red-necked Grebe *Podiceps grisegena* `LC`
Hypothetical (**Bahamas**).
L 40–50 cm; **WS** 77–85 cm. Two subspecies, of which one (*holbollii*) occurs in region. New Providence, Dec 1967 (Bond 1968, 1982, Paterson 1968). In winter on sheltered coastal waters, but also inland lakes. **ID** Mid-sized and fairly long-billed grebe. **Ad non-br** has yellow bill with dark culmen and tip, dark cap, white face, dusky ear-coverts, white throat and breast, dusky flanks, and dark brown upperparts. **Ad br** has black bill with bright yellow base, white face and throat, contrasting dark cap, and rufous neck and breast. **Juv** resembles ad non-br but has rufous neck-sides and dark stripes on face. **Voice** Unlikely to be heard. Alarm call a sharp "tek". **SS** Probably none. **TN** Race *holbollii* has been proposed as separate species, Holböll's Grebe (Bochenski 1994).

Black-necked Grebe / Eared Grebe `LC`
Podiceps nigricollis
Vagrant (**Bahamas**).
L 28–34 cm; **WS** 41–60 cm. Three subspecies, of which one (*californicus*) occurs in region. Grand Bahama, Oct 2015 (*NAB* 70: 129). Will use both inland waterbodies and inshore waters. **ID** Smaller grebe with uptilted appearance to bill. **Ad non-br** has dark cap with dusky feathers extending below eye, dark upperparts and white underparts with dark-streaked flanks. **Ad br** (unlikely in region) is unmistakable due to black head, neck and upperparts, with yellow head plumes and rusty underparts. **Juv** resembles ad non-br but has weak yellowish-buff wash on face. **Voice** Generally silent in winter. **SS** Red-necked Grebe (also known from single record in Bahamas) obviously larger, longer-necked and longer-billed, among other features, whereas both resident grebes are smaller and shorter-necked, and lack obviously two-toned winter plumage of present species, among other features. Bear in mind possibility that similar-sized but extralimital Horned Grebe *P. auritus* (widespread in N Holarctic) could occur in winter (has shorter, straighter bill, cap finishes level with eye, narrower dark area on hindneck and pale loral spot).

PHOENICOPTERIDAE
Flamingos
6 extant species, 1 in region

American Flamingo *Phoenicopterus ruber* `LC`
Abundant (**S Bahamas**) to locally common or rare resident (**GA**), vagrant (**LA**).
L 120–145 cm; **WS** 140–165 cm. Monotypic. Breeds in large, dense colonies on Great Inagua (Bahamas), Cuba and Hispaniola. Reintroduced to Virgin Is. Vagrants recorded St Kitts to Barbados and Grenada (eBird). Saline lagoons and estuaries. In flight neck and legs held fully extended. **ID** Size, brightly coloured plumage and overall profile unique in region. **Ad** is bright pinkish red with black remiges and distinctly curved, black-tipped, white-based bill. ♀ up to 20% smaller and shorter-legged. **Juv** is grey-brown with some pink in underparts, wings and tail; legs, feet and bill mainly brown. **Subad** gradually attains pink over three years. **Voice** Commonest call a goose-like double honk, "ka-hank". Also low-pitched "kok-kok-kok..." and nasal "nyaah"; large feeding flocks can maintain a continuous "kucking". **SS** None. **TN** Previously considered conspecific with extralimital Greater Flamingo *P. roseus* (Old World). **AN** Caribbean Flamingo.

Least Grebe
dominicus

breeding non-breeding

Pied-billed Grebe

podiceps
breeding

non-breeding

antillarum
breeding

Red-necked Grebe
holbollii

breeding non-breeding

Black-necked Grebe / Eared Grebe
californicus

breeding non-breeding

American Flamingo

47

PHAETHONTIDAE
Tropicbirds
3 extant species, 2 in region

Red-billed Tropicbird *Phaethon aethereus* `LC`
Common (**Virgin Is**) to uncommon and local resident (**off E Puerto Rico**, **LA**), vagrant elsewhere.

L 91–107 cm (including 46–56 cm tail-streamers); **WS** 99–106 cm. Three subspecies, of which one (*mesonauta*) occurs in region. Highly pelagic. Visits coast only to breed in small loose colonies on cliffs. **ID** Mostly white with black outer primaries. Back, rump and scapulars heavily barred black. **Ad** has long, pointed central tail-streamers and bright red bill. Black mask extends well beyond eyes to sides of nape. **Juv** lacks streamers and has black tail tip. Bill usually yellow tipped black. Eyestripes often meet on nape. **Voice** Usually silent away from colonies. When breeding, gives loud, raspy, tern-like screams, "kreeeee-kreeeee-kri-kri-kri-kri…". **SS** White-tailed Tropicbird (commoner in Bahamas and Greater Antilles) is smaller, has yellow to orange bill and shorter eyestripe, lacks barring on back and has prominent black carpal bars. Juv has coarser barring on back than any plumage of Red-billed and shorter eyestripe.

White-tailed Tropicbird *Phaethon lepturus* `LC`
Fairly common but local resident (**GA**, **O**), rare (**LA**).

L 70–82 cm (including 33–45 cm tail-streamers); **WS** 90–95 cm. Six subspecies, of which one (*catesbyi*) occurs in region. Highly pelagic. Visits coast only to breed in small loose colonies on cliffs. **ID** Smaller and more delicately built than Red-billed Tropicbird. Mainly white with black outermost primaries. Bill yellow to reddish orange. **Ad** has long, pointed central tail-streamers. Distinctive black carpal bars form V on spread wings. Back unbarred. Black mask does not reach nape. **Juv** lacks tail-streamers and has white forecrown, but rest of upperparts, uppertail- and wing-coverts barred black, plus black outermost primaries, black-tipped tail and black-tipped yellow bill. **Voice** Usually silent away from colonies, where utters harsh screams, like calls of Red-billed Tropicbird but faster, shorter, drier and higher-pitched, e.g. "ki-rik…ki-rik…ki-rik…". **SS** Red-billed Tropicbird (mostly E & S of Puerto Rico) lacks black carpal bars. Juv larger, with finer, denser barring on back, longer and more obvious mask, and more prominent black tail tips.

COLUMBIDAE
Pigeons and Doves
350 extant species, 18 in region + 4 introduced

Rock Dove / Rock Pigeon *Columba livia* `LC`
Introduced, locally common resident (**throughout**).

L 31–34 cm. Nine subspecies, but population in region entirely feral. Widespread in urban areas, where abundant and tame. **ID** The archetypal city pigeon. **Ad** (wild morphotype) is to varying degrees grey with slightly darker head and breast, green- and purple-glossed neck patches, two broad black wingbars and dark tail-band; underwing and rump whitish. Extensive plumage variability in feral birds, which can show patches of, or be almost entirely, white, brown or black (the norm). **Juv** in pure individuals has plumage duller with gloss on neck reduced or absent. **Voice** Repeated short "rrrrooh" "r-r-rrrooh", in display gives a more elaborated mournful "cu-cu-crroooh". **SS** Largely pale grey plumage with black wingbars unique among pigeons in region. Be aware of much variation; in particular, compare White-crowned and Scaly-naped Pigeons, but present species is largely confined to towns and cities. **AN** Feral Pigeon.

Eurasian Collared-dove *Streptopelia decaocto* `LC`
Introduced, common to uncommon resident (**GA**, **LA**, **O**).

L 30–32 cm. Monotypic. Introduced to New Providence in 1974, from where spread through Bahamas (Smith 1987) and colonized Cuba and Hispaniola in 1989/90 and 2007, respectively (Garrido & Kirkconnell 1990, Dhondt & Dhondt 2008). Separate introduction to Guadeloupe, in 1976, promoted colonization of almost all Lesser Antilles as far S as Martinique, St Lucia (Smith 1995, Barré *et al.* 1996, Feldmann *et al.* 1999) and Barbados (eBird). Favours urban areas and adjacent farmland; shuns forest. **ID** Mid-sized, long-tailed dove with largely uniform plumage. **Ad** has black neck-ring bordered white, pale grey forehead and crown, grading into pinkish-grey neck/breast, pale grey belly and flanks, pale brown upperparts with grey leading edge to wing and dark primaries. In flight grey-brown uppertail with broad pale tips to all but central rectrices, undertail dark-based with broad pale terminal band. **Juv** lacks collar and is sandier overall, feathers of upperparts with narrow buff fringes. **Voice** A repeated rhythmic, monotonous phrase of three rather low-pitched coos, "who.HOOO..hu". Call a nasal, whining, drawn-out "wheeeh". **SS** African Collared-dove (sympatry known only in Puerto Rico, but vagrants elsewhere possible) distinguished by slightly smaller size and shorter, rounded or square-ended tail, paler underparts (belly and undertail-coverts whitish), darker primaries and different vocalizations.

African Collared-dove *Streptopelia roseogrisea* `LC`
Introduced, fairly common resident (**Puerto Rico**, **Cayman Is**).

L 26–29 cm. Two subspecies, but birds in region presumably all pertain to domestic form (informally known as '*S. risoria*'). Present in feral state in Puerto Rico (mainly around La Parguera) since at least 1970s, and Cayman Is since 1983 (Bradley 2000); perhaps also on Virgin Is (St Croix, St John); but Guadeloupe (Levesque *et al.* 2005, Levesque & Delcroix 2015) and Martinique populations not self-sustaining. Bond (1980) mentioned it for New Providence and Andros (Bahamas). Urban areas and adjacent farmland; shuns forest. **ID** Similar to slightly larger Eurasian Collared-dove. **Ad** has pale sandy-brown upperparts, pinkish-grey underparts fading to whitish belly and undertail-coverts. Outermost tail feathers grey above with white tips, but black below and white-tipped. **Juv** is paler and duller with less demarcated black half-collar. **Voice** Song a short emphatic phrase, followed by short pause, then a drawn-out coo that starts as a rolling trill: "coo, rrrrrr-oooh". Call a bleating descending laughter, "ceh-heh-heh-heh-he", fading at end. **SS** See Eurasian Collared-dove, with which it co-occurs on Puerto Rico, Cayman Is and possibly on Martinique (Feldmann *et al.* 1999). Much confusion exists as to whether both occur in Bahamas (on at least New Providence and perhaps Andros) and most literature suggests that only present species occurs in Puerto Rico. **AN** Barbary Dove (in reference to domestic form).

Red-billed Tropicbird
mesonauta

adult

juvenile

White-tailed Tropicbird
catesbyi

adult

juvenile

Rock Dove / Rock Pigeon
feral

Eurasian Collared-dove

African Collared-dove

White-crowned Pigeon *Patagioenas leucocephala* `NT`
Common to uncommon resident (**GA, LA, O**).

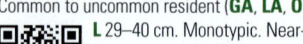

L 29–40 cm. Monotypic. Near-endemic; also S Florida and locally on Caribbean coast of Middle America between Yucatán and NC Colombia. Coastal woodland and mangroves, sometimes in mountains when not breeding (to 1500 m in Jamaica). Wanders freely between islands, often up to 45 km from nesting sites to feed. Regular as far S as Guadeloupe, vagrant further S. **ID** Very distinctive, arboreal pigeon. **Ad** is dark slate-grey with contrasting white crown, and bronze-green iridescence on hindneck. ♀ is slightly duller and paler, with greyer crown. **Juv** is much duller, browner and less bluish throughout. Entire hindneck dull greyish brown and crown greyish white. **Voice** Song a rhythmic series of coos, "whooo..gwhu-wu.. whooo..gwhu-wu..whooo...". Call a repeated, slightly lower-pitched, burry, three-note phrase "puh-whrru-poorrrr" (faster and less deliberate than Scaly-naped Pigeon). Also distinct low purring. **SS** Scaly-naped Pigeon looks similarly all dark at distance, but larger and lacks white crown. **AN** White-headed Dove.

Scaly-naped Pigeon *Patagioenas squamosa* `LC`
Common to uncommon resident (**GA, LA**).

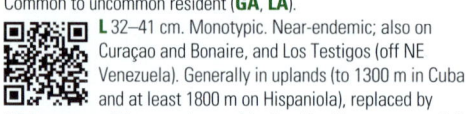

L 32–41 cm. Monotypic. Near-endemic; also on Curaçao and Bonaire, and Los Testigos (off NE Venezuela). Generally in uplands (to 1300 m in Cuba and at least 1800 m on Hispaniola), replaced by White-crowned Pigeon at lower altitudes. Humid areas, especially rainforest and moist broadleaf forest, but may descend to arid lowlands. In some areas, especially Bridgetown, Barbados (where introduced early 20th century), has adapted to gardens and urban areas. Movement between islands recorded; vagrant to Anguilla and Jamaica. **ID** Rather wary, arboreal species, most frequently seen in flight. **Ad** is slate-grey with drab purple head and neck, purple-brown nape patch, metallic purplish hindneck with dark scaling, and broad orange orbital ring. ♀ is duller than ♂. **Juv** has feathers fringed buff; purple and red areas of ad reddish brown. Bare parts duller and browner. **Voice** Song a rhythmic series of low-pitched coos, typically a long monotonous note followed by repeated triple notes: "whooo...pu-whu-whuh...pu-whu-whuh.. pu-whu-whuh...". Call a low-pitched burry "rhurrrrrrr". **SS** See White-crowned Pigeon. **AN** Red-necked Pigeon.

Ring-tailed Pigeon *Patagioenas caribaea* `VU`
Locally common resident (**Jamaica**).

L 38–48 cm. Monotypic. Endemic. Mostly wet highland or limestone forests, especially in John Crow Mts, E Blue Mts and Cockpit Country. Occasionally second growth and agricultural areas. Breeds to 2000 m; regularly descends to 100–300 m in autumn/winter. Flight appears heavy, but still fast and direct. **ID** Rather drab, long-tailed pigeon. **Ad** has pinkish-grey head and underparts, metallic green hindneck and grey upperparts. Tail grey with dark slate band across centre. ♀ has back, scapulars and wing-coverts suffused olive or brown. **Juv** has grey of head and neck suffused brown; foreneck to breast greyish brown then fawn or cinnamon on belly. **Voice** A series of 4–5, usually bisyllabic, comparatively low-pitched coos achieving highest pitch and amplitude over second half, e.g. "cooOOh ... cooOOh ... cooOOh ...". **SS** Plain Pigeon lacks metallic hindneck and reddish orbital skin, has white leading edge to wing-coverts, reddish-brown wings and breast, and different tail pattern. **AN** Jamaican Band-tailed Pigeon.

Plain Pigeon *Patagioenas inornata* `NT`
Locally common to rare resident (**GA**).

L 39–41 cm. Monotypic. Endemic. In Cuba, I of Pines and some offshore cays (rare) mostly in lowland forest and open country with tall trees, but also lower mountains and, very rarely, in mangroves; on Hispaniola (where more numerous) in highland pine and broadleaf forests to 1200 m, dry scrub and cactus thorn woodland; in Jamaica (scarce) nests in montane rainforest, but visits lowlands to feed and in non-br season; and Puerto Rico (very rare) now basically restricted to second growth and agriculture (but formerly used range of forested habitats). Usually in small groups, larger when not breeding. Easily overlooked unless song known. **ID** Rather nondescript pigeon. **Ad** has vinaceous-pink head and underparts, greyish-brown upperparts, reddish-brown breast and wings, latter with narrow white leading edge, bluish-grey rump, and blackish tail. **Juv** is greyish brown all over, paler on belly; wing-coverts and breast feathers with narrow pale fringes. **Voice** Song a series of coos, typically a single note followed by repeated double notes, "whoOo...pUh-whoOOo...pUh-whoOOo...pUh-whoOOo...". **SS** Compare Ring-tailed Pigeon (Jamaica alone) and *Zenaida* doves. **TN** Various single-island races sometimes recognized, e.g. *exigua* (Jamaica), *wetmorei/exsul* (Puerto Rico) and *proxima* (I of Pines) but differences considered to reflect individual variation.

Blue-headed Quail-dove *Starnoenas cyanocephala* `EN`
Rare and patchily distributed resident (**Cuba**).

L 29–34 cm. Monotypic. Endemic. Semi-deciduous, evergreen and swamp woodland, to c. 1790 m. Prefers areas with dense canopy but relatively open understorey, on stony substrate with substantial leaf litter; often favours drier areas than Grey-headed Quail-dove. Usually in pairs or small groups, and prefers to run rather than fly when startled. **ID** Largest quail-dove on Cuba. **Ad** has bright blue head bordered by black eyestripe, and white stripe below this. Black throat extends to form white-bordered black bib. Rest of body buffy brown, with darker olive-brown upperparts and wings. Bill red with blue tip. **Juv** is duller; back, scapulars and wing feathers narrowly edged buff-brown, while breast feathers also display some buff fringes. Blue on head duller, and legs and feet dull red. **Voice** Song a series of short, deep "whooo-up" or "ooowup" notes, with distinct upward inflection on second syllable, repeated every 2–4 seconds, ending abruptly. **SS** None. **AN** Blue-headed Partridge-dove.

Crested Quail-dove *Geotrygon versicolor* `NT`
Locally fairly common resident (**Jamaica**).

L 27–31 cm. Monotypic. Endemic. Wet montane limestone forest in John Crow Mts, Blue Mts, Port Royal Mts, Mt Diablo and Cockpit Country; occurs as low as 100 m near Windsor in Cockpit Country, occasionally up to 1800 m. Feeds on forest floor, sometimes along trails and even road edges. **ID** Typical *Geotrygon* with small, often indiscernible crest on nape. **Ad** has grey head with duskier forehead and pale orange-buff patch below eye. Back and wings iridescent reddish purple with contrasting chestnut primaries and upper back. Underparts grey with chestnut rear flanks. ♀ usually paler and browner than ♂ on neck and belly. **Juv** is duller than ad, most feathers edged rusty. **Voice** A short phrase of two coos (lasting c. 1.5 seconds), "whuuOh..whuuw"; second note more subdued and sometimes barely audible. **SS** No similar species in Jamaica, but beware briefly seen Caribbean Dove or Ruddy Quail-dove, as colours are not always obvious in quick view. **AN** Jamaican Quail-dove.

White-crowned Pigeon

Scaly-naped Pigeon

Ring-tailed Pigeon

Plain Pigeon

Blue-headed Quail-dove

Crested Quail-dove

51

Grey-headed Quail-dove *Geotrygon caniceps* VU
Rare resident (**Cuba**).

L 26–30 cm. Monotypic. Endemic. Tropical lowland forest and old second growth (mainly Zapata Peninsula and Sierra del Rosario), both on dry land with permanent small pools and semi-inundated forest abutting swamps. Also damper mid-elevation forests, to 1200 m on Pico Cuba. Secretive, but often seen along tracks. Flight swift and direct. **ID** Attractive, but often appears rather plain in brief views. **Ad** has grey head and underparts, faded whitish forehead, glossy purple mantle becoming bluish purple on lower back, rump and uppertail-coverts. In flight, chestnut underwing. ♀ is slightly duller with greyer head and reduced iridescence. **Juv** is apparently undescribed. **Voice** Song—given year-round, though less so May–Sept—a long series of short, low-pitched "uup" notes, either given in rapid succession ('fast song') or more slowly (every c. 0.5 seconds). Some evidence suggests changes only from slow to fast songs (see White-fronted Quail-dove). **SS** Blue-headed, Ruddy and Key West Quail-doves occur sympatrically, but none liable to confusion with present species in a clear view. Compare Zenaida Dove, which shares same areas and is commoner. **TN** Formerly treated as conspecific with White-fronted Quail-dove, but now usually split on morphology and some vocal differences (Garrido *et al.* 2002). **AN** Grey-fronted Quail-dove.

White-fronted Quail-dove *Geotrygon leucometopia* EN
Uncommon resident (**Dominican Republic**).

L 28 cm. Monotypic. Endemic. Historically probably also Haiti. Wet montane forest with dense undergrowth, and shade-coffee plantations (Cordillera Central, Sierra de Baoruco and Sierra de Neiba, possibly extirpated in last); 745–1685 m (Kirwan *et al.* 1999), perhaps to 1800 m. Very locally at sea level in much drier areas characterized by tall cacti, emergent palms and 1–3 m-tall scrub. **ID** Very similar to Grey-headed Quail-dove (Cuba). **Ad** has dark grey head and underparts, obviously contrasting white forehead, bluish-purple upperparts reaching breast-sides, and chestnut undertail-coverts and in wings. **Juv** is browner than ad, lacking reddish or purple sheen on mantle and neck-sides. **Voice** Song like Grey-headed Quail-dove—a continuous series of low "uup" notes that can become a prolonged "coo-o-o"—but said to change only from fast to slow song (Kirwan & Kirkconnell 2002). **SS** No other quail-dove on Hispaniola has white forecrown, and others have stripes on face and browner upperparts. **TN** See Grey-headed Quail-dove. **AN** Hispaniolan Quail-dove.

Ruddy Quail-dove *Geotrygon montana* LC
Common to rare resident (**GA**, **LA**), vagrant (**Virgin Is**).

L 20–27 cm. Two subspecies, of which both (nominate[A] Greater Antilles, *martinica*[B] Lesser Antilles) occur in region. Commonest in dense forest and shade-coffee plantations at mid elevations (to 850 m in Cuba and 750 m on Hispaniola); locally on coasts. Only on large, forested islands in Lesser Antilles. Like other quail-doves, forages in undergrowth. **ID** Plump, ground-dwelling dove. **Ad** ♂ is predominantly reddish brown with broad buff streak below eye, pale throat, and purple-red iridescence on nape, mantle, wing-coverts and rump. Buff patch on breast-sides visible as streak at edge of folded wing. **Ad** ♀ is mostly olive-brown where ♂ chestnut, with greenish sheen. Bare parts paler. **Juv** is paler overall with cinnamon-tipped upperparts and wing feathers. **GV** Race *martinica* larger and darker. Birds on Grenada appear intermediate between the two races. **Voice** Song a single monotonous, extremely ventriloquial low-pitched "whoOOou" (0.7–1 second), repeated every 3–5 seconds. Very similar to Key West Quail-dove, but dies off more rapidly and is slower and more mournful-sounding. **SS** Reddish-brown plumage unique among quail-doves. Key West and Bridled Quail-doves both have white stripe below eye, some iridescence on head and neck, and former has grey breast.

Key West Quail-dove *Geotrygon chrysia* LC
Fairly common resident (**GA**, **O**).

L 27–31 cm. Monotypic. Endemic; currently only vagrant to Florida (Howell *et al.* 2014). Semi-arid woodland, lowland scrub and wet montane forest with less disturbed understorey, generally below 500 m, locally to at least 1000 m in Dominican Republic. Scrubby second growth on I of Pines and regularly near habitation in Bahamas. **ID** Ground-dwelling dove with broad white facial stripe. **Ad** is chestnut-brown above with bronze-green or amethyst iridescence on crown and nape, purplish mantle, inner wing-coverts, and rump. Underparts greyish, paler on throat and dusky brown on flanks. ♀ has greyish-brown wing-coverts, rear scapulars and edges of rectrices; darker below, notably on breast. **Juv** has rufous-cinnamon upperparts with distinct buff tips to scapulars and wing-coverts, duller head and neck, and dull, cinnamon-edged breast. **Voice** Call a repeated, single low-pitched cooing note (c. 1–1.5 seconds), gradually increasing in amplitude before dropping slightly in pitch and fading rather abruptly (see Ruddy Quail-dove). **SS** See Bridled Quail-dove in Puerto Rico. Otherwise, white facial stripe unique in range, except Cuba, where compare very different Blue-headed Quail-dove.

Bridled Quail-dove *Geotrygon mystacea* LC
Uncommon to rare (**LA**) or very rare resident (**Puerto Rico**).

L 24–30 cm. Monotypic. Endemic. Puerto Rico and Virgin Is through most of Lesser Antilles to St Lucia (absent Anguilla, Barbados, St Vincent and Grenadines, Grenada); probably commonest Virgin Is. Lowland forest to at least 700 m; in many (but not all) areas, apparently prefers drier localities than Ruddy Quail-dove. In Puerto Rico, where perhaps mainly a recent colonist, regularly seen only on Vieques I and in Guajataca State Forest. **ID** Similar to Key West Quail-dove. **Ad** has conspicuous white stripe below eye, dark olive-brown upperparts, chestnut primaries and outer tail feathers, purple iridescence on nape and upper mantle, and buffy-brown underparts. ♀ is usually slightly duller. **Juv** is duller above and on breast, with no iridescence on hindneck, and darker and duller bare parts. **Voice** Call a repeated, single, low-pitched cooing note (c. 1–1.5 seconds long), often preceded by short introductory note, e.g. "hu.. whuuUUuw". **SS** Key West Quail-dove occurs sympatrically only on Puerto Rico (where commoner than Bridled); present species best separated by darker underparts, and duller crown and upperparts with less iridescence. Compare Ruddy Quail-dove.

Grey-headed Quail-dove

White-fronted Quail-dove

♀ montana

♂

martinica ♂

Ruddy Quail-dove

Key West Quail-dove

Bridled Quail-dove

53

Caribbean Dove *Leptotila jamaicensis* `LC`
Common to uncommon resident (**GA, O**).

L 29–33 cm. Four subspecies, of which three (nominate[A] Jamaica and introduced Bahamas, *collaris*[B] Cayman Is, *neoxena*[C] San Andrés) occur in region. Near-endemic; also SE Mexico, NE Belize, and Honduran islands of Barbareta, Roatán and Little Hog. Locally common Jamaica and San Andrés; uncommon Grand Cayman and very rare (status uncertain) Cayman Brac. Introduced in 1920s to New Providence (Bahamas), where currently uncommon. Typically in semi-arid areas, preferably with some cover, usually in lowlands. In Jamaica, in dry limestone forest, but commoner in secondary forest in foothills; also gardens and orchards; in Blue Mts reaches 2000 m. **ID** Plump, terrestrial dove. **Ad** has white face, rosy-vinaceous neck-sides with green-blue iridescence on hindneck and mantle, vinaceous breast grading to white belly and vent, and olive-brown upperparts with conspicuous white band on front of folded wing. In flight, chestnut wing-linings and white-tipped tail feathers (except central pair). **Juv** is duller than ad with most coverts edged rufous; neck and breast barred pale reddish brown in iridescent areas. **GV** Race *neoxena* more olivaceous with less bright neck iridescence and marginally darker vinaceous breast (in field recalls extralimital White-tipped Dove *L. verreauxi* of Middle and South America); *collaris* like nominate but averages slightly smaller. **Voice** Song a rather rhythmic series of four mournful monotonous notes, with emphasis on last, "wo-o-o-oooooo", repeated every 6–10 seconds. **SS** None. Other ground-dwelling woodland doves in range have darker underparts. **AN** Jamaican Dove, White-bellied Dove.

Grenada Dove *Leptotila wellsi* `CR`
Rare resident (**Grenada**).

L 28–31 cm. Monotypic. Endemic. Strongholds Mt Hartman and Perseverance estates. Just 100–200 birds survive; formerly more widespread. Dry scrub woodland with closed canopy and large patches of bare ground, to 150 m. Never outside forest. **ID** Similar to allopatric Caribbean Dove. **Ad** has pale pinkish face, grey crown, dull reddish-brown upperparts, pinkish-buff neck and upper breast, and white belly. In flight, cinnamon wing-linings and white-tipped outertail feathers. **Juv** is darker with buff-fringed scapulars and wing-coverts, and cinnamon tips to breast feathers. Bare parts duller. **Voice** A single soft mournful note every 5–8 seconds, "hooooo......hooooo"; low-pitched and reminiscent of sound produced by blowing across top of bottle. **SS** None.

White-winged Dove *Zenaida asiatica* `LC`
Common to rare resident (**GA, LA, O**).

L 25–31 cm. Three subspecies, of which nominate occurs in region. Common S Bahamas, Grand Cayman and most of Greater Antilles, uncommon N Bahamas, Virgin Is and other Cayman Is. In Lesser Antilles, undergoing range expansion (Larsen & Levesque 2008, Francis 2012, Levesque 2013), perhaps mirroring colonization of Puerto Rico (in 1943) and some of Virgin Is (McNair *et al*. 2006, Madden *et al*. 2015), reaching Anguilla (2000), Barbuda, Guadeloupe, Dominica and Antigua (2005), and Nevis (2011). Range of habitats, from mangroves to evergreen and pine forests, but mostly coastal lowlands. **ID** Fairly distinctive by large central white wing patch. **Ad** has buff head and neck, blue orbital skin, black spot below eye and iridescent bronze patch behind it. Belly and undertail-coverts greyish. Outertail feathers grey with black subterminal bar and broad white terminal band. **Juv** is paler and greyer with indistinct pale shaft-streaks on foreneck and upper breast; metallic gloss on neck lacking. **Voice** Song a low-pitched, rather hoarse, rhythmic cooing, "whohoo...hu...hoo" lasting c. 2 seconds, usually rather louder than virtually all other pigeons; also often gives longer phrases. **SS** Extensive white in wing distinguishes from other *Zenaida* and other pigeons and doves.

Zenaida Dove *Zenaida aurita* `LC`
Common resident (**throughout**).

L 28–31 cm. Three subspecies, of which two (nominate[A] Lesser Antilles, *zenaida*[B] Bahamas and Greater Antilles) occur in region. Near-endemic; also SE Mexico including some offshore islands. Less frequent in S Lesser Antilles, where Eared Dove more common. Usually lowlands, in open woodland, forest edge and clearings, plus cultivation, second growth, scrubby areas and mangroves. **ID** Typical *Zenaida* with fairly short, rounded tail. **Ad** has cinnamon-coloured head and neck with iridescent purple spot below and behind eye (usually appears blackish), metallic purple gloss on lower hindneck. Upperparts reddish brown with white-margined black spots on wings, and diagnostic white trailing edge to secondaries. Outertail feathers white-tipped with black subterminal bar. ♀ is paler with back less reddish. **Juv** is like ♀ but lacks iridescence on neck and has buff fringes to all back and wing feathers. **GV** Race *zenaida* darker, with terminal band on outermost rectrices bluish grey rather than white, and no white margins to wing spots. **Voice** Song a mournful cooing, "whoo'oOO...hu...hu...hu"; very similar to Mourning Dove but slightly higher-pitched, faster and with more rhythmic cadence. **SS** White in secondaries distinguishes it from partially sympatric Eared and Mourning Doves, and latter also has much longer, graduated tail.

Eared Dove *Zenaida auriculata* `LC`
Locally common resident (**LA**).

L 22–28 cm. Eleven subspecies, of which one (*stenura*) occurs in region. Relatively recent colonist in S Lesser Antilles. Fairly common St Lucia (probably since 1950s), Barbados (arrived 1950s), St Vincent and Grenadines, and Grenada. Breeds S Martinique. Small number of records Guadeloupe. Semi-arid brushy areas, primarily in lowlands. **ID** Proportions similar to Zenaida Dove. **Ad** has pinkish-grey head and neck with black markings behind and below eye, and iridescent pink or bronzy patch on neck-sides. Upperparts olive-brown with black markings. Outertail feathers grey with black central bar and white terminal patch. ♀ has head, neck and underparts less pinkish, and iridescence on neck-sides less obvious. **Juv** has no iridescence on neck-sides, buffy-brown neck and breast, and pale-fringed wing-coverts and scapulars. **Voice** Low-pitched, rather hoarse mournful cooing "whoo'oOO...hu...hu...hu"; very similar to Mourning and Zenaida Doves, but shorter than latter. Not very vocal and song not far-carrying. Occasionally a low-pitched growling "ohrrrr". **SS** White-winged and Zenaida Doves have white wing markings. Mourning Dove (note range) has long graduated tail.

Caribbean Dove
jamaicensis

Grenada Dove

White-winged Dove
asiatica

Zenaida Dove
aurita

Eared Dove
stenura

55

Mourning Dove *Zenaida macroura* `LC`
Locally common resident and rare winter visitor (**GA**, **O**), vagrant (**LA**).
L 23–34 cm. Five subspecies, of which two (nominate[A] Greater Antilles, *carolinensis*[B] mainly Bahamas) occur in region. Rare visitor Cayman Is, and reported Guadeloupe (Levesque & Delcroix 2015). Race *carolinensis* also a winter visitor (Oct–Apr) to Cuba and Hispaniola, but abundance unknown (Keith *et al.* 2003, Kirkconnell *et al.* in press). Perhaps expanding range, e.g. first recorded Puerto Rico in 1935. Wide range of habitats, but commonest in open and agricultural areas of lowlands. **ID** Mid-sized dove with long, graduated tail. **Ad** has buff face and foreneck, black spot on ear-coverts, greyish crown and nape, and metallic purple or bronze gloss on neck-sides. Wings greyish brown, with black spots on inner wing-coverts and scapulars. Outertail feathers grey with black bar and white tips. ♀ is paler, with less grey on head and less iridescence on neck. **Juv** resembles ♀ but has pale buff fringes to most feathers, and blackish spots on some head and breast feathers. **GV** Race *carolinensis* difficult to separate, but is overall paler and larger. **Voice** Song a low-pitched, mourning cooing "whoo'hoo...hu...hu..hu". Also several single hoots, e.g. "whoooo". **SS** Long, graduated tail readily distinguishes this species, and also note lack of white in wings. See Eared and Zenaida Doves.

Common Ground-dove *Columbina passerina* `LC`
Very common resident (**throughout**).
L 15.5–18 cm. Nineteen subspecies, of which ten (*bahamensis*[A] Bahamas, *exigua*[B] Mona I, *insularis*[C] Cuba, I of Pines, Cayman Is, Hispaniola and satellites, *jamaicensis*[D] Jamaica, *umbrina*[E] Î de la Tortue, *navassae*[F] Navassa I, *portoricensis*[G] Puerto Rico, Culebra, Vieques and Virgin Is, except St Croix, *nigrirostris*[H] St Croix and Lesser Antilles S to Dominca, *trochila*[I] Martinique, *antillarum*[J] S Lesser Antilles) occur in region. Wide range of habitats, including cities, shunning only heavily wooded areas. Largely ground-dwelling; if flushed, flies short distance on distinctively quick, shallow wingbeats. **ID** The smallest dove in region, not much larger than House Sparrow. **Ad** is sandy brown with scaled pinkish-brown breast, silvery-pink crown and nape, greyish-brown upperparts, black-spotted wing-coverts and secondaries, and chestnut primaries, axillaries and underwing. Outertail feathers black with white tips. ♀ has less grey on head, wing spots purplish brown or dull chestnut. **Juv** is similar to ♀ but has pale buff edges and fainter dusky breast markings. Young ♂ has hint of pink on neck. **GV** Races very similar, differ in bill colour or general shade of brown upperparts, and extent of white on tail tips. **Voice** Song an even-spaced, low-pitched, slightly upslurred series of cooing notes, "whuuu....whuuu....whuuu" (0.8–1.2 notes/second). More rarely, soft "whop" notes heard. **SS** Size and scaly appearance unmistakable.

Pied Imperial-pigeon *Ducula bicolor* `LC`
Introduced resident (**Bahamas**).
L 35–42 cm. Monotypic. Apparently well-established population (with single counts of at least 50 individuals) on New Providence, stemming from deliberate release of three individuals in 1996 (*NAB* 63: 664). More or less wooded habitats. **ID** Large black-and-white pigeon, uniquely plumaged in a regional context. **Ad** has white head and body (often tinged yellow, sometimes heavily) with largely black flight feathers and tail, but extensive white in outer rectrices. **Juv** has white areas suffused grey and broad yellow-buff fringes to most feathers. **Voice** Five descending and steadily shorter notes "whoo whoo whoo hoo hoo". **SS** None.

NYCTIBIIDAE
Potoos
7 extant species, 1 in region

Northern Potoo *Nyctibius jamaicensis* `LC`
Fairly common to rare resident (**GA**).
L 38–44 cm. Five subspecies, of which two (nominate[A] Jamaica, and probably this race Cuba, *abbotti*[B] Hispaniola and Gonâve I) occur in region. Forest edge and nearby areas with scattered trees, including parks, farmland, golf courses and mangrove edge (Jamaica), also xerophytic vegetation characterized by tall columnar cacti, emergent palms and 1–3 m-tall scrub on Hispaniola and in E Cuba; to 900 m on Hispaniola. Day-roosts in trees or atop poles. At night forages from posts and snags. Vagrant to Puerto Rico, on islands of Mona (Dec 1974) and Desecheo (Oct 1987). **ID** Unique nightbird that perches vertically at day-roost. **Ad** appears dark and intricately patterned, especially breast and wings, although overall plumage varies from more reddish brown to decidedly grey-brown; eyes yellow or orange. **Juv** is paler and usually exhibits dark gape line. **GV** Race *abbotti* has black markings on crown reduced in extent and size, throat more ashy (less cinnamon-buff) and blackish post-ocular stripe almost obsolete. **Voice** Song (mostly heard on moonlit nights) a harsh, guttural, unmusical "bwaaahw, bwa-bwa-bwa", lasting c. 2 seconds, first part often inaudible, frequently likened to a wailing or laughing sound; sometimes up to ten final syllables in *abbotti*. Calls include a hoarse "waark-cucu". **SS** Obviously larger and longer-tailed than nightjars, and upright stance very different. Huge-looking eye but virtually invisible bill. Even flight silhouette identified by unusually broad wings and somewhat cross-like shape. Voice too very distinctive.

Mourning Dove
macroura

Common Ground-dove
insularis

Pied Imperial-pigeon

Northern Potoo

jamaicensis

abbotti

57

CAPRIMULGIDAE
Nightjars
98 extant species, 10 in region + 1 vagrant

Common Nighthawk *Chordeiles minor* — LC
Uncommon to rare passage migrant (**GA**, **LA**, **O**).
L 22–25 cm. Nine subspecies, of which at least four (nominate, *sennetti*, *howelli*, *chapmani*) occur in region, but relative abundance and distribution unknown, as only specimens generally assignable. Passage late Mar–mid May and late Aug–early Nov, with occasional winter records (Dec–Jan) mainly in S (e.g. Barbados). Behaviour and habitat usage as Antillean Nighthawk, and also regularly observed on wing in daylight. **ID** Somewhat hawk-like: comparatively slender, pointed wings and buoyant flight with sometimes striking changes of direction. At rest, white throat conspicuous to hardly noticeable depending on posture. **Ad** ♂ has broad white band on five outermost primaries and broad white subterminal band on all but central rectrices. Generally dark greyish, spotted whitish and buff, with brown and black spots and streaks. **Ad** ♀ has narrower, less bold white wing-bands, no white in tail and buffier throat. In flight, both sexes have white wing-band broader at rear, inner primaries above band appear blackish and unmarked. **Imm** is similar to ad but more heavily barred below, throat patch paler and spotted, no white in tail, and flight feathers narrowly tipped white. **GV** Races vary mainly in size, coloration, barring of underparts and, especially in ♂, extent of white wing and tail markings. **Voice** Territorial/courtship call of ♂ a distinctive nasal "peent", usually in flight (heard on migration). **SS** Slightly smaller Antillean Nighthawk near-identical; certain identification usually reliant on voice. At rest, wingtips of Antillean sometimes fall short of tail (reach well beyond it in Common) and Antillean can appear paler and browner above, and buffier below, but all these average differences cannot be evaluated in flight. From below in flight, Antillean has tan-coloured wing linings (blackish in Common), but exceptional views required for use. In practice, any nighthawk that does not call is probably best left unidentified, but midsummer records are only likely to be Antillean. See Lesser Nighthawk.

Antillean Nighthawk *Chordeiles gundlachii* — LC
Fairly common to common summer visitor and passage migrant (**GA**, **O**), mainly passage migrant or vagrant (**LA**).
L 20–23 cm. Monotypic. Near-endemic breeder; also nests S Florida. Wintering grounds presumably N South America. Breeds Bahamas, Cuba, Cayman Is, Jamaica, Hispaniola, Puerto Rico and Virgin Is; recently expanded range SE to Guadeloupe (Levesque *et al.* 2005). Mainly late Mar–early Oct, but at least three winter records (Puerto Rico, late Dec/Jan; Cuba, mid Jan). Can appear early Feb, but rare Greater Antilles before mid Apr. Claimed Providencia Apr 2018 (Donegan & Huertas 2018) but too unsatisfactorily for acceptance. Arid, open, slightly rocky terrain, including coasts and pine barrens, from sea level to higher elevations; even breeds on rooftops in cities (e.g. Havana). Migrants can appear virtually anywhere. **ID** Almost identical to Common and Lesser Nighthawks. Two colour morphs, greyish and tawny. **Ad** ♂ has broad white band on five outermost primaries and white subterminal band on all but central rectrices; throat white. **Ad** ♀ has less distinct white wing-bands and no white in tail; throat buffish. In both sexes inner primaries above white wing-band appear blackish and unmarked. **Imm** is similar but more heavily barred below, throat patch paler and spotted, no white band on tail, and flight feathers narrowly tipped white. **Voice** Territorial/courtship call of ♂ a repetitive, nasal "chitty-chit", "penk-dik", "pity-pit-pit" or "killikadick", occasionally a longer "killi-kadick-dick-dick-dick", usually in flight. Similar to Common Nighthawk, but rhythm different. Mechanical booming wing noise during display flight. **SS** Only nighthawk likely in region mid May to late Aug; see Common and Lesser Nighthawks. **TN** Race *vicinus* sometimes recognized for those breeding in Bahamas and S Florida.

Lesser Nighthawk *Chordeiles acutipennis* — LC
Hypothetical (**San Andrés**).
L 19–23 cm. Seven subspecies, of which one (presumably *texensis*) claimed in region (McNish 2003, Pacheco 2012), albeit in error. Most likely during passage periods when habitat probably similar to other nighthawks. Behaviour like Common Nighthawk, but tends to fly very close to ground; however, differences probably of little use since migrating birds fly higher. **ID** Very similar to Common and Antillean Nighthawks. Outer wing shorter than Common and inner wing broader. Wingtip more rounded or notched. Broad white band on four outermost primaries obviously closer to wingtip than carpal joint. At rest, wings reach tip of tail, not beyond; diagnostic tawny bars often visible at base of closed primaries. **Ad** ♂ has white band on four outermost primaries and narrow white subterminal band on all but central rectrices. **Ad** ♀ has buffish or buffish-white wing-band and no white on tail. **Imm** is greyer or buffier than ad, with flight feathers tipped greyish white or tawny. **Voice** Almost cat-like "weah" occasionally given in flight, sometimes in rapid series. May utter "chuck" notes from ground. **SS** Common Nighthawk has more pointed wingtips and white patch midway between carpal joint and wingtip. Antillean Nighthawk (claimed on Providencia) best separated by voice.

White-tailed Nightjar *Hydropsalis cayennensis* — LC
Rare resident (**Martinique**), vagrant (**Puerto Rico**).
L 20–22.5 cm. Six subspecies, of which one (*manati*) occurs in region. Grassy fields on Martinique. Vagrant Puerto Rico (Nov 1974). Occurrence on Barbados (claimed as vagrant by Raffaele *et al.* 1998) extremely doubtful (Buckley *et al.* 2009). **ID** Mainly greyish-brown nightjar with broad tawny-buff nuchal collar, large white spots on wing-coverts, buffish breast heavily spotted white, and white belly. **Ad** ♂ has white band on four outermost primaries, a mainly white tail and large white throat patch. **Ad** ♀ lacks white on wings and tail, and has buffish throat. **Juv** is undescribed. **Voice** Song of ♂ a series of short, high-pitched whistles: "pt-cheeeeeee". Sings from low perches, mostly at dusk and dawn. Flight call of high-pitched "see-see" notes; also soft "tic-tic" or "wut-wut-wut". **SS** The only nightjar on Martinique. Compare nighthawks.

Jamaican Poorwill *Siphonorhis americana* — CR(PE)
Former resident (**Jamaica**).
L 23–25 cm. Monotypic. Endemic. Probably extinct. Precise habitat unknown, but possibly dry limestone forest, semi-arid woodland or open country. Known from four specimens taken mid-19th century, last dated 1860. **ID** Fairly small but long-tailed, dark nightjar with pale tail tips but no white in wings. **Ad** has rufous-brown upperparts streaked blackish brown and indistinct buffish nuchal collar. Wing-coverts rufous-brown marked blackish brown, with small, brown-centred, pale buff spots. Large white throat patch and rufous-brown underparts, boldly spotted whitish on upper belly, becoming buff barred brown towards tail. ♂ has all but central tail feathers narrowly tipped white, ♀ has yellowish-buff tips. **Imm** plumages unknown. **Voice** Unknown. **SS** Antillean and Common Nighthawks are larger with obvious white band on wing, and presumably very different behaviour. Chuck-will's-widow is larger still with shorter tail. **AN** Jamaican Pauraque.

Common Nighthawk
minor

Antillean Nighthawk

Lesser Nighthawk
texensis

White-tailed Nightjar
manati

Jamaican Poorwill

59

Least Poorwill *Siphonorhis brewsteri* NT
Rare and local resident (**Hispaniola**).

L 17–21.5 cm. Two subspecies, of which both (nominate[A] mainland Hispaniola, *gonavensis*[B] Gonâve I, off Haiti) occur in region. Endemic. Arid or semi-arid lowlands, especially scrubby limestone woodland and cactus and thorn scrub; also deciduous, coniferous or mixed forest, to 800 m. **ID** The smallest resident nightjar on Hispaniola. **Ad** has greyish-brown upperparts, broad buff nuchal collar, dark brown below boldly spotted white on breast and becoming white, barred and vermiculated brown, on belly. No white on wings; ♂ has all but central tail feathers narrowly tipped white, ♀ possibly has tips pale buff. **Juv** is similar to ad, but upperparts distinctly spotted (rather than streaked) blackish brown, and belly and flanks pale buff faintly barred brown. **GV** Race *gonavensis* claimed to have palest parts of plumage much whiter, almost devoid of any buff, tip of tail purer white and flanks barring less conspicuous and somewhat narrower; also said to be smaller. **Voice** Song of ♂ a whistled, rising "toorrrrri", or a warbled "tworrri"; calls include short, whistled "toorric" or "to-ic" and dove-like scratchy sounds. **SS** Northern Potoo much larger with very different voice; nighthawks fly much higher and often during daylight. Most likely to be confused with commoner Hispaniolan Nightjar, which is also larger and darker, and note voice. Compare Chuck-will's-widow (winter visitor to Hispaniola). **TN** Possibly best considered monotypic (Cleere 2010). **AN** Least Pauraque.

Eastern Whip-poor-will *Antrostomus vociferus* NT
Vagrant (**Bahamas**, **Cuba**, **Jamaica**).

L 22–27 cm. Monotypic. Could occur in any forested or semi-open area, especially dry, open woodland. Most reports sight only. San Salvador, Bahamas, Feb–Mar 2013 (presumably this species, rather than Mexican Whip-poor-will *A. arizonae*; eBird); Cuba, Jan 1932, Apr 2001, Mar 2013, plus two undated specimens, one now destroyed (Kirkconnell *et al.* in press); and Jamaica, Mar 1993 (*Gosse Bird Club Broadsh.* 61: 19). **ID** Indistinct tawny-buff nuchal collar, wing-coverts greyish brown, speckled and spotted tawny, buff, greyish white and blackish brown, with large blackish spots on scapulars, white or buffish-white band on lower throat; underparts brownish, heavily spotted whitish, greyish white and cinnamon. **Ad** ♂ has broad white tips to three outermost tail feathers. **Ad** ♀ has narrow buff tips to three outermost rectrices. Both sexes lack white markings on wings. **Juv** is very similar to ad. **Voice** Song of ♂ (could be heard in region) a whistled "whip, pr-will" or "whip, pr-weeea", last note usually rising at end, and often preceded by or interspersed with several short, sharp "quit" notes. Calls include soft "coo" or "couk", melodious "coo-eu" or "coo-eu-ah". **SS** Chuck-will's-widow usually more reddish brown and lacks black on throat, also larger, while ♂ of present species has much more white on outermost tail feathers and ♀ more readily apparent buff tips to tail feathers; Cuban and Hispaniolan Nightjars are larger, darker and greyer, with less white on face, neck and tail. **TN** Formerly considered conspecific with extralimital Mexican Whip-poor-will (SW USA to N Central America).

Puerto Rican Nightjar *Antrostomus noctitherus* EN
Rare resident (**Puerto Rico**).

L 21.5–22.5 cm. Monotypic. Endemic. S & SW Puerto Rico, where most frequently observed around Guánica. Semi-deciduous forest with little or no ground vegetation, usually on dry, limestone soils. In disturbed areas only if canopy intact. Sea level to 230 m. Forages, sings and nests under forest canopy. **ID** Medium-sized, generally greyish-brown nightjar, with indistinct tawny-buff collar on hindneck. **Ad** ♂ has broad white tips to three outermost tail feathers and white throat-band. **Ad** ♀ has narrow, buffish tips to outertail and possibly a more buff-coloured throat. Both sexes lack white in wings. **Juv** is paler than ad. **Voice** Song of ♂ a rapid series of short, liquid whistles, "whlip, whlip, whlip, whlip, whlip", often preceded by faint "quert" notes. Calls include short growls, "quert" and "gaw" notes, and soft, clucking or guttural sounds. Sings from perches throughout night. **SS** Only regularly occurring nightjar in range. Chuck-will's-widow (rare winter visitor) distinctly larger, more reddish brown overall with much less white in tail of ♂. Antillean Nighthawk has quite different behaviour, foraging high above rather than inside forest, and has white wing patch. **AN** Puerto Rican Whip-poor-will.

Chuck-will's-widow *Antrostomus carolinensis* NT
Fairly common resident (**N Bahamas**) and winter visitor (**GA**), vagrant (**LA**).

L 27–34 cm. Monotypic. Reasonably common breeder N Bahamas (Hayes *et al.* 2010) and might occasionally nest Cuba (Kirkconnell *et al.* in press). Fairly common (Sept–May) Hispaniola, less so (Aug–May) Cuba and Saba, rare (Sept–May) Jamaica, Puerto Rico, Virgin Is and Cayman Is. Vagrant St Martin, St Barthélemy, Barbuda and reported Guadeloupe (Feb 2013). Semi-deciduous and evergreen woodland, scrub, swampy borders, coastal vegetation and marshy ground, to at least 750 m. **ID** Comparatively large, well-variegated, brownish nightjar, with narrow buffish nuchal collar and whitish throat-band; no white in wing. **Ad** ♂ has white inner webs to three outermost tail feathers, outer webs tawny, speckled or barred blackish. **Ad** ♀ lacks white in tail. **Juv** is similar to ad ♀. **Voice** Song of ♂ a loud, repetitive "chuk, weeo, weeo"; can be faster, higher-pitched, or more muted. Sings from perch, mainly at dusk, dawn and throughout moonlit nights, sometimes during day. Flight calls, by both sexes, include clucks, growls and deep "quok" notes. **SS** Compare especially Cuban and Hispaniolan Nightjars, and Puerto Rican Nightjar.

Rufous Nightjar *Antrostomus rufus* LC
Five subspecies divided into three subspecies groups, of which one occurs in region.

St Lucia Nightjar *Antrostomus (rufus) otiosus*
Locally common resident (**St Lucia**).

L 25–30 cm. One subspecies in group. Endemic. Prefers arid brush country at low elevations, especially in NE coastal areas from Grand Anse S to Dennery; might also occur on Anse la Raye. **ID** Medium-sized rather dark-coloured nightjar, with reddish-brown edges to flight feathers, bold white throat-band and comparatively short, rounded wings without any white. **Ad** ♂ has large white spot on inner web of three outermost tail feathers, outer webs tawny with black bars. **Ad** ♀ lacks white in tail. **Juv** is paler than ad, with crown, scapulars and some wing-coverts pale buff or whitish, spotted black. **Voice** Song of ♂ a loud "chuk, wee, wee, weeo" given from branches or rocks, sometimes ground. Call a guttural croak. **SS** No other nightjar occurs on St Lucia. **TN** St Lucian population has been considered a separate species, but Robbins & Parker (1997) relegated it to subspecies status.

Least Poorwill
brewsteri

Eastern Whip-poor-will
♂ ♀

Puerto Rican Nightjar
♂ ♀

Chuck-will's-widow
♂ ♀

St Lucia Nightjar
(Rufous Nightjar)
♂ ♀

Cuban Nightjar *Antrostomus cubanensis* LC
Fairly common resident (**Cuba**).

L 25–29.5 cm. Two subspecies, of which both (nominate[A] mainland Cuba and many larger cays of Sabana-Camagüey archipelago off N coast, *insulaepinorum*[B] I of Pines) occur in region. Endemic. Semi-deciduous, evergreen, riparian and swampy woodland, coastal thickets, pine forest and second growth, to 600 m. **ID** Fairly large, generally rather dark grey and blackish-coloured nightjar with no nuchal collar, but a broad buff throat-band. **Ad** ♂ has white tips to three outermost tail feathers. **Ad** ♀ has narrow buff tips to outer rectrices. Both sexes lack white in wings. **Juv** is apparently undescribed. **GV** Race *insulaepinorum* smaller and darker, but perhaps only doubtfully valid. **Voice** Song of ♂ a short, even-pitched, trilled whistle, "terrrrrrro"; also a four-syllable phrase, "gua bai ah ro", which gives rise to the species' Cuban vernacular name. **SS** Compare larger and somewhat browner Chuck-will's-widow (which lacks spots on breast). **TN** Usually considered conspecific with Hispaniolan Nightjar by New World authorities, under name Greater Antillean Nightjar.

Hispaniolan Nightjar *Antrostomus ekmani* LC
Locally fairly common to rare resident (**Hispaniola**).

L 26–30 cm. Monotypic. Endemic. Woodland including pines, from near sea level to 1825 m, but mainly known from middle elevations in W Dominican Republic; elsewhere rare and local, and current status in Haiti very poorly understood. **ID** Overall fairly large and dark-coloured nightjar, very similar to allopatric Cuban Nightjar. **Ad** ♂ has white tips to three outermost tail feathers. **Ad** ♀ has narrow buff tips to outer rectrices. Both sexes lack white in wings. **Juv** is very dark, almost blackish. **Voice** Song of ♂ a short "clik" or "tuc" followed by trilled whistle that rises in pitch: "tuc, terrrreo". Calls include soft "quat" and deep, crow-like "gaaw" sounds. **SS** Compare larger and somewhat browner (even redder) Chuck-will's-widow, which lacks spots on breast of present species, and is usually found at lower elevations on Hispaniola. **TN** See Cuban Nightjar.

APODIDAE
Swifts
96 extant species, 7 in region + 2 vagrants

Black Swift *Cypseloides niger* VU
Rare to fairly common resident or partial migrant (**GA**, **LA**), vagrant (**O**).

L 15–18 cm. Three subspecies, of which nominate occurs in region (but see GV). Usually over forested mountains and hills to at least 1800 m, but also lowland and more open areas, perhaps most frequently in non-breeding season or during inclement weather, but nests in sea-caves in Barbados. Potentially only visitor to islands such as St Lucia, even in breeding season. Temporal status open to clarification; often considered resident, at least in most of Greater Antilles, but some (juveniles?) may depart post-breeding, numbers perhaps varying annually. Presumed migrants observed in Virgin Is and parts of Lesser Antilles, where present Mar/Apr–Aug/Oct. Vagrant Cayman Is. **ID** Well-named swift, appearing virtually featureless at distance. **Ad** has shallow but distinctly forked tail, long, broad wings, and stocky body. All-dark plumage with white fringes to underparts, sometimes lacking. ♀ has broader white fringing to underparts and more fan-shaped tail (slightly notched or square-ended in ♂). **Juv** shows broader, white, rather spotty, fringes reaching to vent. **GV** Race *borealis* (might have occurred extreme W Cuba; Navarro Pacheco 2018) larger with strikingly notably paler, grey head, especially forehead, and more prominent white fringing below. **Voice** Calls include a short "chip" or "tip", often repeated, sometimes becoming a fast rattle and terminating in short notes "tip...tip...tip...trirrirrrrrrrrrr...trip...tip...tip", alternatively ending in a rather squeaky note, "tip...tip...tip...tri-ri-ri-rrrrrrrr-sqeew". Rather silent away from colonies. **SS** Larger, longer-tailed and bigger-headed than other resident swifts, other than White-collared (which see). Vagrant Alpine Swift obviously larger with paler brown upperparts and striking underparts pattern, whereas vagrant Common Swift (which see), being similarly sized, could pose more problems, especially if unfamiliar with one or both relevant genera. **TN** Relationship between North and Middle American populations of Black Swift, and those in West Indies, requires study.
AN American Black Swift.

White-collared Swift *Streptoprocne zonaris* LC
Fairly common resident (**GA**), more local or vagrant (**LA**).

L 20–22 cm; **WS** 48–53 cm. Nine subspecies, of which at least two (*pallidifrons* resident, *minor* vagrant or visitor) occur in region. Montane, hilly and wooded areas, to lowlands (if bad weather forces them downslope?). Note mapped region in Cuba includes all areas with regular records, not just known breeding sites. Often in quite large, rather compact flocks that fly purposefully and in unison in travelling flight, but mix with other swifts when foraging. Some colonies shared with Black Swift. Vagrant Puerto Rico, Saba, St Kitts, Guadeloupe, Martinique, Barbados and Grenadines. Status unclear Grenada, perhaps only visitor but sometimes in quite large numbers (Frost & Massiah 2003); a specimen (NHMUK 1898.2.8.174) was said (Bond 1957) to be race *albicincta* and he later speculated that one in Jul 1971 on Vieques I (Puerto Rico) was same race (Bond 1973). Our examination of the specimen, his publications and correspondence held at COP indicate that the population to which Bond was referring would be *minor* (N Venezuela coastal ranges and Trinidad) under our taxonomy, rather than *albicincta* (lowland Venezuela and Guianas). **ID** Large black swift, with long, broad but rakish, well-angled wings, plus a shallowly (but usually obviously) forked tail. **Ad** (*pallidifrons*) has slightly paler foreface (obvious only when very close) and ragged, variable, white collar, broadest on foreneck. **Juv** has extensive white fringes to body feathers and a mottled, less distinct or obvious collar. **GV** Race *minor* small, with narrow, often mottled, breast-band and greyish throat; extralimital *albicincta* also small, with broad breast-band and restricted buff fringing to underparts. **Voice** Quite vocal, with several loud calls somewhat reminiscent of parakeets. Commonest is a loud, bright "cleeee" or "peeew", oft-repeated and sometimes continuing as a burst of similar notes (usually by several individuals simultaneously). Also a more mellow "wheee" and burry "prreeew". **SS** Largest swift in region, other than vagrant Alpine. Only likely to be confused (and only if white collar cannot be discerned) with frequently sympatric Black Swift, which is smaller with a less obviously forked tail and shorter, less scimitar-like wings (head shape and vocalizations also differ). **TN** Restall *et al.* (2006) considered race *minor* a synonym of *albicincta* based on specimens and live birds trapped (M. Lentino pers. comm.).

Cuban Nightjar
cubanensis

Hispaniolan Nightjar

Black Swift
niger

adult

juvenile

White-collared Swift
pallidifrons

63

Lesser Antillean Swift *Chaetura martinica* `LC`
Fairly common resident (**LA**).
L 10.5–11 cm. Monotypic. Endemic. Breeds Guadeloupe, Dominica, Martinique, St Lucia and St Vincent. Vagrant Nevis. In flocks of up to c. 40, over forests in montane areas, but also feeds over lowlands and open zones, perhaps especially during wet weather in mountains. **ID** Relatively small swift with short square tail; rectrix spines comparatively short (similar to Short-tailed Swift). **Ad** has black-brown upperparts with fairly narrow but obvious grey rump-band, and brown underparts with indistinctly paler throat. **Juv** is reported to differ only by showing small white tips to inner primaries, secondaries and tertials. **Voice** Probably very similar to Grey-rumped Swift. Main call a high-pitched twittering trill "prrrrrrrr" or "prrrrrrr-titi" given at intervals. **SS** Compare Short-tailed, Grey-rumped and (migrant) Chimney Swifts. Former has obviously different flight silhouette and much paler rump/uppertail, whereas Chimney Swift lacks obvious contrast between rump and rest of upperparts. Grey-rumped shows much more contrast than Chimney, and compared to present species pale area appears broader, while Lesser Antillean shows more brown in upperparts.

Grey-rumped Swift *Chaetura cinereiventris* `LC`
Seven subspecies divided into two subspecies groups, of which one occurs in region.

Ashy-rumped Swift *Chaetura (cinereiventris) sclateri*
Locally fairly common resident or migrant breeder (**Grenada**).
L c. 10.5 cm. Six subspecies in group, of which one (*lawrencei* Grenada) occurs in region. Most frequent over forest, mainly in hills, breeding in hollows in trees or caves, perhaps occasionally in buildings. Seasonal status (migrant breeder or resident) requires confirmation. Claimed twice in Dominican Republic (Jan 1997, Feb 2000), by same observer, but records undocumented and not acceptable. **ID** Smallish, relatively slender-bodied swift with short square tail and comparatively long rectrix spines (like Chimney Swift, and longer than Short-tailed). **Ad** has glossy blue-black upperparts including tail, broad grey rump and proximal uppertail-coverts (offering some contrast with mantle). Underparts generally blackish grey, becoming paler from upper breast to chin. **Juv** differs only in showing indistinct pale-fringed tertials, secondaries and greater coverts. **Voice** Thin, high-pitched, metallic twitter, often extended into a liquid trill. **SS** Short-tailed Swift has quite different shape in flight, and the obviously very short tail appears much paler from above. Lesser Antillean Swift is brownish black and has less contrasting rump. **TN** Ashy-rumped Swift may prove specifically distinct from nominate Grey-rumped Swift (Atlantic Forest of E Brazil to NE Argentina), but more information on voice and genetics required.

Chimney Swift *Chaetura pelagica* `VU`
Uncommon to rare passage migrant (**GA**, **O**), vagrant (**LA**).
L 12–14 cm; **WS** 29–31 cm. Monotypic. Observed Aug–early Nov and late Feb–May (one Jun report), perhaps most frequently during or following bad weather, sometimes in large flocks (up to 1000 reported) but usually in singles or very small numbers, over all manner of habitats, including towns, fields and woodland. **ID** Comparatively large and long-tailed *Chaetura*, with long rectrix spines. **Ad** has very uniformly dark plumage, showing almost no contrast between rump/uppertail-coverts and rest of dark grey-brown upperparts. Underparts dark grey-brown, usually with slightly (sometimes obviously) paler throat. **Juv** differs from ad only in having narrow white tips to inner primaries, secondaries and tertials. **Voice** Main calls a rapid series of dry "chip" notes and twittering descending trills "t-t-rrrr...t-t-rrrr...t-t-rrrr...". **SS** The only *Chaetura* definitely recorded in Greater Antilles, Cayman Is and Bahamas; elsewhere, other congenerics require elimination, but note exceptional lack of contrast between tail/rump vs. rest of upperparts, whereas Lesser Antillean, Short-tailed and Grey-rumped Swifts all have obviously but variably pale tail/rump areas.

Short-tailed Swift *Chaetura brachyura* `LC`
Four subspecies divided into two subspecies groups, of which one occurs in region.

Short-tailed Swift *Chaetura (brachyura) brachyura*
Fairly common breeding visitor (**St Vincent**), uncommon resident (**St Lucia**), vagrant elsewhere.
L 10 cm. Three subspecies in group, of which one (*praevelox*) occurs in region. Over towns, open and forested areas, in lowlands and hills. Status unclear; it is suggested that breeders on St Vincent (late Feb–Oct, mainly Mar–Jul) are otherwise absent, but apparently present year-round St Lucia. Vagrants Puerto Rico (May, sight), St Croix (Aug 1936), Guadeloupe (May 2005, May 2006), Martinique (May–Jun), Barbados (Jun–Oct) and Grenada. **ID** The most bat-like *Chaetura*, with protruding head, remarkably short square tail (wings appearing disproportionately long), and distinctive wing shape, with hooked outer wing and deeply bulging mid-wing, cutting in at body. **Ad** is largely brownish black, with slightly paler throat and obviously pale grey-brown rump and vent. **Juv** has narrow white tips to inner primaries, secondaries and tertials (usually invisible in field). **Voice** Mainly a series of dry, unmelodious, single notes "chip...chip...chip...". Song a bright melodious twitter "wee-d-drr...wee-d-drr...". Voice lower-pitched than most *Chaetura*. **SS** Very short tail and distinctive wing shape produce instantly recognizable silhouette unique among swifts in region. Lesser Antillean Swift is shorter-winged. Black Swift obviously larger and lacks pale rump and vent.

Antillean Palm-swift *Tachornis phoenicobia* `LC`
Locally common resident (**GA**).
L 9–11.5 cm; **WS** 23.5–25 cm. Two subspecies, of which both (nominate[A] Jamaica, Hispaniola and satellites of Saona, Beata and Vache, *iradii*[B] Cuba and I of Pines) occur in region. Endemic. Nearly always in reasonably close proximity to palms (especially *Roystonea* and *Washingtonia*), including treed savannas, cultivated zones and urban environments; occasionally breeds in sea caves and other sites, but typically uses palms, even fronds used as roofing. To c. 1100 m. Flickering, fast-paced flight, rarely very high above ground. Often in small groups, regularly with hirundines. Vagrant Puerto Rico (Jul 1969, Oct 2008, Aug 2011, Aug and Nov 2017), Bahamas (Great Inagua, Aug 2018), Turks & Caicos (Sept 2016) and Cayman Is (Sept 2015). **ID** Very slim-bodied, narrow-winged swift. **Ad** has long, obviously forked tail, black-brown upperparts broken by large white rump band with dark central stripe, and white throat and central belly on otherwise brown underparts. **Juv** has white of underparts duller and overall pattern less contrasting. **GV** Race *iradii* has deeper tail-fork (no overlap), sootier upperparts and more extensively pale flanks and face; in addition, sexes perhaps sometimes separable, with ♂ being darker with white throat, ♀ overall paler above with greyish-white throat. **Voice** Weak twittering calls given persistently in flight, rendered "tooee-tooee". **SS** Only regularly occurring sympatric swifts are much larger with far less or no white in plumage (White-collared and Black Swifts) or slightly larger and wholly lack white (Chimney Swift).

Lesser Antillean Swift

Ashy-rumped Swift
(Grey-rumped Swift)
lawrencei

Chimney Swift

Short-tailed Swift
praevelox

Antillean Palm-swift
phoenicobia

65

Alpine Swift *Tachymarptis melba* LC
Vagrant (**GA**, **LA**)

L 20–22 cm; **WS** 54–60 cm. Ten subspecies, of which one or more (all those identified have been *tuneti*) occurs in region. Barbados, Sept 1955, Jun–Jul 2003, Jul 2005, Jul 2015 (Frost & Burke 2005, *NAB* 69: 511); Guadeloupe, Apr 1987 (Feldmann & Pavis 1995); Desecheo I, Puerto Rico, Jul 1987 (Meier *et al.* 1989); and St Lucia, Aug 1992 (Burke 1994). Could appear almost anywhere in region. **ID** Very large swift with broad wings and shallowly forked tail. **Ad** has olive-brown upperparts becoming blacker on outer wing, white throat and highly distinctive large oval patch encircled by olive-brown breast-band, flanks and undertail-coverts. **Voice** Probably unlikely to be heard. Commonest call a loud, metallic-sounding trilling phrase lasting c. 2–4 seconds, accelerating then decelerating, "peee-ti-ti-titititititi-ti-ti-ti--ti", and increasing/decreasing in pitch and amplitude. **SS** Unmistakable if seen well. Only White-collared Swift approaches it in size, but underparts pattern unique and usually has obviously paler body plumage. **TN** Some authorities place in genus *Apus*.

Common Swift *Apus apus* LC
Vagrant (**Puerto Rico**), hypothetical (**Grenada**).

L 16–18 cm; **WS** 42–48 cm. Two subspecies, of which presumably nominate occurs in region. Faro de Rincón, Nov 2015 (Ławicki & van den Berg 2016). Bond (1973) speculated that a claim, from Grenada, Aug 1971 (Lack & Lack 1973) was erroneous, although the observers had spent years studying *A. apus*. Could appear almost anywhere, over open country and forests. **ID** Larger rakish swift with long, deeply forked tail and sharply pointed wings. **Ad** has black-brown mantle, outer wing, lesser coverts, tail, crown and underparts, except small off-white throat patch and pale forehead. Paler inner wing, greater underwing-coverts and rump, but contrast varies individually and temporally; fresh feathers darker with narrow white fringes. **Juv** is very black with little contrast; white fringes most obvious on forehead, and larger, more defined white throat patch. **Voice** Very unlikely to be heard, but commonest call a shrill, piercing, wheezy or reedy scream "srreeeeerrr", with variations, especially a much shorter and less reedy but still monotone call. **SS** Regularly occurring swifts are mainly either smaller and shorter-winged (all *Chaetura*), or larger and more powerful (White-collared Swift). Antillean Palm-swift much slighter and narrower-bodied, with more winnowing flight. Greatest risk of confusion almost certainly with Black Swift, which lacks pale throat (but difficult to evaluate at long range), making structural differences key (Black has narrower and less tapered wings, shallower tail-fork and heavier-looking body against which wings appear less set forward due to shorter tail).

TROCHILIDAE
Hummingbirds
365 extant species, 20 in region + 2 vagrants

White-necked Jacobin *Florisuga mellivora* LC
Vagrant (**Grenadines**).

L 11–12 cm. Two subspecies, of which one (*flabellifera*) occurs in region. Imm ♂, Carriacou, Aug 1904 (Clark 1905: 275; MCZ 113001), involved the Tobago race per Bond, but other authorities have listed it under nominate. Prefers reasonably well-wooded habitats. **ID** Mid-sized hummingbird with short, slightly decurved bill. **Ad** ♂ has blue head and breast, broad white crescent on nape, bright green upperparts, white belly and most of tail, the latter narrowly edged and tipped black. **Ad** ♀ varies, but mostly has blue-green breast heavily scaled whitish, dull white belly, green upperparts and mainly green tail with dark blue tip and white outermost feathers. **Juv** ♂ varies from essentially ♀-like, but more white in tail, to ♂-like with more black in tail. **Juv** ♀ is equally variable but usually has less white in tail, and more bronzy throat and chest. Both sexes at this age can show buffy malar and central rump. **Voice** Rather silent. Calls include a short "tsik", sometimes doubled "tsi-sik", also longer, high-pitched "sweet". **SS** At all ages should be unmistakable.

Rufous-breasted Hermit *Glaucis hirsutus* LC
Fairly common resident (**Grenada**).

L 10–12 cm. Two subspecies, of which one (*insularum*) occurs in region. Forest, edges, and banana, cocoa, and nutmeg plantations, principally above 450 m. Rare lower. **ID** Unmistakable, the only hermit in West Indies, with long, somewhat decurved bill (mandible yellow). **Ad** has dark green upperparts and brownish-rufous underparts. Outer rectrices rufous with broad black subterminal band and white tip. ♀ has paler throat and upper breast, shorter wings and tail, and more decurved bill. **Juv** has ochraceous upperparts and pale tips to remiges. Young ♂ usually has some prominent dark throat feathers. **Voice** Song a high-pitched descending "tsee-tsee-tsi-tsi-tsi", irregularly alternated with high-pitched "seee" notes. Flight call a sharp high-pitched upslurred "tseeet!". **SS** None.

Ruby-topaz Hummingbird *Chrysolampis mosquitus* LC
Vagrant (**Grenada**).

L 8–9.5 cm. Monotypic. One definite record, at Point Salines, Sept 1962, plus possible ♂ on Bequia, Grenadines (Bond 1963, 1982). Favours savanna-like vegetation, open country, gardens and cultivation, in lowlands. **ID** Striking hummingbird with short straight black bill. **Ad** ♂ is dark brown with brilliant ruby-red crown and nape, iridescent golden throat and breast, and rich chestnut, black-tipped tail. **Ad** ♀ has copper-green back and pale grey underparts; birds in Trinidad & Tobago (from where vagrants probably originate) sometimes have greenish stripe on chin to central breast. Central tail feathers green, others chestnut with broad black subterminal band and white tip. Bill averages slightly longer than ♂. **Juv** is like ad ♀; larger white post-ocular spot and dark violet outertail. **Voice** Song reportedly a high-pitched, doubled "tliii...tliii...tliii". **SS** Ad ♂ unmistakable. Other plumages easily confused but for distinctive and unusual head shape (nape, especially, appears shaggy), and combination of short bill and presence of rufous in outertail feathers.

Green-breasted Mango *Anthracothorax prevostii* LC
Common resident (**Providencia**, **San Andrés**).

L 11–12.5 cm. Four subspecies, of which one (*hendersoni*) occurs in region. Open coastal areas with scattered trees and bushes. Mar 1976 claim in Puerto Rico now generally not accepted. **ID** Fairly large with long, decurved bill. Regularly fans tail when feeding. **Ad** ♂ is bronze-green, with brighter green throat and blackish central throat-stripe bordered deep blue. Outertail feathers deep purple. **Ad** ♀ is mostly white below with green sides and broad central stripe black mixed with green. Outer rectrices have subterminal blue-black band and narrow white tip. Small percentage are like ad ♂, albeit usually with much less black and violet-blue on underparts. **Juv** has dark central stripe bordered or smudged rufous, and buffy fringes to much of back. **Voice** Rather silent. Song, usually from high perch, a buzzy, twanging "kazick-kazee", usually repeated rapidly 3–4 times. Calls include a liquid "tsup" or "tseep", and buzzy "pzzzt".

Alpine Swift
tuneti

Common Swift
apus

Rufous-breasted Hermit
insularum

White-necked Jacobin
flabellifera

Green-breasted Mango
hendersoni

Trinidad form

Ruby-topaz Hummingbird

67

SS No confusion species on Providencia or San Andrés, where it is the only regular hummingbird (compare rare migrant Ruby-throated Hummingbird). From other mango species in Greater Antilles by dark median stripe, and distinctive tail pattern in ♀. **TN** Proposed race *pinchoti* (San Andrés) considered inseparable from *hendersoni*.

Hispaniolan Mango *Anthracothorax dominicus* `LC`
Common resident (**Hispaniola**).

L 11–12.5 cm. Monotypic. Endemic. Throughout Hispaniola, and offshore islands of Tortue, Gonâve, Vache and Beata. Clearings, gardens, shade-coffee plantations, second growth and arid shrubby hillsides up to 2600 m, but rare above 1500 m and unusual in pine forest above 1100 m. Most abundant in semi-arid regions. **ID** Largest hummingbird on Hispaniola, with fairly long, slightly decurved bill. **Ad** ♂ has shiny green-bronze upperparts, metallic green throat and velvet-black underparts. Tail violet with blue-black tips. **Ad** ♀ has small white spot behind eye and grey underparts with white vent. Tail reddish brown with conspicuous white tip. **Juv** is like ad ♀ with velvet-black median line and no white spot behind eye. **Voice** Song apparently undescribed. Calls include a repeated short "tsip" and a high-pitched liquid, twittering trill. **SS** Smaller Hispaniolan Emerald has pale-based mandible, straighter bill and less striking tail pattern. **TN** See Puerto Rican Mango.

Puerto Rican Mango *Anthracothorax aurulentus* `LC`
Locally common to rare resident (**Puerto Rico, Virgin Is**).

L 11–12 cm. Monotypic. Endemic. Gardens and at forest edges, mainly in lowlands. Common in coastal areas with scattered trees, where the most numerous Puerto Rican hummingbird. Probably extinct on Anegada (Virgin Is) and Vieques (last seen Dec 1994), and rare to very rare in E Puerto Rico and rest of N Virgin Is, apparently all due to competition from Green-throated Carib. **ID** Very similar to allopatric Hispaniolan Mango. **Ad** ♂ is shiny green-bronze, with metallic green throat and black breast to mid-belly. Tail violet with blue-black tips and central rectrices bronzy green. **Ad** ♀ has grey underparts and white spot behind eye. Dull brownish-grey outertail, shading to darker subterminal tips and white tips. **Juv** is like ♀ with black median line on underparts and no white spot behind eye. **Voice** Call a repeated short "tsip" and high-pitched descending twittering trill. **SS** ♀ superficially resembles much smaller ♀ Puerto Rican Emerald, but latter has shorter, straighter bill and different tail pattern. See mainly highland Green Mango. **TN** Usually considered conspecific with Hispaniolan Mango by New World authorities, under name Antillean Mango.

Green Mango *Anthracothorax viridis* `LC`
Locally common resident (**Puerto Rico**).

L 11–14 cm. Monotypic. Endemic. Common in shade-coffee plantations and forests of C & W mountains, mainly at 800–1200 m. Less common in Luquillo Mts (E Puerto Rico); rare and perhaps seasonal on coasts (where largely replaced by Puerto Rican Mango). **ID** Fairly large hummingbird with little sexual dimorphism. **Ad** is emerald-green with slightly metallic blue tinge to underparts, and metallic blue-black, rounded tail. ♀ has tiny white post-ocular spot. **Juv** has brown-fringed feathers on head and back. **Voice** Rather silent. Song a repeated high-pitched phrase commencing with a drawn-out buzz, "szzzzz-szi-szi-chup-tsz-tsz.....". Calls include a repeated short "tsik", high-pitched twittering trill, and harsh rattle in alarm. **SS** Uniform emerald-green underparts in both sexes distinctive; ♂ Puerto Rican Mango has black breast and greyish vent, while ♀ is very different. See Green-throated Carib.

Jamaican Mango *Anthracothorax mango* `LC`
Common resident (**Jamaica**).

L 11–12 cm. Monotypic. Endemic. Widespread, commonest on N coast. Open lowlands including arid areas, gardens and plantations, commonest below 800 m; scarce in mangroves. Much dispersal to mid-altitudes (Cockpit Country) and to Blue and John Crow Mts in Jun–Aug, reaching c. 1500 m, following flowering season at higher altitudes. **ID** Unmistakable, all-dark hummingbird. **Ad** ♂ has green-brown crown, magenta-purple head-sides, velvet-black underparts and dull bronze upperparts. Central rectrices dusky bronze to dull black, rest metallic violet with narrow dark blue band. **Ad** ♀ is like ♂ but faded green velvet on flanks and white-tipped outertail feathers. **Juv** ♂ has deep blue throat, becoming black after 2nd-y. **Voice** Rather silent. Song apparently undescribed. Call a sharp, raspy "tic..tic..tic..". **SS** None, but beware streamertail without long rectrices, which is similarly sized.

Green-throated Carib *Eulampis holosericeus* `LC`
Common resident (**Puerto Rico, Virgin Is, LA**).

L 11–12.5 cm. Two subspecies, of which both (nominate[A] Puerto Rico, Virgin Is and most of Lesser Antilles, *chloroleamus*[B] Grenada) occur in region. Endemic. Open secondary vegetation, cultivation, parks, semi-deciduous and wet forest at all elevations; primarily coastal in Puerto Rico and prefers drier areas on Martinique. Commonest below 500 m. In some areas, disperses in Jun–Sept to forest borders at higher altitudes (c. 800–1000 m). **ID** Large hummingbird with little sexual dimorphism. **Ad** ♂ is metallic bronzy green with contrasting deep violet-blue central breast, velvet-black belly and blue-green uppertail-coverts grading to blue tail. **Ad** ♀ is very similar; bill slightly longer and more decurved. **Juv** is like ad ♀ but head feathers tinged brown. **GV** Race *chloroleamus* has darker green throat and broader violet-blue patch on centre of breast. **Voice** Calls include a short "tsip" and sharp "chewp", repeated rapidly in agitation. **SS** In Puerto Rico, larger Green Mango is similar but rarely encountered in E lowlands, and lacks dark patch on underparts.

Purple-throated Carib *Eulampis jugularis* `LC`
Locally common resident (**LA**).

L 11–12 cm. Monotypic. Endemic. Fairly common St Barthélemy, Saba, Guadeloupe, Dominica, Martinique, St Lucia, St Vincent and Grenada; uncommon St Eustatius, St Christopher, Nevis, Antigua, Barbuda and Montserrat. Vagrant Anguilla, Barbados (Jun–Aug), Puerto Rico including Vieques I (Jun, Oct) and Culebra I, and Virgin Is (St Croix, St John). On some islands (St Lucia, St Vincent) occurs at sea level at end of May, otherwise mainly in forest edge and canopy of secondary and primary forest at 800–1200 m; also swamp forest on Guadeloupe. **ID** Spectacular hummingbird with fairly short, decurved bill. **Ad** is dark velvet black with fiery purplish-red throat and breast, bright golden-green wings, and metallic greenish-blue tail. ♀ has longer, more decurved bill. **Juv** has throat and chest orange, speckled red. **Voice** Calls a strident "tsip" and sharp "chewp", repeated rapidly when agitated. **SS** None.

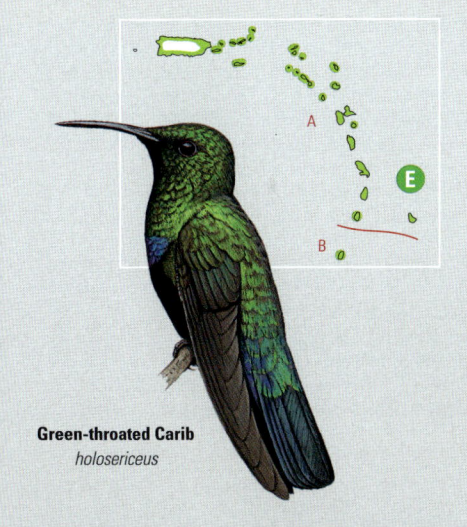

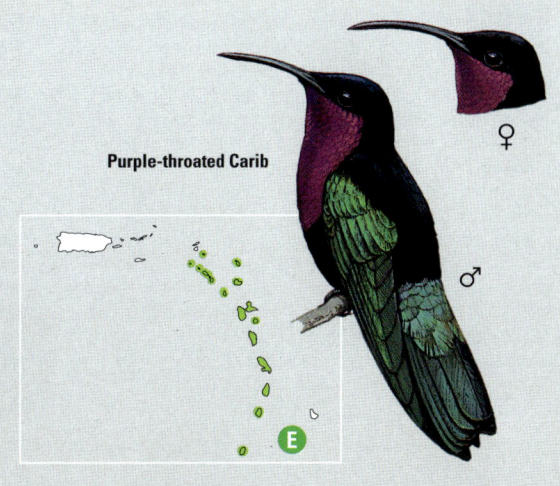

Hispaniolan Mango

Puerto Rican Mango

Green Mango

Jamaican Mango

Green-throated Carib
holosericeus

Purple-throated Carib

69

Cuban Emerald *Chlorostilbon ricordii* LC
Common resident (**Cuba**, **Bahamas**).

L 10–11.5 cm. Monotypic. Endemic. Cuban range includes Cayo Coco, Cayo Largo, I of Pines and many smaller cays. In Bahamas, regular Grand Bahama, Great Abaco, Andros and Green Cay; casual elsewhere (South Bimini). Varied wooded habitats to at least 1970 m, but most frequent below 1300 m, in humid and arid open forest. On Grand Bahama, common in bushy undergrowth of open pine woodland, and coastal scrub forest, gardens, parks and plantations. **ID** Largest hummingbird in range. Bill slightly decurved with red-based mandible, and tail deeply forked. **Ad** ♂ is emerald-green with short greyish-white spot behind eye, whitish vent and metallic bronze tail. **Ad** ♀ has longer post-ocular mark, pale brownish-grey central underparts, and less forked tail. **Juv** has duller green upperparts than ad ♀. **Voice** ♂ song high-pitched, a rapid rolling series of "slee" notes and sputtering metallic sounds, e.g., "zzi zzi zzi-zzi zzih", while ♀ gives high-pitched "seeeee" flight call. Other calls include a buzzy "zzzir" or "chi-di-dit". **SS** In Cuba, only sympatric species is much smaller (and rarer) Bee Hummingbird, with which confusion is unlikely, especially due to white-tipped blue tail.

Hispaniolan Emerald *Chlorostilbon swainsonii* LC
Fairly common resident (**Hispaniola**).

L 9.5–10.5 cm. Monotypic. Endemic. Not definitely recorded on satellite islands, but recently reported without documentation from I Saona (eBird). Dense forest, including pines, shade-coffee plantations, edges and scrub at 300–2500 m, occasionally to sea level and exceptionally 3075 m. **ID** Small with fairly straight bill, basally red on mandible. Tail forked. **Ad** ♂ is dark bronze-green with brownish forehead and crown, large black breast patch and dark brown tail. **Ad** ♀ has slightly more decurved bill, grey underparts, darker belly and undertail-coverts. Outer rectrix has broad dark brown subterminal band and white tip. **Juv** resembles ad ♀. **Voice** Sharp metallic chipping notes, sometimes in long series. **SS** Red base to mandible distinguishes it from all sympatric hummingbirds. ♀ further separated from same sex of much smaller Vervain Hummingbird by considerably longer bill, while Ruby-throated Hummingbird, which is only a vagrant to Hispaniola, has smaller white post-ocular spot.

Puerto Rican Emerald *Chlorostilbon maugaeus* LC
Common resident (**Puerto Rico**).

L 7.5–9.5 cm. Monotypic. Endemic. Coastal mangroves, open forest, woodland and coffee plantations, to 800 m; commonest in mountains. Generally forages in lower strata. **ID** Short- and straight-billed hummingbird with forked tail. **Ad** ♂ is metallic green with large iridescent blue-green throat, and shining steel-blue tail. Mandible red. **Ad** ♀ has all-black bill, white post-ocular spot, and pale grey underparts. Tail less forked with white tips to outer rectrices. **Juv** resembles ad ♀. **Voice** Presumed song a repeated twittering phrase of high-pitched descending notes, followed by a few even-pitched notes, "tseereetseetseetsee-tslew-tslew-tslew-tslew-tslew" lasting c. 2 seconds. Calls include a constantly repeated high-pitched "tsik" and irregular "si…si…sik-sik…tsik..". **SS** Red-based mandible and blue-green throat of ♂ distinctive. ♀ could be mistaken for same sex of Antillean Crested Hummingbird, which largely replaces present species in NE Puerto Rico and has much shorter bill, rounded tail, duskier underparts and no post-ocular spot, or ♀ Puerto Rican Mango, which is larger, with longer, more decurved bill.

Blue-headed Hummingbird *Cyanophaia bicolor* LC
Locally common resident (**Dominica**, **Martinique**).

L 9–11 cm. Monotypic. Endemic. Patchy; favours high-elevation humid forest, edges and second growth, especially at 800–1000 m. Often hawks insects over streams. **ID** Mid-sized hummingbird with fairly long, forked tail and long straight bill. **Ad** ♂ has metallic violet-blue head, pinkish-red mandible, shining green back and underparts, and blue uppertail. **Ad** ♀ is shining green above with bronze sheen on back, dusky ear-coverts, whitish-grey underparts, tail has blue band, outer rectrices narrowly tipped white. Bill all black. **Juv** is like ad ♀ but head feathers pale green with brown fringes. **Voice** Shrill, metallic notes, rapidly descending in pitch, and metallic "click-click-click". **SS** ♂ unmistakable. ♀ easily separated from ♀ Antillean Crested Hummingbird by longer tail and bill, and whiter underparts.

Antillean Crested Hummingbird LC
Orthorhyncus cristatus
Common resident (**Puerto Rico**, **Virgin Is**, **LA**).

L 8–9.5 cm. Four subspecies, all of which (nominate[A] Barbados, *exilis*[B] Puerto Rico S to St Lucia, *ornatus*[C] St Vincent, *emigrans*[D] Grenadines and Grenada) occur in region. Endemic. Open areas, parks, plantations and forest borders at all elevations, but commonest below 500 m. Rare W Puerto Rico. Possible dispersal to higher altitudes Jul/Aug. **ID** Short-billed hummingbird. **Ad** (*exilis*) has green head with obvious crest (rarely tipped blue), dull metallic bronze-green upperparts, and black underparts, usually with paler throat and upper breast. Tail black and rounded. **Ad** ♀ lacks crest (crown concolorous with upperparts) and has dull whitish-grey underparts. Outertail feathers broadly tipped white. **Juv** is like ♀ but head feathers tinged cinnamon. **GV** Races differ in crest colour of ♂♂: nominate golden to emerald with broad violet tip; *emigrans* more bluish violet, throat paler grey; *ornatus* has terminal portion abruptly blue. ♀♀ largely similar but differ in paleness of underparts. **Voice** Short "tsip" or "tzip" notes and longer "tslee-tslee-tslee-tslee". **SS** Seen well, ♂ should be unmistakable. On Martinique and Dominica, see ♀ Blue-headed Hummingbird, and Puerto Rico see ♀ Puerto Rican Emerald.

Red-billed Streamertail *Trochilus polytmus* LC
Common resident (**Jamaica**).

L ♂ 22–30 cm (including 13–17 cm tail); ♀ c. 10.5 cm. Monotypic. Endemic. At all altitudes and in all habitat types; especially common in Blue Mts at c. 1000 m. **ID** Remarkable ♂ is longest-tailed hummingbird. Always some red on bill. **Ad** ♂ has black cap, lateral crown feathers and ear-coverts extend beyond nape, iridescent emerald-green body, and black tail. Bill all red with small black tip. Red-chinned morph occasionally reported in St Andrews (EC Jamaica). **Ad** ♀ has crown concolorous with upperparts, and white underparts lightly spotted on sides and belly. Lacks streamers; tail centrally green, otherwise dark blue with broad white tips. Bill blackish at tip, always with extensive red base. **Juv** has black maxilla, red only at bill base. **Subad** ♂ is like ad ♂, but lacks tail-streamers (also absent in moulting ad ♂). **Voice** Loud, repetitive "chink, chink", "tsee, tsee" or "teet, teet". ♂ also emits distinctive whirring sound in flight caused by modified outer primaries. **SS** Confusion only likely with Black-billed Streamertail (note range), which always has all-black bill. **TN** Sometimes considered conspecific with Black-billed Streamertail, with which it occasionally hybridizes in W John Crow Mts. **AN** Streamertail (when lumped with Black-billed Streamertail).

Cuban Emerald

Hispaniolan Emerald

Puerto Rican Emerald

Blue-headed Hummingbird

Red-billed Streamertail

Antillean Crested Hummingbird

emigrans

exilis

cristatus

Black-billed Streamertail *Trochilus scitulus* `LC`
Common resident (**Jamaica**).

L ♂ 22–35 cm (including 16–24 cm tail); ♀ c. 10.5 cm. Monotypic. Endemic. Restricted to E, from sea level (Port Antonio) to elfin forest (John Crow Mts). Most abundant in semi-open areas like parks, banana plantations and on ridges; rare in dense vegetation. **ID** Very similar to Red-billed Streamertail. Entirely black bill at all ages. **Ad** ♂ is emerald-green, with black head, elongated feathers extending beyond nape, and black tail. **Ad** ♀/**Juv** are both identical to Red-billed Streamertail, but bill all black. **Voice** Song a long series of loud chips, "tsink...tsink...tsink..", at rate of c. 1.5 notes/second. Calls include a buzzing "zeet" or longer "zeeerrr" in agitation. ♂ emits distinct whirring sound in flight. **SS** See Red-billed Streamertail. **TN** See Red-billed Streamertail. **AN** Streamertail (when lumped with Red-billed Streamertail).

Lyre-tailed Hummingbird *Nesophlox lyrura* `LC`
Fairly common resident (**Bahamas**).

L 7.8–8.2 cm. Monotypic. Endemic. Only on Great and Little Inagua, in all habitats including dune scrub, riparian areas, gardens and parks, but perhaps not mangroves; habitat use appears to vary seasonally. **ID** Similar to Bahama Hummingbird. **Ad** ♂ differs from latter by metallic reddish-purple forehead (as well as gorget), lyre-shaped outer rectrices and more strongly forked tail. **Ad** ♀/**Juv** ♂ are not known to differ from same plumages of Bahama Hummingbird except in morphometrics, with young ♂ of present species having significantly shorter r1 but longer r5. **Voice** Song quiet and relatively simple (sounding like wet, squeaking shoes), also typically much shorter (lasting < 5 seconds) than Bahama Hummingbird (which see), either from perch or as part of shuttle display. Calls vary in both number and rate of syllables given in series, from single "chip" to a lengthy, rapid-fire sequence. **SS** The only hummingbird in range. **TN** Recently split from Bahama Hummingbird based on morphology (especially of ♂), vocalizations, courtship display and genetics (Feo *et al.* 2015). Some authors place this and next species in genus *Calliphlox*. **AN** Lyre-shaped or Inagua Woodstar, Inagua Sheartail, Inaguan Lyretail.

Bahama Hummingbird *Nesophlox evelynae* `LC`
Common resident (**Bahamas**), hypothetical (**Cuba**).

L 8–9.5 cm. Monotypic. Endemic. Coastal vegetation to pine forests, gardens and parks. Virtually throughout Bahamas (including Caicos Is). One report (sight only, presumably this species) Cayo Paredón Grande off N Cuba, Apr 2001 (Kirkconnell & Kirwan 2008). Usually less common where range overlaps with Cuban Emerald. **ID** Small, distinctive hummingbird with short, slightly decurved bill. **Ad** ♂ has iridescent green head and upperparts, small white post-ocular spot, glittering violet-purple gorget, and mottled rufous-and-green belly. Tail deeply forked: green central feathers, others with cinnamon inner webs. Post-breeding, iridescent throat replaced by pale grey plumage. **Ad** ♀ has duller upperparts, white throat and underparts with contrasting rufous flanks. Tail rounded. **Juv** is like ad ♀ but has buff-tipped upperparts; young ♂ has dusky throat, slightly forked tail and mostly blackish outer rectrices with little cinnamon. **Voice** Song a rhythmic "prítitidee, prítitidee, prítitidee" or sharper "tit, titit, tit,

titit" that often becomes a rattling sound (fast, high-pitched and relatively long vs. Lyre-tailed Hummingbird, lasting 3–27 seconds, also louder and audible further away). Calls include a high, fairly sharp, often persistently repeated, chipping "tih" or "chi". **SS** Rufous flanks distinguish both sexes from other regular hummingbirds in range. See Rufous Hummingbird (hypothetical) for comparison. **TN** See Lyre-tailed Hummingbird. **AN** Bahama Woodstar (when lumped with Lyre-tailed Hummingbird).

Vervain Hummingbird *Mellisuga minima* `LC`
Common resident (**GA**).

L 6–7 cm. Two subspecies, of which both (nominate[A] Jamaica, *vielloti*[B] Hispaniola and satellites, and presumably this race Puerto Rico) occur in region. Endemic. Common and widespread Jamaica and Hispaniola; Puerto Rico (Mona I, Mar 1976 and records 2015–2018 including ♀ on nest Sept 2016). Broadleaf deciduous and pine woodland, to shrubbery and gardens, even nesting inside houses in Jamaica; absent from dense, moist forest. Recorded to 1600 m (Hispaniola). More often heard than seen, usually gives penetrating song from canopy of tall trees, where hard to locate. **ID** Tiny, fractionally larger than Bee Hummingbird, thus smallest hummingbird in range. Bill straight and dull. **Ad** has dull metallic bronze-green upperparts and dull white underparts; chin and throat often flecked green. ♂ has notched tail, ♀ a rounded and white-tipped tail. **Juv** is similar to ♀, but young ♂ has chin and throat flecked grey and poorly defined white tips to rectrices. **GV** Race *vielloti* darker, ♂ with longer wings and more deeply forked tail. **Voice** Song a loud, long sequence of high-pitched, metallic squeaks. Also sings during diving display-flight. Calls: a throaty buzz "bzzzr" and soft twittering rattles. **SS** ♀ Ruby-throated Hummingbird (a rare migrant) has longer bill, more elongated profile and white spot behind eye.

Bee Hummingbird *Mellisuga helenae* `NT`
Rare and local resident (**Cuba**), hypothetical (**Bahamas**).

L 5–6 cm. Monotypic. Endemic. Woodland, shrubbery and gardens; occasionally in fairly open country, but usually requires mature growth with lianas and many epiphytes, to c. 1790 m. Now restricted to Guanahacabibes Peninsula, Zapata Swamp, and various sites in far E of mainland Cuba. ♀, Providenciales, Turks & Caicos, Feb 1986 (Aldridge 1987). **ID** Smallest bird in world, with short, rounded tail and black bill. **Ad** ♂ has fiery red head and throat, iridescent gorget with elongated lateral plumes, bluish upperparts, and mostly greyish-white underparts. In eclipse plumage resembles ♀ but has black-tipped tail. **Ad** ♀ is slightly larger, with green crown, black lores, indistinct white spot behind eye, and white tips to outer rectrices. **Juv** is like ♀, but young ♂ has deeper blue upperparts. **Voice** Song of high-pitched warbling phrases and a very high-pitched, drawn-out buzzy note, typically given from high, bare vertical twig, above canopy of low forest. Chase call a series of high-pitched buzzy scratchy notes. Also a high-pitched "tsit" when feeding. **SS** No other hummingbird in range has blue upperparts. Much smaller than Cuban Emerald (which see). When size not apparent, ♀ could be mistaken for ♀ Ruby-throated Hummingbird (rare migrant), but latter has longer bill and indistinct green throat spots.

Black-billed Streamertail

Lyre-tailed Hummingbird

Bahama Hummingbird

Vervain Hummingbird
minima

Bee Hummingbird

73

Ruby-throated Hummingbird *Archilochus colubris* LC
Rare migrant or vagrant (**GA**, **O**).

L 8–9.5 cm. Monotypic. Rare migrant N Bahamas, Cuba (mainly W) and Cayman Is, especially Mar–Apr but records Sept–Nov and Jan–May. Vagrant Jamaica (Jan 1971), Hispaniola (three records, one in mid Jan) and Puerto Rico (a few records). Could occur in almost any open habitat, parks, gardens etc. **ID** Small hummingbird, with long, thin dark bill. **Ad** ♂ is iridescent green with black eyestripe, white post-ocular spot, iridescent red throat bordered white below; rest of underparts dusky grey with greenish sides. Tail forked: central feathers green, outers greyish brown. **Ad** ♀ has chin and throat dusky or speckled white; tail less forked, with outer rectrices rounded and white-tipped. **Juv** is similar to ♀; most young ♂♂ show one or more ruby feathers on throat. **Voice** A soft "chup", also doubled, or longer chatters. Song (unlikely to be heard) a repeated, faint, insect-like, high-pitched sputtering rattle. **SS** ♂ always distinctive. ♀ could be mistaken for Cuban and Hispaniolan Emeralds, but they have paler bills and large white post-ocular mark. ♀ Rufous Hummingbird (hypothetical) usually has rufous flanks and distinctive tail pattern. Compare Bee and Vervain Hummingbirds (Cuba and Hispaniola, respectively).

Rufous Hummingbird *Selasphorus rufus* NT
Hypothetical (**Bahamas**).

L c. 8.5 cm. Monotypic. Juv ♂, Grand Bahama, Oct 1966 (Kale *et al.* 1969) could have been an extralimital Allen's Hummingbird *S. sasin* (breeds Pacific coast of USA, winters SC Mexico), but latter probably very unlikely. Could appear in any disturbed habitat rich in foodplants; now fairly regular in Florida in autumn/winter. **ID** Fairly small with rather short, straight black bill. **Ad** ♂ has bronze-green crown, scarlet-bronze gorget, white breast and rufous underparts. Upperparts and tail rufous, sometimes with green feathers on back and black-tipped tail feathers. **Ad** ♀ is bronze-green with pale underparts (throat often has green or bronze spots/streaks) and rufous flanks. Central rectrices metallic green, outers rufous, subterminally green and black, tipped white. **Juv** resembles ad ♀, but young ♂ may show reddish-brown back by winter, before acquiring gorget. **Voice** Dull "chup" notes, often doubled or in short chatters. **SS** Tail pattern in both sexes distinctive. Compare Ruby-throated Hummingbird.

CUCULIDAE
Cuckoos
149 extant species, 11 in region + 4 vagrants

Greater Ani *Crotophaga major* LC
Vagrant (**Virgin Is**, **Barbados**).

L 46–49 cm. Monotypic. One photographed, St Croix, Oct 2010 (*Cotinga* 33: 159), remarkably on same date as two photographed on Curaçao, Leeward Antilles, just outside region (*Cotinga* 33: 158); sight, Barbados, Nov 2012 (eBird). Typically in well-vegetated areas near water (standing or flowing), whereas Smooth-billed Ani occurs in all manner of more open country. **ID** Large ani with massive, black, two-tier bill and long flat tail. **Ad** has sleek, glossy blue-black plumage with bronzy-green cast. Obvious white eyes. **Juv** has smaller bill and brown eyes. **Voice** Unlikely to be heard, but utters a variety of croaks, grates and hisses. **SS** Smooth-billed Ani much smaller, with smaller bill and more matt plumage. Juv potentially more confusing, but remains larger than Smooth-billed. Grackles have much slenderer bills and different behaviour.

Smooth-billed Ani *Crotophaga ani* LC
Common to uncommon resident (**GA**, **LA**, **O**).

L 33–35 cm. Monotypic. Generally common resident (Bahamas, Greater Antilles, Virgin and Cayman Is, Guadeloupe, Dominica, St Vincent, Grenada, Providencia), but less common or even rare (Martinique) elsewhere in Lesser Antilles. Open habitats with scattered trees or bushes, including moist pastures, farmland, clearings, scrub and gardens, mostly in lowlands, rarely foothills and mountains. Also on small satellite islands. Sociable and conspicuous, typically in small, noisy flocks. Flight straight, rapid, shallow wing flaps alternating with longer glides. Typically flies in single file and often lands clumsily. Anis build bulky nests used communally by several ♀♀. **ID** Large, all-black bird with heavy, prominent bill and long, flat tail. **Ad** is black with slight bluish gloss. Bill laterally compressed into thin keel and swollen at base. Eyes brown to black. **Juv** lacks gloss and has smaller bill. **Voice** Loud, squawking, whistled "ah-nee" or "a-leep", rising in pitch and often given in flight. Also varied short growls and barks. **SS** Bill shape unique in region. Greater Antillean Grackle smaller with shorter tail and smaller bill.

Yellow-billed Cuckoo *Coccyzus americanus* LC
Common to rare breeding and non-breeding visitor and passage migrant (**throughout**).

L 28–32 cm. Monotypic. Scrub and dry forests in lowlands. Occasionally pine forest to at least 700 m. Also small satellite islands. Uncommon breeding visitor (May–Aug) Cuba, Hispaniola and Puerto Rico; rare Grand Bahama, Jamaica and Virgin Is. Passage migrant mainly Aug–Nov and Mar–May, but large numbers clearly overfly region between North & South America: fairly common S Bahamas, Cuba, Hispaniola, Puerto Rico and Guadeloupe; generally uncommon N Bahamas, Cayman Is and Jamaica; rare Virgin Is; uncommon to rare Lesser Antilles. Occasional in midwinter (at least Cuba, Cayman Is and Puerto Rico). **ID** Slender cuckoo with clean white underparts and decurved bicoloured bill (blackish maxilla and yellow mandible). **Ad** is brownish olive above with primaries edged rufous, conspicuous in flight. Long, graduated, blackish tail has conspicuous, large white oval spots on underside. Eye-ring yellow. **Juv** has narrower tail feathers, duller whitish tail spots and grey eye-ring. Bill black above and grey below. **Voice** A throaty "ka-ka-ka-ka-ka-ka-kow-kow-kowlp-kowlp", first notes loud and guttural, then slower and lower-pitched at end. **SS** Mangrove Cuckoo has black ear-coverts and buff-coloured underparts. Black-billed Cuckoo lacks yellow in bill and reddish brown in wing, with smaller, duller and less conspicuous spots on undertail. See Pearly-breasted Cuckoo. Other cuckoos and lizard-cuckoos are much larger.

Pearly-breasted Cuckoo *Coccyzus euleri* LC
Vagrant (**Anguilla**).

L 28 cm. Monotypic. Sombrero (Hat) I, Oct 1863 (AMNH 44495; Lawrence 1864, Greenway 1978, Banks 1988). Usually in well-wooded habitats. **ID** Slender with clean white underparts, very similar to but slightly smaller than Yellow-billed Cuckoo. **Ad** differs by uniform brown wings (no rufous in primaries). **Juv** has some rufous in wings, but inner webs of primaries white (brown in Yellow-billed). **Voice** Slow series of 5–20 sad "kuoup" notes, at c. 1 note/second, repeated several times. Also short rattle followed by 4–9 accented notes similar to Yellow-billed Cuckoo. **SS** Easily confused with Yellow-billed Cuckoo (see above). Mangrove Cuckoo has prominent black ear-coverts and buff-coloured underparts. Black-billed Cuckoo lacks yellow on bill and has smaller, duller and less conspicuous spots on undertail. Compare Dark-billed Cuckoo.

Ruby-throated Hummingbird
♂ ♀

Rufous Hummingbird
♂ ♀

Greater Ani

Smooth-billed Ani

Yellow-billed Cuckoo

Pearly-breasted Cuckoo

75

Mangrove Cuckoo *Coccyzus minor* `LC`
Fairly common to uncommon resident (**throughout**).
L 28–34 cm. Monotypic. Mangroves (but rarely principal habitat), thickets near water, dry scrub and forest. Also small satellite islands. Mostly lowlands, but also montane rainforest in Lesser Antilles, locally to 1300 m. Mainly understorey, where best located by voice. **ID** Variable, slender cuckoo with black mask, long graduated tail and decurved bicoloured bill. **Ad** is greyish brown above, greyer on crown. Wings brown. Underparts rich cinnamon to whitish or buff, sometimes with pale greyish throat and breast. Tail blackish with prominent white oval spots below. Indistinct yellowish eye-ring. **Juv** is generally duller with wing-coverts and flight feathers edged rufous. Mask and tail spots indistinct, breast buffy white and bill duller. **Voice** Similar to Yellow-billed Cuckoo, but slower, lower-pitched and more nasal. **SS** Black mask unique (but see vagrant Dark-billed Cuckoo). Black-billed and slightly smaller Yellow-billed Cuckoos whitish below. Juv Yellow-billed recalls juv Mangrove, but smaller with whiter underparts. Lizard-cuckoos much larger. **TN** Sometimes considered polytypic, with following races recognized in West Indies: *maynardi* (S Florida and Bahamas), *teres* (most of Greater Antilles), *caymanensis* (Cayman Is), *nesiotes* (Jamaica), *abbotti* (Providencia and San Andrés), *rileyi* (Barbuda, Antigua), *dominicae* (Montserrat, Guadeloupe, Dominica, *vincentis* (Martinique, St Lucia, St Vincent) and *grenadensis* (Bequia, Grenada).

Dark-billed Cuckoo *Coccyzus melacoryphus* `LC`
Vagrant (**Grenada**).
L 25–28 cm. Monotypic. One collected in mangroves, May 1963 (Schwartz & Klinikowski 1965). In native range, varied forested habitats from dry to humid, also farmland, clearings, thickets and mangroves, mainly in lowlands. **ID** Rather small, slender cuckoo with long, tapered tail and decurved black bill. **Ad** is greyish brown above and buffy below, with greyer crown and nape, black mask and narrow grey band on neck-sides. Tail bronzy above, blackish below with broad white spots. Eye-ring grey to olive-yellow. **Juv** is duller, with crown and nape brown, wings sometimes rufous and tail without distinct white spots below. **Voice** A "cu-cu-cu-cu-cu-kolp, kolp, kulop". **SS** Smaller than most other cuckoos. Grey band on cheeks and neck-sides diagnostic. Mangrove Cuckoo obviously larger, ad Yellow-billed Cuckoo lacks mask and has yellow mandible, and Black-billed Cuckoo lacks mask and has browner upperparts.

Black-billed Cuckoo *Coccyzus erythropthalmus* `LC`
Very rare passage migrant (**Bahamas**, **Cuba**, **Cayman Is**), vagrant elsewhere.
L 27–31 cm. Monotypic. Scrub, mangrove and dry and moist forests in lowlands, where typically hard to see. Records Sept–Dec and Mar–May. Vagrant Jamaica, Dominican Republic, Puerto Rico, Virgin Is, Barbuda, Antigua, Guadeloupe, Dominica, St Lucia and Barbados. **ID** Slender with a long, tapered tail and decurved dark bill. Similar to Yellow-billed Cuckoo. **Ad** has all-black bill, red eye-ring, grey-brown wings without rufous, tail greyer and duller, with less contrasting whitish spots. **Juv** is like ad but warmer above and buffier below, with yellowish to greyish eye-ring. Undertail duller with less distinct spotting. Bill black above, greyish below, with yellowish eye-ring.

Voice Silent on migration and in winter. **SS** Dark bill always distinguishes from Yellow-billed and Mangrove Cuckoos. Mangrove also differs by larger size and dark mask. Lizard-cuckoos much larger. Compare vagrant Dark-billed Cuckoo.

Chestnut-bellied Cuckoo *Coccyzus pluvialis* `LC`
Fairly common resident (**Jamaica**).
L 48–56 cm. Monotypic. Endemic. Moist to wet evergreen and montane forest, open woodland, thickets in open areas, brush-covered limestone hills and gardens, to c. 1500 m. Forages in midstorey and canopy, runs along branches and glides on outstretched wings between trees. **ID** Very large cuckoo with a thick, blackish, decurved bill. **Ad** is dull blackish brown above. Face and breast light grey with paler throat. Belly chestnut. Long, broad black tail has purple gloss and broad white spots on underside. Orbital skin black and eyes red to brown. **Juv** lacks purple gloss on tail. **Voice** Throaty, hoarse "quak-quak-ak-ak-ak", slow, then accelerating towards end. **SS** Nothing similar in Jamaica.

Bay-breasted Cuckoo *Coccyzus rufigularis* `EN`
Rare and local resident (**Hispaniola**).
L 46–51 cm. Monotypic. Endemic. Rare and declining; perhaps extinct in Haiti. Various forest types, from tropical deciduous to montane rainforest, to c. 900 m, rarely higher. Occasionally overgrown pastures and agricultural areas. Secretive. Forages in midstorey and canopy, leaping from branch to branch. **ID** Very large cuckoo with chestnut breast. **Ad** has grey upperparts. Chin to breast and patch on primaries dark chestnut, and belly pale rufous. Long, broad black tail has broad white spots on underside and white tip. Eye-ring yellow. Thick, downcurved black bill has yellow mandible. **Juv** is basically identical. **Voice** Guttural crow-like "ú-wack-ú-wack-ú-wack-", also bleats like a lamb. **SS** Nothing similar on Hispaniola. Smaller Hispaniolan Lizard-cuckoo has pale grey breast.

Cuban Lizard-cuckoo *Coccyzus merlini* `LC`
Common resident (**Cuba**).
L 45–55 cm. Three subspecies, all of which (nominate[A] throughout Cuba, Cayo Conuco and Cayo Saetía, *santamariae*[B] Archipiélago de Sabana-Camagüey, off NC Cuba, *decolor*[C] I of Pines) occur in region. Endemic. Varied forest types, from tropical deciduous to moist, evergreen and pine. Also thickets, overgrown pastures and limestone hills. To 1250 m. Forages in midstorey and canopy, clumsily, occasionally on ground. **ID** Very large with long tail and long straight bill. **Ad** is olive-brown above, with rufous in primaries. Throat whitish, breast and cheeks pale grey, and rear underparts rufous. Tail grey with black subterminal band and broad white tip. Bare ocular skin red to orange. **Juv** has narrower tail feathers, more pointed and without distinct terminal spots. Bare ocular skin yellow. **GV** Race *decolor* paler below, greyish brown above with shorter bill than nominate; *santamariae* smaller than nominate, with paler upperparts and longer bill. **Voice** A long, loud throaty series, "ka-ka-ka-ka-ka-ka-kau-kau-ko-ko", lasting c. 9 seconds, with second part gradually increasing in volume and slightly in speed. **SS** All other cuckoos in Cuba much smaller, with shorter, decurved bills. **TN** Usually considered conspecific with Bahama Lizard-cuckoo, under name Great Lizard-cuckoo.

Mangrove Cuckoo

variant

Black-billed Cuckoo

Dark-billed Cuckoo

variant

Bay-breasted Cuckoo

Chestnut-bellied Cuckoo

decolor

merlini

Cuban Lizard-cuckoo

77

Bahama Lizard-cuckoo *Coccyzus bahamensis* `NT`
Uncommon resident (**Bahamas**).

L c. 50 cm. Monotypic. Endemic. On Andros, New Providence and Eleuthera, in wide variety of well-vegetated habitats, especially with many vines, including coffee plantations, scrub and pine forest. Habits probably similar to Cuban Lizard-cuckoo. **ID** Very similar to allopatric Cuban Lizard-cuckoo. **Ad** differs in slightly smaller size, much smaller (usually concealed) rufous wing panel, greyer upperparts, much paler underparts and black tip to central uppertail. **Juv** presumably differs from Cuban mostly in smaller size. **Voice** Similar to Cuban Lizard-cuckoo, but not subject to detailed study. **SS** Large and distinctive. Nothing similar within limited range. **TN** See Cuban Lizard-cuckoo. Population of Andros sometimes separated as race *andria*.

Jamaican Lizard-cuckoo *Coccyzus vetula* `LC`
Common and widespread resident (**Jamaica**).

L 38–40 cm. Monotypic. Endemic. Tropical evergreen forest, woodland, semi-arid country with trees and shrubs, wet montane areas and brush-covered limestone hills, to 1200 m. Forages in midstorey and canopy. **ID** Fairly large, with long tail and long, straight bill. **Ad** is greyish above with forehead to nape dark brown, and rufous patch in primaries. Throat white, breast grey and rest of underparts rufous, with paler vent. Central tail feathers grey, with black subterminal band and broad white tips, others black with broad white tips. Orbital skin red. **Juv** is basically indistinguishable from ad in field. **Voice** A rapid, low "cak-cak-cak-ka-ka-ka-k-k". **SS** Chestnut-bellied Cuckoo is larger, has chestnut rear underparts and shorter, decurved bill. Mangrove Cuckoo is smaller, has black mask and shorter, bicoloured bill. Other cuckoos in range smaller and paler below.

Hispaniolan Lizard-cuckoo *Coccyzus longirostris* `LC`
Two subspecies placed in separate subspecies groups, of which both occur in region.

Hispaniolan Lizard-cuckoo
Coccyzus (longirostris) longirostris
Common and widespread resident (**Hispaniola**).
L 41–46 cm. One subspecies in group. Endemic. Widespread in all manner of wooded areas including suburban gardens and shade-coffee plantations, to 1700 m, rarely 2200 m. Forages from understorey to canopy. Occurs on all major satellites (except Gonâve I). **ID** Large cuckoo with long tail and long, straight bill. **Ad** is grey above, with rufous patch in primaries. Throat and rear underparts rufous contrasting with pale grey breast. Tail black with broad white tips. Bare orbital skin red. **Juv** has whitish throat and generally browner tail with buff tips. **Voice** A throaty, descending "ka-ka-ka-ka-ka-ka-ka-kau-kau-ko-ko". **SS** Bay-breasted Cuckoo larger with dark reddish-brown breast and throat. Other cuckoos in range much smaller. **TN** Population of Saona I sometimes separated as race *saonae*.

Gonave Lizard-cuckoo *Coccyzus (longirostris) petersi*
Common (**Gonâve I, off W Hispaniola**).
L 41–46 cm. One subspecies in group. Endemic. Habitat and habits as Hispaniolan Cuckoo, but restricted to much narrower range of forest types. **ID** Very similar to Hispaniolan Lizard-cuckoo. **Ad** differs in having generally paler upperparts and underparts (especially belly), with whitish (rather than very pale rufous) throat, and apparently slightly longer bill and tail. **Juv** plumage is apparently undescribed. **Voice** Not known to differ from Hispaniolan Lizard-cuckoo. **SS** Unlikely to be confused in tiny range.

Puerto Rican Lizard-cuckoo *Coccyzus vieilloti* `LC`
Fairly common resident (**Puerto Rico**), vagrant (**Virgin Is**).

L 40–48 cm. Monotypic. Endemic. Single mid-19th century specimen from Vieques I (Gemmill 2015). Vagrant (also historically) on St Thomas (Shelley 1891) sometimes considered doubtful (Peters 1940). Dense tropical deciduous and evergreen forest, shade-coffee plantations and brush-covered limestone hills, mainly to c. 800 m. Forages in midstorey and canopy. **ID** Large cuckoo with long straight bill, long tail and two-toned underparts. **Ad** is greyish brown above, with primaries concolorous with back (no rufous in wing). Tail grey with very broad subterminal bars and broad white tips. Throat whitish, breast grey and lower underparts rufous. Bare ocular skin red. **Juv** is brown above with rufous feather edges. Chin and throat light grey, belly paler rufous. Tail narrower and paler with broad black bands, and tips more pointed. **Voice** An emphatic "ka-ka-ka-ka…", accelerating and becoming louder. **SS** Other cuckoos in range much smaller and shorter-billed.

Common Cuckoo *Cuculus canorus* `LC`
Vagrant (**Barbados**).

L 32–36 cm. Four subspecies, of which at least one (probably nominate) occurs in region. 1st-y ♀, Nov 1958 (Buckley *et al.* 2009); 1st-y, Nov 2014 (*NAB* 69: 170). Possible in wooded or semi-open habitats with trees or bushes. Flight on regular, shallow wingbeats and occasional short glides. **ID** Medium-sized cuckoo with barred underparts. **Ad** ♂ has grey head, throat, upper breast and upperparts. Lower breast and belly white, barred black. Undertail spotted and tipped white. Bill black with yellow base; legs yellow. **Ad** ♀ has two colour morphs. **Grey morph** resembles ♂ but has rufous-washed upper breast. **Rufous morph** (nominate race) has upperparts barred chestnut and blackish, and rufous rump. Tail chestnut barred black. Underparts white barred pale chestnut and blackish, with rufous-tinged breast. **Juv** has white nape spot and white-tipped crown and upperparts. **Voice** Unlikely to be heard in region. **SS** Only cuckoo in region that has barred underparts, but other features of plumage almost equally distinctive. However, extralimital African Cuckoo *C. gularis*, a potential vagrant, would require very careful separation. The African species has much more extensive and obvious yellow at base of bill (including maxilla) and barred outer tail feathers (spotted in Common Cuckoo). Young African Cuckoo is greyer above than Common, with paler, more narrowly barred throat, broader pale barring above, and larger white spots and bars on tail. Hepatic juvenile is much less rich brown than Common.

Bahama Lizard-cuckoo

Jamaican Lizard-cuckoo

Hispaniolan Lizard-cuckoo

Gonave Lizard-cuckoo
(Hispaniolan Lizard-cuckoo)

Puerto Rican Lizard-cuckoo

Common Cuckoo
canorus

grey morph

♀ rufous morph

♂

79

RALLIDAE
Rails, Gallinules and Coots
142 extant species, 11 in region + 3 vagrants

Yellow Rail *Coturnicops noveboracensis* LC
Vagrant (**Bahamas**).

L 16–19 cm. Two subspecies, of which nominate has occurred in region. Grand Bahama, mid-Sept 1971 (Bond 1972). Drier margins of marshes, meadows and fields. Very secretive, runs mouse-like under vegetation if disturbed. Perhaps under-recorded. **ID** Small, quail-like rail with short bill, neck and tail. **Ad** has buffish head and underparts faintly barred brown, and whitish throat and belly. Upperparts broadly streaked blackish and yellow. In flight, white secondaries. ♂ has corn-yellow bill, dusky brown in ♀ and non-br. **Juv** is poorly known, but perhaps much darker than ad. **Voice** Usually silent away from nesting areas. **SS** Compare Spotted and Yellow-breasted Crakes.

Black Rail *Laterallus jamaicensis* NT
Rare resident or visitor (**GA**), vagrant (**LA**, **O**).

L 12–15 cm. Four subspecies, of which nominate occurs in region. Rare resident Hispaniola, Jamaica and probably Cuba, but winter visitor Puerto Rico (perhaps formerly resident). Vagrant Bahamas (though heard singing on Andros, Mar 2000, and Abaco, Jun 2013), Antigua and Barbuda. Marshes and wet grassland. Principally nocturnal, very shy and reluctant to leave cover. **ID** Tiny blackish rail with short black bill. **Ad** is dark slate-grey from head to lower breast with distinctive black-and-white barring on belly to vent. Chestnut upperparts spotted white. **Juv** has white chin, throat and central belly; flanks and vent browner. Legs darker. **Voice** ♂ gives "kic-kic-kerr" ending in downward slur; ♀ a low "croo-croo-croo". Growls and barking notes. Rarely vocal by day. **SS** Fairly distinctive, other ad rails considerably larger. Most likely to be mistaken for downy young of other rallids, which are all black and lack bars or spots.

King Rail *Rallus elegans* NT
Fairly common resident and rare winter visitor (**Cuba**), vagrant (**Jamaica**).

L 38–48 cm. Two subspecies, of which both (*ramsdeni* resident, nominate winter visitor) occur in region. Freshwater wetlands with tall, dense emergent vegetation and, more rarely, brackish marshes throughout mainland Cuba, I of Pines and perhaps some offshore cays (risk of confusion with Clapper Rail high). At least small numbers of nominate race winter in W & C Cuba (Llanes Sosa *et al*. 2016, Kirkconnell *et al*. in press). Largely diurnal, feeding near cover. **ID** Largest *Rallus* in region, the size of American Coot. **Ad** (*ramsdeni*) has long slender bill, slightly decurved at tip and orange at base. Black upperparts feathers broadly fringed tawny-brown, grey on face only around eye or absent altogether, underparts pale cinnamon-buff, with bold black and white bars on flanks. **Juv** is largely fuscous-black above with brown feather edges, bars on flanks indistinct and ground colour of underparts generally paler. **GV** Nominate race larger and typically more saturated below. **Voice** Both sexes give series of even-spaced dry "kek" notes, 2–3 per second, slower and more regular than Clapper Rail. Also a soft, rapid "tuk" call, a deep boom (♂ alone), a purring call (♀) and "rak-k-k" or "chur-ur-ur" in distress. **SS** Larger than congeners (especially vs. Virginia Rail), without extensive grey on head-sides. Clapper Rail also has less contrasting black and white bars below, and brown rather than rufous wing-coverts (visible in flight). Virginia Rail tends to have redder bare parts and marginally brighter rufous wing-coverts.

Clapper Rail *Rallus crepitans* LC
Locally common to rare resident (**GA**, **LA**, **O**).

L 32–41 cm. Eleven subspecies, of which three (*coryi*[A] Bahamas, *leucophaeus*[B] I of Pines, *caribaeus*[C] elsewhere) occur in region. Locally common Bahamas, Greater Antilles and Barbuda, rare St Kitts, and Guadeloupe, vagrant elsewhere. Feeds in saltmarshes and among mangrove roots; generally near shoreline. Active at dawn and dusk. **ID** Rather pale, dull, long-necked rail with long dark-tipped reddish bill. **Ad** (*caribaeus*) is somewhat variable. Typically has greyish head, reddish-brown upperparts with olive-grey to brown feather fringes, and cinnamon-washed underparts with indistinct greyish breast-band; flanks barred blackish and white. **Juv** is like ad, but back more uniform olive-brown, wing-coverts with pale tips and subterminal bars, and more extensively whitish below. **GV** Race *coryi* much paler, with whitish underparts, and upperparts feathers always fringed brown; *leucophaeus* slightly less pale, breast with pale cinnamon-buff wash. **Voice** Both sexes give c. 10-second bellowing series of dry "kek" notes, 3–4/second, sometimes accelerating at end, often in duet; ♀ often follows these with a "burr". Also a grunt and very low "hoo". **SS** Virginia Rail much smaller and darker with richer rufous underparts and wing-coverts. See King Rail. **TN** Race *limnotis* (Puerto Rico) sometimes recognized, but subsumed here within *caribaeus*.

Virginia Rail *Rallus limicola* LC
Four subspecies divided into two subspecies groups, of which one occurs in region.

Virginia Rail *Rallus (limicola) limicola*
Rare visitor (**Grand Bahama**) or vagrant (rest of **Bahamas**, **GA**).

L 20–25 cm. Two subspecies in group, of which nominate occurs in region. Present Sept–Apr. In Greater Antilles, records only in Cuba and Puerto Rico. Freshwater marshes with emergent vegetation, but will use brackish wetlands. Usually remains in or near cover, but may feed in open at dusk. **ID** Substantially smaller than King and Clapper Rails. **Ad** has extensive blue-grey head-sides, rather dull upperparts with contrasting rufous wing-coverts, cinnamon-buff underparts and black-and-white barred rear. **Juv** has mainly sooty-black upperparts, duller sepia-tinged wing-coverts, blackish underparts mottled white, poorly defined flanks barring and whitish central breast and belly. **Voice** Loud series of "kik kik kidik kidick" notes followed by "queeah" is song and mainly given when breeding (thus unlikely to be heard); at other times, a descending series of pig-like grunts. **SS** Combination of small size, grey head-sides, dark upperparts and rich rufous wing-coverts distinguishes it from congeners.

Corncrake *Crex crex* LC
Vagrant (**Guadeloupe**).

L 27–30 cm. Monotypic. Exhausted juv, Sept/Oct 2003 (Levesque & Saint-Auret 2007). Several other New World records, thus potentially a more regular vagrant. In native range, favours wet meadows and marshy fringes; very secretive. **ID** Fairly large, plump-bodied crake with chestnut wing-coverts. **Ad** has streaked crown, dull grey supercilium, foreneck and upper breast, and rufous-brown upperparts spotted blackish. Lower breast, belly and flanks barred brown and buff. Legs and feet dark grey. **Juv** has narrower, more buff-yellow-tinged upperparts, less barred wing-coverts, and buff-brown face and breast. **Voice** Unlikely to be heard. **SS** None. Compare smaller Spotted Crake.

Yellow Rail
noveboracensis

Black Rail
jamaicensis

elegans

King Rail
ramsdeni

Clapper Rail
caribaeus

Virginia Rail
limicola

Corncrake

81

Zapata Rail *Cyanolimnas cerverai* `CR`
Extremely rare and local resident (**Cuba**).
L c. 29 cm. Monotypic. Endemic. Extremely rare, little-known and very infrequently seen endemic of Zapata Swamp, in dense sawgrass (*Cladium jamaicense*) and *Typha* marshes with some bushes and low trees. Last definite sighting Nov 2014. **ID** Apparently almost flightless; very short wings and tail. Relatively plain with few obvious markings. **Ad** has fairly straight, red-based yellowish-green bill, brown upperparts, grey head and underparts, very faintly pale-barred lower belly, and white undertail-coverts. **Juv** is reportedly duller, without red on bill and has olivaceous legs. **Voice** Very poorly known. Alarm a loud "kwowk", much like Limpkin. Song unknown; previously attributed recordings are of Spotted Rail. **SS** Plain grey underparts and brown upperparts distinctive. Spotted Rail (shares same marshes) has strikingly variegated black-and-white underparts.

Spotted Rail *Pardirallus maculatus* `LC`
Two subspecies placed in separate subspecies groups, of which one occurs in region.

Southern Spotted Rail *Pardirallus (maculatus) maculatus*
Rare to very rare resident (**GA**).
L 25–28 cm. One subspecies in group. Freshwater swamps with dense emergent vegetation; occasionally rice fields. No records Puerto Rico, and very few until last decade on Hispaniola. In Cuba and Jamaica, most records in W of both islands. Usually secretive, but will feed in open early morning and evening. **ID** Strikingly variegated plumage. Pinkish legs and long, slightly decurved yellowish bill with red spot at base. **Ad** is black with head and brown upperparts spotted, and underparts strongly barred, white. **Juv** has duller bill and legs, with less well-marked plumage. **Voice** Loud, repeated, rasping screech, usually preceded by grunt or pop. Also an accelerating series of deep, gruff, pumping notes, like distant motor starting up, and sharp "gek" in alarm. **SS** Compare Zapata Rail.

Yellow-breasted Crake *Hapalocrex flaviventer* `LC`
Locally uncommon to rare resident (**GA**).
L 12.5–14 cm. Five subspecies, of which two (*gossii*[A] Cuba and Jamaica, *hendersoni*[B] Hispaniola and Puerto Rico) occur in region. Freshwater marshes and canals with grassy edges. Forages among emergent plants. Emerges early morning and evening; runs across floating plants and climbs easily in tangled reeds. When flushed, flies weakly a short distance, legs dangling. **ID** Small crake, with long toes and short, dark bill. **Ad** (*gossii*) has dark line through eye, whitish supercilium, blackish crown, rich buffy neck- and breast-sides, white-streaked upperparts, and black and white bars on flanks. **Juv** has indistinct dusky barring on neck and breast. **GV** Race *hendersoni* slightly smaller and paler above. **Voice** Low, harsh, churring "k'kuk kurr-kurr", plaintive squealing, single or repeated "kreer" or "krreh", and high-pitched, whistled "peep" notes. **SS** Facial pattern diagnostic vs. Yellow Rail (a potential vagrant to same range) and combination of overall colour and size distinctive compared to other crakes.

Sora *Porzana carolina* `LC`
Regular passage migrant and winter visitor (**throughout**).
L 19–25 cm. Monotypic. Common Cuba and Bahamas, uncommon (but doubtless overlooked) elsewhere. Present Sept–May, in marshes, rice fields and mangroves. **ID** Rotund crake with short, straight yellow bill. **Ad** has blackish mask, white spot behind eye, grey breast, buff undertail-coverts; belly barred black and white. ♀ has duskier bill. **Imm** is like ad but black on head less extensive and mottled grey; cheeks and neck-sides paler grey, and breast often tinged olive. **Juv** has upperparts and wings with more white streaks; mainly buff on head, neck and breast where ad is grey, with duller flanks barring. Black restricted to lores and around eye, with white chin, throat and central crown-stripe. **Voice** Usually silent on wintering grounds. High-pitched descending whinny sometimes heard, and short, sharp "keek" in alarm. **SS** See Spotted Crake.

Spotted Crake *Porzana porzana* `LC`
Vagrant (**St Martin, Guadeloupe**).
L 22–24 cm. Monotypic. 1st-y ♀, St Martin, Oct 1956 (Voous 1957; specimen now at Naturalis); Désirade I, Guadeloupe, Feb 2014 (Chabrolle & Levesque 2015). Habitat as Sora. **ID** Similar to Sora. **Ad** appears spotted, with extensive white and grey bars and speckles on underparts, plus buff vent. Yellow bill with greenish tip and orange-red base at base of maxilla. White leading edge to wing in flight. ♂ has grey face and breast; ♀ a grey face, heavily spotted. **Imm** has more white spots on head-sides, whitish throat and duller yellow-brown bill. **Juv** is like imm but has streak over eye brown or cream, with tiny white spots, neck mottled grey-brown and off-white, breast olive-brown to bright brown with white or buff markings, and flanks less contrastingly barred. **Voice** Usually silent. Advertising call (heard away from breeding sites) a short, sharp ascending "whitt", suggesting whiplash, repeated about once per second for several minutes, mainly at dusk through night. Also quiet "hui" notes, a loud repetitive ticking, hard "eh" in alarm and warning "tshick". **SS** Ad Sora readily distinguished by blackish mask and grey breast. Juv Sora has unspotted breast and head-sides, and distinct central crown-stripe (vs. streaked crown).

Purple Gallinule *Porphyrio martinicus* `LC`
Fairly common (**GA**) to rare resident (**LA**), uncommon migrant (**O**).
L 27–36 cm. Monotypic. In Lesser Antilles, rare resident St Barthélemy, Montserrat, Guadeloupe, Martinique and Barbados, rare winter visitor Dominica; vagrant elsewhere. In Bahamas (rare on smaller islands), present Aug–Oct and Mar–May. Freshwater habitats including marshes and rivers. Walks on floating vegetation and swims, like American Coot. **ID** Attractive, with vivid yellow legs and feet. Red bill has yellow tip and is topped by pale frontal shield. **Ad** is bluish purple with iridescent greenish-bronze back and wings, and white vent. Bare parts duller in non-br plumage. **Juv** is warm buffy brown with darker upperparts, and wings have slight iridescent green sheen. Bare parts much duller and browner. **Voice** Typical is a sharp, high-pitched "kyik", sometimes with booming undertone, loud "kur", often preceded by "cook" notes. Also a wailing scream and rapid "ka-ka-ka". **SS** Ad unmistakable. Juv browner than Common Gallinule and lacks flanks patch.

Zapata Rail

Southern Spotted Rail
(Spotted Rail)

Yellow-breasted Crake
gossii

Sora

Spotted Crake

Purple Gallinule

Common Gallinule *Gallinula galeata* LC
Common resident (**throughout**).

L c. 35–36 cm. Seven subspecies, of which at least two (*cerceris*[A] Greater and Lesser Antilles, *barbadensis*[B] Barbados) occur in region, with those wintering on San Andrés possibly *cachinnans*. Wetlands and marshes with emergent vegetation. Conspicuous and often tame. **ID** Mid-sized gallinule. **Ad** (*cerceris*) is essentially black at any distance, with prominent yellow-tipped red bill and squared-off red frontal shield, prominent white line at top of flanks, and white lateral undertail-coverts made obvious by constant flicks of tail. **Juv** has crown, hindneck and upperparts duller brown, underparts paler, becoming whitish on throat and belly. Bill and shield dull yellowish brown, legs and feet yellowish grey. **GV** Race *barbadensis* has brighter body plumage and paler head and neck; *cachinnans* has browner wings and back than other races. **Voice** Wide range of clucks and chatters. Commonest a series of gradually slowing clucks terminating in long whining notes, e.g. "kukkuk-kuk-kuk kuk, kuk, kuk, peeehr peehr peehr peehr". **SS** None; white line on flanks distinguishes from juv Purple Gallinule. **TN** Previously considered conspecific with extralimital Common Moorhen *G. chloropus* (Old World). **AN** Laughing Moorhen.

American Coot *Fulica americana* LC
Fairly common resident and common winter visitor (**GA**, **Bahamas**), less common (**LA**).

L 34–43 cm. Two subspecies, of which nominate occurs in region. Freshwater lakes, ponds and marshes; less frequent in coastal brackish lagoons. Gregarious but territorial when breeding. Bobs when swimming. **ID** Unmistakable, duck-like blackish rail with white vent. **Ad** has white bill with small frontal shield, broken chestnut subterminal band, and red-brown horny callus. Yellowish-orange legs and feet. **Caribaea morph** (local or rare resident on most of Greater Antilles, except Cuba; also Virgin Is and Guadeloupe; sporadic breeder elsewhere in Lesser Antilles) has larger, swollen white shield, often lacks horny callus, and only sometimes shows reddish subterminal band on bill. **Juv** has crown, hindneck and upperparts dark brown to dark olive-brown, pale ash-grey on head-sides and underparts; bill and shield ivory-grey with small, pale red callus. Most attain ad colours at c. 4 months. Legs pale blue-grey. **Voice** Varied cackling and clucking notes. Aggressive ♂ gives explosive "hic". Loud "puhlk" or "poonk" in alarm. **SS** None. **TN** So-called Caribbean Coot ('*F. caribaea*'), previously given species status by some, now considered a morph.

ARAMIDAE
Limpkin
1 extant species, 1 in region

Limpkin *Aramus guarauna* LC
Common to uncommon resident (**GA**, **Bahamas**), vagrant (**Cayman Is**).

L 56–71 cm; **WS** 101–107 cm. Four subspecies, of which two (*pictus*[A] Bahamas, Cuba and Jamaica, *elucus*[B] Hispaniola and Puerto Rico) occur in region. Generally most numerous N Bahamas, Cuba and Jamaica; rare to very rare Hispaniola but increasing Puerto Rico; vagrant Cayman Is (Apr–Jun 2010). Said to have been recorded Guadeloupe (Oct 1982). Open freshwater wetlands, humid forest, swampy borders and short-grass wet savannas, exceptionally to 1675 m in Dominican Republic. Wades in shallow water, sometimes in small groups. **ID** Large, ibis-like bird with thick, yellow, slightly decurved bill. **Ad** (*pictus*) is darkish oily brown with extensively white-streaked head and neck, and white speckles on breast and wings. ♀ is slightly smaller. **Juv** is more streaked than speckled. **GV** Race *elucus* very similar, but slightly smaller with fewer white streaks and wing markings. **Voice** Most vocal at dusk and by night. Most frequently heard a somewhat wild-sounding, far-carrying "karrao". Various other loud, guttural screams and yapping sounds. **SS** Unmistakable.

GRUIDAE
Cranes
15 extant species, 1 in region

Sandhill Crane *Antigone canadensis* LC
Uncommon and local resident (**Cuba**).

L 95–120 cm; **WS** 160–210 cm. Six subspecies, of which at least one (*nesiotes*) occurs in region. Open wetlands with emergent vegetation, marshes and natural savannas in Cuba and I of Pines, where declining and populations fragmented. Vagrant Bahamas, on Andros Nov–Dec 2003 (*Cotinga* 22: 111) and Abaco, Dec 2018 to Mar 2019 at least; presumably more likely to have involved one of mainland North American breeding populations. **ID** Fairly small crane with all-grey plumage and red forecrown. **Ad** has variable and irregular rusty feathering above, whitish ear-coverts and chin, and grey bill and legs. **Juv** has reddish-brown fringes to wing-coverts, nape and back. **GV** Races only truly differ in size, making subspecific identification of vagrants probably impossible. **Voice** Song a distinctive loud, far-carrying, rattled, trumpeting "karrou, karrou…", perhaps most reminiscent of Limpkin. Calls include loud, lower-pitched rattles. **SS** None.

GAVIIDAE
Loons/Divers
5 extant species, 1 vagrant in region

Common Loon *Gavia immer* LC
Vagrant (**Cuba**).

L 69–91 cm; **WS** 122–148 cm. Monotypic. Inshore waters, in Nov (twice), Dec, Apr, May and Jul (once each), all since 1971 (Kirkconnell *et al.* in press). **ID** Large and thick-necked with steep forehead, flat crown and deep bill, distal half of mandible angled up and culmen curved. **Ad non-br** has dark brown head and hindneck fairly cleanly separated from white chin and foreneck. Lower neck with white V-shaped indentation above dark half-collar. Dark brown upperparts and white underparts with brown-streaked upper breast/flanks. Pale grey-white bill with dark culmen and tip. **Ad br** has black head and neck, necklace of short white streaks on upper foreneck and conspicuous collar of parallel white lines forming large oval on neck-sides. Upperparts blackish grey, each feather with small white spot, and much larger white markings forming rows of squares on scapulars. **Juv** recalls ad non-br but browner, with pale scaling on upperparts. **Voice** Largely silent in winter. **SS** None. **AN** Great Northern Diver.

Common Gallinule
cerceris

American Coot
americana

caribaea morph

Limpkin
pictus

Sandhill Crane
nesiotes

adult non-breeding

adult breeding

Common Loon

85

OCEANITIDAE
Southern Storm-petrels
9 extant species, 1 in region

Wilson's Storm-petrel *Oceanites oceanicus* LC
Fairly common non-breeding visitor (**Bahamas**, **Guadeloupe**), rare (**elsewhere**).

L 15–20 cm; **WS** 34–42 cm. Three subspecies, of which at least one (nominate, perhaps *exasperatus*) occurs in region. Highly pelagic. Flutters over sea as if walking on water, legs dangling and wings held up. Can follow vessels. Mostly Apr–Jun (Bahamas), but also mid Feb–Aug (Guadeloupe) and Mar–May (Barbados), with exceptional records (backed by specimens) in Dec/Jan (Cuba) and sight only Dec (Dominican Republic). **ID** Small blackish-brown storm-petrel with longish legs projecting slightly beyond square-ended tail in flight. **Ad** has conspicuous horseshoe-shaped white rump that almost wraps around to vent. Wings rather short and broad with rounded tips, blackish flight feathers and paler greyish panel on greater coverts. Underwing dark, sometimes with paler tips to greater coverts forming inconspicuous pale bar. **GV** Race *exasperatus* separable only mensurally (larger), but not always recognized (Howell 2012). Specimens not identified to race. **Voice** Usually silent at sea, but low squeaks sometimes heard while feeding. **SS** Leach's and Band-rumped Storm-petrels have slightly longer, more pointed wings, with more prominent bend at carpal. Shorter legs do not project beyond tail and white rump does not wrap around to vent. Also, Leach's has slightly forked (not square) tail.

HYDROBATIDAE
Northern Storm-petrels
18 extant species, 1 in region + 1 vagrant

Band-rumped Storm-petrel *Hydrobates castro* LC
Vagrant (**Bahamas**, **Cuba**, **Jamaica**, **Antigua**, **Martinique**).

L 18.5–20.5 cm; **WS** 42.5–48 cm. Monotypic. Pelagic, rarely near land. Flight buoyant and direct, occasionally erratic. Probably more regular in oceanic waters Apr–May than records suggest, being overlooked among Wilson's Storm-petrels. Records (not all definite) in Bahamas, Abaco, May 2001 (*NAB* 55: 370), Exumas, Apr 2011 (*NAB* 65: 534), Grand Bahama, May 2012 (*NAB* 66: 567), Eleuthera, May 2014 (*NAB* 68: 441); Cuba, Apr 1964 (specimen, with two other records involving either this species or Leach's Storm-petrel; Kirkconnell *et al.* in press); Jamaica, three records Nov 2009 (*Cotinga* 32: 174); Antigua; Martinique (Belfan & Conde 2016). **ID** Medium-large storm-petrel with square-ended or slightly notched tail and long, angled, blunt-tipped wings held horizontally. **Ad** is mostly dark sooty brown with pale greyish-brown crescent-shaped wingbar not reaching forewing. Underparts slightly paler. In flight, legs do not project beyond tail. White rump noticeably broader than it is long and does not wrap around to vent. Worn plumage paler with more obvious wingbar. **Juv** in fresh plumage has greater coverts greyer. **Voice** Unlikely to be heard. **SS** Commoner Leach's Storm-petrel slightly longer-winged, with forked tail, more prominent crescent-shaped wingbar that usually extends to carpal, and rump is longer than it is broad. See also commoner Wilson's Storm-petrel. **TN** Populations breeding in NE Atlantic have been suggested to comprise three species, two breeding at different seasons in some of same island groups (Robb *et al.* 2008), and same pattern exists or probably does at archipelagos elsewhere (Bennett *et al.* 2009, Howell 2012). Cuban specimen tentatively ascribed to winter-breeding so-called Grant's Storm-petrel (Kirkconnell *et al.* in press). **AN** Madeiran Storm-petrel.

Leach's Storm-petrel *Hydrobates leucorhous* VU
Rare to uncommon non-breeding visitor (**throughout**).

L 19–22 cm; **WS** 43.5–49.5 cm. Two subspecies, of which nominate occurs in region. Pelagic. Does not follow fishing vessels or 'walk' over sea surface like Wilson's Storm-petrel. Flight somewhat bounding. Mostly Nov–Jun (Howell 2012), with large numbers regular off Guadeloupe in Mar–Jun (Levesque & Yésou 2005), but documented records in Cuba in Oct and Jul (Kirkconnell *et al.* in press) and sightings off Barbados (Buckley *et al.* 2009) and Puerto Rico in Oct, and Dominican Republic in Jul (Keith *et al.* 2003). **ID** Mid-sized storm-petrel with forked tail (often held closed) and distinctly hooked wingtip. Legs do not project beyond tail. Bill long and fairly thin. **Ad** has large and often contrasting pale wing panel reaching carpal on forewing, and slightly greyer head. Underwing usually sooty black. U-shaped white rump, typically divided by smudged median stripe, is longer than it is broad and does not reach vent. **Voice** Unlikely to be heard. **SS** Wilson's and Band-rumped Storm-petrels have square-ended or slightly notched (not forked) tail, with less prominent and shorter wing panel (not reaching carpal). Much easier to separate Wilson's Storm-petrel also has longer legs that extend beyond tail, shorter, broader and round-tipped wings, and flutters over sea as if walking. In contrast, Band-rumped very difficult to separate from present species, even if photographed, especially in brief or distant views (see Howell 2012, and other specialist literature). Some Leach's can appear or are dark-rumped, and might be confused with similar-sized Swinhoe's Storm-petrel *H. monorhis* (possible record St Martin, May 2015; *NAB* 68: 441), which has smaller head, slimmer appearance, notched rather than forked tail, narrower wing with often less contrasting (albeit variable) ulnar bar that usually does not reach wingbend, and pale shafts at bases of outer five (or six) primaries. **TN** Taxonomy complex and confused, especially Pacific populations.

DIOMEDEIDAE
Albatrosses
22 extant species, 2 vagrants in region

Atlantic Yellow-nosed Albatross EN
Thalassarche chlororhynchos
Vagrant (**Guadeloupe**).

L 71–82 cm; **WS** 180–215 cm. Monotypic. Juv photographed, May 2016 (eBird). **ID** Comparatively small albatross, with relatively light structure, longish bill and white underwing with obvious black leading edge and tips to primaries. **Ad** has somewhat grey-hooded effect, white crown, and black bill with yellow stripe on culminicorn and orange nail. **Juv** initially has grey confined to hindneck and all-dark bill that gradually gains culminicorn stripe after 1–2 years; black underwing margins distinctly ragged. **Voice** Very unlikely to be heard. **SS** Only other albatross recorded in region is Black-browed, which is stockier-looking, with broader wings (always with much more extensive dark feathering on underside) and a shorter bill (initially dark, but eventually becoming extensively pale orange). In closer views, Black-browed has narrow dark browline (vs. eye patch in present species) and ad has clean white head (but note, grey hood of Atlantic Yellow-nosed not always obvious, depending on light). **TN** Until recently considered conspecific with extralimital Indian Yellow-nosed Albatross *T. carteri* (breeds subantarctic Indian Ocean islands), under name Yellow-nosed Albatross.

Wilson's Storm-petrel
oceanicus

Band-rumped Storm-petrel

adult

Leach's Storm-petrel
leucorhous

adult

juvenile

Atlantic Yellow-nosed Albatross

87

Black-browed Albatross *Thalassarche melanophris* `LC`
Vagrant (**Bahamas, Martinique**).

L 79–93 cm; **WS** 205–240 cm. Monotypic. One collected (from group of c. 10!), Martinique, Nov 1956 (Pinchon 1976), two at sea, 240 nm NNE of Los Roques, May 1968 (de Bruijne 1970), off Conception I, Dec 1997 and (same bird) off Exuma, Bahamas, Jun 1998 (*FN* 52: 507), imm, off Abaco, Bahamas, Jul 2013 (*NAB* 67: 659; photo) and ad, off San Salvador, Bahamas, Oct 2018 (eBird). **ID** Very large and heavy build with very long wings compared to most seabirds. **Ad** is white with brownish-black upperwing, and greyish-black back and tail. Black streak above eye. Stout orange-yellow bill with pinkish wash, especially at tip. Underwing has broad blackish leading and trailing edges. **Juv** has crown and hindneck greyish. Broad dark smudgy margins to underwing. Bill greyish black. **Voice** Unlikely to be heard. **SS** Large size, long wings, white underparts and (in ad) orange-yellow bill distinctive. Compare Atlantic Yellow-nosed Albatross. **TN** Until recently considered conspecific with extralimital Campbell Albatross *T. impavida* (breeds subantarctic islands of New Zealand).

PROCELLARIIDAE
Petrels and Shearwaters
95 extant species, 7 in region + 3 vagrants + 1 extinct

Northern Fulmar *Fulmarus glacialis* `LC`
Hypothetical (**Bahamas, Virgin Is**).

L 45–50 cm; **WS** 102–112 cm. Three subspecies, of which one (either *auduboni* or nominate) occurs in region. Pelagic, with characteristic gliding flight on stiff, level wings. New Providence, Bahamas, Feb 1988 (*NAB* 42: 327); at sea in British Virgin Is, Aug 1975 (Norton *et al.* 1989). Neither documented. **ID** Robust, bull-headed and broad-winged petrel. Heavy, mainly yellow bill topped by large, usually darker nostril-tube. Polymorphic, with wide range of intermediates between two morphs. **Ad pale morph** (most likely in region) has whitish head, neck and underparts, with small dusky patch in front of large black eye. Rest of upperparts mid-grey with slightly paler rump and tail. Upperwing like mantle with paler panel at base of primaries. Underwing white with narrow dark margins. **Ad dark morph** ('blue Fulmar') is entirely dark smoky grey (sometimes almost blackish) except slightly paler panel at base of primaries. **Voice** Unlikely to be heard. **SS** Distinguished from similar-sized gulls by nostril-tube, longer wings and stiff-winged glides. From shearwaters by large head and neck, and thicker, mostly yellow bill. **TN** Some authors propose splitting into separate Atlantic and Pacific species (Kerr & Dove 2013), with Atlantic Fulmar *F. glacialis* in region. Validity of race *auduboni* often questioned.

Trindade Petrel *Pterodroma arminjoniana* `VU`
Vagrant (**Bahamas, Puerto Rico, Guadeloupe**).

L 35–40 cm; **WS** 88–102 cm. Monotypic. First record Jul 1986, a dark-morph bird prospecting on land, Culebra I, off NE Puerto Rico (Gochfeld *et al.* 1988); an intermediate or dark morph, Elbow Cay, Bahamas, Nov 2013 (*NAB* 68: 162); two dark morphs, Providenciales, Turks & Caicos, Nov 2014 (*NAB* 69: 169); Guadeloupe, May 2014 (eBird). Records (mainly dark morph) in E USA waters mid May to late Sept (Howell 2012). **ID** Mid-sized, slim-bodied, narrow-winged and fairly small-headed petrel with two morphs plus intermediates. **Ad dark morph** is overall dark brown, usually with white flashes on underside of primaries and wing-coverts. **Ad pale morph** appears dark-hooded; underparts whitish, often with extensive white on underwings. Intermediates can be variably pale brown below.

Voice Unlikely to be heard. **SS** Especially at longer range, dark morph (and intermediates) could be mistaken for Sooty Shearwater or virtually any jaeger. Latter have shorter, broader wings and more gull-like flight, plus white on upperside of wings. Sooty Shearwater smaller-headed and shorter-tailed than Trindade Petrel, with a slenderer bill and no white in primary bases; its flight is characterized by quicker, stiffer wingbeats, whereas present species is more effortless-looking and buoyant.

Black-capped Petrel *Pterodroma hasitata* `EN`
Rare and very local resident (breeds **Hispaniola, Dominica**).

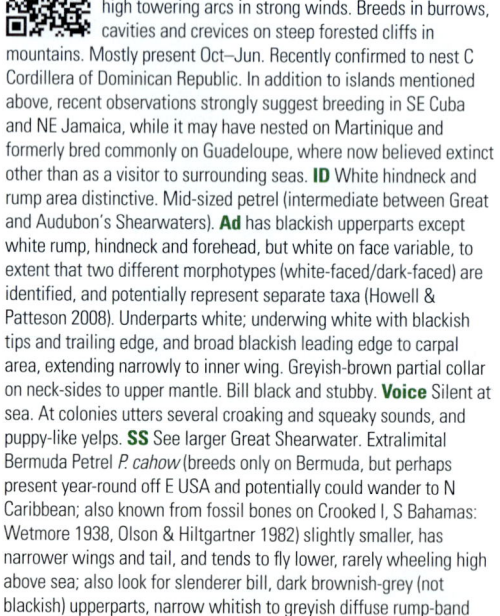

L 38–45 cm; **WS** 98–105 cm. Monotypic. Endemic breeder. Pelagic; flight buoyant and strong, performing high towering arcs in strong winds. Breeds in burrows, cavities and crevices on steep forested cliffs in mountains. Mostly present Oct–Jun. Recently confirmed to nest C Cordillera of Dominican Republic. In addition to islands mentioned above, recent observations strongly suggest breeding in SE Cuba and NE Jamaica, while it may have nested on Martinique and formerly bred commonly on Guadeloupe, where now believed extinct other than as a visitor to surrounding seas. **ID** White hindneck and rump area distinctive. Mid-sized petrel (intermediate between Great and Audubon's Shearwaters). **Ad** has blackish upperparts except white rump, hindneck and forehead, but white on face variable, to extent that two different morphotypes (white-faced/dark-faced) are identified, and potentially represent separate taxa (Howell & Patteson 2008). Underparts white; underwing white with blackish tips and trailing edge, and broad blackish leading edge to carpal area, extending narrowly to inner wing. Greyish-brown partial collar on neck-sides to upper mantle. Bill black and stubby. **Voice** Silent at sea. At colonies utters several croaking and squeaky sounds, and puppy-like yelps. **SS** See larger Great Shearwater. Extralimital Bermuda Petrel *P. cahow* (breeds only on Bermuda, but perhaps present year-round off E USA and potentially could wander to N Caribbean; also known from fossil bones on Crooked I, S Bahamas: Wetmore 1938, Olson & Hiltgartner 1982) slightly smaller, has narrower wings and tail, and tends to fly lower, rarely wheeling high above sea; also look for slenderer bill, dark brownish-grey (not blackish) upperparts, narrow whitish to greyish diffuse rump-band with very little or no white (instead of broader and pure white), lack of white collar on hindneck (although not all Black-capped show this) and more extensive dark head markings (like a cowl). Some Bermuda Petrels possess virtually all-dark upperparts. **AN** Capped Petrel.

Jamaican Petrel *Pterodroma caribbaea* `CR(PE)`
Former resident (**Jamaica**).

L 35–40 cm; **WS** 93 cm. Monotypic. Endemic. Only recorded with certainty in Jamaica, but dark-plumaged petrels were observed on Dominica and speculated also to have bred on Guadeloupe. Reported to have bred in burrows in forest above 1600 m. Behaviour at sea unknown. Despite targeted searches on land and at sea, last recorded in 1879, when as many as 22 specimens collected; feared extinct. **ID** Dark, sooty-brown petrel with horseshoe-shaped pale rump and uppertail-coverts. **Ad** is smaller and darker than Black-capped Petrel, with a smaller bill; some have slightly demarcated black cap (Howell 2012). White, rather ill-defined and often blotchy, horseshoe between lower rump and uppertail-coverts, and longer tail-coverts have brown tips; white mark hardly visible in some birds, being always narrower than exposed dark uppertail and its longest coverts (broader in Black-capped Petrel). Hindneck usually brown, but pale or narrowly whitish in some. **Voice** Unknown. **SS** Plumage combined with presumably typical *Pterodroma* behaviour should be distinctive.

Black-browed Albatross

adult

juvenile

Northern Fulmar
glacialis

pale morph

dark morph

dark morph

pale morph

Trindade Petrel

Black-capped Petrel

dark-faced

white-faced

Jamaican Petrel

89

Sooty Shearwater *Ardenna grisea* NT
Uncommon to very rare non-breeding visitor (**throughout**).
L 40–51 cm; **WS** 94–109 cm. Monotypic. Pelagic, but more regularly seen from land than many petrels. Flight swift and direct, with rapid flapping ascents and long glides, usually low over water. Mostly May–Aug, especially Apr–Jun, but can occur in any month, for example, specimens from Cuba also in Mar and Nov, and sight record Sept (Kirkconnell *et al.* in press), plus sight record St Martin in Jan, and other records from Guadeloupe in Jan, Feb and Nov. **ID** Medium-large sooty-grey shearwater with long, narrow pointed wings, fairly short tail, cigar-shaped body, and long, slim dark bill. **Ad** has whitish underwing-coverts contrasting with rest of sooty-brown to greyish plumage, usually darkest on head and upperside of primaries and tail. **Juv** as ad, but fresh May–Jul when older birds mostly in wing moult. **Voice** Mostly silent at sea. **SS** Only medium-sized all-dark seabird with pale underwing-coverts. Skuas and jaegers can look superficially similar, but shape different and they show whitish on primaries, not on underwing-coverts.

Great Shearwater *Ardenna gravis* LC
Locally common to rare non-breeding visitor (**throughout**).
L 43–51 cm; **WS** 100–118 cm. Monotypic. Pelagic, but sometimes near shore. Generally glides low on stiff wings. Most records in Lesser Antilles in Jun (large numbers off Guadeloupe), with others in seas between Puerto Rico and Bahamas in Apr–Jul, occasionally from Mar, exceptionally Jan (Keith 1997, Feldmann *et al.* 1999, Bradley 2000, Levesque & Yésou 2005, Buckley *et al.* 2009, García-Quintas & Marichal 2016). **ID** One of largest shearwaters, with a two-toned (dark-and-white) and capped appearance. **Ad** has cap including forehead blackish. Rest of upperparts and wings mostly sooty brown except narrow but conspicuous white collar (sometimes incomplete) and uppertail-coverts. At close quarters, pale fringes to upperparts give slightly scaled appearance. Underparts mainly white with diffuse and sometimes weak dark smudge on belly. Underwing white with conspicuous dark trailing edge, broader at tip and variable dark on axillaries and narrow leading edge. Bill blackish. **Juv** lacks white hindcollar. **Voice** Occasional bleating and squawking calls while foraging. **SS** Compare Cory's, Scopoli's and very rare Cape Verde Shearwaters. Smaller Black-capped Petrel has white forehead and conspicuous black leading edge, and lacks dark belly smudge.

Scopoli's Shearwater *Calonectris diomedea* LC
Uncommon non-breeding visitor (**Bahamas**), very rare elsewhere.
L 44–49 cm; **WS** 117–135 cm. Monotypic. Pelagic. Behaviour similar to Cory's Shearwater (which see). Mostly May–Jun, but records until Sept. Most observers did not distinguish between it and Cory's Shearwater until recently. **ID** Very similar to Cory's Shearwater (sometimes still treated as same species); flight faster and less heavy. Probably only safely distinguished in flight. **Ad** differs by slightly smaller size and lighter build, narrower wings, smaller and paler head, and less robust bill. Underwing has dark trailing edge and tips, but inner webs of primaries show as white 'tongues' projecting into dark wingtip. **Voice** Rarely vocalizes at sea. **SS** See very similar Cory's and Cape Verde Shearwaters. Great Shearwater shows more contrasting two-toned appearance, is darker above, with blurry dark smudge on belly, usually conspicuous white collar and rump, and has dark (not pale yellow) bill. **TN** Until recently considered conspecific with Cory's and Cape Verde Shearwaters (as Cory's Shearwater), and Scopoli's and Cory's are often still united to exclusion of Cape Verde.

Cory's Shearwater *Calonectris borealis* LC
Uncommon to fairly common non-breeding visitor (**Bahamas**, **LA**), rare (**GA**).
L 48–56 cm; **WS** 113–124 cm. Monotypic. Pelagic. Flight more leisurely than Great Shearwater, with slow, rather deliberate or even languid wingbeats. Mainly May–Jul, but also observed in much smaller numbers Nov–Apr. **ID** Largest shearwater in region, with heavy build, prominent pale yellowish bill and little contrast above. **Ad** has upperparts rather uniform greyish brown. Head and neck blend smoothly into all-white underparts. Inconspicuous or no white rump area. Underwing white with broad and solid black tip and trailing edge. **Voice** Unlikely to be heard. **SS** Great Shearwater has more contrasting two-toned appearance, is darker above, with dark smudge on belly, often conspicuous white collar and rump, and dark (not pale yellow) bill. Very similar Scopoli's Shearwater is slightly smaller and less robust, with lighter bill and white tongues at base of primaries. See vagrant Cape Verde Shearwater. Other shearwaters much smaller. **TN** See Scopoli's Shearwater.

Cape Verde Shearwater *Calonectris edwardsii* NT
Vagrant (**Guadeloupe**).
L 42–47 cm; **WS** 101–112 cm. Monotypic. At sea, Apr 2014 (*NAB* 68: 446). Pelagic. Behaviour similar to Cory's Shearwater, but wingbeats slightly shallower and wings held slightly more forward. **ID** Smallest of *C. diomedea* complex. Similar to Cory's Shearwater. **Ad** differs by smaller size, lighter build, slightly narrower wings, darker upperparts, with more contrasting two-toned appearance (dark above, white below), somewhat capped look and thinner greyer bill (not thick and yellow), with blackish subterminal band. **Voice** Rarely vocalizes at sea. **SS** See Cory's Shearwater. Scopoli's Shearwater has yellowish bill and paler upperparts with less contrast. Larger Great Shearwater has narrow white hindneck and rump band, and dark smudge on belly. Black-capped Petrel has darker (blackish) mantle and wings, and broad and contrasting white forehead, hindneck and rump band. **TN** See Scopoli's Shearwater.

Sooty Shearwater

Great Shearwater

Scopoli's Shearwater

Cory's Shearwater

Cape Verde Shearwater

91

Manx Shearwater *Puffinus puffinus* LC
Rare to locally common non-breeding visitor (probably **throughout**).
L 30–38 cm; **WS** 75–89 cm. Monotypic. Pelagic and probably overlooked, e.g. just one record (Mar 2005) in Barbados waters. Mostly Lesser Antilles (especially off Guadeloupe, where sometimes in large numbers and records in every month except Sept) mainly in Nov–May, but records until at least Jun (Keith *et al.* 2003, Levesque & Yésou 2005) and probably oversummers in N of region at least occasionally (Howell 2012). Very rare Greater Antilles, where never recorded Cuba (*contra* Raffaele *et al.* 1998) and very few times Hispaniola, Puerto Rico, Bahamas (*NAB* 57: 272, 69: 306) and San Andrés (Donegan *et al.* 2010). **ID** Rather small, contrasting blackish-and-white shearwater with long thin bill. Flight (in calmer conditions) typically 4–5 distinctive, stiff-looking, wingbeats and short glide on bowed wings. **Ad** differs from browner-looking Audubon's Shearwater in marginally larger size, slightly blacker upperparts, longer, narrower wings, shorter tail, pale notch or crescent behind ear-coverts, white undertail-coverts, and white underwing with dark trailing edge and narrower dark leading edge and tip, often with ill-defined dark patch at base, sometimes connected to rear wing base via more sharply defined narrow blackish band (never shown by Audubon's). Beware, in worn plumage (and certain lights) looks duller and browner, especially 1st-y birds in summer. **Voice** Generally silent at sea, but loud whistles reported. **SS** White undertail-coverts very distinctive. See much commoner but slightly smaller Audubon's Shearwater. Compare other shearwaters and petrels.

Audubon's Shearwater *Puffinus lherminieri* LC
Three subspecies divided into three subspecies groups, of which one occurs in region.

Audubon's Shearwater *Puffinus (lherminieri) lherminieri*
Locally common to very rare resident (**throughout**).
L 27–33 cm; **WS** 64–74 cm. One subspecies in group. West Indian population much-reduced (Mackin 2016). At sea except when breeding (Jan/Mar–May/Jul, in colonies on rocky offshore islets, cliffs and earth slopes). Rarely follows fishing boats, but joins mixed-species feeding frenzies. Flight several quick flaps followed by short glide, low over waves. Still fairly common Bahamas mostly Mar–Jul (uncommon at other times); generally uncommon and local elsewhere when nesting, and rare (due to lack of pelagic trips) outside this. **ID** Relatively small shearwater, with short broad wings and comparatively long rounded tail. **Ad** has blackish-brown upperparts and head to just below eye. White underparts, with a slightly capped and obviously two-toned appearance. Usually exhibits some indistinct white around eyes (can form spectacles), but no white hook around rear ear-coverts. Undertail-coverts blackish brown (albeit variably, being typically solidly dark in Lesser Antilles but sometimes extensively white in Bahamas). Underwing white, with broad (but variable) dark margins. In worn plumage looks browner above. **Voice** Generally silent at sea. At colonies utters squeaky or bleating calls; ♀ lower-pitched and harsher than ♂. Mostly nocturnal at colonies, calling only at night. **SS** In many areas the most frequently encountered shearwater year-round. Slightly larger Manx Shearwater (mainly during boreal winter in region) differs structurally and in head pattern (appearing more dark- than white-faced), has white undertail-coverts and cleaner (whiter) underwing with thinner margins. Compare other, larger (and usually less common) shearwaters and petrels.
TN Sometimes recognized race *loyemilleri* (breeds on islands off NW Panama and W Venezuela), which presumably wanders to our region, is treated here as synonym of nominate (Austin *et al.* 2004). No evidence to suggest that those nesting around Providencia are *loyemilleri* (Howell 2012), *contra* van Halewyn & Norton (1984). Howell also suggested that birds nesting in Bahamas might warrant subspecific status as *P. l. auduboni* (Bahama Shearwater). **AN** Antillean Shearwater.

Bulwer's Petrel *Bulweria bulwerii* LC
Vagrant or very rare non-breeding visitor (**Guadeloupe, Dominica**).
L 26–29 cm; **WS** 61–73 cm. Monotypic. Pelagic. Usually seen in low, weaving flight with quick wingbeats, but in stronger winds can wheel higher. Holds head higher than body in pigeon-like fashion. Records in late Apr (first off Dominica in 2003), Jun and Jul (Levesque & Yésou 2005, *NAB* 60: 589, eBird). **ID** Dark-plumaged seabird with long, narrow and pointed wings, and long, tapered tail. **Ad** is dark sooty brown to blackish brown, often slightly darker on head and diffusely paler on chin. Conspicuous pale crescent on upperwing-coverts (hardly noticeable in worn plumage or at long range). Underwing all dark, sometimes with paler silvery reflection in strong light. Short black bill. **Voice** Unlikely to be heard. **SS** All-dark plumage with pale crescent on upperwing, long wedge-shaped tail and short black bill distinctive.

CICONIIDAE
Storks
20 extant species, 1 in region + 1 vagrant

Wood Stork *Mycteria americana* LC
Rare resident (**Cuba**), vagrant (**elsewhere**).
L 83–102 cm; **WS** c. 150 cm. Monotypic. Patchily distributed Cuba (including I of Pines, plus records on several cays); formerly Dominican Republic (now extirpated). Vagrants reported Grand Bahama, Jamaica, Puerto Rico, St John and Dominica. Wetlands, mangroves, coastal mudflats, rice fields and inland waterbodies. Wades in shallow, open water with little emergent vegetation. Flies with slow wingbeats; often soars. **ID** The only regular stork; large and heavy-billed. **Ad** has unfeathered black head and neck, white body and black flight feathers. **Imm** has brownish head and neck with some feathering, and yellow bill. **Voice** Mostly silent. Occasional grunts and bill-claps when feeding in groups. At nest, nasal goose-like calls and grunts; also bill-clattering in courtship. **SS** Jabiru has all-white wings, deep-based bill and reddish throat sac. **AN** American Wood Stork.

White Stork *Ciconia ciconia* LC
Vagrant (**Antigua, Martinique**).
L 100–102 cm; **WS** 155–165 cm. Two subspecies, of which nominate occurs in region. Singles, SW Antigua, Aug 1993–Mar 1994, perhaps seen Barbuda earlier (Gricks 1994a,b), and Martinique, Feb 2007 (Leblond 2007). **ID** Large, mostly white stork with conical red bill. **Ad** is pure white with contrasting black flight feathers. **Juv** has duller, browner plumage, and bare parts. **Voice** Generally silent. **SS** Combination of white, feathered head and neck, and red bill and legs distinctive.

Jabiru *Jabiru mycteria* LC
Hypothetical (**Grenada**).
L 122–140 cm; **WS** 230–260 cm. Monotypic. Sight record, May 1955 (Devas 1970, Groome 1970), was questioned by Bond (1973). Favours freshwater marshes and swamps. Regularly soars. **ID** Huge stork with massive, bulging, slightly upturned bill. **Ad** has all-white plumage, bare black head and neck, reddish throat sac. ♀ is smaller with less massive bill. **Juv** has pale grey upperparts with grey-brown feather fringes and white primaries washed pale brown on inner webs. **Voice** Unlikely to be heard. **SS** See Wood Stork.

Manx Shearwater

Audubon's Shearwater

Bulwer's Petrel

White Stork
ciconia

juvenile

adult

Wood Stork

Jabiru

93

THRESKIORNITHIDAE
Ibises and Spoonbills
35 extant species, 4 in region + 2 vagrants

Roseate Spoonbill *Platalea ajaja* `LC`
Locally common to uncommon resident (**Bahamas, Cuba, Hispaniola**), vagrant elsewhere.

L 68–87 cm; **WS** 120–130 cm. Monotypic. Locally common Cuba and Hispaniola, common Great Inagua, uncommon Andros, rare Caicos. Vagrant elsewhere in Bahamas, plus Cayman Is, Jamaica, Puerto Rico, Virgin Is, St Martin, St Barthélemy, Guadeloupe and St Lucia (*Cotinga* 31: 166). Saltwater lagoons, mangrove swamps and other coastal sites with salt or brackish water; rare inland. Forages alone or in small groups. Flies with outstretched neck. **ID** Combination of pinkish coloration and spatulate bill unique. **Ad** has bare facial skin, black nape-band, whitish neck, and pinkish body with deep reddish-pink wing-coverts. **Juv** has head feathered, white plumage variably tinged pink, outer primaries tipped and fringed brown, and yellowish bill. **Voice** Calls include a rapid slightly descending series, "rruh-ruh-ruh-ruh-ruh-ruh", and low-pitched croaks and grunts. Bill-clacking at colonies and roosts. **SS** See Eurasian Spoonbill.

Eurasian Spoonbill *Platalea leucorodia* `LC`
Vagrant (**Antigua, St Lucia, Barbados**).

L 70–95 cm; **WS** 115–135 cm. Three subspecies, of which presumably nominate occurs in region. Ad, St Lucia, intermittently Nov 2007–Apr 2009 (Buckley *et al.* 2009; *Cotinga* 31: 166, 32: 174); Antigua, at least Feb–Apr 2009, perhaps one of Barbados birds (*NAB* 63: 340); up to three, all 1st-y, Barbados, Nov 2008–Jan 2011, up to two (one 1st-y), Nov 2017–Mar 2019, at least (Buckley *et al.* 2009; eBird). Shallow wetlands, including marshes, rivers, lakes, flooded areas and, occasionally, tidal flats. **ID** Distinctive egret-like bird with broad, spatulate bill. **Ad br** is all white, with yellowish nuchal crest and variable yellow area at base of neck. Bare skin of throat yellow. Bill black with yellow tip. **Ad non-br** lacks yellow, and crest also absent. **Juv** is similar to ad non-br, but bill initially pinkish, and has black tips to outermost primaries. **Voice** Usually silent. **SS** Juv Roseate Spoonbill has pinkish hue.

White Ibis *Eudocimus albus* `LC`
Common resident (**GA**), rare visitor (**Bahamas**), vagrant elsewhere.

L 56–71 cm; **WS** 94–105 cm. Two subspecies, of which at least nominate occurs in region. Fresh and saltwater wetlands and marshes, rice fields and flooded land. Roosts and feeds in large flocks, and nests colonially. Breeds NC coast of Puerto Rico. Vagrants recorded Cayman Is, Dominica and Barbados (the latter perhaps equally likely to have involved South American race *ramobustorum*). **ID** Typical (white) ibis with pinkish-red facial skin and bill. **Ad br** has distal half of bill blackish, black wingtips and reddish legs. **Ad non-br** has bare facial skin reduced. **Juv** has olive-brown upperparts with variable white; head and neck with brown or grey spots. **GV** Race *ramobustorum* differs mainly in smaller size, and during courtship bill all or mostly dark (black in ♂, brown in ♀), while both sexes have well-developed gular sac when breeding. **Voice** Main call a nasal grunting "uhnk" or "uurr". **SS** Compare Scarlet Ibis. **TN** Occasionally considered a race, or even morph, of Scarlet Ibis. **AN** American White Ibis.

Scarlet Ibis *Eudocimus ruber* `LC`
Very rare visitor (**Grenada**), vagrant (rest of **LA, GA**).

L 56–71 cm; **WS** 94–105 cm. Monotypic. Mostly Jan–Jun, but seen all months Grenada, and also Oct–Nov in Cuba. Elsewhere in Greater Antilles, sightings attributed to wild birds on Jamaica. Presumably reaches West Indies from both North & South America, but free-ranging and unbanded population (c. 45 birds) recently introduced on Necker I (Virgin Is) seems likely to also wander. Swamps, estuaries and tidal mudflats. **ID** Dazzling scarlet plumage unmistakable. **Ad** has black wingtips. Bill typically reddish, only briefly black when breeding. **Juv** has head and neck streaked or mottled brown and white; underparts and rump white. Dark grey brown above, can have pink mottling on white parts. **Voice** Rather silent. May give nasal grunt, "uh-runk", in flight. **SS** Juv White Ibis identical to juv of present species, but older imm Scarlet separable by patchy pinkish plumage. Imm White Ibis never shows pinkish on rump or back. Beware possibility of hybrids.

Glossy Ibis *Plegadis falcinellus* `LC`
Fairly common to locally common resident (**GA**), uncommon to rare visitor (**O**), mainly vagrant (**LA**).

L 48–66 cm; **WS** 85–95 cm. Monotypic. Uncommon visitor Bahamas, formerly rare visitor but now breeds Puerto Rico, and usually rare Cayman Is (can appear in large flocks). Increasingly regular in some numbers in Lesser Antilles, at least a few apparently transatlantic vagrants, as evidenced by bird ringed in Spain controlled on Barbados (*NAB* 65: 181). Mudflats, marshes and flooded areas. **ID** Medium-sized, dark brown ibis with glossy-green wings. **Ad br** has head, neck and body rich chestnut, but appears merely dark at long range. **Ad non-br** is dull dark brown, with head and neck finely but densely streaked whitish. **Juv** is similar to ad non-br, with oily green sheen. Head and neck browner, with variable white on forehead, throat and foreneck. **Voice** Usually silent. At colonies, calls include grunts, guttural coos, croaks and rattles. **SS** See vagrant White-faced Ibis.

White-faced Ibis *Plegadis chihi* `LC`
Vagrant (**Cuba**).

L 46–66 cm. Monotypic. Ad photographed, May 2018 (Kirkconnell Posada *et al.* 2018). Should be searched for at wetlands, including rice fields, as might be far more regular than supposed. **ID** Very similar to Glossy Ibis with which often confused (and considered conspecific) in past. **Ad br** is principally deep chestnut, tinged bronzy, with green-glossed wings, especially coverts. Note white border (extending behind eye) to pinkish-red loral skin. Legs and eyes red; bill variable, but can be reddish. **Ad non-br** has pale pink facial skin. **1st-w** has head and neck strongly streaked white, underparts dull brown, faintly tinged purple; starts to acquire reddish eyes and loral skin. **Voice** Flight call a rapid series of nasal quacks "nrah..nrah..nrah..", higher-pitched and more nasal than Glossy Ibis. **SS** Careful separation from Glossy Ibis required. In all plumages focus on bare-parts coloration, especially loral skin (always grey-green in Glossy), legs (grey-green with red joints in Glossy) and eyes (brown in Glossy). Ad br White-faced shows complete white border to lores (bluish-tinged and incomplete in Glossy), while in other plumages both species show pale border only above lores (but its colour remains useful).

Roseate Spoonbill

Eurasian Spoonbill
leucorodia

White Ibis
albus

Scarlet Ibis

Glossy Ibis

White-faced Ibis

95

ARDEIDAE
Herons
64 extant species, 14 in region + 6 vagrants

American Bittern *Botaurus lentiginosus* LC
Uncommon winter visitor (**Bahamas**, **Cuba**), very rare (rest of **GA**), vagrant (**LA**).

L 56–85 cm; **WS** 105–125 cm. Monotypic. Favours freshwater marshes with dense emergent vegetation. Secretive. In alarm, points bill up to aid camouflage. Records mid Aug to Apr, S to Barbados. **ID** The only *Botaurus* recorded in region, but extralimital Eurasian Bittern *B. stellaris* is potential vagrant to Barbados and Pinnated Bittern *B. pinnatus* (Middle and South America) also expected, especially in S Lesser Antilles. Large, stocky buff-brown heron with broad neck; dull yellow bill with dark culmen. **Ad** has pale buffish supercilium, brown upperparts finely flecked blackish, underparts coarsely streaked brown and white, with distinct black stripe on neck-side. In flight, unbarred blackish remiges contrast with much paler wing-coverts and buff trailing edge. **Imm** lacks black markings on neck-sides. **Voice** May give hoarse "kok-kok-kok" in alarm and flight. **SS** Imm night-herons darker and shorter-necked, without dark markings on neck-sides and dark remiges. Eurasian Bittern is larger, with a black cap, indistinct moustachial, more crisply and intricately patterned upperparts and, in flight, broader wings without contrasting darker remiges. Pinnated Bittern is smaller, has a black-barred crown, overall browner (less rufous) ground colour, upperparts patterned more like Eurasian Bittern, and virtually lacks a black moustachial.

Least Bittern *Ixobrychus exilis* LC
Fairly common to uncommon or rare resident (**GA**, **O**, **C LA**), vagrant elsewhere.

L 28–35 cm; **WS** 41–46 cm. Six subspecies, of which nominate occurs in region. Dense freshwater marshes with tall emergent vegetation. Climbs nimbly through reeds and cattails; secretive and often perches motionless. Most easily detected in flight (quick, jerky wingbeats distinctive) or if vocalizing. Probably an uncommon resident also Guadeloupe, Dominica and Martinique, where discovered only in early 1980s. **ID** Tiny heron, smaller than Common Gallinule. **Ad** ♂ has blackish cap, rich tawny-buff face and neck-sides, whitish throat and underparts, and blackish upperparts and rump; white line bordering wing conspicuous at rest. In flight, distinctive combination of black flight feathers, chestnut secondary-coverts and buff wing-coverts. Primary-coverts and secondaries tipped chestnut. **Ad** ♀ averages larger with less obvious dark crown, chestnut upperparts, dark streaks on foreneck and breast, and paler wing patch. **Juv** is like ad ♀ but even paler, with browner mantle and crown, dusky-streaked lesser coverts and throat, and breast streaks browner. **Voice** Rather vocal, sometimes throughout day. ♂ advertisement a low-pitched, quiet, dove-like "uh, uh, uh, oo, oo, oo, oo, ooah". Also "gack, gack" in contact, and "kuk, kuk" in flight or alarm. **SS** Vagrant Little Bittern has paler head- and neck-sides, whitish (vs. chestnut) secondary-coverts, all-black flight feathers (lacking chestnut tips), and no conspicuous white border to upperwing. Creamy wing-coverts and jerky flight distinguish from all other herons.

Common Little Bittern *Ixobrychus minutus* LC
Vagrant (**Barbados**).

L 27–38 cm; **WS** 49–58 cm. Three subspecies, of which one (nominate or *payesii*) occurs in region. Ad ♂, Dec 1995; from photos, general coloration seemed closest to *payesii* (Buckley *et al.* 2009). Habitat and behaviour as Least Bittern. **ID** Very similar to Least Bittern. **Ad** ♂ has blackish crown, tail and remiges, greyish face-sides, buff-white wing-coverts, white underwing and buffish-white underparts, sometimes with slight streaking; bill yellow or yellow-green, with dark brown maxilla. **Ad** ♀ is typically smaller and duller, with browner tone to dark parts, smaller and slightly streaked wing patches, and streaked underparts. **Juv** is similar to ♀, but more heavily streaked, with brown-streaked crown and heavily mottled brown-and-buff wing-coverts. **Voice** Rather silent when not breeding. **SS** See Least Bittern. **TN** Previously considered conspecific with extralimital Australian Little Bittern *I. dubius*. **AN** Little Bittern.

Black-crowned Night-heron *Nycticorax nycticorax* LC
Locally uncommon resident and winter visitor (**GA**, **O**), mainly rare non-breeding visitor or resident (**LA**).

L 56–65 cm; **WS** 105–112 cm. Four subspecies, of which one (*hoactli*) occurs in region. Marshes, swamps, brackish lagoons and saltwater bays. Gregarious. Largely nocturnal, easiest to observe dawn and dusk. Large numbers of North American breeders winter in region. Common to uncommon resident St Martin, Antigua, Guadeloupe, Martinique, St Lucia, Barbados and Grenada. Recently colonized Guadeloupe and Barbados; perhaps formerly bred St Lucia. **ID** Mid-sized, stocky, short-necked, large-headed, short-billed heron. **Ad** has distinctive black cap and mantle, long white nuchal plumes, and dark grey body. In flight, black mantle contrasts with dark grey upperwing, rump and tail; feet just project beyond tail. **Juv** is spotted and streaked brown, grey, buff and whitish. Upperwing grey-brown with paler spots. Acquires ad plumage gradually over 2–3 years. **Voice** Commonest call (typically in flight) a "quok", "kwark" or "kowak", slightly lower-pitched than Yellow-crowned Night-heron. **SS** Ad Yellow-crowned Night-heron has different head pattern. Imm Yellow-crowned has stouter all-black bill, longer legs (projecting further beyond tail in flight), slightly greyer upperparts with smaller white flecks, and narrower, more distinct streaking on underparts. Compare Limpkin.

Yellow-crowned Night-heron *Nyctanassa violacea* LC
Fairly common to uncommon resident and winter visitor (**throughout**).

L 51–70 cm; **WS** 101–112 cm. Six subspecies, of which two (nominate winter visitor, *bancrofti* resident almost throughout) occur in region. Favours mangroves, but also mudflats, savannas and sometimes dry thickets away from water. Usually crepuscular and nocturnal, but can feed by day depending on tides. Rarer in S Lesser Antilles. Numbers augmented in winter. **ID** Chunky with unmistakable head pattern, orange eyes and chunky, all-black bill. **Ad** (*bancrofti*) has cream-white crown and crest plumes, black head with broad white streak on face, and grey neck and body. In flight, legs project far beyond tail. **Juv** is brownish, with underparts distinctly streaked cream and brown, upperparts with small pale spots, and blackish crown streaked buffy white. **GV** Nominate race darker overall (at all ages), with broader dorsal stripes and smaller bill. **Voice** Rather vocal, commonest a "squawk" similar to Black-crowned Night-heron but higher-pitched. Also guttural "ahhh, ahhh" in aggression. **SS** See Black-crowned Night-heron. **TN** Not all authors accept *bancrofti* as resident race in region; arguments in favour of this presented by Watts (2011).

American Bittern

Least Bittern
exilis

Common Little Bittern
minutus

Black-crowned Night-heron
hoactli

Yellow-crowned Night-heron
violacea

97

Green-backed Heron / Green Heron LC
Butorides striata

Thirty-three subspecies divided into three subspecies groups, of which two occur in region.

Green Heron *Butorides (striata) virescens*
Common resident and less common winter visitor (**throughout**). **L** 41–48 cm; **WS** 52–60 cm. Four subspecies in group, of which two (*bahamensis*[A] Bahamas, *virescens*[B] resident elsewhere) occur in region. Wide range of waterbodies, but commonest in coastal lowlands. Typically alone. No definite evidence that North American breeders occur in Lesser Antilles in winter (see Buckley *et al.* 2009). **ID** Small, stocky, dark-coloured heron with hunched posture, relatively long bill and bright yellow legs. **Ad** (*virescens*) has green-black crown and nuchal plumes, thick-based maroon neck, and cream line from bill below eye. Throat and upper breast streaked white, forming vertical stripe. Back dark with green hue. **Juv** is streaky brown with plain dusky-brown cap; cream line under eye lined black. **GV** Race *bahamensis* (often not recognized) averages smaller and paler. **Voice** Characteristic "skeow" given in alarm, flight and advertisement. **SS** Least Bittern smaller, with conspicuous creamy wing patches. **TN** Nominate race and *bahamensis* often split from Striated Heron *B. striata* as *B. virescens*. Some authors recognize name *maculata* for resident Antillean populations.

Striated Heron *Butorides (striata) striata*
Vagrant (**Puerto Rico**, **LA**, **San Andrés**).
L 35–48 cm. Twenty-eight subspecies in group, of which one (*striata*) occurs in region. Puerto Rico, Feb–Sept 2010 and May 2016; St John, May 2003; St Thomas, May 2003; Guadeloupe, Dec 2007–Jan 2008; St Vincent, Jul 1924 (♂ in breeding condition); Grenada, Feb–Jul 2013; and Barbados, Jun–Jul 2013, Sept 2015–Jun 2018 at least (Bond 1964, Hayes & Hayes 2006, Buckley *et al.* 2009, *NAB* 57: 417, eBird). Apparently vagrant on San Andrés (McNish 2003, 2011, Pacheco 2012), and has occurred Nicaragua (Sandoval & Arendt 2011). Breeds as close as Trinidad & Tobago. **ID** Much like conspecific Green Heron. **Ad** has mid-grey neck-sides, with variable rusty-brown border to white stripe on central foreneck (never approaches extent or intensity of maroon neck of Green Heron). Greenish-grey back plumes. **Juv** lacks dark chestnut streaking on throat and neck of same-age Green Heron, with browner upperparts and generally darker underparts. **Voice** No differences known vs. Green Heron. **SS** Green Heron has rich maroon neck-sides.

Squacco Heron *Ardeola ralloides* LC
Vagrant (**Guadeloupe**).

L 42–47 cm; **WS** 80–92 cm. Two subspecies, of which nominate presumably occurs in region. Photographed, La Désirade, Nov 2018 (AL). Prefers well-vegetated wetlands, but could appear in wet fields including rice. **ID** Appears mainly brownish buff at rest, but largely white wings, rump and tail become apparent on take-off. **Ad br** has yellow-buff or straw-coloured crown, golden to cinnamon-buff plumes on lower neck and back, reddish-gold foreneck and breast, pinkish-brown back, with long golden plumes, and rest of plumage white. Bare parts brighter during courtship; legs can be reddish (soon fade to pink), bill bright blue with dark tip. **Ad non-br** is generally dull brown with dark and pale streaks; legs dull yellow-green, bill pale greenish yellow with black tip and maxilla. **Juv** is like ad non-br but drabber, with more heavily streaked neck, no plumes on crest or back, dark-tipped bill, grey (rather than white) underparts and brown-tinged flight feathers. **Voice** Mainly silent, but a shrill, harsh-sounding "karr" sometimes heard at dusk, especially in alarm; also frog-like croaks or grunts, and a rasping "kek-kek-kek-kek" in aggression. **SS** Smaller and slighter than Cattle Egret, which always looks much whiter, has less pointed bill and an obvious jowl.

Cattle Egret *Bubulcus ibis* LC

Three subspecies divided into two subspecies groups, of which one occurs in region.

Western Cattle Egret *Bubulcus (ibis) ibis*
Common resident (**throughout**).
L 46–56 cm; **WS** 88–96 cm. Two subspecies in group, of which nominate occurs in region. Open grassy areas, meadows, freshwater and brackish swamps, rice fields and mangroves. Often with cattle, feeding on insects. Colonized region mid 1900s, e.g. first noted Providencia in 1933, Puerto Rico in 1948, Eleuthera (Bahamas) in 1953, Cuba in 1954, and Barbados and Grenada in 1956. **ID** Fundamentally all-white egret with yellow bare parts. Stubby aspect, promoted by short tibia, and shorter, stouter bill than other egrets, with relatively short neck, hunched posture and comparatively large head. Short, rapid wingbeats in flight. **Ad br** has reddish-buff crown, nape, breast and mantle. **Ad non-br** is entirely white. **Juv** as non-br, but sometimes slightly greyer, with very dark to blackish legs. **Voice** Call "rick, rack", first syllable louder and higher-pitched. **SS** Other all-white egrets are taller, slimmer, longer-legged and longer-billed.

Grey Heron *Ardea cinerea* LC

Four subspecies divided into two subspecies groups, of which one occurs in region.

Grey Heron *Ardea (cinerea) cinerea*
Rare resident (**Barbados**, **Guadeloupe**), vagrant elsewhere (**LA**).
L 84–102 cm; **WS** 175–195 cm. Three subspecies in group, of which nominate occurs in region. First recorded 1959 on Montserrat, then 1963 on Barbados, where resident since 1997 and small population still augmented by new arrivals (Buckley *et al.* 2009), but no evidence of breeding. Vagrant N to St Kitts, with French-ringed birds recovered on Martinique and Montserrat (Bond 1962). All types of shallow fresh, brackish or saltwater areas, standing or flowing. **ID** Slightly smaller, with shorter neck, bill and legs than similar Great Blue Heron. **Ad non-br** has grey upperparts and wings, paler whitish face and neck, and broad black head-stripes extending as long plumes. Flanks and thighs grey. **Ad br** has all bare parts flushed deep orange to red. **Juv** has duller and less contrasting plumage, with darker grey crown, shorter nape plumes, greyer neck and grey maxilla; can show buff on thighs. **Voice** A loud, far-carrying "frarnk", regular in flight. **SS** Great Blue Heron is 10–40% larger with darker grey upperparts, chestnut-brown flanks, thighs and tips to wing-coverts, and dark vinous neck-sides. In flight, legs and feet project less beyond tail than in Great Blue, and at rest note whiter face and foreneck, with more contrasting black stripe.

juvenile

adult

Green Heron
(Green-backed Heron / Green Heron)
virescens

adult

juvenile

Striated Heron
(Green-backed Heron / Green Heron)
striata

breeding

non-breeding

Squacco Heron
ralloides

non-breeding

non-breeding

breeding

breeding

Grey Heron
cinerea

Western Cattle Egret
(Cattle Egret)
ibis

Great Blue Heron *Ardea herodias* `LC`
Common (**GA**, **O**) to uncommon winter visitor (**LA**), rare resident (mainly **GA**).

L 91–137 cm; **WS** 170–190 cm. Five subspecies, of which two (nominate winter visitor and passage migrant, *occidentalis* resident) occur in region. Race *wardi* suspected in Cuba (Kirkconnell *et al.* in press) but potential for confusion with so-called Würdemann's Heron high. River and lake margins, wet meadows, marshes, swamps, fields, mangroves and tidal mudflats. Solitary, often stands motionless. Nominate race mainly occurs Oct–Apr. Resident *occidentalis* breeds at very few localities (e.g. on Cuba, Hispaniola and Virgin Is). **ID** Largest heron; long-necked and long-legged. Dimorphic. **Ad typical morph** (nominate) is essentially grey-blue with white crown, cheeks and throat, black eyestripe and long crest plumes, grey neck, often with chestnut or vinous tones on back and sides, chestnut thighs and 'headlights' on wingbend. **Ad white morph** is rare (and apparently only in Bahamas and Cuba); all white with yellow bill and creamy-yellow legs and feet. Intermediates occur, but are not common. **Juv typical morph** is darker, lacks crest but has variably dark crown, belly streaked, bare parts duller and thighs grey to white. **Juv white morph** is off-white with buffish-grey legs and feet. **GV** Race *occidentalis* larger with distinctly more massive bill, and much greater percentage are white morph; *wardi* also markedly larger and paler than nominate. **Voice** A harsh, deep "frawnk" in alarm, when disturbed or in flight. **SS** See Grey Heron and (for white morph) American White Egret. **TN** Very complex. In past, both white *occidentalis* and Würdemann's Heron (usually treated as colour morph of *occidentalis* or extralimital race *wardi*) were frequently treated as separate species, and very recently McGuire *et al.* (2019) again recommended that *occidentalis* be afforded recognition at specific level. Würdemann's Heron is genetically indistinguishable from white birds (McGuire 2002, who regarded it as a hybrid swarm). **AN** Great White Heron (white morph).

Cocoi Heron *Ardea cocoi* `LC`
Vagrant (**Grenadines**).

L 95–127 cm. Monotypic. Mustique, Jan–Feb 2014 (*NAB* 68: 290). Likely in habitats similar to those used by Grey and Great Blue Herons. **ID** Large, slender heron with long neck and legs, and white thighs. **Ad non-br** has crown to below eyes black. Grey above with black shoulder patch. Neck, vent and upper breast white, latter with long plumes and usually thin vertical row of black streaks on foreneck. Belly black. Long bill dull yellow, legs blackish. Eye yellow and lores bluish. In flight, pale grey upperwing with blackish flight feathers. **Ad br** has black nuchal plumes, brighter yellow bill and dusky-pink legs. **Juv** is like ad non-br but greyer, especially neck, and crown is duller black. Thighs pale greyish or whitish. **Voice** Has "rraahbm rraabb" call typical of other large grey herons. **SS** Ad Great Blue Heron has grey neck, less extensive black on crown and rufous thighs. In flight, Great Blue has marginally darker upperwing, with less contrasting flight feathers. Juv Great Blue is even more similar to juv Cocoi, but note rufous thighs. **AN** White-necked Heron.

Purple Heron *Ardea purpurea* `LC`
Four subspecies divided into two subspecies groups, of which one occurs in region.

Purple Heron *Ardea (purpurea) purpurea*
Vagrant (**Barbados**).

L 78–90 cm; **WS** 120–150 cm. Three subspecies in group, of which nominate occurs in region. Shallow freshwater marshes with dense vegetation, especially reeds. Six records, all since autumn 1998, most involving 1st-y and all between Sept and Apr. **ID** Large, slim-bodied heron with long slender neck. **Ad** is chestnut and grey, with prominent black stripes on head and neck. In flight, grey wings with darker flight feathers and rufous underwing-coverts. **Juv** is duller, predominantly rufous-brown with darker streaks. In flight appears brown with blackish flight feathers. **Voice** Call a simple "frank", higher-pitched than Great Blue and Grey Herons, given mainly in flight. **SS** Generally smaller-bodied, but slimmer and longer-necked than congeners, which against light, or in otherwise dark conditions, with experience can serve to identify alone. In flight feet look larger and retracted neck bulges more obviously.

Great White Egret / Great Egret *Ardea alba* `LC`
Four subspecies divided into four subspecies groups, of which one occurs in region.

American Great Egret *Ardea (alba) egretta*
Common to uncommon resident (**GA**, **N LA**, **O**), uncommon non-breeding visitor (**S LA**).

L 90–104 cm; **WS** 140–170 cm. One subspecies in group. Wetlands (fresh, brackish and salt), rivers, marshes and flooded fields. Usually nests colonially in mangroves. Breeds S as far as Guadeloupe. In Lesser Antilles commonest Sept–Apr. During same period, migrants from North America certainly reach Cuba and probably beyond. **ID** Large all-white heron, with long S-shaped neck. **Ad br** has long feathers on lower neck and breast. Briefly develops bright orange bill, fading to yellow post-courtship, bright green lores and red eye-ring. **Ad non-br** lacks long plumes and has all-yellow bill and facial skin. **Juv** as ad non-br, with blackish tip to yellow bill. **Voice** A low-pitched "kraak" when disturbed and in flight. **SS** White-morph Great Blue Heron (rare) larger with heavier bill, and yellow legs. Other all-white egrets smaller and shorter-necked, among other features.

Whistling Heron *Syrigma sibilatrix* `LC`
Vagrant (**San Andrés**).

L 50–64 cm. Two subspecies, of which one (presumably *fostersmithi*) occurs in region. One photographed on an unknown date early in the 21st century (McNish 2011). Considered remarkable by the author, but records from Bonaire (Jan 2003) and S Central America illustrate the species' propensity to wander. Vagrancy to Lesser Antilles likely. **ID** Very distinctively plumaged heron, observed in drier habitats than many relatives. In flight, wings appear rather broad and bulky neck is often largely retracted. **Ad** has slate-coloured crown, blue skin around eye, long head plumes, cinnamon-buff neck and breast, and largely grey upperparts with buffy and dark-streaked smaller wing-coverts. Legs dark, but bill bright pinkish red with solid black tip. Facial skin and bill colours duller in non-br season. **Juv** is overall duller, with streaking on neck and wings, and narrower streaking on lesser wing-coverts. **Voice** A far-carrying, melodious whistle, "kee" or well-spaced "kee, kee, kee", typically in flight, a slower, drawn-out rendition on take-off and a higher-pitched, rapidly repeated, metallic call. **SS** Plumage and bare-parts colorations render the species unmistakable.

typical morph

Great Blue Heron
herodias

typical morph

white morph

Cocoi Heron

Purple Heron
purpurea

non-breeding

American Great Egret
(Great White Egret / Great Egret)

Whistling Heron
fostersmithi

not to scale

breeding

101

Reddish Egret *Egretta rufescens* NT
Locally common resident (**Bahamas**, **Cuba**), uncommon to rare non-breeding visitor (rest of **GA**, **Cayman Is**), vagrant (**LA**). **L** 66–81 cm; **WS** c. 116–121 cm. Two subspecies, of which nominate occurs in region. Sheltered coasts, lagoons and brackish ponds. An active heron; pursues prey by walking quickly or running energetically, often with wings spread to form a canopy. Vagrants reported St Martin, Antigua, Barbuda, Dominica and Montserrat. **ID** Dimorphic heron with fairly long legs and neck. Legs and feet dark. **Ad br** has longer, shaggy plumes on hindneck and lower foreneck (covering breast). Bill pinkish with black tip; develops bright violet lores in courtship, becoming flesh-coloured during incubation and chick-rearing, darkening later. **Ad non-br** lacks long plumes and has all-blackish bill. **Dark morph** has dark cinnamon head and neck, blue-grey body and wings. **White morph** is all white. **Juv** is like respective morph but typically has all-blackish bill, dark lores and lacks neck plumes; dark morph has cinnamon tips to wing-coverts. **Voice** A soft "awh" while feeding and "crog-crog" in flight displays. **SS** Ad dark morph larger and paler than ad Little Blue Heron; ad white morph larger than imm Little Blue Heron or Snowy Egret; in both cases note heftier bill. In ad br, two-toned bill and shaggy neck useful in combination. Also distinctive, very active feeding behaviour and all-dark legs and feet.

Tricolored Heron *Egretta tricolor* LC
Fairly common resident (**GA**, **O**), uncommon resident or vagrant (**LA**). **L** 50–76 cm; **WS** c. 90 cm. Three subspecies, of which one (*ruficollis*) occurs in region. Usually near coasts, using lagoons, marshes, mangroves or small satellite islands. Active feeder, running and chasing prey, but also wades belly deep. Alone or in small groups. Mainland South American nominate race, or Trinidadian *rufimentum*, perhaps reach S Lesser Antilles periodically, although a specimen from Grenada is neither (Bond 1959). **ID** Notably slender heron with long legs and thin, snake-like neck. **Ad br** has slate-coloured head, neck and upperparts, white plumes on nape and buff ones on mantle. Foreneck white with red-brown stripe reaching breast. Belly and underwing-coverts white. Bright blue bill with black tip, cobalt-blue lores, scarlet-red eyes, and pinkish-maroon legs and feet. **Ad non-br** has bill, lores and legs yellow. **Juv** has warm brown neck-sides and wings, chestnut-grey streaks on breast, dull black-tipped yellow bill, and greenish-yellow legs and feet. **GV** Nominate race and *rufimentum* (both potential vagrants) have, to varying degrees, dark chestnut or rufous (not white) chin and throat-stripe. **Voice** An "aah" in aggression and display, with a short guttural "aahrr" in alarm. **SS** White foreneck stripe and belly diagnostic. **AN** Louisiana Heron.

Little Blue Heron *Egretta caerulea* LC
Fairly common resident (**throughout**). **L** 51–76 cm; **WS** 95–106 cm. Monotypic. Shallow fresh, brackish and saltwater wetlands, including slow-flowing streams and rivers, at all elevations. Relatively passive and methodical feeder, moves slowly or stands motionless. In groups of up to 20+. **ID** Fairly small with quite heavy-based and slightly decurved bill. **Ad br** typically has grey-blue body and wings, and purple-maroon head. Neck plumes become red-brown and develops long, slate-blue, lanceolate plumes on back. Bill base and lores turquoise-blue, legs black. **Ad non-br** has duller grey bill base and lores, greenish-grey legs, and lacks plumes. **Juv** is all white with slate-grey primary tips; bare parts as ad non-br. **Imm** gradually acquires dark grey feathers. **Voice** Rather quiet, but a "scah" or "sken" in threat or aggression, "gwa" in flight, and repeated "gerr" if disturbed. **SS** Compare Reddish Egret and (vagrant) Western Reef-egret. Juv/imm resembles Snowy Egret, but has grey-based (vs. all-black) decurved bill, blue-grey (vs. yellow) lores, dusky green-grey (vs. black) legs, dusky-tipped primaries, and slower feeding action.

Snowy Egret *Egretta thula* LC
Common to uncommon resident (**throughout**). **L** 47.5–68 cm; **WS** 84–91 cm. Two subspecies, of which nominate occurs in region. Coasts and wetlands. Very active feeder. Not known to breed on most of Lesser Antilles, but nests as far S as Barbados and possibly St Lucia. **ID** Thin-bodied, mid-sized white heron, with long, thin neck and bill. **Ad br** has yellow eyes and lores, long dorsal plumes, and bushy crest. Black legs and yellow feet, often with yellowish-green rear to lower tarsus. **Ad non-br** lacks ornamental plumes and typically has duller bare parts. **Juv** has shorter crest and paler bill, lacks dorsal plumes and has greyish-yellow lores. **Voice** Rather vocal, giving characteristic "rah" during antagonistic encounters. **SS** See Little Egret and Little Blue Heron.

dark morph
breeding

white morph
breeding

dark morph
non-breeding

white morph
non-breeding

Reddish Egret
rufescens

breeding

Tricolored Heron
ruficollis

non-breeding

adult
breeding

Little Blue Heron

adult
non-breeding

juvenile

non-breeding

breeding

Snowy Egret
thula

103

Little Egret *Egretta garzetta* `LC`
Uncommon resident (**Barbados**, **Antigua**, **Guadeloupe**), vagrant elsewhere.

L 55–65 cm; **WS** 86–104 cm. Three subspecies, of which nominate occurs in region. First recorded (in W Hemisphere) Apr 1954, Barbados, where first bred in 1994 (now up to 25 pairs), followed by Antigua in 2008 (Kushlan & Prosper 2009), with records elsewhere in Lesser Antilles and N to Bahamas (first, Aug 1999), Hispaniola and Puerto Rico (first, May–Jun 1986) probably originating from there, at least in part, but Spanish-ringed bird recovered Martinique. Ponds and swamps. **ID** Very similar to Snowy Egret. **Ad br** during courtship develops head, breast and back plumes, and lores become pink then red. **Ad non-br** has blue-grey lores and yellow feet contrasting with all-dark legs. **Juv** as ad non-br, with some brown feathers. Green bill with black markings, pale green lores, and dull black-and-green legs. **Voice** A "kark, kark, kark" in aggression and flight, and "aaah" when flushed. **SS** Care required to distinguish from Snowy Egret. Latter slightly smaller with proportionately shorter bill, neck and legs; in breeding plumage, bushier and shorter nape plumes, bright yellow lores (vs. blue-grey), and often has yellow rear tarsi. See (vagrant) Western Reef-egret.

Western Reef-egret *Egretta gularis* `LC`
Vagrant (**GA**, **LA**).

L 55–65 cm; **WS** 86–104 cm. Three subspecies, of which nominate occurs in region. Rocky shores, estuaries, mangroves, mudflats, lagoons, sometimes sandy beaches, but could appear at inland wetlands. Often chases prey at water's edge. First record Barbados (in 1975) and subsequently Puerto Rico (first Sept 1999), St Lucia (Feb 1984, Jan 1985), the Grenadines (Mustique, Feb–Jun 2004), with possible sight record in Cuba (*Cotinga* 28: 87). Could colonize, given number of records in Barbados, some for long periods in heronries. **ID** Mid-sized, thin-bodied egret with long, thin neck and yellowish-green bill, dark (sometimes green) legs and yellow feet. Dimorphic. **Ad br** develops two nuchal plumes, and others on breast and back. **Dark morph** is overall slate-grey with white throat (all regional records in this plumage). **White morph** is all white, but almost always shows some dark feathers. **Juv** is similar to respective morph of ad non-br, but white morph can exhibit scattered dark flight feathers. **Voice** No differences known from Little Egret. **SS** Little Blue Heron (ad confusable with dark morph, juv with white morph) has grey-based, slightly decurved bill with no yellow, and foraging much more laboured; ad lacks white throat. White morph very similar to Little Egret, but bill usually slightly longer, with subtly fatter tip and more extensive yellow; legs often more extensively greenish. **TN** Often considered conspecific with Little Egret, with which it hybridizes. **AN** Western Reef-heron.

PELECANIDAE
Pelicans
8 extant species, 2 in region

Brown Pelican *Pelecanus occidentalis* `LC`
Common to uncommon resident and winter visitor (**throughout**).

L 105–152 cm (including bill 28–35 cm); **WS** 203–228 cm. Five subspecies, of which two (nominate resident, *carolinensis* mainly winter visitor) occur in region. Coastal: harbours, estuaries, lagoons, bays and other calm waters. Fishes by plunging into sea, often from height. Forms colonies on mangroves or other vegetation, sometimes cliffs. Nest sites and numbers in region both few, with no colonies known S of Guadeloupe and perhaps just 1500 pairs overall (Collazo *et al*. 2000). Generally rare and declining S Lesser Antilles, although apparently increasing on Guadeloupe and Martinique. Race *carolinensis* common visitor, at least in N, but potentially throughout; records year-round. **ID** Very large, with massive pouched bill. **Ad br** is mostly greyish above and dark brown below, with blackish flight feathers. Hindneck chestnut, rest of head yellowish to white (infrequently mostly white). **Ad non-br** is similar but most of head and neck white, with yellowish forehead. **Juv** has head, neck and upperparts dusky brown, underparts greyish white. **GV** Race *carolinensis* slightly paler below. **Voice** Rarely heard sounds include harsh grunts and bill-claps. **SS** American White Pelican (very rare in most of region) larger and mostly white, with black flight feathers.

American White Pelican *Pelecanus erythrorhynchos* `LC`
Rare to locally fairly common winter visitor (**Cuba**), vagrant (elsewhere in **GA**, **O**, **LA**).

L 127–178 cm (bill 26.5–37 cm); **WS** 244–299 cm. Monotypic. Coastal bays, but also estuaries, brackish and freshwater lakes. Does not plunge to catch fish like Brown Pelican, but scoops them up while swimming. Mainly Nov–Apr but records virtually year-round; numbers in region, especially Cuba, have increased dramatically since 1990s. One record in Lesser Antilles (Antigua, Jan 2019). **ID** Huge, with massive bill. **Ad br** mostly white with black primaries and outer secondaries. Nuchal crest and hindneck creamy. Bill orange with knob on maxilla. **Ad non-br** is similar but hindneck greyish to dusky, without crest, and bill usually yellower, without knob. **Juv** is even duller, with dull brownish upperwing and pinkish-grey bill. **Voice** Silent in region. **SS** Much commoner Brown Pelican smaller, mostly brown overall and never has knob on maxilla.

FREGATIDAE
Frigatebirds
5 extant species, 1 in region

Magnificent Frigatebird *Fregata magnificens* `LC`
Fairly common but local resident (**throughout**).

L 94–104 cm; **WS** 217–244 cm. Three subspecies, of which one (*rothschildi*) occurs in region. Bays, harbours, inshore waters and offshore cays. Snatches fish from surface. Never touches water other than with bill and never intentionally settles on it. Breeds in colonies, usually on mangroves but also other coastal vegetation. Non-breeders can wander widely. **ID** Long forked tail and long, pointed wings, sharply bent at wrist, give distinctive aerodynamic profile in flight. **Ad** ♂ is all black with bright red inflatable gular pouch, unobtrusive when not breeding, but conspicuous at colonies. **Ad** ♀ is brownish black with white breast and pale brownish band on upperwing. **Juv** is similar to ♀ but has head to belly white. **Imm** shows intermediate plumage. **Voice** Usually silent away from nests, where gives gull-like grating and gurgling calls, and bill-clacking. **SS** Unmistakable.

Little Egret
garzetta

breeding

breeding courtship

non-breeding

Brown Pelican
occidentalis

breeding

Western Reef-egret
gularis

dark morph

white morph

American White Pelican

adult non-breeding

Magnificent Frigatebird
rothschildi

adult ♂

adult ♂

adult ♀

juvenile

105

SULIDAE
Gannets and Boobies
10 extant species, 4 in region

Northern Gannet *Morus bassanus* `LC`
Rare to uncommon non-breeding visitor (**Bahamas**), mainly vagrant elsewhere.

L 87–100 cm; **WS** 165–180 cm. Monotypic. Coastal and pelagic. Visits region Sept–May, occasionally groups of up to c. 20. Very rare or vagrant Cuba, Puerto Rico, Guadeloupe and Martinique. **ID** Largest of family. **Ad** is white with apricot-yellow tinge on head/neck. Wingtips black. Bill pale grey. ♀ is less bright on head. **Juv** is mostly dark brownish grey, flecked white on wings and mantle; paler below. **Imm** gradually acquires ad plumage. Juv and imm most likely in region, but three (of five) Cuban records were ad (Kirkconnell *et al.* in press). **Voice** Silent in region. **SS** Masked Booby is smaller; ad has black mask and flight feathers (not just wingtips), with no apricot tinge on head. Dark juv Northern Gannet much larger than any dark juv boobies.

Red-footed Booby *Sula sula* `LC`
Fairly common but very local resident (**throughout**).

L 66–77 cm; **WS** 124–152 cm. Three subspecies, of which nominate occurs in region. Pelagic. Seldom seen from shore. Plunges diagonally into water to catch fish, may follow vessels and often forages at night. Nests in large colonies on offshore islands. Non-breeders can wander widely. **ID** Small, highly polymorphic booby, with longish tail. Ad always has bright red feet, and pale blue bill and facial skin. **Ad white morph** is all white, often more or less tinged apricot-yellow on head and neck, with black flight feathers, greater and primary-coverts. Underwing has blackish-grey carpal. **Ad brown morph** is usually chocolate-brown with flight feathers darker than rest. Tail usually white but can be brown (or rarely black). Rump, belly and undertail-coverts often whitish or pale brown. **Juv** is all dark brown, often with slightly paler head and neck. Underwing-coverts pale brown to blackish brown, but never strongly patterned. Bill dusky brown and feet dull yellowish or bluish. **Imm** varies with morph, but blotchier and less uniform than respective ad. **Voice** Silent at sea. In colonies utters fast "rah-rah-rah-rah-rah"; also harsh rasping calls. **SS** Smallest of boobies. Red feet in ad and white tail (if present) diagnostic. Ad Brown Booby has white belly, axillaries and median coverts; always darker than Red-footed. Juv Brown Booby recalls dark morph, but often has distinctly paler brown/whitish underwing-coverts, and belly densely spotted white. Ad Masked Booby has black (not white or brown) tail.

Brown Booby *Sula leucogaster* `LC`
Fairly common to rare resident (**throughout**).

L 71–76 cm; **WS** 132–150 cm. Four subspecies, of which nominate occurs in region. Bays and coasts, including harbours and ports, but also at sea. Plunges diagonally into sea to catch fish. Breeds colonially on rocky cliffs, mangroves or other coastal vegetation. Fairly common near colonies or at fishing areas throughout, rare only in N Bahamas. **ID** Dark chocolate-brown booby with pale yellowish-green legs. **Ad** has head and upper breast contrasting sharply against white mid-breast to undertail-coverts. In flight, underwing has white axillaries and median coverts, with darker leading edge. Bill greenish yellow (but varies). Feet yellow. ♀ is larger and typically has cream-yellow face (bluish in ♂) but much variation. **Juv** is similar to ad but breast to undertail-coverts mottled brown. Bill dull greyish. **Imm** has more white feathers on abdomen and underwing approaches ad pattern. **Voice** Silent away from nests, where it utters low hisses and hoarse squawks. **SS** Ad similar to juv/imm Masked Booby, which is larger, has white collar and upper breast, and appears greyer on upperparts. Juv Red-footed Booby is dark brown overall.

Masked Booby *Sula dactylatra* `LC`
Rare and local resident (virtually **throughout**).

L 81–92 cm; **WS** 152–170 cm. Four subspecies, of which nominate occurs in region. Prefers deeper water than other boobies, but behaviour similar. Nests colonially on rocky offshore islands. Non-breeders can wander widely. **ID** Largest booby, with relatively heavy yellow bill (variably tinged olive) and dark dull pinkish-orange legs. **Ad** is white with black face, flight feathers and tail. Broad black trailing edge to wing. **Juv** has blackish-brown head, neck, back, wings and tail. White hindcollar and underparts; wing-coverts and mantle pale-fringed. White underwing has dark margins. **Imm** slowly becomes progressively whiter on head and upperparts. **Voice** Silent at sea. At colony, reedy whistles, plus hisses and quacking notes. **SS** Larger and more pelagic than other boobies. Juv similar to juv Brown and Red-footed Boobies, which are mostly brown and lack white hindcollar. Ad similar to white-morph Red-footed Booby, which has white or brown (not black) tail. See larger Northern Gannet. **TN** Until recently considered conspecific with extralimital Nazca Booby *Sula granti* (E tropical Pacific).

juvenile

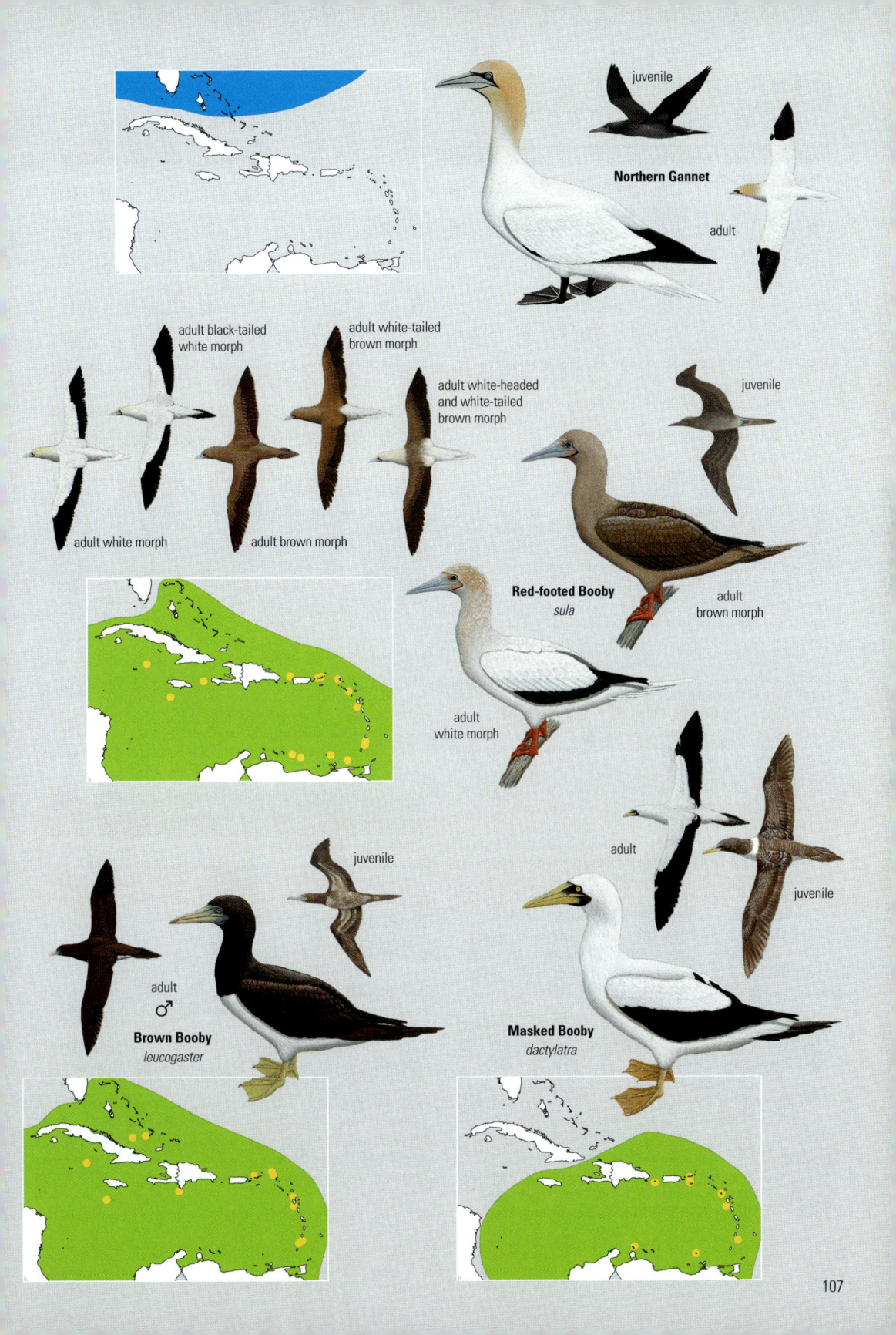

Northern Gannet

adult

adult black-tailed white morph

adult white-tailed brown morph

adult white-headed and white-tailed brown morph

juvenile

adult white morph

adult brown morph

Red-footed Booby
sula

adult brown morph

adult white morph

adult

juvenile

juvenile

Brown Booby
leucogaster

adult ♂

Masked Booby
dactylatra

107

PHALACROCORACIDAE
Cormorants
34 extant species, 2 in region

Double-crested Cormorant *Nannopterum auritus* LC
Common (**Cuba**, **San Salvador**) to uncommon resident (rest of **Bahamas**), rare or vagrant elsewhere.
L 76–91 cm; **WS** 114–137 cm. Five subspecies, of which three (nominate vagrant, *floridanus*[A] S to Cuba, *heuretus*[B] breeds San Salvador and perhaps Eleuthera, in Bahamas) occur in region. Inland waterbodies and sheltered coastal waters; rarely far from coast. Nominate race and *heuretus* at least vagrants to Cuba; latter perhaps reasonably common and even breeds there, while nominate potentially much more numerous (Kirkconnell *et al.* in press); same potentially true in Bahamas. Rests with wings spread. **ID** Large, proportionately long-necked cormorant with long shaggy crest of upcurved feathers behind eye (rarely visible). **Ad br** (*floridanus*) has mostly glossy blue-green plumage, duller on mantle, scapulars and wing-coverts. Head/neck sometimes display whitish filoplumes; gular bright yellow-orange. **Ad non-br** is crestless and duller black. **Juv** is mostly dark grey-brown, underparts mottled whitish brown.
GV Races differ mainly in size: *heuretus* smaller and darker than nominate, which is largest and palest of three races recorded, but in-hand examination required, making knowledge of their respective ranges rudimentary at best. **Voice** Clucking sounds and low-pitched gargling at nest, otherwise occasional low guttural calls.
SS Neotropical Cormorant smaller with white line around gular region (often indistinct in ad non-br and juv), longer tail especially noticeable in flight, duller lores (mainly in non-br plumage) and has different-shaped naked skin on gular. **TN** Previously placed in genus *Phalacrocorax*.

Neotropical Cormorant *Nannopterum brasilianus* LC
Common resident (**Bahamas**, **Cuba**), non-breeders wander more widely.
L 58–73 cm; **WS** c. 101 cm. Two subspecies, of which one (*mexicanus*) occurs in region. Inland and shallow coastal waters; more likely on freshwater bodies than Double-crested Cormorant. Sometimes feeds cooperatively. **ID** Smaller, slenderer and longer-tailed than Double-crested Cormorant. **Ad br** is mostly black with bluish or purplish sheen, except line of white feathers around gular pouch. Often has very short white filoplumes on superciliary, sometimes longer filoplumes elsewhere on head/neck. Bare skin on gular dull yellow. **Ad non-br** is duller, white line around gular often less defined. **Juv** is dark dull brown, paler on throat and breast. Whitish line around gular noticeable after first moult. **Voice** Low grunts and croaks. **SS** See Double-crested Cormorant. **TN** Previously placed in genus *Phalacrocorax*.

ANHINGIDAE
Darters
4 extant species, 1 in region

Anhinga *Anhinga anhinga* LC
Fairly common (**Cuba**) to scarce resident (**Cayman Is**), vagrant elsewhere.
L 81–91 cm; **WS** c. 120 cm. Two subspecies, of which potentially both (*leucogaster* breeds in Cuba, nominate probably wanders to Lesser Antilles) occur in region. Mainly on still, shallow inland waters; rarely coasts.

Often swims with virtually entire body submerged, head and neck very upright, like a periscope. In flight, almost dagger-shaped, and can appear raptor-like when soaring. Vagrants reported N to Bahamas (has bred) and S to Barbados and Grenada (those in S Lesser Antilles, occasionally present long periods, are presumably wanderers from South America). **ID** Large but attenuated, cormorant-like bird, with long dagger-shaped bill and very long, narrow, snake-like neck. **Ad** ♂ is black with long white scapulars and extensively silvery-grey scapulars and wing-coverts. **Ad** ♀ has brown throat to lower breast. **Juv** is like ♀, but even browner, and lacks most of silvery grey on upperparts. **GV** Nominate race (not confirmed) differs only in being larger with broader pale tip to tail. **Voice** Rather quiet away from nest. May give varied single rasping or croaking notes, plus longer descending "krr..krr..krr..krr" and prolonged grating rattle. **SS** Dagger-like bill, long, slim neck and long tail afford diagnostic profile. Further distinguished from cormorants by whitish centres to upperpart feathers.

BURHINIDAE
Thick-knees
10 extant species, 1 in region

Double-striped Thick-knee *Burhinus bistriatus* LC
Uncommon and local resident (**Great Inagua**, **Hispaniola**), vagrant (**Barbados**).
L 38–43 cm. Four subspecies, of which one (*dominicensis*) occurs in region. Open dry savannas and agricultural land. Mainly crepuscular or nocturnal, and terrestrial; by day, typically encountered in pairs resting in shade of bushes or fences. Bred Great Inagua, May 2003 (*Cotinga* 21: 82) where two seen Jul 2014 (*NAB* 68: 562). Vagrant Barbados, Sept 1937 and Oct 1978. **ID** Large, mainly streaky brown, plover-like shorebird, with large yellow eyes, short stout bill and long yellow legs. **Ad** has broad white supercilium bordered above by blackish lateral crown-stripe. **Juv** is duller and paler, with head, neck and breast tinged buff. **Voice** A chattering, often strident series, e.g. "ke-ke-ke..." or clucking "kah-kah-kah...", often in duet, sometimes becoming a noisy cacophony. **SS** Unmistakable.

HAEMATOPODIDAE
Oystercatchers
9 extant species, 1 in region

American Oystercatcher *Haematopus palliatus* LC
Two subspecies placed in separate subspecies groups, of which one occurs in region.

American Oystercatcher *Haematopus (palliatus) palliatus*
Fairly common (**Bahamas**, **Puerto Rico**, **Virgin Is**) to rare and local resident (**Cuba**, **Hispaniola**, **LA**).
L 40–44 cm; **WS** 76 cm. One subspecies in group. Stony beaches, rocky headlands and occasionally sandy beaches. Typically in pairs, or alone; occasionally small flocks. **ID** Unmistakable. **Ad** is a robust, black-and-white wader with stout red bill and pink legs. **Juv** has more drab-coloured bare parts and buff-fringed black feathers. **Voice** Far-carrying shrill piping whistle, "kleep", typically repeated in series. **SS** None. **TN** Bahamas population formerly treated as separate race, *pratti*, based on its larger bill than continental North American birds.

Double-crested Cormorant
auritus

Neotropical Cormorant
mexicanus

Anhinga
leucogaster

Double-striped Thick-knee
dominicensis

American Oystercatcher

RECURVIROSTRIDAE
Avocets and Stilts
7 extant species, 2 in region

American Avocet *Recurvirostra americana* LC
Mainly rare winter visitor (**Cuba**, **Cayman Is**, **Puerto Rico**), vagrant elsewhere.
L 41–51 cm; **WS** 68 cm. Monotypic. Freshwater and brackish lagoons. Often in small flocks. Mainly Nov–Apr, but recently (Jun 2007) bred Cuba (Labrada & Blanco 2011) where numbers have increased dramatically in recent decades. Records S to Barbados.
ID Unmistakable, large mainly pied wader with quite long, upturned bill. **Ad br** has head, neck and breast orange-brown. ♂ has longer, less recurved bill than ♀. **Ad non br** has head, neck and breast grey. **Juv** as ad, but crown pale brownish or greyish, with dull chestnut nape and hindneck merging into pinkish upper mantle. Whitish on inner primaries. Mantle and wing-coverts have contrasting pale tips when fresh. **Voice** Far-carrying "kleek". **SS** None.

Black-winged Stilt *Himantopus himantopus* LC
Five subspecies divided into five subspecies groups, of which two occur in region.

Black-winged Stilt *Himantopus (himantopus) himantopus*
Vagrant (**Guadeloupe**).
L 35–40 cm; **WS** 67–83 cm. One subspecies in group. Habitat and behaviour as Black-necked Stilt. Two, Aug 2014, in wake of Hurricane Bertha (AL); another, believed to be hybrid between this and Black-necked Stilt, Jan 2016 (eBird). **ID** Very similar to Black-necked Stilt but has white hindneck and upper back in all plumages. **Ad** has much-reduced black on head; some are entirely white-headed. **Juv** is paler and greyer than Black-necked Stilt. **Voice** Very similar to Black-necked Stilt, but discernibly shriller "kyi kyi kyi…" or "kik kik kik…". **SS** See Black-necked Stilt.

Black-necked Stilt *Himantopus (himantopus) mexicanus*
Resident (**throughout**).
L 35–40 cm; **WS** 67–83 cm. One subspecies in group. Shallow, open, brackish and freshwater wetlands. Usually in flocks. Commonest Greater Antilles and Bahamas, becoming uncommon in N Lesser Antilles (but expanding on Guadeloupe), and rare further S. **ID** Structure distinctive; very long pink legs trail noticeably in flight, and slim black bill. **Ad** has contrasting black upperparts (including hindneck and upper back) and white underparts. **Juv** is brownish above with pale feather edges. **Voice** Commonest call a repeated nasal, staccato, strident, yelping "kek..kek..kek…". **SS** Scarcely possible to confuse with American Oystercatcher (usually in different habitat) or American Avocet (likely to share habitat), despite their superficial similarity. **TN** All races of Black-winged Stilt often elevated to species level.

CHARADRIIDAE
Plovers
71 extant species, 9 in region + 4 vagrants

Grey Plover / Black-bellied Plover LC
Pluvialis squatarola
Common non-breeding visitor (**throughout**).
L 27–31 cm; **WS** 71–83 cm. Three subspecies, of which one (*cynosurae*) occurs in region. Intertidal mudflats, beaches and coastal pools. Typically in loose flocks. Mostly Aug–May, but recorded all months.
ID Large, slender, upright, long-legged plover with largely monochrome plumage. Black axillaries and white rump diagnostic in all plumages. **Ad non-br** is grey, mottled above, with broad whitish supercilium, dirty white underparts and dusky breast-band. **Ad br** has spangled black and silver upperparts, with contrasting black face and underparts. **Juv** has darker upperparts with pale yellow spots, faint streaking on breast and flanks. **Voice** Call a plaintive, drawn-out, pure, trisyllabic "peee-oo-ee", downslurred in middle. **SS** American Golden Plover smaller and more delicate, with thinner bill and noticeably more rounded head with more conspicuous white supercilium. Axillaries greyish, concolorous with rest of underwing. Rarely in same habitat, latter preferring open fields or savannas. Compare Pacific Golden Plover.

Pacific Golden Plover *Pluvialis fulva* LC
Vagrant (**GA**, **LA**, **O**).
L 23–26 cm; **WS** 60–72 cm. Monotypic. Has usually occurred beside artificial waterbodies (water treatment or shooting ponds) near coasts. Eleuthera, Bahamas (Mar 2004), Cayman Is (Dec 2015), Dominican Republic (Dec 2016), St Kitts & Nevis (Nov 2006 and 2012), Guadeloupe (Dec 2018 to Jan 2019) and Barbados (Apr 1993, Feb 2000, Feb 2008, May 2017–Nov 2018 at least). **ID** Very similar to American Golden Plover. **Ad non-br** is rather nondescript, with golden-spangled upperparts (appear pale brown at distance) and dark patch on ear-coverts. Underparts off-white. **Ad br** has spangled black, gold and silver upperparts with black face and underparts demarcated by broad white forehead, white supercilium reaching breast-sides and flanks. **Juv** is more noticeably golden-spangled above and appears barred below. **Voice** Whistled "klu-eeuh" or shorter and more emphatic "ku-eet!" with stress on second, rising syllable. **SS** Grey Plover larger, chunkier and much greyer, with square head, heavier bill and comparatively shorter legs. In flight, white rump and black axillaries diagnostic. Tertials of very similar American Golden Plover fall well short of tip of tail, revealing 4–5 primaries exposed beyond longest tertial, and 12–22 mm primary projection beyond tail-end. Tibia shorter and bill often shorter and finer. Plumage is greyer-toned.

American Golden Plover *Pluvialis dominica* LC
Rare passage migrant (**throughout**).
L 24–28 cm; **WS** 65–72 cm. Monotypic. More often on fields and golf courses than coastal flats. Can occur in large flocks with hundreds shot annually in Guadeloupe and Martinique, but elsewhere records often of singles or small groups. More frequent in autumn (Aug–Nov) than spring (Jan–Apr). **ID** Medium-large, upright, long-legged, long-winged plover with overall grey-brown plumage and brownish-grey underwing. **Ad non-br** has grey-spangled upperparts faintly washed brownish, extensively grey on head and neck, with broad pale supercilium and darker ear-coverts. Underparts off-white. **Ad br** has spangled black, gold and silver upperparts with black face and underparts demarcated by white forehead extending above eye to form broad white lateral breast patch. **Juv** is like ad non-br, but

breeding

American Avocet

non-breeding

Black-winged Stilt

variant

Black-necked Stilt
(Black-winged Stilt)

juvenile

Grey Plover / Black-bellied Plover
cynosurae

non-breeding

breeding

Pacific Golden Plover

non-breeding

breeding

American Golden Plover

non-breeding

111

almost totally lacks golden feathering, and has heavily marked greyish underparts. **Voice** Call a plaintive "klu-eet". **SS** Grey Plover larger, chunkier and much greyer, with large square head, heavier bill and relatively shorter legs. In flight, white rump and black axillaries diagnostic. Vagrant Pacific Golden Plover is more golden-toned, shorter-winged, with longer tertials reaching almost to tail tip, revealing 2–3 primaries exposed beyond longest tertial and 0–9 mm primary projection beyond tail-end. Tibia longer, and bill often longer and more bulbous.

Common Ringed Plover *Charadrius hiaticula* `LC`
Vagrant (**Guadeloupe**, **Barbados**).

L 18–20 cm; **WS** 48–57 cm. Three subspecies, but that in region not determined (all seem possible). Intertidal mudflats, occasionally sandy beaches. Juv specimen, Barbados, Sept 1888 (plus possibles, Apr 1993 and Nov 2008); Guadeloupe, Sept 2010 (*NAB* 65: 181), Jan 2019 (eBird). **ID** Smallish, rather plump plover closely recalling Semipalmated Plover. **Ad non-br** has head and upperparts greyish brown, white forehead and prominent supercilium, blackish lores to ear-coverts, and olive-brown breast-band. Black lores as broad as entire bill base. Yellow-orange legs. **Ad br** has black lores to ear-coverts, frontal bar and breast-band, short white post-ocular stripe. Base of bill orange. **Juv** recalls pale ad, with buffy fringes above (soon lost). No black. Narrower or broken breast-band. **Voice** Commonest call, usually given in flight, a mellow whistled "too-eep", with emphasis on longer first syllable. **SS** Semipalmated Plover best distinguished by voice. Slightly smaller with distinctly shorter bill that has less extensive orange base. Narrow eye-ring except in juv plumage. White eyebrow less prominent, breast-band narrower in centre. Narrower black lores reach bill above mandible. See Little Ringed Plover.

Semipalmated Plover *Charadrius semipalmatus* `LC`
Common passage migrant and winter visitor (**throughout**).

L 17–19 cm; **WS** 43–52 cm. Monotypic. Intertidal mudflats, occasionally sandy beaches. Usually in dispersed flocks, often numbering dozens. Commonest Aug–Apr, but records all months. **ID** Smallish, fairly plump, round-headed plover with mid-brown upperparts and dark breast-band. **Ad non-br** has head and upperparts dirty brown, white forehead and prominent supercilium, blackish lores to ear-coverts, and narrow olive-brown breast-band. Yellow-orange legs. **Ad br** has black lores to ear-coverts, frontal bar and breast-band, thin white post-ocular stripe. Base of bill orange. Narrow yellow orbital ring. **Juv** is like pale ad non-br, upperpart feathers with buff fringes, breast-band narrower (sometimes rather obscure), legs duller yellow-brown. **Voice** A clear whistled "kli-weeet" with emphasis on second syllable. **SS** See very similar vagrant Common Ringed (best separated by call) and Little Ringed Plovers.

Little Ringed Plover *Charadrius dubius* `LC`
Vagrant (**Martinique**).

L 14–17 cm; **WS** 42–48 cm. Three subspecies, not separable in field, but presumably *curonicus* occurs in region. Ad ♀, Apr 2005 (Lemoine 2005). Sand and shingle beaches, and artificial pools. Alone or in pairs. **ID** Small, dainty 'ringed' plover with tiny rounded head, attenuated rear with long tertials and short primary projection. Tail projects beyond primaries. **Ad br** has head and upperparts greyish brown, black frontal bar separated from brown crown by narrow white line, white forehead, black lores to ear-coverts, and black breast-band.

Bright yellow eye-ring, dull yellowish-pink legs. **Ad non-br** has reduced brownish breast-band and black on head replaced by brownish. Lacks white eyebrow. **Juv** recalls ad non-br but olive-brown upperparts have buff fringes. **Voice** Call, given especially in flight, a slightly burry, piping, downslurred "preeu". **SS** Semipalmated and Common Ringed Plovers are larger and bulkier with longer primary projection, wings projecting beyond tail and prominent white eyebrow. Common Ringed lacks prominent eye-ring.

Wilson's Plover *Charadrius wilsonia* `LC`
Fairly common resident (**Bahamas**, **GA**, **Virgin Is**, **N LA**), rare or vagrant elsewhere.

L 16.5–20 cm; **WS** 36 cm. Four subspecies, of which two (nominate[A] breeds Bahamas S to Guadeloupe, *cinnamominus*[B] breeds Grenadines and Grenada) occur in region; the two almost certainly overlap in distribution at other times of year. Sandy beaches and (especially) tidal mudflats. Typically in loose pairs, often with Semipalmated Plovers. In some areas perhaps only or mainly a summer visitor. **ID** Distinctive *Charadrius* with large, thick black bill and broad breast-band. Legs pinkish. **Ad br** ♂ has sooty forecrown, lores and breast-band. **Ad br** ♀ has black of breast-band and forecrown replaced by brown. **Ad non-br** has mouse-brown upperparts (without any rufous tones) and mask. **GV** Nominate has medium-brown mask concolorous with upperparts. Race *cinnamominus* darker above, with shorter supercilium and more rufous on crown; mask broader and pale to bright cinnamon. **Voice** A distinctive, emphatic, musical "pit!", often given when flushed. **SS** Size, bulk and colour of bill immediately separate it from congeners. All, except more rakish Killdeer, are smaller. Semipalmated Plover is commonest confusion species, but clearly smaller with much stubbier bill that typically shows some orange at base. **TN** Precise range of races probably demand clarification; e.g., in Barbados (where does not breed) most records indeterminate, but both subspecies have been confirmed. Birds ascribed to nominate nest as far S as Guadeloupe (where *cinnamominus* has perhaps been observed), and probably occur in winter throughout basin. **AN** Thick-billed Plover.

Killdeer *Charadrius vociferus* `LC`
Common to fairly common resident (**Bahamas**, **GA**, **Virgin Is**), uncommon to rare passage visitor and winter visitor elsewhere.

L 20–28 cm; **WS** 59–63 cm. Three subspecies, of which two (*ternominatus* resident, nominate winter visitor) occur in region. Open fields and flats with short vegetation, not always near water: agricultural land, lawns, beaches, ponds. Perhaps resident over a wider area than currently perceived, S to at least St Martin. Residents augmented by migrants Sept–Jun, when uncommon on Cayman Is, Lesser Antilles S to St Barthélemy, Barbados, and San Andrés. **ID** Distinctive large *Charadrius* with double blackish breast-bands and long tail (reaching beyond wingtips). Cinnamon rump distinctive in flight. **Ad br** ♂ has partial black mask and forecrown bar, whitish post-ocular supercilium, white hindneck collar, and orange-red eye-ring. Legs pale yellowish or pinkish. **Ad br** ♀ tends to have browner mask and breast-bands. **Ad non-br** has rufous and buffish-brown fringes to upperparts. **Juv** is like dull ad non-br, but has off-white legs. **GV** Race *ternominatus* smaller, paler and greyer (less brown) than nominate. **Voice** Calls include plaintive or strident piping notes, e.g. "keee", or "kil-deee" (hence its name). **SS** Other 'ringed plovers' smaller with single breast-bands.

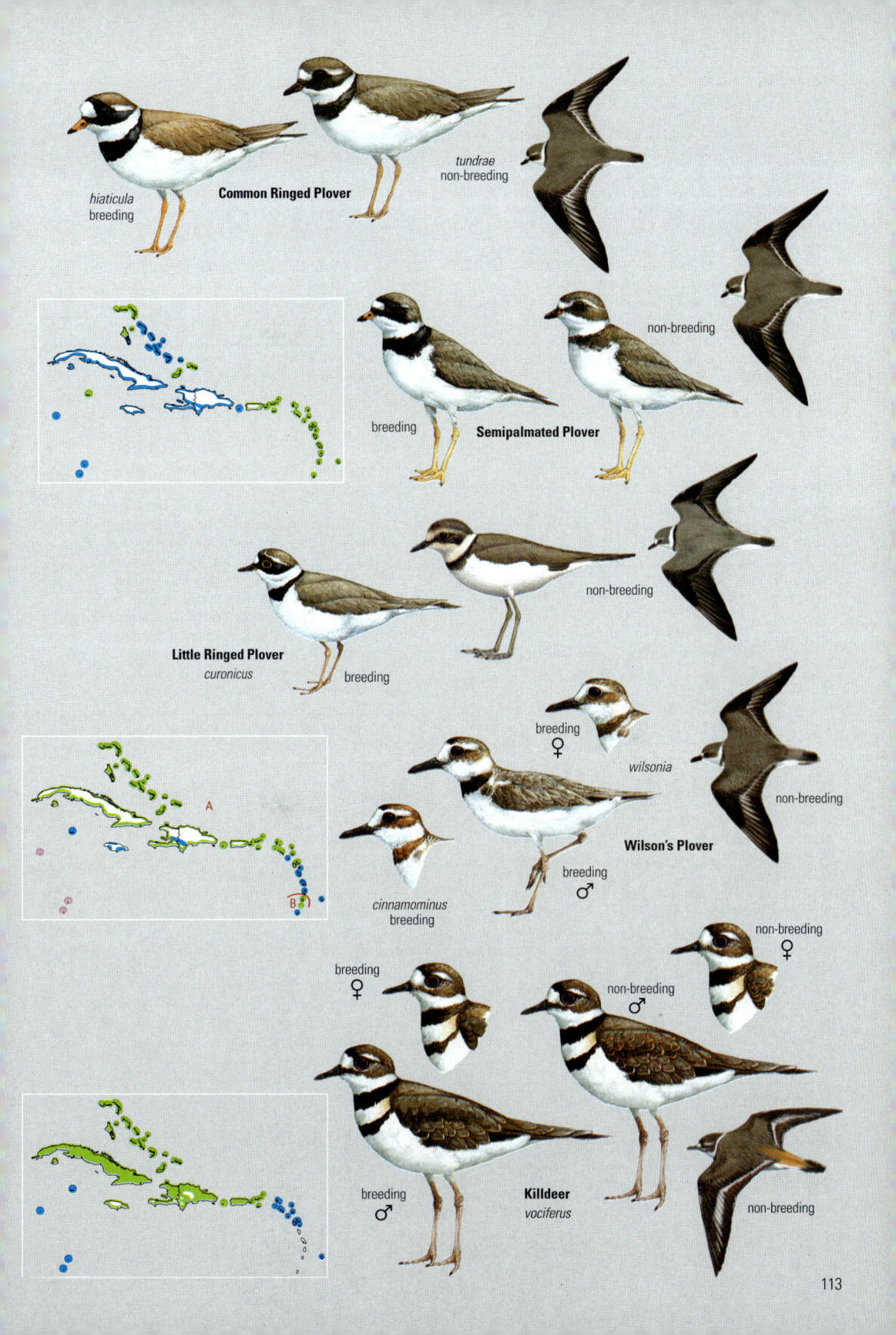

hiaticula breeding
Common Ringed Plover
tundrae non-breeding

breeding
Semipalmated Plover
non-breeding

Little Ringed Plover
curonicus breeding
non-breeding

breeding ♀
wilsonia
non-breeding
Wilson's Plover
cinnamominus breeding
breeding ♂

breeding ♀
non-breeding ♂
non-breeding ♀
breeding ♂
Killdeer
vociferus
non-breeding

Piping Plover *Charadrius melodus* `NT`
Fairly common (**Bahamas**) to uncommon winter visitor (**GA**), vagrant (**LA**).

L 17–18 cm; **WS** 36 cm. Two subspecies, of which both (nominate, *circumcinctus*) occur in region. Respective ranges in West Indies, if indeed separate, unknown, given relative impossibility of distinguishing them in non-breeding plumage. Sandy beaches, usually with piles of seaweed, often with Ruddy Turnstones or Sanderlings. Fairly common in Bahamas, uncommon and local in Cuba, Hispaniola and Puerto Rico, and vagrant in Jamaica, Virgin Is and further S (to Martinique and Barbados). Mainly recorded Oct to late Feb, but overall seen mid Jul–late Apr. **ID** Small, compact plover, with pale, almost silvery upperparts, stubby orange bill with black tip and short, bright orange legs. **Ad br** has black frontal bar and breast-band, latter often broken. ♀ has black of forehead and breast browner. **Ad non-br** has greyish forehead and breast-band, and often all-black bill. **Juv** is like ad non-br, but upperparts fringed pale buff; bill black. **GV** Races clearly differ genetically (Miller *et al.* 2010): nominate tends to have all-white lores and narrow or incomplete breast-band, and *circumcinctus* a fairly broad complete breast-band and some dark markings on lores; non-breeders apparently inseparable (Chandler 2009). **Voice** Calls a short mellow "peep" and plaintive double-noted "peee-lew". Rarely heard in region. **SS** Semipalmated Plover has darker, browner upperparts, slightly longer bill and dark lores. Snowy Plover has black legs.

Snowy Plover *Charadrius nivosus* `NT`
Fairly common to very rare resident (**S Bahamas** to **N LA**), mainly vagrant elsewhere.

L 15–17 cm; **WS** 42–45 cm. Two subspecies, of which nominate occurs in region. Common S Bahamas, Hispaniola and Anguilla, uncommon Puerto Rico, Virgin Is and St Barthélemy, rare but regular Guadeloupe, and very rare Cuba and Jamaica. Vagrant S to Barbados, but also breeds Grenadines. Extinct as breeder on St Martin (Brown 2012). Rare on autumn passage and, even more occasionally, in winter in Lesser Antilles. Sandy beaches, alkaline flats and saltpans. Pale coloration renders it almost invisible until moves, characteristically making fast runs, punctuated by brief stops. **ID** Small pale plover with slender black bill and black legs. White hindneck and breast-band always restricted to sides. **Ad br** ♂ has black crown, ear-coverts and breast-side patches, and may develop pale cinnamon wash to crown. **Ad br** ♀ tends to have less black on head. **Ad non-br** has brown head markings. **Juv** lacks black on head and has pale-fringed mantle feathers. **Voice** Call from ground an upslurred "curr-weeEET!". Flight call a low burry "prripp". **SS** Only similarly pale species is Piping Plover, which is larger with orange legs and almost always has some orange on its stubby bill. Collared Plover (note range) more richly coloured, with pinkish legs. **TN** Until recently considered conspecific with extralimital (but potential vagrant) Kentish Plover *C. alexandrinus* (Old World). Caribbean race *tenuirostris* was based on supposedly paler coloration, and genetics, but now included in nominate.

Collared Plover *Charadrius collaris* `LC`
Rare non-breeding visitor or vagrant (**LA**).

L 14–16 cm; **WS** 36–39 cm. Monotypic. Sandy beaches and ponds. Suggestion it breeds on Grenada and perhaps Grenadines (Raffaele *et al.* 1998) now believed erroneous; instead apparently mainly a post-breeding visitor to S Lesser Antilles (May–Oct), N to St Kitts and St Martin (Voous 1983, Steadman *et al.* 1997, Smith & Smith 1999b). **ID** Delicate small plover with narrow black bill and dull pinkish legs. **Ad** has narrow black breast-band, frontal bar and lores, and pale chestnut central crown. ♀ can have less extensive chestnut and black areas are sometimes brownish. **Juv** lacks breast-band, but has tawny-brown lateral patches; no black on head; buff fringes on upperparts. **Voice** Call a short emphatic "pit" or "chit", which can become a metallic series. Not very vocal. **SS** Piping Plover (mainly N of region) much paler with thicker, stubby bill and orange legs. Semipalmated Plover stockier with yellow or orange legs. Snowy Plover much paler with black legs. Beware of confusion with larger and far more robust Wilson's Plover, thought to be source of previous confusion concerning status in region.

Northern Lapwing *Vanellus vanellus* `NT`
Vagrant (**Bahamas**, **Puerto Rico**, **Martinique**, **Barbados**).

L 28–31 cm; **WS** 82–87 cm. Monotypic. Grassy fields, agricultural land and coastal mudflats. Singles: Hog (= Paradise) I, Bahamas, Nov 1900 (Fleming 1901), Puerto Rico, Dec 1978–Jan 1979 (Bond 1984), Martinique, Feb 1976 (Pinchon 1976) and Dec 2015 (AL); Barbados, Dec 1886 and Dec 1963 (Buckley *et al.* 2009). An observation on Tortola, Virgin Is, Dec 1996, assumed to be this species might have been a Southern Lapwing (Ebels 2002). **ID** Very distinctive. Large with metallic glossy green upperparts, blackish crest, bronze scapulars and very broad wings. **Ad non-br** has buff face, short crest, white chin and throat, and broad black breast-band. Upperwing-coverts and scapulars fringed buff. **Ad br** ♂ has black chin, throat and breast; ♀ is similar to non-br. **Juv** has shorter crest, more extensive buff feather fringes, and narrower, browner breast-band. **Voice** A shrill "cheew" or more plaintive "cheew-ip" or "wee-ip" in flight, though vagrants unlikely to vocalize. **SS** None. Only vaguely resembles paler, browner Cayenne Lapwing.

Southern Lapwing *Vanellus chilensis* `LC`
Four subspecies divided into two subspecies groups, of which one occurs in region.

Cayenne Lapwing *Vanellus (chilensis) cayennensis*
Rare resident (**Barbados**, **Grenada**), vagrant (**St Vincent and Grenadines**), vagrant? (**San Andrés**).

L 32–38 cm. Two subspecies in group, of which nominate occurs in region. Pastures, agricultural land, marshes. Noisy, conspicuous shorebird, usually in singles or pairs. Resident in very small numbers Barbados (just one bird remains) and Grenada (a few pairs), with first record in 1998 and first breeding in 2007; three, Mustique (Grenadines), May 2011 (*NAB* 65: 535), one photographed, St Vincent, Jan 2017 (eBird), pair on San Andrés, May 2018. **ID** Large, brownish, crested lapwing with blackish breast-band and bronze sheen to upperparts. **Ad** has black forehead and chin outlined white, green-grey upperparts, more rufous and iridescent wing-coverts, black-tipped reddish bill, dark red legs and deep red eye-ring. **Juv** has crown feathers tipped buff, white facial band much reduced and infused buff, diffuse breast-band, short crest and spurs, duller legs; buff fringes and barring to upperparts. **Voice** Various metallic, cackling or shrill notes and yapping sounds, e.g. "keek, keek, keek…", "kee, kee, kee…" or "chero-chero-chero". **SS** See vagrant Northern Lapwing.

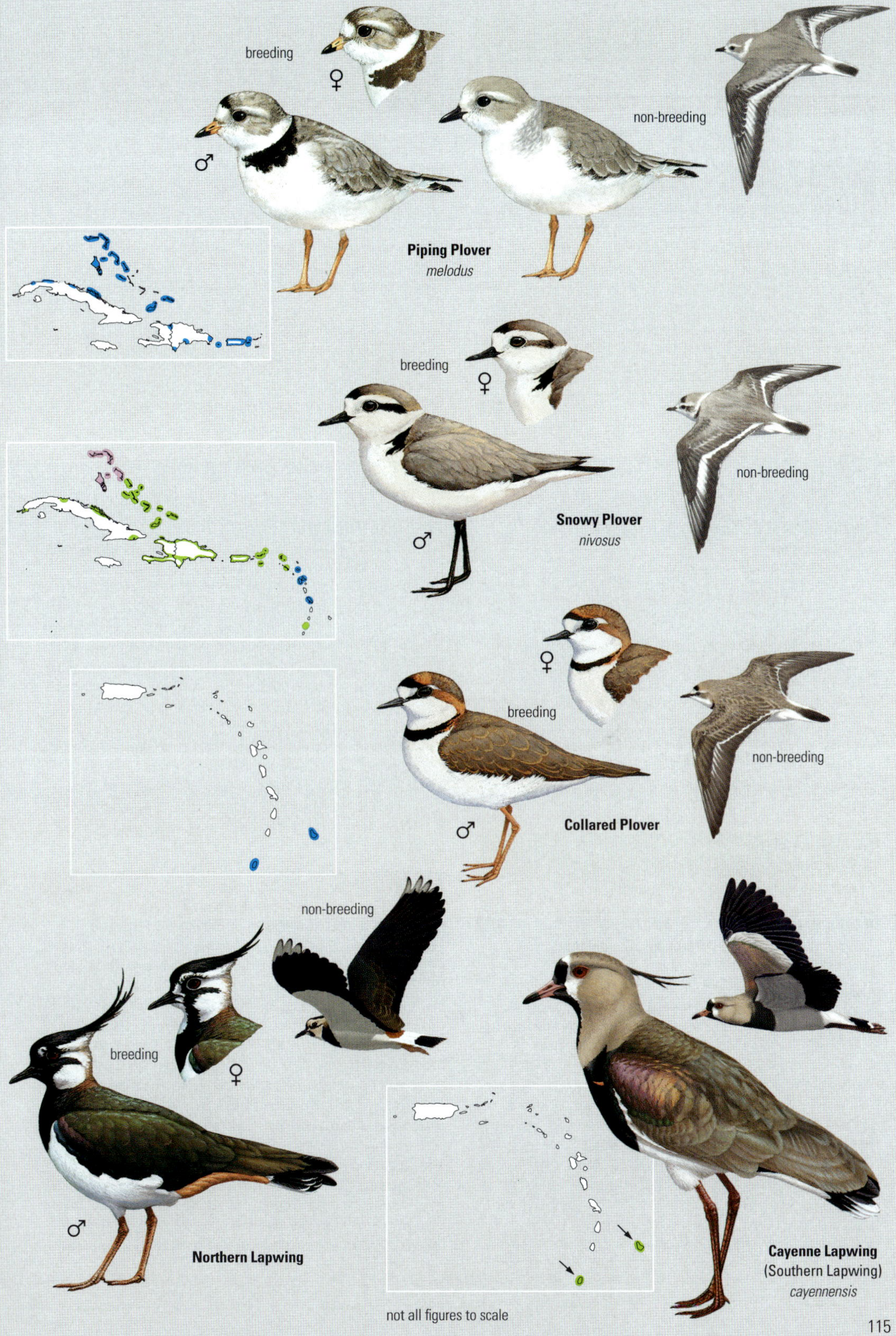

not all figures to scale

JACANIDAE
Jacanas
8 extant species, 1 in region + 1 vagrant

Northern Jacana *Jacana spinosa* `LC`
Common resident (most of **GA**), vagrant (**Puerto Rico**).
L 17–23 cm; **WS** 36 cm. Three subspecies, of which one (*violacea*) occurs in region. Breeds on floating and floating-emergent vegetation in permanent and seasonal shallow wetlands, even roadside ponds. Also grassy marshes. Walks on floating vegetation. Single old (1870s) record from Puerto Rico. **ID** Unmistakable, colourful chicken-like bird. **Ad** has black head, neck and foreparts contrasting with chestnut upperparts. Bright yellow-green flight feathers only conspicuous when wings spread. **Juv** is quite different: upperparts medium-brown, underparts white, black eyestripe and white supercilium. **Voice** Varied repertoire of loud, harsh calls, "chek" or "kak", and raucous chatter. **SS** See Wattled Jacana.

Wattled Jacana *Jacana jacana* `LC`
Six subspecies divided into two subspecies groups, of which one occurs in region.

Chestnut-backed Jacana *Jacana (jacana) jacana*
Vagrant (**Antigua**).
L 21–25 cm; **WS** 40 cm. Five subspecies in group, of which one (presumably *intermedia* on geographical grounds) occurs in region. Jan 2019 (eBird). Typically found at freshwater wetlands with floating and emergent vegetation, but will use wet farmland. **ID** Rather similar to Northern Jacana. **Ad** has red bi-lobed frontal comb and conspicuous rictal lappets contrasting with yellow bill and black head; like Northern Jacana upperparts largely reddish chestnut-brown, with yellow to pale greenish-yellow flight feathers. **Juv** is very similar to same-age Northern Jacana, but small frontal shield is bi-lobed (vs. tri-lobed), without rictal wattles. **Voice** Vocalizations very similar to Northern Jacana. **SS** Any jacana in Lesser Antilles should be carefully scrutinized, as Northern Jacana could also wander there.

SCOLOPACIDAE
Sandpipers and allies
91 extant species, 29 in region + 8 vagrants

Upland Sandpiper *Bartramia longicauda* `LC`
Fairly common to uncommon (**Barbados**, **Guadeloupe**) or very rare passage migrant, or vagrant, elsewhere.
L 26–32 cm; **WS** 64–68 cm. Monotypic. Rough grass, pastures, golf courses and airfields. Sometimes in flocks (mostly Barbados). Mostly autumn, late Aug–late Oct, much less frequent in Feb–May. **ID** Small, somewhat curlew-like shorebird with small head, short bill and legs, and long tail. Erect posture. **Ad** is buff-brown overall with extensive barring, white belly to vent. Straw-coloured legs and bill. **Juv** is similar, but fresh, even-aged feathers usually evident. Extensive pale feather fringes; tertials exhibit pale notches. **Voice** Flight call a rolling, whistled, Whimbrel-like "quip-ip-ip-ip". **SS** Drier habitat aids identification, shared only with smaller Buff-breasted Sandpiper (also tends to overfly West Indies), which lacks streaking on central breast and belly. Lesser and Greater Yellowlegs have longer yellow legs. Curlews larger with decurved bills.

Whimbrel *Numenius phaeopus* `LC`
Seven subspecies divided into two subspecies groups, of which both occur in region.

Eurasian Whimbrel *Numenius (phaeopus) phaeopus*
Vagrant (mainly **LA**).
L 40–46 cm; **WS** 76–89 cm. Five subspecies in group, of which nominate (presumably) occurs in region. At least eight records Barbados, all since 1962, Aug–Oct, Dec and Apr (Buckley *et al.* 2009); also W Cuba, Apr 2013 (Kirkconnell *et al.* in press), St Thomas and St Croix, Aug–Sept 1983 (*AB* 38: 252), St Martin, Apr 2004–Jan 2005 (Brown & Collier 2007), Guadeloupe, four records, with one bird spending 15 consecutive winters at Petite Terre Nature Reserve (Levesque & Saint-Auret 2007, *NAB* 62: 171, 492, 65: 182) and Mustique, Grenadines, Feb 2011 (*NAB* 65: 357). Perhaps most likely to be noticed with American Whimbrels, and given annual records in W Hemisphere clearly could occur more regularly. **ID** Medium-sized curlew, sharing structural characteristics of American Whimbrel. **Ad/juv** are like American Whimbrel, but greyer, with triangular white rump and lower back, and whitish underwing. **Voice** Apparently indistinguishable from American Whimbrel. **SS** See American Whimbrel.

American Whimbrel *Numenius (phaeopus) hudsonicus*
Uncommon to rare passage migrant and rare winter visitor (**throughout**).
L 40–46 cm; **WS** 76–89 cm. Two subspecies in group, of which nominate occurs in region. Coastal wetlands, saltmarshes and intertidal mudflats, especially near mangroves. Often in small flocks, sporadically larger groups. Records all months, mainly Jul–Nov and Feb–Apr. **ID** Fairly large shorebird with distinctively striped head and longish, decurved bill. In flight appears entirely buffy brown, with brownish underwing. **Ad** has blackish-brown crown with pale central stripe, conspicuous pale supercilium and blackish eyestripe, and grey-brown upperparts with buff spotting. **Juv** is very similar, with better-defined buff markings and shorter bill. **Voice** Flight call a rapid whinnying or rippling series of 6–8 evenly-pitched "queep" notes. **SS** See conspecific Eurasian Whimbrel. Much rarer Long-billed Curlew substantially larger with considerably longer bill. **TN** Previously accorded species rank by some authorities (e.g. Sangster *et al.* 2011).

Eskimo Curlew *Numenius borealis* `CR(PE)`
Formerly a rare passage migrant (**Barbados**) or vagrant.
L 29–38 cm; **WS** 70 cm. Monotypic. Grasslands, ploughed fields, saltmarshes, occasionally mudflats. Rare but regular autumn passage migrant (late Aug–early Nov) on Barbados amongst American Golden Plover flocks. Also recorded Puerto Rico, Martinique (Bond 1964), Guadeloupe, Grenadines and Grenada. Last record in 1963 on Barbados. Probably extinct. **ID** Tiny curlew, little larger than Upland Sandpiper, with rather short bill and legs. Wings extend well beyond tail at rest. In flight, cinnamon underwing contrasts with plain grey flight feathers. **Ad** has dark eyestripe (reaching to base of bill) and black crown streaked pale buff. Upperparts brown with brown-buff notches; underparts buff-cinnamon with streaked breast and Y-shaped marks on flanks. **Juv** is similar, but underparts more buff and upperparts have neat pale buff fringes and spotting. **Voice** Very poorly known; reported to give an 'oft repeated, soft, mellow, though clear whistle' in flight. **SS** American Whimbrel significantly larger (albeit difficult to judge in the field), with longer bill and (especially)

adult
juvenile
Northern Jacana
violacea

Chestnut-backed Jacana
(Wattled Jacana)
intermedia

Upland Sandpiper

Eurasian Whimbrel
(Whimbrel)
phaeopus

American Whimbrel
(Whimbrel)
hudsonicus

Eskimo Curlew

117

legs. Beware short-billed juveniles. Primaries reach to or just beyond tail at rest, and underwing-coverts plain brown. Upland Sandpiper has shorter, straighter bill, plain brown face and more attenuated wings and tail.

Long-billed Curlew *Numenius americanus* LC
Vagrant (**throughout**).

L 50–65 cm; **WS** 87 cm. Two subspecies, of which both (nominate, *parvus*) occur in region. Intertidal mudflats, lagoons, wetlands and beaches. Strides slowly over mud, stopping to pick at surface or probe deeply. Singles recorded on widely scattered islands, from Exumas (Bahamas) and Cuba S to Barbados, principally in spring and autumn, but also winter. **ID** Large cinnamon-coloured curlew with very long decurved bill and blue-grey legs. **Ad** has virtually unmarked (finely streaked) cinnamon underparts and in flight shows cinnamon underwing-coverts. Bill of ♀ longer than ♂. **Juv** is very similar, with shorter bill. **Voice** Call a loud "CUR-lee" and variations. **SS** Marbled Godwit shares cinnamon coloration, but has straight bill (often invisible on roosting birds) and black legs. American Whimbrel considerably smaller with shorter bill, dark head-stripes and brownish underwing. Vagrant Eurasian Curlew most similar, being greyer overall with more heavily streaked underparts; in flight white underwing and white lower back and rump separate.

Eurasian Curlew *Numenius arquata* NT
Hypothetical (**Bahamas**).

L 50–60 cm; **WS** 80–100 cm. Three subspecies, of which one (presumably nominate) occurs in region. Sight record: Eleuthera, Jan–Mar 1972 (Connor & Loftin 1985). Usually on muddy coasts, bays and estuaries. **ID** Large greyish-brown curlew with long bill and plain head. In flight, triangular white rump and lower back, plus pale unmarked whitish underwing. **Ad non-br** has head, neck, breast and upperparts grey-brown streaked dark, white belly and streaked flanks. **Ad br** has breast and upperparts buffy brown. ♀ averages larger than ♂, with longer bill. **Juv** has buffier breast and less streaked flanks, upperparts with buff spots and fringes. **Voice** Call a far-carrying "curLEE", often repeated as "curLEE-curLEE-curLEE-curLEE …". **SS** Long-billed Curlew and American Whimbrel lack white rump, and former has more cinnamon coloration. Vagrant Eurasian Whimbrel smaller, with strongly marked head pattern.

Bar-tailed Godwit *Limosa lapponica* NT
Hypothetical (**Virgin Is**).

L 37–41 cm; **WS** 70–80 cm. Five subspecies, of which nominate might occur in region. St Croix, Mar–May 1987 (Sladen 1989). Reports on Barbados not accepted locally (Buckley *et al.* 2009). Mainly intertidal areas, preferring mudflats, inlets, mangrove-fringed lagoons, sheltered bays. **ID** Mid-sized godwit with slightly upcurved bill. In flight, triangular white rump and lower back, plus pale unmarked whitish underwing. **Ad non-br** has pale grey-brown upperparts with whitish feather edges, grey breast with fine dark streaking, and white underparts. In flight, rather plain wings and white underwing. **Ad br** has upperparts fringed chestnut. ♂ has face, neck and underparts reddish chestnut. ♀ is larger with longer bill, face, neck and breast washed peach, belly whitish. **Juv** has neck and breast washed buff, with scattered streaks, upperparts dark with bright buff fringes, belly whitish, bill darker. **Voice** Rather silent: short nasal calls, e.g. single "kek", double "kek-kek" or "kEh-rik". **SS** Triangular white rump and lower back shared only with (also vagrant) Eurasian Whimbrel and Eurasian Curlew, both of which have decurved bills.

Marbled Godwit *Limosa fedoa* LC
Very rare migrant or vagrant (**throughout**).

L 42–48 cm; **WS** 70–80 cm. Two subspecies, of which nominate occurs in region. Intertidal mudflats, coastal pools and saltmarshes, preferably with adjoining wet savannas or grassy borders, as regularly uses grassland at high tide. Deliberate, with 'leggy' gate, probing deeply in mud from one angle, then another. Late Jul–Apr. **ID** Largest godwit, with overall cinnamon-buff plumage and very long, slightly upturned bicoloured bill. **Ad br** has cinnamon wingbar and underwing; buff underparts finely barred black; upperparts with sooty-black and reddish-buff spots and bars. **Ad non-br** has buff underparts irregularly streaked brown. **Juv** as ad non-br, but has buff fringes to upperparts and cinnamon-buff underparts with only a little streaking. **Voice** Nasal, slightly crowing or laughing calls, e.g. "ah-ha" or "ah-ahk" and single "ahk". **SS** Other godwits show white in rump or tail. Larger Long-billed Curlew similar in plumage but has decurved bill.

Hudsonian Godwit *Limosa haemastica* LC
Very rare migrant or vagrant (**throughout**).

L 36–42 cm; **WS** 67–79 cm. Monotypic. Muddy estuaries, tidal pools, coastal lagoons, flooded grassland, less often sandy beaches. Walks deliberately, probing deeply in mud. The most commonly encountered godwit, especially in Puerto Rico, Virgin Is, Guadeloupe, Martinique and Barbados and in autumn, but only vagrant elsewhere and virtually unknown in spring (Apr); mainly Sept–Oct (Jul–Nov). **ID** Darkish, fairly short-tailed godwit with slightly uptilted bill. In flight, white tail with broad black terminal band, narrow white wingbar and very dark underwing with black coverts. **Ad non-br** is plain grey-brown on breast and upperparts, white on lower belly, white supercilium. **Ad br** has whitish face and throat, deep chestnut breast, belly and undertail-coverts, irregularly barred. Dark brown upperparts with variable buff or cinnamon fringes and notches. **Juv** has dark upperparts fringed buff, neck and breast washed brownish buff, lower belly whitish. **Voice** Rarely heard away from nesting areas. **SS** Marbled Godwit larger and cinnamon-coloured with much longer bill; in flight lacks distinctive (black or white) markings. See even rarer Bar-tailed and Western Black-tailed Godwits.

Black-tailed Godwit *Limosa limosa* NT
Three subspecies divided into two subspecies groups, of which one occurs in region.

Western Black-tailed Godwit *Limosa (limosa) limosa*
Hypothetical (**St Kitts**).

L 36–44 cm; **WS** 70–82 cm. Two subspecies in group, of which one (nominate or *islandica*) claimed to occur in region. Sept 1988 (*AB* 43: 175, Steadman *et al.* 1997); no verifiable documentation available, but said to be possibly *islandica* in fading summer plumage. Records in Barbados rejected (Buckley *et al.* 2009). Could appear on intertidal mudflats, coastal lagoons, saltmarshes, saltflats, sandy beaches and freshwater habitats. Clearly likely to occur, given records in E USA (Howell *et al.* 2014) and on Trinidad (Hayes & Kenefick 2002). **ID** Tall, elegant godwit with long, rather straight bill, flesh-pink at base, and long legs. In flight, striking white-based black tail, broad white wingbar and flashing white underwing. **Ad non-br** is grey-brown above, greyish white below, with pale greyish supercilium. **Ad br** develops chestnut breast and upper belly, with dark brown bars on belly. Mantle and scapulars blotched pale red, black and grey.

Long-billed Curlew
americanus

Eurasian Curlew
arquata

♂ breeding ♀ non-breeding
Bar-tailed Godwit
lapponica

non-breeding

breeding ♂ breeding ♀ non-breeding
Hudsonian Godwit

breeding
Marbled Godwit
fedoa

non-breeding

♂ breeding ♀ non-breeding non-breeding
Western Black-tailed Godwit
(Black-tailed Godwit)
limosa

119

Juv as ad non-br, but upperparts dark grey-brown with pale chestnut-and-buff fringes, neck and breast pale cinnamon. **GV** Races not reliably distinguished in field in non-breeding plumage; *islandica* is darker red in breeding plumage with shorter bill than nominate. **Voice** Short nasal "kek" or "kek-kek", and more grating subdued notes. **SS** Non-breeder very similar to same plumage of Hudsonian Godwit, which is shorter-legged with noticeably uptilted bill, less prominent white wingbar and black underwing. Bar-tailed Godwit has triangular white rump and lower back. Marbled Godwit larger and cinnamon-coloured, with much longer bill; in flight lacks distinctive (black or white) markings.

Ruddy Turnstone *Arenaria interpres*   LC
Common to fairly common non-breeding visitor (**throughout**).
L 21–26 cm; **WS** 50–57 cm. Two subspecies, of which one (*morinella*) occurs in region. Chiefly coastal, on rocky and shingly shores, short-grass saltmarshes, seaweed-strewn sandy beaches, and less frequently mudflats. Turns over seaweed and pebbles, or excavates prey. Commonest as passage migrant and winter visitor, fewest in Jun. **ID** Robust shorebird with short orange legs and wedge-shaped bill. Distinctive pattern in flight: white back, uppertail-coverts, wingbar and patch on inner wing, contrasting with otherwise dark upperparts. **Ad non-br** is dark greyish brown and blackish above; head and breast mostly dark grey-brown, white underparts. **Ad br** has head to breast black and white, upperparts bright chestnut-orange with black-brown patches, underparts clean white. **Juv** is like ad non-br, but browner above with buff fringes and paler head. **Voice** Contact and flight call a short chuckling staccato rattle, "tuk-tuk-i-tuk-tuk". **SS** None.

Red Knot *Calidris canutus*   NT
Uncommon passage migrant and rare winter visitor (**throughout**).
L 23–25 cm; **WS** 45–54 cm. Six subspecies, of which three (*rufa*, *roselaari*, *islandica*) either do or could occur in region. Prefers large intertidal mudflats or sandflats. Usually in groups of 1–5, sometimes tens. Walks slowly, feeds mainly by shallow probing or picking. Records all months, mainly Greater Antilles and Barbados (probably involving both *rufa* and *islandica*); mostly Aug–Nov and, to lesser extent, Mar–early May. **ID** Distinctive, large, compact yet long-winged *Calidris*, with shortish green legs and straight mid-length bill. In flight shows whitish rump. **Ad non-br** is plain grey above with narrow white fringes to larger feathers, underparts white with grey bars on breast and flanks. **Ad br** has underparts rich rusty chestnut, upperparts mainly blackish mixed pale to rufous-chestnut. **Juv** is like ad non-br, but breast washed buffish, upperparts brownish grey, and wing-coverts and scapulars have buff fringes and dark subterminal bars. **GV** Races indistinguishable in field. **Voice** Rather silent, occasionally gives short subdued "kuk" or louder nasal upslurred "kweh-kweh". **SS** Non-br Stilt Sandpiper is superficially similar in flight and at rest, but has much longer legs and longer, decurved bill.

Ruff *Calidris pugnax*   LC
Very rare (almost annual) passage migrant (**Puerto Rico**, **Guadeloupe**, **Barbados**), vagrant elsewhere.
L ♂ 26–32 cm, ♀ 20–25 cm; **WS** ♂ 54–58 cm, ♀ 48–52 cm. Monotypic. Muddy margins of lakes and pools, damp grasslands. Probes in shallow water and mud. Powerful flier. Usually alone in region. Most records Aug–Oct (occasional late Jun/Jul), much less common in spring, Mar–May, and even less so in winter (Guadeloupe, Puerto Rico, Jamaica). **ID** Large, erect wader with mid-length, slightly curved bill and long legs. ♂ is larger than ♀. Bold V-shaped white rump in all plumages. **Ad non-br** is pale grey-brown above, usually with whiter face and throat, dusky-buff breast. Bill dark with orange base, legs dull orange. **Ad br** shows extreme sexual dimorphism. ♂ has head-tufts and ruff variably coloured buff, chestnut, dark purple, black or white, often barred or flecked. Mantle and scapulars black to brown, buff, chestnut, ochre or white. Underparts usually dark, lower belly and vent white. Yellow to brown facial warts. Bill brown to dull orange and legs yellow-green to dark orange. ♀ is like ad non-br with dark brown, scaled upperparts. **Juv** (most frequently observed plumage) is similar to ad non-br but has buff-fringed dark brown upperparts and rufous-buff foreneck, breast and belly, face mainly buff with pale throat. Bill black, legs pinkish orange to green or grey. **Voice** Largely silent. **SS** A distinctive wader. Dark central rump surrounded by white V is unique and obvious in flight.
AN Reeve (♀).

Stilt Sandpiper *Calidris himantopus*   LC
Uncommon to locally common non-breeding visitor (**throughout**).
L 18–23 cm; **WS** 38–47 cm. Monotypic. Typically at freshwater wetlands, flooded fields, shallow ponds and pools, sewage lagoons and brackish marshes. On coast, will use intertidal mudflats. Sometimes in hundreds, but mostly in smaller groups, often with dowitchers. Records all months, especially Aug–Apr, with largest numbers passing Greater Antilles and Virgin Is, and winter records mainly in first-named group. **ID** Larger, rather distinctive *Calidris*, with long, slightly decurved bill and long, yellow-green legs clearly reaching beyond tail in flight. **Ad non-br** has plain grey upperparts with white-fringed wing-coverts, and white supercilium. Underparts white with grey-streaked lower neck, breast and flanks. **Ad br** has chestnut stripe on head-sides, broad white supercilium, dark upperparts fringed rufous and whitish. Neck and upper breast streaked dark brown, below barred. **Juv** has upperparts dark brown fringed rufous or whitish buff, wing-coverts greyer with buff fringes, foreneck and breast washed buff and streaked, belly white. **Voice** Often silent in winter, but commonest call, in flight, a rattling "grrrt" or "querp". Also a whistled "kueu". **SS** Non-br and juv superficially resemble chunkier dowitchers (often together), which have longer, straight bills and shorter legs. Both yellowlegs have straight bills. See rarer Dunlin and Curlew Sandpiper, which also shows white rump in flight, but has white wingbar and on ground appears less leggy with a blunter-tipped, straighter bill.

Ruddy Turnstone
morinella

breeding
♀
non-breeding
♂

Red Knot
rufa

breeding
non-breeding

Ruff

breeding ♂
grey bird
breeding ♀
rufous bird
non-breeding

Stilt Sandpiper

breeding
non-breeding

121

Curlew Sandpiper *Calidris ferruginea* NT
Very rare passage migrant (**Barbados**), vagrant elsewhere.
L 18–23 cm; **WS** 38–46 cm. Monotypic. Intertidal mudflats, coastal lagoons and saltmarshes; also inland, at muddy edges of wetlands. All records involve singles (mainly juv), Sept–Oct and Apr–Jun, but also Nov–Jan. No definite reports N & W of Puerto Rico; details of two sight records from W Cuba involving large numbers (Navarro Pacheco 2018) that would be quite unprecedented anywhere in the region are awaited. **ID** Mid-sized sandpiper with longish neck and legs, and long decurved bill. Black bill and legs. White rump. **Ad non-br** is plain grey above, with contrasting white supercilium, white underparts, breast-sides washed grey. **Ad br** has head, neck and all underparts rusty rufous to deep chestnut-red, with dark-streaked crown. Mantle and scapulars dark brown with chestnut and whitish fringes, wing-coverts greyer. **Juv** has brownish upperparts with whitish fringes, breast washed peachy buff with fine dark streaking. **Voice** A pleasant, soft, twittering "chirrup" in flight. **SS** Stilt Sandpiper averages marginally larger with longer greenish legs. Dunlin smaller and more compact, with shorter legs and bill, and dark-centred rump.

Sanderling *Calidris alba* LC
Fairly common winter visitor, rare in summer (**throughout**).
L 19–21 cm; **WS** 35–39 cm. Two subspecies, of which one (*rubida*) occurs in region. Open sandy beaches exposed to sea. Characteristically scurries to and fro at edge of waves, feeds by rapid probing and pecking. Records year-round, mostly in autumn/winter. **ID** Chunky, mid-sized *Calidris* with shortish, thick bill. Strikingly pale with black bill and legs. In flight, white wingbar in all plumages. Unique among Scolopacidae, lacks a hallux. **Ad non-br** is very white, with pale grey upperparts, white face and underparts, lesser coverts darker. **Ad br** (rarely seen) has head, upperparts and breast rufous, with dark-streaked head and breast, and bold black marks on upperparts. **Juv** has buffy-white head and breast, streaked crown and breast-sides, upperparts marked black and grey, with almost black shoulders. **Voice** Typical call a very short steeply upslurred "twick". **SS** None share behaviour and habitat plus very white plumage.

Dunlin *Calidris alpina* LC
Very rare passage migrant/winter visitor or vagrant (**throughout**).
L 16–22 cm; **WS** 33–40 cm. Ten subspecies, of which at least one (presumably *hudsonia*) occurs in region, although a specimen (from Barbados) is labelled (perhaps incorrectly) *pacifica*. One of Palearctic races might easily occur. Tidal mudflats, lagoons and brackish wetlands. Usually alone or in flocks (of up to 200), often with other *Calidris*. Records Jan–May in Cuba, but also autumn in Lesser Antilles and could occur any month. Mainly Bahamas, Greater Antilles, Cayman and Virgin Is. Much scarcer Lesser Antilles. **ID** Smallish, short-necked sandpiper with hunched posture, and longish, distinctly decurved bill, especially at tip. Resembles a large Western Sandpiper. **Ad non-br** has plain brownish-grey head, breast and upperparts, white chin, throat and rear underparts. Short, pale eyebrow. **Ad br** has upperparts rufous, breast white with heavy black streaks, belly black. **Juv** has streaked pale buff breast. Flanks and sides of white belly display lines of bold brownish spots. **Voice** Commonest call a short reedy or gravelly trill, often given in flight, "kreet". **SS** Significantly larger than 'peeps' (Least, Semipalmated and Western Sandpipers), but lone non-br hard to separate from extremely long-billed Western, which is otherwise smaller and greyer.

Baird's Sandpiper *Calidris bairdii* LC
Very rare passage migrant or vagrant (**throughout**).
L 14–17 cm; **WS** 36–40 cm. Monotypic. Favours upper, drier margins of wetlands, where picks insects from mud. Most records of singles, Aug–Oct, from Bahamas and Cayman Is S to Barbados, but basically vagrant except on those islands mapped. Presumably mainly overflies region, especially in spring. **ID** Medium-small sandpiper with very long wings projecting well beyond tail at rest. Fairly short, slightly decurved black bill and short, black legs. Lower back, rump and uppertail dark. **Ad non-br** has dull brownish-grey upperparts, buffy head and brown-streaked breast. White chin and belly. Face rather expressionless. **Ad br** has black-brown crown, nape, mantle and scapulars with broad buff and brown fringes, ear-coverts and breast buff with brown streaks. **Juv** is like ad non-br, but has narrow white fringes to feathers of upperparts. **Voice** Typical call when flushed a rather low-pitched, slightly rolled or trilled "prr-reet" or "kree". **SS** Main confusion species, White-rumped Sandpiper, separable by rump colour or, if latter unseen, by having pale base to slightly longer bill, and streaked flanks. Smaller peeps, which are superficially similar, all have shorter wings that do not project beyond tail. Larger Pectoral Sandpiper has longer, yellowish legs, and Dunlin has heavier bill, shorter wings and more conspicuous wingbar, while juv lacks scaly appearance of juv Baird's.

Little Stint *Calidris minuta* LC
Vagrant (**Montserrat**, **Barbados**).
L 12–14 cm; **WS** 28–31 cm. Monotypic. Could occur at any freshwater or brackish wetland. Four records Barbados: Apr–May 1997, May 1997 (another), May 1999 and May 2002 (Buckley *et al.* 2009); Montserrat, Aug 2003 (*NAB* 58: 159). Reports of three, Antigua, Nov 1975 (Holland & Williams 1978) and one Jul 1976 (Morrison 1980) unacceptable, given lack of descriptions (Iliff & Sullivan 2004). **ID** Tiny, compact stint with short black bill and legs. **Ad non-br** has brownish-grey upperparts mottled dark and fringed pale. Crown grey, streaked dark, eyestripe and breast-sides dull grey, rest of face and underparts white. **Ad br** has upperparts dark brown with pale rufous fringes, mantle with yellowish edges forming distinct V. Head, neck and breast rufous streaked brown, rest of underparts to chin white. Split white supercilium characteristic. **Juv** has tertials, lower scapulars and wing-coverts centred dark brown, edged rufous and white; white edges of mantle form distinct V. Head and breast suffused pale buff, crown, neck and broad eyestripe pale grey with fine streaking, and split supercilium white. **Voice** Commonest call a short staccato "tip". **SS** Only wader of similar size (a trifle smaller) is Least Sandpiper, which has yellowish or greenish legs and longer tail (projecting slightly beyond wings at rest). Very slightly larger Semipalmated Sandpiper has shorter primary projection, partially webbed feet, blunter-tipped bill, and in non-br plumage has streaking on breast most obvious at sides (never across entire chest) and lacks obvious dark feather centres above; juv lacks obvious rufous fringes above and pale tramlines.

Curlew Sandpiper

Sanderling
rubida

Dunlin
hudsonia

Baird's Sandpiper

Little Stint

123

Least Sandpiper *Calidris minutilla* `LC`
Common passage migrant and uncommon to rare winter visitor (**throughout**).

L 13–15 cm; **WS** 33–35 cm. Monotypic. Muddy inland and coastal wetlands. Usually in small dispersed flocks, often with Semipalmated and Western Sandpipers, albeit favouring drier areas. Year-round, but least common Jun–Jul. **ID** Smallest wader. Tiny, compact, short-winged, mouse-like stint with short, slightly decurved black bill and yellowish legs. **Ad non-br** has brown head and upperparts with smudgy dark centres to scapulars. Faint grey eyestripe. Brownish breast, rest of underparts white. **Ad br** has upperparts dark brown with rufous fringes, rufous cheeks. Breast finely streaked brown, rest of underparts white. **Juv** is brighter, appearing rufous-brown overall; even-aged feathers create near-scaled upperparts. Brownish streaks on breast sharply demarcated from otherwise clean white underparts. **Voice** Flight call a shrill, reedy "kreeep". **SS** Leg colour and dark-breasted appearance separate from all other small 'peeps'.

White-rumped Sandpiper *Calidris fuscicollis* `LC`
Uncommon to rare passage migrant and very rare winter visitor (**throughout**).

L 15–18 cm; **WS** 36–38 cm. Monotypic. Beaches, mudflats, saltmarshes, rice fields and marshes. Migrants often alone, but sometimes in flocks of up to 100, especially Barbados and Guadeloupe, Aug–Oct; has occurred all months. Clearly, most overfly region between North and South America. **ID** Small sandpiper with long wings projecting beyond tail at rest. White uppertail-coverts contrast with dark rump and tail. Fairly short, slightly decurved black bill with very pale orange base to mandible. Legs dusky green to black. **Ad non-br** has plain grey head and upperparts with faint streaks and paler fringes, pale grey breast with faint dark streaks. Broad whitish supercilium. **Ad br** has crown, cheeks, mantle and scapulars centred dark brown and edged rufous-pink and grey, wing-coverts paler. Neck- and breast-sides and flanks spotted and streaked brown. **Juv** is like ad br, but upperparts brighter with more white tips. Breast streaked and washed buff-grey. **Voice** When flushed, gives a thin, high-pitched "jeet" or "tzeep". **SS** Baird's Sandpiper (which see) shares long wings, but has dark-centred rump; calls differ. Smaller peeps (superficially similar) all have shorter wings that do not extend beyond tail.

Buff-breasted Sandpiper *Calidris subruficollis* `NT`
Very rare passage migrant (**LA**), vagrant (**GA**, **O**).

L 18–20 cm; **WS** 43–47 cm. Monotypic. Fields, pastures, golf courses and other short-grass areas. Usually observed alone or in small flocks. Reasonably regular Guadeloupe and Barbados, very scarce elsewhere, especially Greater Antilles. Mostly Sept–Nov, fewer in Aug, much scarcer Mar–Apr (once late Jun on Barbados). **ID** Elegant, erect *Calidris* with small, round head, fairly long neck, warm buff face and underparts, and blackish spots on crown, hindneck and mantle. Rear belly white. Underwing mostly silvery white with black band on coverts. Short, thin black bill, and bright yellow legs. **Ad non-br** has dark brown centres with broad brown fringes to upperparts. **Ad br** has narrower, buffier fringes above. **Juv** is paler below, especially towards vent, with narrower fringes and dark subterminal line to feathers of upperparts forming neat scalloping. **Voice** Rather quiet; the most characteristic and frequently heard sound a series of "tick" notes, like stones being knocked together. A low "cheep" when flushed or in flight. **SS** Obviously larger Ruff has V-shaped patch of white around rump. Pectoral Sandpiper plumper and pot-bellied, with longer bill and greenish legs.

Pectoral Sandpiper *Calidris melanotos* `LC`
Uncommon passage migrant and rare winter visitor (**throughout**).

L 19–23 cm; **WS** 37–45 cm. Monotypic. Freshwater and brackish wetlands, coastal and inland, as well as grassy margins and flooded grassland. Usually alone or in small flocks. Commonest Aug–Nov, less so Jan–Apr. **ID** Mid-sized, long-necked and pot-bellied sandpiper, with heavily streaked buff breast sharply demarcated from white belly. **Ad non-br** has black-brown crown, mantle and scapulars with buff fringes, paler wing-coverts. Greenish-yellow legs. **Ad br** is brighter with chestnut fringes and white tips above. **Juv** is like ad br but upperparts have paler, narrower fringes, forming clear V on mantle and scapulars. Finer streaking on breast. **Voice** Alarm or flight note a reedy "kreet" or "churrk, and sharp "trit-trit". **SS** See Buff-breasted Sandpiper and Ruff.

Semipalmated Sandpiper *Calidris pusilla* `NT`
Fairly common passage migrant and winter visitor (**throughout**).

L 13–15 cm; **WS** 34–37 cm. Monotypic. Tidal mudflats, sometimes sandy beaches, occasionally inland. Usually in large flocks; picks food from surface. Year-round, but scarcest in Jun–Jul. **ID** Small, plumpish stint with short black bill and black legs. Toes partially webbed. Relatively drab in all plumages. Often the default 'peep', being the commonest and of 'average' characters. **Ad non-br** has smooth grey-brown upperparts and smudgy breast-sides. Grey cap, faint whitish supercilium. **Ad br** has dark brown upperparts with pale rufous fringes, but very few of latter on head. Overall less rufous than congeners. **Juv** is similar to ad br but even-aged feathers produce neater, more scalloped look; white supercilium more prominent. Upperparts mostly grey with some rufous on crown and scapulars. **Voice** A low-pitched "chrrp". **SS** See Western Sandpiper, which in non-breeding plumage is very similar and short-billed individuals can be indistinguishable. Least Sandpiper has yellowish legs. White-rumped and Baird's Sandpipers have long wings, extending beyond tail at rest. See Little Stint.

Western Sandpiper *Calidris mauri* `LC`
Uncommon passage migrant and winter visitor (**throughout**).

L 14–17 cm; **WS** 35–37 cm. Monotypic. Tidal mudflats, favouring wetter areas than Semipalmated Sandpiper. Often in large flocks; picks food from surface, but more prone to probe than Semipalmated. Mostly Aug–Apr, but records year-round. **ID** Small, chest-heavy stint with long bill typically drooped at tip. Toes partially webbed. **Ad non-br** has uniform pale grey upperparts. Head, neck and breast-sides finely streaked grey, rest of underparts white. **Ad br** has rufous-chestnut crown and ear-coverts, chestnut-and-black scapulars with pale fringes. Neck, breast and flanks heavily streaked dark brown, with black chevrons on breast-sides. **Juv** has rufous on upper scapulars and grey on lower scapulars. Crown buff-grey and streaked, mantle blackish fringed rufous and white. Underparts white with breast washed very pale orange-buff and finely streaked. **Voice** A thin, high-pitched "jeet". **SS** Semipalmated Sandpiper very similar but usually shorter-, blunter- and thicker-billed (longest in eastern ♀, overlapping with ♂ Western), with swollen tip, and streaking on crown and breast in non-br plumage is coarser. Posture at rest generally more 'balanced' vs. front-heavy Western, which latter also appears to have a larger, squarer head. Ad br Semipalmated Sandpiper lacks obvious rufous on head and upperparts, and br Western also has diagnostic chevrons on breast and upper flanks. Voice also useful in separating them. Non-br/juv Dunlin is larger and longer-billed, as well as appearing longer-bodied but smaller-headed. For differences vs. other small *Calidris* see Semipalmated Sandpiper.

breeding | non-breeding
Least Sandpiper

breeding | non-breeding
White-rumped Sandpiper

breeding | non-breeding
Buff-breasted Sandpiper

♂ breeding | ♀ breeding | non-breeding
Pectoral Sandpiper

breeding | non-breeding
Semipalmated Sandpiper

breeding | non-breeding
Western Sandpiper

125

Short-billed Dowitcher *Limnodromus griseus* LC
Fairly common (**Bahamas**, **GA**, **Barbados**) to uncommon passage migrant and winter visitor (**LA**).
L 25–29 cm; **WS** 45–51 cm. Three subspecies, of which two (nominate, *hendersoni*) occur in region. Tidal mudflats bordered by mangroves. Records overwhelmingly coastal. Gregarious, often in large flocks, sometimes with Stilt Sandpiper and other longer-legged waders. Walks in thigh-deep water and probes with characteristic 'sewing-machine' action. Records all months. Race *hendersoni* perhaps little more than vagrant (very few records in Lesser Antilles especially), but difficulties of identifying unquestionably suggest it is more numerous than records currently indicate. **ID** Stocky wader with snipe-like bill and comparatively short greenish legs. Wedge-shaped white rump and lower back obvious in flight. **Ad non-br** has plain grey upperparts and breast, pale supercilium, dark eyestripe and whitish underparts. **Ad br** has upperparts blackish with chestnut fringes. Underparts rusty brown, fading to white on belly to vent, spotted black on neck and densely barred black on breast-sides. **Juv** has broad chestnut-buff edges and bars on crown, upperparts and tertials. Obvious capped effect. Face, foreneck and breast buffish, white at rear. Tertials irregularly barred and notched buff. **GV** Race *hendersoni* has all-rufous underparts, with similar black markings to nominate, and broad fringes to mantle and scapulars; indistinguishable in winter plumage. **Voice** Call, given frequently in flight, and useful in separating Long-billed Dowitcher, a clean and fluid, "tu-tu-tu", usually uttered as trio of notes. **SS** See Long-billed Dowitcher, which should be separated with great care.

Long-billed Dowitcher *Limnodromus scolopaceus* LC
Uncommon passage migrant and rare winter visitor (**Bahamas**, **Cuba**, **Cayman Is**, **Puerto Rico**), vagrant elsewhere.
L 24–30 cm; **WS** 46–52 cm. Monotypic. Freshwater or brackish habitats. Behaviour similar to Short-billed Dowitcher. Identification difficulties have compromised understanding of its status. Records Aug–Apr. **ID** Stocky, snipe-like wader with long, straight, green-based bill and greenish-yellow legs. In flight, white lower back contrasts with rest of upperparts. **Ad non-br** has grey upperparts and breast, white supercilium, dark eyestripe and whitish underparts. **Ad br** has upperparts blackish with chestnut feather fringes; when fresh, scapulars boldly tipped white. Below (including belly) rusty brown, spotted black on neck and densely barred black on breast-sides. **Juv** has neat rusty-fringed upperparts, notably scapulars. Tertials dark grey with narrow chestnut edges. Obvious capped effect. Face, foreneck and breast buffish. Underparts white at rear. **Voice** Commonest call, given from ground and in flight, a shrill penetrating "keek", initially doubled or tripled then repeated singly. **SS** Short-billed Dowitcher very similar, especially race *hendersoni* in breeding plumage; best separated by voice. Pale-based bill tends shorter, especially in ♂ (overlap in ♀), foreneck with few or no spots, tail pattern variable, but white bars normally broader than black ones. Juv has dark brown tertials irregularly barred and notched buff.

Wilson's Snipe *Gallinago delicata* LC
Fairly common winter visitor (**throughout**).
L 26–28 cm; **WS** 40–47 cm. Monotypic. Open fresh or brackish marshes, wet meadows. Relatively immobile, probing deeply in shallow water or soft mud. Often seen only when flushed at short range, escaping in rapid zigzag flight. Most records Oct–Apr (Aug–May) in Bahamas and Greater Antilles, less frequent Virgin Is and Lesser Antilles (other than Guadeloupe and Barbados, where reasonably numerous). **ID** Long slender bill and cryptic plumage. Narrow white trailing edge to secondaries. Underwing-coverts mostly dark with narrow whitish barring. **Ad** has crown striped black and buffy. Above a mix of brown, black and grey, forming spots and bars. Pale-coloured spots tend to form four lines on back. Tail appears russet. Underparts mostly white, but neck and breast heavily streaked or spotted brown. **Juv** has mantle and scapular lines narrower, and wing-coverts (especially inner median coverts) fringed pale buff with black subterminal lines. **Voice** A rasping "scaap" when flushed (or in flight on migration), plus a hard sharp "jick" (commonly from ground, also in flight). **SS** None.

Jack Snipe *Lymnocryptes minimus* LC
Vagrant (**Barbados**).
L 17–19 cm; **WS** 34–42 cm. Monotypic. Juv ♂ shot, Nov 1960 (Buckley *et al.* 2009). In Eurasia uses brackish and freshwater habitats. Often bobs while foraging. Sits tight until almost trodden upon, then flutters off, often circling and landing close by; behaviour more reminiscent of a rail than a snipe. **ID** Smallest snipe, with large head but relatively short bill. Short rounded wings with white trailing edge. From all other snipes by wedge-shaped tail without white. Split supercilium, but no central crown-stripe. **Ad** has brown stripes on undertail-coverts. **Juv** is very similar, but undertail-coverts white. **Voice** Generally silent in winter; occasionally a low, weak "etch" when flushed. **SS** Wilson's Snipe is considerably larger, with much longer bill.

Wilson's Phalarope *Steganopus tricolor* LC
Very rare passage migrant or vagrant (**throughout**).
L 22–24 cm; **WS** 35–43 cm. Monotypic. Shallow ponds and salt lagoons. Typically swims, often spinning in tight circles, but regularly wades or walks. In small flocks of up to four, but can be highly gregarious, mixing with other shorebirds, e.g. Stilt Sandpipers and Lesser Yellowlegs. Records Aug–May, but most on southbound passage and very few in winter; mapped range shows only those territories with more or less regular records. **ID** Largest, most slender phalarope, with longest legs and bill. Bill needle-like. Reversed sexual dimorphism. **Ad non-br** is uniform pale grey above, white below, with white forehead and supercilium. White uppertail-coverts and rump conspicuous in flight; no wingbar. Legs yellow. **Ad br** has greyish-white cap and nape, black band from bill to breast-sides, blue-grey mantle and wing-coverts, reddish-chestnut edge to mantle and scapulars. Foreneck and upper breast rich orange. Legs dark. ♂ is much darker and generally duller above. **Juv** has dark brown upperparts with buff fringes, producing somewhat scaled appearance. Breast-sides washed buff. Legs yellow. **Voice** Not very vocal; occasional nasal grunts or quacking "week". **SS** Other phalaropes are darker with shorter, stouter bills; in flight they have dark rumps and white wingbars. Stilt Sandpiper superficially similar to non-br, but has thicker, decurved bill. Lesser Yellowlegs has longer legs. **TN** Some authorities place in genus *Phalaropus*.

hendersoni breeding

griseus breeding

non-breeding

Short-billed Dowitcher

breeding

non-breeding

Long-billed Dowitcher

Wilson's Snipe

Jack Snipe

♀ breeding

♂

non-breeding

Wilson's Phalarope

127

Red-necked Phalarope *Phalaropus lobatus* `LC`
Very rare passage migrant or vagrant (**throughout**).
L 18–19 cm; **WS** 32–41 cm. Monotypic. Coastal lagoons. Almost always on water, sometimes in small flocks with other shorebirds. Records Sept–Jan and Apr–May; most Cuba, Hispaniola and Bahamas. **ID** Smallest phalarope with needle-like bill and slender neck. In flight dark rump and white wingbar. Reversed sexual dimorphism. **Ad non-br** has dull blue-grey upperparts with white fringes forming obvious V on mantle. Head white with black around eye extending to form post-ocular patch. White below with faintly streaked lower flanks. **Ad br** has slate-grey head, neck and breast-sides, with bright orange-red horseshoe collar, white throat, and golden-buff fringes to upperparts forming lines on mantle-sides. ♂ is much duller with browner head, neck and upperparts. Often has narrow supercilium. **Juv** has brown crown, hindneck and eye patch, black-brown upperparts with bright buff fringes, buff-washed breast, and rest of underparts white. **Voice** A sudden chirping "prek" or "chep", sometimes an almost disyllabic, harsh "cherrp". **SS** Wilson's Phalarope larger with longer, finer bill; non-br much paler. See Red Phalarope.

Red Phalarope *Phalaropus fulicarius* `LC`
Vagrant (**Bahamas, Cuba, Puerto Rico, LA**).
L 20–22 cm; **WS** 37–40 cm. Monotypic. Usually at sea, but singles or pairs can occur on ponds and lagoons. Almost always on water, often spinning in tight circles. San Salvador, Bahamas, Mar 2009 (*NAB* 63: 520), Old Bahama Channel, Jul 2010 (*NAB* 64: 657), Cuba, Dec 1963, Jan 1967, Dec 2010 (Kirkconnell *et al.* in press), Puerto Rico, Apr 2011, Jan 2016 (eBird), Virgin Is (Raffaele *et al.* 1998), Antigua, Oct 1983 (Bond 1984) and Guadeloupe, Nov 2012 (eBird). **ID** Chunky with relatively broad, heavy and mainly yellow bill. In flight dark rump and white wingbar. Reversed sexual dimorphism. **Ad non-br** is pale unstreaked blue-grey above, white below, with black hindcrown and black eye patch. **Ad br** ♀ is mainly chestnut-red with blackish-brown crown and foreface, grey central nape, white face-sides, blackish-brown upperparts with cinnamon and buff fringes. **Ad br** ♂ is duller, with streaked crown and mantle, white on face less pure, duller underparts often mixed white. **Juv** has upperparts like ad br ♂ and head pattern as ad non-br, but browner. Face, neck and breast-sides pink-buff, belly white. **Voice** Call a short, sharp, high-pitched "pit" or "wit". **SS** Red-necked Phalarope smaller and daintier with thinner, all-black bill; non-br has white fringes to upperparts, obvious white V on mantle and less black on hindcrown. Non-br Wilson's Phalarope is larger and paler with needle-like bill. **AN** Grey Phalarope.

Terek Sandpiper *Xenus cinereus* `LC`
Vagrant (**Barbados**).
L 22–25 cm; **WS** 57–59 cm. Monotypic. May 2000 (*NAB* 54: 335). Uses intertidal mudflats, estuaries and shallow freshwater wetlands. Bobs rear body like Spotted Sandpiper and flies low, skimming over water with flicking downbeats to wings, also like latter. **ID** Distinctive by long upcurved bill and short orange to greenish-yellow legs. **Ad br** has grey-brown upperparts with almost black centres to feathers, particularly on scapulars, streaked crown, hindneck, cheeks and breast-sides. Broad white trailing edge to wings, but no wingbar. **Ad non-br** is plainer, brownish grey above with pale fringes, paler head. **Juv** has upperparts darker and browner with narrow buff fringes, black scapular lines less prominent. **Voice** A rippling, melodious "hühühühü" in contact and flight, and fluty "to-li" or "wee-we" in alarm. **SS** Spotted Sandpiper has shorter, decurved bill. Both yellowlegs have much longer legs.

Spotted Sandpiper *Actitis macularius* `LC`
Common passage migrant and winter visitor (**throughout**).
L 18–20 cm; **WS** 37–40 cm. Monotypic. Freshwater pools, lakes, rivers and marshes, also sandy beaches, muddy lagoons and mangroves. Constantly bobs rear body. Flies low over water with burst of shallow wingbeats followed by glide on stiff wings. Records every month, but rare late May–early Jul. **ID** Distinctive, small, long-tailed, short-legged sandpiper. Darker eyestripe and narrow whitish supercilium. Bill flesh to orange with black tip. Legs yellow-ochre. In flight dark rump and weak white wingbar. **Ad non-br** has plain olive-brown upperparts and breast-sides, plus white underparts. **Ad br** has neatly dark-spotted underparts, brown upperparts with dark barring. **Juv** is like ad non-br, but upperparts greyish with narrow buff fringes, wing-coverts have distinct buff-and-brown fringes. Tertials plain brown, greater coverts barred at tips. **Voice** A clear, ringing "peet-weet" or "peet-weet-weet". **SS** Tail of much darker Solitary Sandpiper does not project beyond wingtips, and eyestripe and supercilium do not reach behind eye. Note quite different wing and tail pattern in flight, and flight action also very different.

Green Sandpiper *Tringa ochropus* `LC`
Hypothetical (**Guadeloupe**).
L 21–24 cm; **WS** 57–61 cm. Monotypic. Sept 2014; brief description (*NAB* 69: 169). Similar habitats to Solitary Sandpiper, almost always fresh water. Behaviour also similar. **ID** Old World equivalent of Solitary Sandpiper. Wings dark green, underwing blackish; rump white, tail white with broad black bars. **Ad non-br** has foreneck and breast streaked grey-brown, underparts white. **Ad br** has neat white spots on upperparts, white breast with sharply defined olive streaks. **Juv** has more prominent buff spots above. **Voice** Call, often given when flushed or in flight, a musical "weet" or "klu-eet", often becoming a more melodious ringing trisyllabic "klueet-weet-weet". **SS** Solitary Sandpiper has dark green rump and is slightly slimmer with longer legs and bill. In flight also appears narrower-winged and at rest more attenuated at rear, with longer primary projection; eye-ring is also usually more conspicuous in Solitary and upperparts typically seem more spotted than in Green. Call more muted, lacking distinctive ringing quality of Green Sandpiper.

Solitary Sandpiper *Tringa solitaria* `LC`
Uncommon to fairly common passage migrant and winter visitor (**throughout**).
L 18–21 cm; **WS** 55–59 cm. Two subspecies, of which both (nominate, *cinnamomea*) occur in region. Mostly at freshwater wetlands, often in isolated ditches and temporary stagnant pools, also reservoirs, wet meadows, quiet streams and swamps. Lives up to its name, almost always alone. Mainly Aug–May, but isolated records year-round. **ID** Slim, medium-sized, dark-looking sandpiper, initially suggesting a small, compact yellowlegs with long, tapering body. Thin, dark, green-based bill; dull green legs. Conspicuous white eye-ring, white loral-stripe and dark lores. Only *Tringa* with both dark rump and underwing. **Ad non-br** has dark olive-green upperparts, finely spotted white. Head, neck, mantle and breast more smudgy olive-brown. **Ad br** has larger white spots above, and clearly streaked head, mantle and breast. **Juv** is like ad br, but more olive-brown head and breast. **GV** Race *cinnamomea* (rarely distinguishable in field) has broader tail barring and weaker loral-stripe. **Voice** Call a piping "peet" or "pleet-weet-weet", higher-pitched than Spotted Sandpiper. **SS** Both yellowlegs larger, with yellow legs and white rump. See superficially similar Spotted Sandpiper.

breeding

♀ breeding

♂ breeding

♀ breeding

non-breeding

non-breeding

Red-necked Phalarope

Red Phalarope

breeding

non-breeding

non-breeding

breeding

Spotted Sandpiper

Terek Sandpiper

non-breeding

juvenile

non-breeding adult

breeding adult

Solitary Sandpiper
solitaria

breeding

Green Sandpiper

129

Willet *Tringa semipalmata* `LC`
Fairly common (**GA**, **O**) to uncommon or rare resident (**LA**).
L 33–41 cm; **WS** 56–66 cm. Two subspecies, of which both (nominate throughout, *inornata*) occur in region. Due to identification issues, their distribution and relative abundance need clarification, but *inornata* usually perceived as rare, with records in Bahamas, Cuba (perhaps regular), Cayman Is, Hispaniola, Guadeloupe, Barbados and Grenadines, at least. Always near seashore, on saltmarshes (especially at high tide), intertidal mudflats especially if bordered by mangroves, and sandy or rocky beaches. Forages in shallow water, or on sand or mud, picking at food. **ID** Large, thickset *Tringa* with striking black-and-white wing pattern. **Ad non-br** has very plain brownish-grey upperparts with narrow white fringes. Head to breast washed grey. White underparts. **Ad br** has black markings above and bars below. **Juv** as non-br, but upperparts grey-brown with dark barring and broad buff fringes. **GV** W race *inornata* averages larger and more godwit-like, has a long, tapered, fine-tipped bill that often appears upturned with a blue-grey base, longer legs and appears more attenuated at rear. Field separation deserves caution. **Voice** A loud "klip" or "kleep" in alarm and harsh "wee-wee-wee" in flight. **SS** None. **TN** Probably best treated as two species, Western and Eastern Willets (Oswald *et al.* 2016).

Lesser Yellowlegs *Tringa flavipes* `LC`
Common passage migrant and uncommon winter visitor (**throughout**).
L 23–25 cm; **WS** 59–64 cm. Monotypic. Freshwater and coastal wetlands, especially mudflats and mangroves. Walks rapidly, picking at surface. Often in small, dispersed flocks. Commonest Aug–Oct and Mar–May, but records year-round. **ID** Slim, mid-sized *Tringa* with relatively short, straight, thin, pointed, typically all-dark bill (about same length as head) and long yellow legs. In flight, square white rump. **Ad non-br** has brownish-grey upperparts with small white spots. Head, neck and breast washed brownish grey with fine streaks. **Ad br** has blackish upperparts with white spots and grey-brown wing-coverts. Head, neck and breast heavily streaked blackish. **Juv** is like ad non-br, but upperparts brown with buff spots, breast washed brownish grey with little streaking. **Voice** Call a quiet "tu", given either singly or doubled, softer than Greater Yellowlegs. **SS** Despite significant difference in size, Greater Yellowlegs can be remarkably tricky to separate, except in direct comparison. It is larger and heavier, with longer (by c. 50%), stout, straight or slightly uptilted, greenish-based bill, and longer legs. Call also useful.

Spotted Redshank *Tringa erythropus* `LC`
Vagrant (**Puerto Rico**, **Guadeloupe**, **Barbados**).
L 29–32 cm; **WS** 61–67 cm. Monotypic. Marshes and lagoons. Typically wades in deep water, may even up-end or swim. Puerto Rico, Aug 2000 (eBird); Guadeloupe, Aug 1999 (Levesque & Jaffard 2002); six Barbados records, Oct–Mar (first 1963). **ID** Elegant wader, with long neck, legs and bill. Legs red to orange. Bill black with red basal half of mandible. In flight, narrow white rump and lower back stripe. **Ad non-br** has contrasting dark eyestripe and white supercilium, ash-grey upperparts with white fringes, plain grey breast and white underparts. **Ad br** is entirely blackish, with white dots above. **Juv** is like ad non-br, but overall darker. Upperparts, head and breast brownish with dense white marks, underparts paler and densely barred. **Voice** Call a clear, rising "chu-it", recalling Semipalmated Plover. **SS** Both yellowlegs have yellow legs and lack red base to mandible.

Common Greenshank *Tringa nebularia* `LC`
Vagrant (**Puerto Rico**, **Barbados**).
L 30–35 cm; **WS** 68–70 cm. Monotypic. Similar habitat and behaviour to both yellowlegs. Puerto Rico, Jul 1993 (*FN* 49: 204, *Cotinga* 5: 76); seven records Barbados, Mar, Oct–Nov, Feb–Apr, first 1980 (Buckley *et al.* 2009, *NAB* 67: 532). **ID** Largest *Tringa* with long, robust, slightly recurved bill, dull green legs, dark wings, white back and rump. **Ad non-br** is rather uniform grey above. Breast, foreneck and face white, and wing-coverts dark. **Ad br** has prominent blackish centres to some scapulars, and heavily streaked crown, neck and breast. **Juv** has upperparts browner with buff fringes, and neck and breast more streaked. **Voice** Call a clear ringing trisyllabic "tyu-tyu-tyu", with last note usually falling. Very similar to Greater Yellowlegs. **SS** Greater Yellowlegs is slightly smaller, with smaller head, less hefty bill and relatively longer legs. In all plumages, upperparts more pale-spotted. In flight white rump is square (not wedge-shaped).

Greater Yellowlegs *Tringa melanoleuca* `LC`
Common passage migrant and uncommon winter visitor (**throughout**).
L 29–33 cm; **WS** 70–74 cm. Monotypic. Similar habitats and behaviour to Lesser Yellowlegs, but less common and tends to probe more. Perhaps commonest Aug–Oct, but records year-round. **ID** Large *Tringa* with long yellow legs and longish, slightly upturned bill (c. 1.5× length of head) with basal third greenish, rest black. In flight, square white rump. **Ad non-br** has grey-brown upperparts with white spots and fringes. Head, neck and breast washed brownish grey, with heavy streaks. **Ad br** has head and neck heavily streaked dark brown. Mantle, scapulars and wing-coverts blackish with narrow white tips. Breast, flanks and upper belly show blackish spots. **Juv** is like ad non-br, but buff spots and brown on breast usually form a band. **Voice** Call a sharp and rather penetrating, trisyllabic, vaguely descending "tew-tew-tew". **SS** See Lesser Yellowlegs.

Wood Sandpiper *Tringa glareola* `LC`
Vagrant (**Guadeloupe**, **Barbados**).
L 19–23 cm; **WS** 54–57 cm. Monotypic. Freshwater pools and marshes. When agitated may bob rear body. Seven records Barbados, Oct–Apr, one long-stayer, first in 1955 (Buckley *et al.* 2009, *NAB* 66: 188); two Guadeloupe, Sept 2000 (Levesque & Jaffard 2002), Sept 2014 (*NAB* 69: 169). **ID** Mid-sized, graceful sandpiper, with medium-length bill and long green legs. Wings extend to tail tip at rest. In flight, dark with pale grey underwing, white rump and barred tail. **Ad non-br** has olive-brown, pale-speckled upperparts, and grey-washed, olive-streaked breast. White throat and belly. White supercilium and black eyestripe. **Ad br** is more dark brown above with bold white speckles and notches, more clearly streaked head and breast. **Juv** is like ad non-br, but warmer brown above with buff spots, breast washed grey-brown and finely streaked brown. **Voice** Call an excited, shrill but rather dry and unmelodious "chiff-iff-iff". **SS** Slightly larger Lesser Yellowlegs has much longer yellow legs; wings extend beyond tail at rest. Solitary Sandpiper darker with dark rump and underwing; lacks post-ocular eyestripe and supercilium.

Willet *semipalmata*
breeding
non-breeding

Lesser Yellowlegs
breeding
non-breeding

Spotted Redshank
breeding
non-breeding

Common Greenshank
breeding
non-breeding

Greater Yellowlegs
breeding
non-breeding

Wood Sandpiper
breeding
non-breeding

131

GLAREOLIDAE
Coursers and Pratincoles
17 extant species, 1 vagrant in region

Collared Pratincole *Glareola pratincola* LC
Vagrant (**Guadeloupe**, **Barbados**).
L 22–25 cm; **WS** 60–70 cm. Three subspecies, of which (probably) nominate occurs in region. Meadows, pools and agricultural land. Walks like a plover, forages in flight like a tern. Juv (perhaps ♀), Barbados, Nov 1996–Jun 1997 (Buckley *et al.* 2009), with another pratincole, presumably this species, Aug 2009 (*NAB* 64: 170); two juv, Guadeloupe, Oct 2015 (Levesque 2016). **ID** Distinctive tern-like wader with long wings and short, stout decurved bill. In flight, like a brown tern with long, forked tail and white rump. **Ad br** is brown tinged olive above, with white rump. Throat ochre-yellow, narrowly bordered black. Breast brown, grading to white belly. Long wings and deeply forked tail black, underwing-coverts and axillaries deep rich chestnut, narrow but distinct white trailing edge to secondaries (can be almost impossible to see, especially in worn plumage). Bill red with black tip, legs blackish. **Ad non-br** has black border to throat indistinct, lores paler and breast mottled grey-brown. **Juv** is mottled black above and on breast; throat whitish without black collar. **Voice** Shrill, staccato and trilled calls, including a sharp, tern-like "stwick" and shrill, rolling trill that rises then falls. Mainly given in flight. **SS** None.

LARIDAE
Gulls and Terns
100 extant species, 21 in region + 13 vagrants

Brown Noddy *Anous stolidus* LC
Locally common resident (virtually **throughout**), vagrant (**Cayman Is**).
L 38–45 cm; **WS** 75–86 cm. Five subspecies, of which nominate occurs in region. Pelagic, except if breeding, when seen around colonies. Feeds over open ocean snatching prey from surface. Settles on buoys, flotsam, ships and even the sea. Nests on cliffs and sometimes the ground, or small bushes on offshore islets, mostly Apr–Aug. **ID** Dark, elegant, long-tailed and long-billed seabird. **Ad** is uniformly dark chocolate-brown with pale greyish-white crown, whiter on forehead and not sharply defined vs. rest of head, and narrow split white eye-ring. Long wedge-shaped blackish tail shows slight notch at tip. ♀ is smaller. **Imm** is like ad but white restricted to thin line on forehead **Juv** has pale-fringed upperparts and no white on head. **Voice** Generally silent except around colonies, where varied guttural barks and braying "keh-eh-eh-err" or "karrk", repeated several times. **SS** See Black Noddy, vs. which in distant flight views is longer- and broader-winged, longer-tailed, overall paler, often with contrastingly paler wing-coverts bar, and has more rakish, less buoyant flight. Juv superficially resembles imm Sooty Tern, but lacks whitish belly and vent, or broad pale fringes above.

Black Noddy *Anous minutus* LC
Vagrant (**GA**, **LA**).
L 34–36 cm; **WS** 65–70 cm. Seven subspecies, of which one (*americanus*) occurs in region. Pelagic. Habits similar to Brown Noddy. Records/sightings (few documented, at least some potentially erroneous) from Providenciales, Turks & Caicos (May 1984), Jamaica (Pedro Cays, 1990s), Puerto Rico (e.g., Jun 1989, large numbers reported 1997), Virgin Is (St Thomas, Jun 1989), Anguilla (Sombrero, May 1988), Dominica (historical) and Barbados (Sept 1991), presumably from small colonies just outside region. **ID** Very dark noddy with long slim bill noticeably longer than head. **Ad** is very similar to Brown Noddy, but smaller and slighter, with thinner and longer bill, blacker lores, darker plumage and more sharply demarcated and brighter white crown, especially on forehead, reaching further behind eyes. Legs and tail slightly shorter. At rest, wingtips fall almost level with tail tip. **Juv** has obvious white on head and more obscure pale fringes above than juv Brown Noddy. **Voice** More vocal than Brown Noddy. A distinctive "tik-tikoree" and a staccato rattle. Also sharp, nasal cackles, chatters, squeaky notes and plaintive, piping whistles. **SS** Compare much commoner Brown Noddy. **AN** White-capped Noddy.

Common White Tern *Gygis alba* LC
Vagrant (**Bahamas**).
L 25–30 cm; **WS** 76–87 cm. Three subspecies, of which nominate occurs in region. San Salvador, Jun 2010 (White *et al.* 2014). Usually at sea, but not especially pelagic and may come close to coast. Buoyant and erratic flight on deep, slow wingbeats. **ID** Small, ghostly white tern. **Ad** is immaculate white with long, slightly upcurved, pointed black bill. Black eye-ring creates appearance of large eyes. In flight, underwing looks translucent with dark primary shafts. Tail short and slightly forked. Legs blackish to dull bluish. **Juv** is white with sooty-grey mark behind eye, grey hindneck, brown-fringed body feathers and wing-coverts, cinnamon-buff tail tip and black-fringed flight feathers. **Voice** Unlikely to be heard. **SS** Unmistakable in good views, but at long range at sea differences from several *Sterna* terns might not be immediately apparent. **TN** Previously considered conspecific with extralimital Little White Tern *G. microrhyncha* (C Pacific), under name White Tern.

Black Skimmer *Rynchops niger* LC
Rare to uncommon non-breeding visitor (**Bahamas**, **GA**), vagrant elsewhere.
L 40–51 cm; **WS** 107–127 cm. Three subspecies, of which nominate occurs in region. Calm coastal bays and lagoons. Unique feeding method of 'skimming' water surface in flight, with mandible partly submerged. Active day and night. Mostly Oct–Apr, rarely from Jul, sometimes in large numbers. **ID** Distinctive, large coastal waterbird. Unique scissor-like bill has mandible longer than maxilla. **Ad non-br** has forehead and underparts white, rest of upperparts, including slightly forked tail, blackish, with dull greyish-white collar. Long bill dull reddish, broadly tipped black. Short legs reddish orange. ♂ obviously larger, with longer bill. **Ad br** has bright red bill tipped black, and is black above without greyish collar. **Juv** (unlikely in region) has upperparts mottled dingy brown and legs and bill dull pinkish, latter tipped black. **Voice** A tern-like nasal "kaah" given singly or in series. **SS** Unmistakable.

Collared Pratincole
pratincola

Brown Noddy
stolidus

Black Noddy
americanus

adult   juvenile

Common White Tern
alba

Black Skimmer
niger

133

Little Gull *Hydrocoloeus minutus* `LC`
Vagrant (**Puerto Rico**, **Barbados**).

L 25–30 cm; **WS** 62–78 cm. Monotypic. Coasts, lagoons and beaches. Behaves more like a tern than other gulls, with dipping flight. Several records Puerto Rico, including 2nd-w, Mar 1992 (Oberle & Haney 2002), 1st-y/2nd-y, Jan 1995 (*FN* 49: 203), Sept 1999 (Oberle 2000); 1st-y, Barbados, Nov–Dec 1998 (Buckley *et al*. 2009). **ID** Tiny (tern-size) three-year gull, with small bill, stubby red legs and rather rounded wings. Bill black or reddish black (but looks black) in all plumages. **Ad non-br** has pale grey mantle and upperwing, blackish ear-spot and crown, and pale grey nape. In flight, black underwing contrasts with white trailing edge. Underparts and tail white. **Ad br** has black hood. **2nd-w** is like ad non-br, but variable black markings at wingtip and paler underwing. **1st-w** is like previous but has prominent black 'M' across mostly grey upperwing, with black outer primaries. Underwing white with black primary tips. Broad black tail-band. **Juv** (unlikely in region) has upperparts densely barred blackish brown and white. **Voice** Unlikely to be heard. **SS** 1st-w Black-legged Kittiwake is similar to 1st-w, but larger, with heavier bill, almost always black legs, white (not dusky) crown and white secondaries. Compare larger Bonaparte's Gull.

Sabine's Gull *Xema sabini* `LC`
Vagrant (**Cuba**, **Puerto Rico**, **Dominica**).

L 27–36 cm; **WS** 81–92 cm. Four subspecies, of which (presumably) nominate occurs in region. Pelagic in winter. Buoyant, tern-like flight. Cuba, Dec 1954, Oct 1999 (Kirkconnell *et al*. in press); Puerto Rico, Jan 1992 (eBird), Mar 2002 (Oberle & Haney 2002); Dominica, May 1993 (Evans & Jones 1997). **ID** Distinctive, small, two-year gull with slightly forked tail and unique upperwing pattern. **Ad non-br** has white neck, tail and underparts, dark grey mantle and partial brownish-grey hood. In flight, tricoloured wing consists of dark grey inner part, white-tipped (inconspicuous) black outer primaries and contrasting white triangle between them. Underwing whitish, outer primaries with black subterminal marks. Legs black and bill black with yellow tip. **Ad br** has dark grey hood, narrowly bordered black at rear; can show pinkish-tinged underparts. **Juv/1st-w** is like ad non-br but has grey-brown mantle and inner wing, with delicate black-and-white tips producing scaly effect. Tail has black terminal band. Bill often all black. **1st-s** develops white primary tips and greyish hood. **Voice** Usually silent. **SS** Forked tail, striking grey-white-black wing pattern in all plumages, black bill tipped yellow (except 1st-y) and tern-like flight unique, but see slightly larger juv Black-legged Kittiwake.

Black-legged Kittiwake *Rissa tridactyla* `VU`
Two subspecies placed in separate subspecies groups, of which one occurs in region.

Atlantic Kittiwake *Rissa (tridactyla) tridactyla*
Vagrant (**throughout**).

L 38–40 cm; **WS** 91–97 cm. One subspecies in group. Usually pelagic, but regularly seen from shore in regular range. Flight buoyant. Records (Dec–Apr) Bahamas and Cuba S to Barbados, but very few in Lesser Antilles, and a significant number were during midwinter 1984/85 and early 2009. **ID** Mid-sized, compact, three-year gull with short black legs and black eyes. **Ad non-br** has white head and underparts, dusky smudge between nape and eyes, yellow bill and grey mantle. Upperwing grey and underwing white, both with sharply defined solid black tips, as if dipped in ink. **Ad br** has head all white. **1st-w** has prominent dark half-collar, black spot behind eye, black bill and broad black band on slightly notched tail. In flight prominent black 'M' on wings. May show reddish or pinkish legs. **Voice** Unlikely to be heard. **SS** Combination of mostly white head, short almost always black legs, yellow bill and sharply defined black wingtips in ad, or black nuchal collar, tail-band and 'M' across wings in 1st-w, unique in region, but compare 1st-w with smaller and even rarer Little Gull.

Bonaparte's Gull *Larus philadelphia* `LC`
Rare non-breeding visitor to vagrant (**throughout**).

L 28–34 cm; **WS** 78–90 cm. Monotypic. Coastal, but also found well offshore. Frequents harbours, bays and lagoons. Flight more buoyant and tern-like than Black-headed Gull. Most records Bahamas and Cuba, but S to Barbados, in Aug–Apr (once early Jun). **ID** Small, elegant, two-year gull with small black bill and short flesh-pink legs. Resembles a small, more delicate version of Black-headed Gull. **Ad non-br** has dark ear-coverts spot and two faint dark grey bars on crown, plus distinct greyish 'shawl'. In flight, upperwing mid-grey with white outer primaries tipped black; underwing very pale, primaries translucent with distinct black trailing edge. At rest wingtips appear all black and project well beyond tail. **Ad br** (unlikely in region) has blackish hood with slightly bolder split white eye-ring. **1st-w** has darker diagonal bar across wing-coverts and paler inner 'hand' above than Black-headed Gull. **Voice** Typical call a nasal, tern-like "keek" or "chirp", very different to Black-headed Gull. **SS** Differs from larger Black-headed Gull by black bill, slightly darker mantle and much paler underwing, with translucent white underside to primaries.

Slender-billed Gull *Larus genei* `LC`
Hypothetical (**Antigua**).

L 38–44 cm; **WS** 94–110 cm. Monotypic. Apr 1976 (Holland & Williams 1978); said to be present c. 1 week, but no documentation. Feeds in shallow inshore waters, coastal wetlands and salt flats, but rarely in harbours. Swimming birds often hold long neck tilted forward. **ID** Medium-sized, elegant, two-year gull with notably long, slightly drooping bill, legs and neck. Iris pale yellowish or white. **Ad non-br** has head white with faint grey ear-spot. Neck, tail and underparts pure white. Mantle and innerwing pale grey. Outer primaries white with black tips and narrow black outer edge. Bill paler orange-red and legs dark red. **Ad br** has pure white head, pink-tinged breast and dark blackish-red bill (looks black). **1st-w** has pale brown markings on upperparts and wing-coverts, black terminal tail-band and pale fleshy orange legs and bill, latter with diffuse darker tip. **Voice** Unlikely to be heard. **SS** Commoner Black-headed Gull smaller, less elegant, with shorter bill and legs, darker ear-spot and dark marks on head at all ages (ad breeding has brown hood). Imm darker and more contrasting than Slender-billed, with darker legs, and duller orange bill has more distinct dark tip. Compare Black-legged Kittiwake and larger Ring-billed Gull.

Little Gull — adult breeding; first-winter; adult non-breeding

Sabine's Gull *sabini* — adult breeding; adult non-breeding; first-winter

Atlantic Kittiwake (Black-legged Kittiwake) — adult breeding; adult non-breeding; first-winter; first-winter

Bonaparte's Gull — adult breeding; adult non-breeding

Slender-billed Gull — adult breeding; adult non-breeding; first-winter

135

Black-headed Gull *Larus ridibundus* LC
Rare and local non-breeding visitor to vagrant (**throughout**).

L 37–44 cm; **WS** 91–110 cm. Monotypic. Harbours and fishing ports. Most frequently reported in Bahamas (e.g. New Providence, Eleuthera, Grand Turk), Puerto Rico, Virgin Is (St Thomas, St Croix, Tortola), Guadeloupe and Barbados (including one ringed as chick in Russia), although still no more than vagrant in first two-named, and with records throughout (also including Cuba, St Lucia and Grenada) between Nov and Jul. **ID** Small, compact, two-year gull with thin reddish bill, red feet and rounded head. **Ad non-br** has pale grey upperparts, dark grey ear-spot with faint grey bars on crown, white primaries with narrow black tips and edging to outer wing, and black-tipped red bill. In flight, upperwing has conspicuous white leading edge, forming wedge on outer wing. Underwing darker, especially inner flight feathers. **Ad br** has dark brown hood and dark red bill. **1st-w** is like ad but has brown-centred coverts and tertials. Dark brown secondary bar becomes dark trailing edge to primaries, which have dark edges to white (outer) or grey (inner) feathers. **Juv** has extensive rich buff to darker brown markings on upperparts and wing-coverts, and black terminal tail-band. Juv/1st-w have dull yellow-orange legs and bill, latter with obvious dark tip.
Voice Downslurred but melodious "krreeaarh" or similar, often repeated. Also a short "kek" or "kekekek". **SS** Commoner Laughing Gull slightly larger and has darker grey mantle and wings, the latter with broader black tips. Ad br has black (not dark brown) hood and intermediate plumages have blackish bill (always reddish or two-toned in Black-headed Gull). Smaller Bonaparte's Gull has black bill, paler underwing and ad br has black (not dark brown) hood. Franklin's Gull shows more prominent eye-ring and darker mantle and wings, the latter with white-black-white pattern at tip. Rare Slender-billed Gull more elegant, with longer bill and legs. Compare rarer Black-legged Kittiwake. **AN** Common Black-headed Gull.

Grey-headed Gull *Larus cirrocephalus* LC
Vagrant (**Barbados**).

L 38–45 cm; **WS** 100–105 cm. Two subspecies, of which (presumably) nominate occurs in region. Ad, May 2009 (Buckley *et al.* 2009). Coastal, but could occur inland. **ID** A two- or three-year lanky gull with long bill, neck and legs. **Ad br** has crown, face and throat grey, with narrow black line from hindcrown to lower throat. Neck, tail and underparts white. Mantle and wings grey, with black outer primaries showing conspicuous white windows. Underwing mostly dark with small white window near tip. Bill red tipped black and legs bright pinkish red. Eyes yellowish white. **Ad non-br** has hint of grey head pattern, without black border. Bill duller red, with more extensive black tip. **Imm** has solid black wingtip, dark underwing and narrow dark tail-band. Mantle, scapulars and wing-coverts brown with pale feather fringes and dark primary-coverts. Eyes dark. **Voice** Unlikely to be heard. **SS** Commoner Black-headed Gull slightly smaller, has narrower and more pointed wings, shorter bill, paler underwing, different wingtip pattern and dark eyes; additionally, imm has paler underwing and broader tail-band. See vagrant Slender-billed Gull. **AN** Grey-hooded Gull.

Franklin's Gull *Larus pipixcan* LC
Vagrant (**GA, LA, O**).

L 32–38 cm; **WS** 86–97 cm. Monotypic. Bays and estuaries. Usually alone but occasionally small groups, often with Laughing or other gulls. Bahamas (Grand Bahama, Dec 2005; New Providence, Sept 2012), Cuba (two reports Apr 1999, photo Feb 2018), Cayman Is (Dec 2013), Hispaniola (Aug 1978, Feb 2000), Puerto Rico (Jan 1969, Jan 1972, Jan 2010, Oct–Nov 2014), St Martin (Nov 2015), St Barthélemy (1800s), Guadeloupe (Nov–Dec 1992, Nov 2015–Jan 2016), Barbados (Nov 2005, Jan and Oct 2011, Nov 2015) and San Andrés (Oct 2017). **ID** Small, rather compact, three-year gull with rounded head, rather blunt wings and small straight bill. All plumages similar to same ages of Laughing Gull. **Ad non-br** has blackish half-hood on rear head and broad white eye-ring. Wingtips show distinctive white-black-white pattern. Tail pale greyish white. **Ad br** has larger eye-ring and slight pinkish cast to breast. Intermediate plumages differ from Laughing Gull by incomplete tail-band (broadest in centre) and bolder eye-ring. **Voice** Nasal (and laughing) but higher-pitched, more hollow-sounding and less penetrating than Laughing Gull. **SS** Compare much commoner, slightly larger, more robust and less compact Laughing Gull. See Black-headed Gull.

Laughing Gull *Larus atricilla* LC
Common to rare resident (**throughout**).

L 36–46 cm; **WS** 95–120 cm. Two subspecies, of which nominate occurs in region. Strictly coastal, in calm bays, islets, harbours and fishing ports. Gregarious and opportunistic, often in large groups. Dips for small fish and steals food from pelicans and others; also takes carrion. Breeds locally on rocky islands, mostly Apr–Jul. Common Apr–Sept; less numerous over most of region rest of year. **ID** Mid-sized, slender-bodied, three-year gull with relatively small head, long slightly drooping bill, and long angled wings. **Ad br** has all-black hood, thin white split eye-ring, and dark red bill and feet. Mantle and wings dark slate-grey, latter with broad black tip and narrow white trailing edge. **Ad non-br** differs by white head with greyish nape-band. **1st-w** is like ad non-br but has grey nape and breast-sides, and brown wing-coverts. Underwing dusky with dark axillaries and mottled grey coverts. Tail white with broad subterminal black band. **2nd-w** is intermediate between 1st-w and ad non-br. **Juv** is mostly greyish brown above, paler below, with whitish belly and broad black subterminal tail-band. All plumages except ad br have blackish bill and feet. **Voice** A loud "kaha" repeated often and a laughing nasal "ka-ka-ka-ka-kaaaaa-kaaaa", rapid at first, then slow, with typical gull quality. **SS** Much rarer Franklin's Gull very similar but slightly smaller, has slighter, straighter bill, less pointed wings with white on tips. White eye-ring more prominent and underwings paler. Ad non-br has dusky half-hood and intermediate plumages an incomplete tail-band (not reaching sides). See uncommon Bonaparte's and Black-headed Gulls, and rare Black-legged Kittiwake.

Black-headed Gull

adult breeding

adult non-breeding

adult breeding

first-winter

Grey-headed Gull
cirrocephalus

adult breeding

adult non-breeding

first-winter

Franklin's Gull

adult breeding

adult non-breeding

adult breeding

first-winter

Laughing Gull
atricilla

adult breeding

first-winter

adult non-breeding

137

Ring-billed Gull *Larus delawarensis* `LC`
Fairly common to rare non-breeding visitor (**throughout**).
L 41–54 cm; **WS** 115–135 cm. Monotypic. Harbours, lagoons, estuaries, lakes, reservoirs and open ground. Fairly common to uncommon and local in Bahamas, Greater Antilles, Caymans and Barbados, but very rare elsewhere. Records in all months, mostly Oct–May. **ID** Mid-sized, heavy-bodied, three-year gull with rather heavy bill. **Ad non-br** has broad black band on yellow bill, yellow eyes, and heavily dark-spotted head and hindneck. Pale grey mantle, and white underparts and tail. In flight, mostly pale grey upperwing shows black primaries with two small white spots. Legs yellowish. **Ad br** has clean white head. **2nd-w** is like ad non-br but has darker wingtips (often just one white spot), dark primary-coverts and often some dark on secondaries; some show dark tail-band. **1st-w** has pale grey upperparts, mottled greyish-brown wings with dark brownish-black primaries and broad black tail-band. Legs and bill pink, latter with dark tip. **Voice** Fairly high-pitched, nasal "kuleeeeuk", and softer "kowk". **SS** American Herring Gull obviously larger, with bulkier head and bill; it lacks bill ring and ad has pink (not yellow) feet.

Kelp Gull *Larus dominicanus* `LC`
Vagrant (**Barbados**).
L 54–65 cm; **WS** 128–142 cm. Six subspecies, of which one (perhaps *vetula*) occurs in region. Ad, Dec 2000 (Hayes *et al.* 2002, Buckley *et al.* 2009). Coasts, but could visit inland lakes and reservoirs. **ID** Very similar but smaller, with lighter build than Great Black-backed Gull. **Ad** differs from latter by yellowish-green or greyish-green (not pink) legs and less white on wingtips. **Imm** plumages have greyish-brown (not pink) legs. Race *vetula* (which Barbados bird was suspected to be) further differs from Great Black-backed by dark eyes. **Voice** A repeated "ee-ah" and various yelps and raucous calls, but perhaps unlikely to be heard. **SS** Yellowish-green or greyish-green legs in ad, or greyish-brown legs in imm, distinguish it in all plumages from other large gulls. Commoner American Herring and Lesser Black-backed Gulls are smaller, with thinner bill. See Great Black-backed Gull.

Lesser Black-backed Gull *Larus fuscus* `LC`
Five subspecies divided into four subspecies groups, of which one occurs in region.

Lesser Black-backed Gull *Larus (fuscus) graellsii*
Fairly common non-breeding visitor (**N Bahamas**), rarer elsewhere.
L 51–61 cm; **WS** 124–150 cm. Two subspecies in group, of which at least one (*graellsii*) occurs in region. Mostly Sept–Apr, but records year-round throughout region, some involving very long-stayers. Increasing records, since first in 1965, especially in N of region, e.g. first recorded Cuba in 1998, but now regular in small numbers, and Dominican Republic in Nov 1997; in Barbados, first seen 1995. Large groups (> 80) occasionally reported in Bahamas. Diverse coastal and inland waters, mainly sandy beaches, bays, estuaries, harbours, lakes and reservoirs. At least one record from Barbados suspected to involve nominate race (Baltic Gull) and race *intermedius* seems also likely to occur. **ID** Rather large, four-year gull with long narrow wings, large bill and dark back. **Ad non-br** has head, neck, tail and underparts white with brown-streaked head and neck. Mantle and wings dark grey with white trailing edge and black primaries showing small white mirror. Legs and bill yellow, latter with red spot near tip. Eyes pale. **Ad br** has head and neck clean white. **3rd-y** is similar to ad non-br but some brown on wings, yellowish bill with dark smudge at tip and yellowish, rarely pinkish, legs. **2nd-y** has bill pinkish with black band near tip. Mantle grey. Brownish-grey wings without white mirrors. Rump white and tail whitish with broad black band. Legs pinkish. **1st-y** is mottled brownish grey, with paler head and paler rump contrasting with blackish tail. Bill and eyes blackish. **GV** Races differ mainly in size, proportions and degree of darkness of mantle and wings in ad: *L. (f.) fuscus* (separate subspecies group, Baltic Gull) darkest, jet-black above, and smallest and longest-winged race; *intermedius* sooty black; and *graellsi* dark slate-grey. **Voice** Calls resemble American Herring Gull but deeper and more nasal. **SS** Great Black-backed Gull much larger with massive bill, and ad has pink legs, black (not dark grey) mantle and wings, with more white on primary tips. Slightly larger ad American Herring Gull has paler mantle and pink legs; 1st-y/2nd-y have less contrasting white rump. Extralimital Yellow-legged Gull *L. michahellis* (not definitely recorded, but claimed several times in Barbados) should be identified with great care, especially given possibility of Arctic Herring Gull × Lesser Black-backed Gull hybrids—see Howell & Dunn (2007), Howell *et al.* (2014).

Arctic Herring Gull *Larus smithsonianus* `LC`
Three subspecies divided into three subspecies groups, of which one occurs in region.

American Herring Gull
Larus (smithsonianus) smithsonianus
Uncommon and local to rare non-breeding visitor (**GA, N LA, O**), vagrant (**S LA**).
L 53–65 cm; **WS** 120–150 cm. One subspecies in group. Coasts, harbours and lagoons. Much less regular inland, at lakes and reservoirs. Present Bahamas, Cuba, Hispaniola and Cayman Is Sept–May, exceptionally Jun–Aug; rarer Jamaica, Puerto Rico and Virgin Is; and very rare Lesser Antilles (or only vagrant in S) Oct–Mar. **ID** Large, thickset, very variable four-year gull with strong bill. Dull pinkish legs. **Ad non-br** has white head, tail and underparts, with dense to sparse brownish streaking on head, neck and upper breast. Mantle and wings pale grey, with black primaries showing small white spots near tips. Eyes pale. Bill yellow with red spot on mandible. **Ad br** has clean white head, neck and underparts. **3rd-w** is like ad non-br but has more extensive brownish streaks and spots. Broad, irregular black tail-band. Bill yellowish with black smudge near tip. **2nd-w** has variable greyish mantle, brown wings with blackish primaries. Tail blackish brown with limited white at base, rump varies from heavily barred to clean white. Bill pinkish with black tip. **1st-w** is rather uniform brown with diffuse streaking or mottling. Primaries dark brownish black with pale window on innermost feathers. Tail almost uniform blackish. Bill dull and dark, with paler base. Eyes dark. **Voice** Fairly high-pitched "kraaaw kraaaw" and deeper "gyow". **SS** Slightly smaller ad, 2nd-y and 3rd-y Lesser Black-backed Gull have darker mantle. Additionally, ad has yellow legs. 1st-y Lesser Black-backed has less uniform brown plumage with more contrasting, paler, rump. See also much larger Great Black-backed Gull and obviously smaller Ring-billed Gull.
TN Often treated as part of European Herring Gull *L. argentatus*, under name Herring Gull.

adult breeding

adult non-breeding

first-winter

Ring-billed Gull

adult breeding

adult non-breeding

juvenile

first-winter

Kelp Gull
dominicanus

adult non-breeding

first-winter

adult breeding

Lesser Black-backed Gull
graellsii

adult non-breeding

first-winter

adult breeding

American Herring Gull
(Arctic Herring Gull)

Iceland Gull *Larus glaucoides* `LC`
Two subspecies placed in separate subspecies groups, of which one occurs in region.

Iceland Gull *Larus (glaucoides) glaucoides*
Vagrant (**Bahamas**, **Guadeloupe**).
L 55–64 cm; **WS** 125–145 cm. One subspecies in group. 3rd-y, Grand Bahama, Feb 2004, New Providence, Bahamas, Feb–Mar 2009 (*NAB* 63: 340–341); Guadeloupe, Feb 2010 (Levesque & Delcroix 2015). Harbours and other coastal areas. **ID** Highly variable, usually very pale, rather large, four-year gull, with short bill, rounded head and long wings. Very similar to American Herring Gull in all plumages. **Ad** differs in having bill slighter and shorter, plumage generally whiter and more uniform, wings paler with translucent white primaries showing some dark grey markings (variable in extent) near tips (instead of black primaries with small white windows). Additionally, has pinker legs and eyes usually dark in all plumages. **Imm** plumages usually lack dark tail-band of imm American Herring. **Voice** High-pitched "klaaaaw". **SS** One of palest of large gulls. Although very variable, combination of large size but relatively slight, short bill, dark eyes, pale plumage, especially wings, with white primaries showing grey marks at tips and pinker legs (in ad), should distinguish all plumages, but see American Herring Gull and Glaucous Gull. **TN** Closely related to both American Herring and extralimital European Herring Gulls *L. argentatus* and, in past, sometimes treated as conspecific. At least some West Indies records ascribed to so-called Kumlien's Gull, which is currently regarded as a hybrid swarm between the two subspecies groups (Thayer's Gull and Iceland Gull).

Glaucous Gull *Larus hyperboreus* `LC`
Vagrant (**Barbados**).
L 60–77 cm; **WS** 132–162 cm. Four subspecies, of which one (unknown) occurs in region. 1st-w, Nov 2009 (eBird). Most likely to be observed in harbours and ports. **ID** Very large four-year gull with front-heavy appearance. Flight heavy. **Ad non-br** is heavily streaked brown from head to upper breast, creating hooded appearance. Otherwise all white except pale grey mantle and wings; all flight feathers tipped white. Long, heavy yellow bill has small red gonydeal spot. **Ad br** has clean white head. **3rd-w** is like ad with some brown patterning on wing-coverts, and dull bill. **2nd-w** resembles 1st-w but whiter; bill has narrow pale tip and eye becomes pale. **1st-w** is rather uniform pale buff with paler wingtips; undertail-coverts barred brown, and upperparts and wing-coverts show narrower brown markings. Tertials pale-centred with whitish edges. Bill pale pink with distal third black; iris dark. Older (spring) birds can be almost white. **Voice** Typical gull-like "kuwaaa"; also higher, two-part "k-lee". **SS** Compare Iceland Gull, which is smaller, with more rounded head, appearance of larger eyes, a weaker bill, deep pigeon-breast, shorter legs, and longer and narrower wings; flight is lighter than Glaucous. 1st-y Glaucous has different bill pattern, being largely pinkish with large dark tip, vs. largely dark with small pale base.

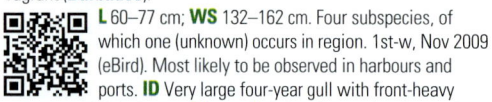

Great Black-backed Gull *Larus marinus* `LC`
Rare non-breeding visitor (**Bahamas**, **GA**), vagrant (**LA**).
L 68–79 cm; **WS** 152–167 cm. Monotypic. Rocky or sandy coasts, bays, estuaries and open sea. Often in harbours, especially fishing ports. Uncommon visitor Bahamas, becoming steadily rarer further S. Mostly Oct–Mar, but exceptionally recorded in Jun/Jul. **ID** Largest gull.

A larger version of Lesser Black-backed Gull, with thicker neck, much heavier bill and broader, less pointed wings. **Ad** has black (not dark grey) mantle and uniformly black upperwing with white trailing edge and more conspicuous white spots near tip of outer primaries. **Ad non-br** has sparser brown streaking on head and neck. **4th-w** resembles ad but may retain some subterminal black on bill and brownish cast above. **3rd-w** has ad-type blackish saddle and median coverts; black primaries have less white on tips than ad, with only small white mirror on p10; tail some dark streaking; bill yellow with black subterminal markings. **2nd-w** has darker mantle and scapulars, showing greater contrast with wing-coverts; bill pinkish with black subterminal markings and pale tip. **1st-w** is similar to juv but head, underparts and rump distinctively whitish. **Juv** is heavily mottled white and pale brown, with an all-dark bill. Legs pink in all plumages. **Voice** All calls very deep, hoarser and slower than American Herring Gull, including a shorter and slower long-call. **SS** In all plumages, huge size and massive bill distinctive. See much smaller and slenderer Lesser Black-backed Gull.

Sooty Tern *Onychoprion fuscatus* `LC`
Common to fairly common resident (**throughout**).
L 36–45 cm; **WS** 82–94 cm. Seven subspecies, of which nominate occurs in region. Highly pelagic, except during breeding season. Breeds in large colonies on cays of sand, coral or rock, usually flat, open and with sparse vegetation, e.g. Culebra (Puerto Rico) and Isla de Aves, in May–Aug, but very rarely observed in other months. **ID** Large, stocky, black-and-white tern with long wings and long, deeply forked tail. **Ad br** has black cap, nape and upperparts, contrasting with white underparts and forehead, which reaches only to eyes. In flight, blackish-grey underwing contrasts with whitish coverts. Tail shows thin white outer edges. Bill and legs black. **Ad non-br** has variable white feather fringes above. **Juv/1st-w** is blackish brown above with white spots and vermiculations, and greyish brown below, becoming paler on lower belly. Tail shorter and less deeply forked. **Voice** Very noisy at colonies, with nasal barking or laughing "wide-a-wake" or "ker-wack-wack" heard constantly. Juv gives high-pitched reedy whistles. **SS** Bridled Tern smaller, with paler upperparts and slightly more pointed wings; white forehead extends as eyebrow behind eyes; white nuchal band, broader white edges to outer tail feathers and different underwing pattern. Juv Bridled has whitish, not all-dark, underparts.

Bridled Tern *Onychoprion anaethetus* `LC`
Fairly common to uncommon resident (**throughout**).
L 35–38 cm; **WS** 76–81 cm. Four subspecies, of which one (*melanopterus*) occurs in region. Pelagic, but less so than Sooty Tern. Otherwise similar to Sooty Tern. Fairly common but local breeder in Apr–Aug; less common in other months. **ID** Mid-sized, dark-backed tern with long, deeply forked tail. **Ad br** has dark grey-brown upperparts, black crown and nape, and white forehead, nuchal band, supercilium and underparts. **Ad non-br** has duller head pattern, with heavily scalloped white crown, white lores and some white fringes to upperparts. **Juv/1st-w** is generally paler than juv Sooty. Head pattern like ad non-br, greyish-brown upperparts with whitish tips and fringes, and whitish underparts. **Voice** Only vocal when nesting. Ad gives a mellow "kowk kowk" or "kwawk kwawk", and a harder "kahrr". **SS** Similar to larger Sooty Tern. Differs in slightly narrower wings, deep brownish-grey (not black) mantle and wings. Whitish collar separates black nape and cap from greyish-brown mantle. White forehead extends as eyebrow slightly beyond eyes. White outer edges to outer tail feathers broader. Underwing mostly white, with black of primaries confined to tips.

adult non-breeding 'kumlieni'

adult non-breeding

first-winter

first-winter

adult breeding 'kumlieni'

Iceland Gull
glaucoides

adult breeding

Glaucous Gull
hyperboreus

adult breeding

adult non-breeding

first-winter

Great Black-backed Gull

juvenile

adult breeding

adult breeding

Sooty Tern
fuscatus

juvenile

adult breeding

adult breeding

Bridled Tern
melanopterus

141

Least Tern *Sternula antillarum* `LC`
Fairly common but local to very rare resident and non-breeding visitor (**throughout**).

L 22–24 cm; **WS** 51 cm. Three subspecies, of which nominate occurs in region. Bays, lagoons, estuaries and harbours. Flight buoyant with quick wingbeats; plunges from mid-heights. Breeds on beaches and sandbars, mostly May–Aug. Local resident Bahamas, to some extent Greater Antilles (but largely summer visitor Cuba), Cayman and Virgin Is, and locally Lesser Antilles S to Guadeloupe, but largely vagrant in S of region (e.g. St Lucia) and generally uncommon Sept–Mar anywhere. **ID** Very small tern with slender pointed bill and short deeply forked tail. **Ad br** has black cap with white wedge-shaped forehead, and pale grey upperparts. Bill yellow tipped black, and legs orange-yellow. In flight, contrasting black outer primaries. **Ad non-br** (less likely in region) has mostly white cap, with only rear crown and thick mark through eyes black, all-black bill and darker legs. **Juv** is similar to ad non-br, but mottled brownish on wings and upperparts, and dusky carpal. **1st-w** is like juv but upperparts greyer and crown whiter. Bill and legs black. **Voice** Four-note "keedee-cui, keedee-cui" mainly in flight at or near colony. Alarm a low-intensity "zweeep" or high-pitched "tsip tsip tsip". **SS** Smallest tern in region. Sandwich and vagrant Large-billed Terns much larger.

Large-billed Tern *Phaetusa simplex* `LC`
Vagrant (**Cuba**, **Grenada**).

L 38–42 cm. Two subspecies, of which nominate occurs in region. Cuba, May 1909, undated, Apr 2019 (Kirkconnell *et al.* in press); Grenada, May–Jun 2010 (White & Jeremiah 2011). Rivers, lakes, marshes, estuaries and other freshwater or brackish wetlands. Feeds mainly by plunge-diving, often from quite high. **ID** Large tern with relatively short, slightly forked tail, heavy yellow bill and striking wing pattern. **Ad non-br** has upperparts and tail mid-grey. Crown blackish and forehead black mottled white (looks greyish). In flight, grey upperwing with black primaries, white secondaries and greater coverts. Legs yellow. **Ad br** has crown solid black. **Juv** has less black on crown, back and wings mottled brown, and bill duller. **Voice** In flight a loud reedy "kree" or "keew", often in series. **SS** Large size, heavy yellow bill and wing pattern distinctive, but see commoner Sandwich Tern and tiny Least Tern.

Common Gull-billed Tern *Gelochelidon nilotica* `LC`
Uncommon resident or summer visitor (**Bahamas** to **Anguilla**), rare migrant (most of **LA**).

L 33–38 cm; **WS** 84–101 cm. Five subspecies, of which two (*aranea* widespread, nominate vagrant) occur in region. Coastal mudflats, marshes, shallow ponds, estuaries, beaches, inland lagoons, but shuns deeper water. More insectivorous than most terns, hawking over fields or water, but rarely dives. Flight buoyant and leisurely. Breeds Apr–Aug; disperses over whole region Sept–Mar, but mainly recorded in autumn across Lesser Antilles. At least one record of nominate; bird ringed as chick in Denmark shot in Barbados, Sept 1935, and another 1st-y, possibly this race, there, Dec 2008 (Buckley *et al.* 2009). **ID** Compact, mid-sized tern with heavy (gull-like), short black bill, rather short, slightly forked tail, and broad wings. **Ad non-br** has white underparts and pale grey upperparts. Blackish-grey wingtips and black on head confined to ear-coverts. Eyes and legs black. **Ad br** has solid black crown and nape. **Juv** is like ad non-br, but some brown markings on scapulars and wing-coverts, and dark-tipped tail. **GV** Nominate race is like *aranea*, but larger and bill has sharp gonydeal angle. **Voice** Commonest call a slightly upslurred "kay-wek", given singly or in series, usually in flight. Also a harsher "aah", sometimes tripled, and "aah-aah-aah" in alarm. **SS** Thickset black bill and shallow-forked tail distinctive. Sandwich Tern slimmer, with a thinner, longer bill, usually black tipped yellow, and longer, more deeply forked tail; also prefers deeper water. **AN** Gull-billed Tern (when lumped with extralimital Australian Gull-billed Tern *G. macrotarsa*).

Caspian Tern *Hydroprogne caspia* `LC`
Fairly common to very rare winter visitor and passage migrant (almost **throughout**), vagrant (**S LA**).

L 48–56 cm; **WS** 127–140 cm. Monotypic. Beaches, sandbars, coastal lagoons, mudflats and estuaries, also inland lakes and reservoirs, rarely pelagic. Much less gregarious than Royal Tern. Hovers and plunge-dives, occasionally surface-dips. Bulky and front-heavy in flight, with shallow wingbeats. More frequent in Bahamas and Greater Antilles. Present year-round, e.g. at Zapata Peninsula (Cuba), but no evidence of nesting. **ID** Largest tern, with heavy coral-red bill and slightly forked tail. **Ad non-br** has black crown and nape mottled white, and variable black tip to bill. In flight, blackish outer primaries on underwing distinctive. **Ad br** has crown and nape all black, small crest, brighter bill. **Juv/1st-w** is like ad non-br, but has some darkish scaling on upperparts, duller orange-red bill, and darker upperwing-coverts and tail. **Voice** Hoarse, deep, heron-like "krree-ahk", lower-pitched than Royal Tern. **SS** From Royal Tern by larger size, heavier build and much thicker blood-red bill, which shows black-and-orange tip up close; underwing has blackish primaries and tail is less deeply forked.

Least Tern
antillarum

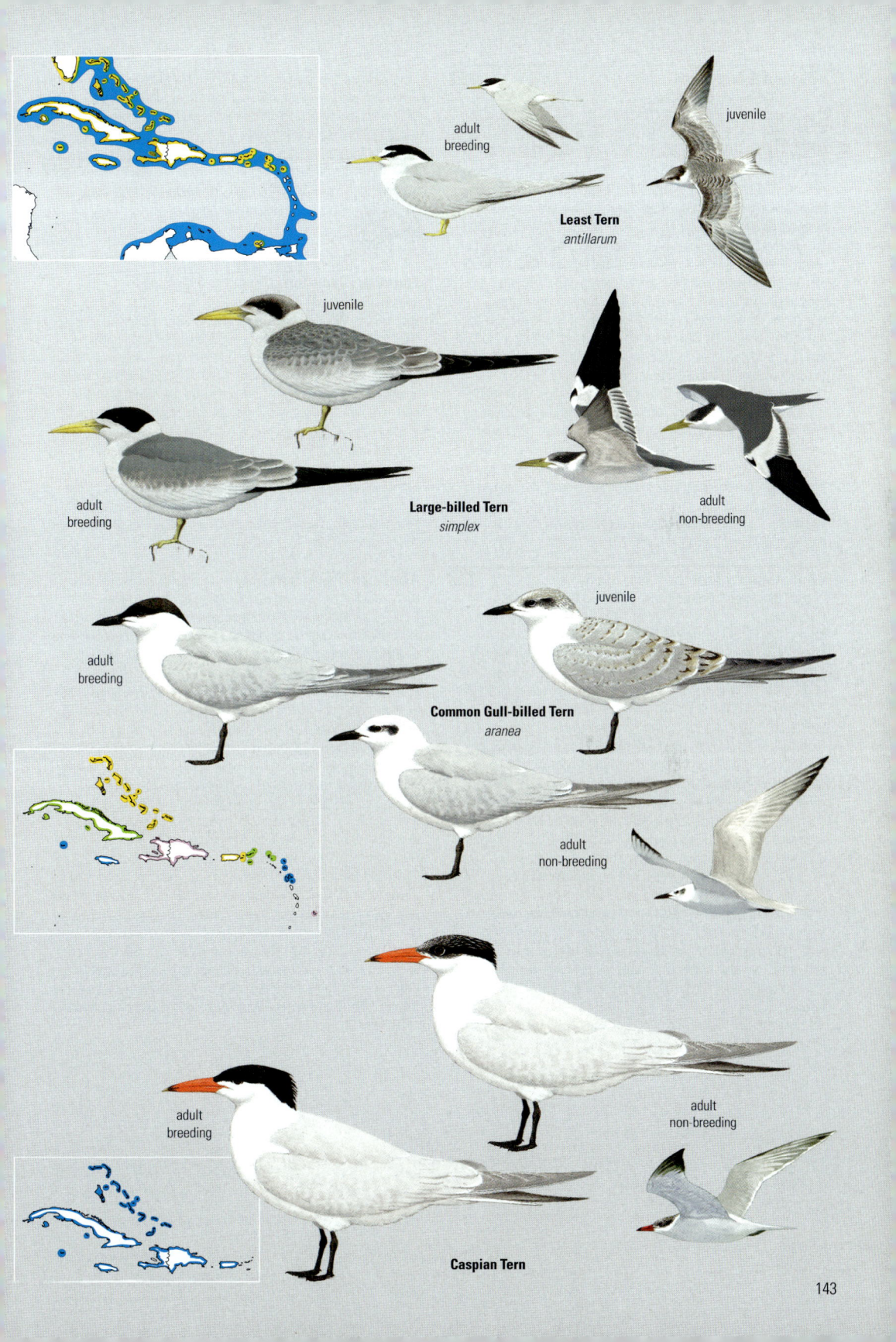

Large-billed Tern
simplex

Common Gull-billed Tern
aranea

Caspian Tern

143

Whiskered Tern *Chlidonias hybrida* `LC`
Vagrant (**Bahamas, Barbados**).

L 23–29 cm; **WS** 64–70 cm. Three subspecies, of which nominate occurs in region. Three in Barbados, juv, autumn 1847, ad, Apr 1994, 1st-y, Nov 2004 (Buckley *et al.* 2009); and one, Great Inagua, Bahamas, Apr–May 2003 (*Cotinga* 21: 82). Inland lakes, marshes and rice fields, also coastal lagoons. Often hawks insects over fields and lake margins. **ID** Slender, rather small tern, with short, slightly notched tail, short bill and relatively short, broad wings. **Ad non-br** has pale grey mantle, white underparts, rump and tail, and white underparts, neck and forecrown, with dark-flecked rear crown, and black patch behind eye merging into dark band on nape. Pale grey underwing and slightly darker flight feathers on upperwing. Bill black and legs dull red. **Ad br** is generally darker, with full black cap, and blood-red bill and legs. Mid-grey underparts, mantle and tail contrast with white cheeks and undertail-coverts. **Juv** is like ad non-br, but has buff-washed face, dark grey-tipped hindneck, and contrastingly darker mantle, scapulars and tertials. **Voice** Alarm a more rasping loud "kerch", often repeated, but unlikely to be heard. **SS** Slightly smaller ad br White-winged Tern has contrasting black-and-white plumage. Non-br differs in having dark restricted to hindcrown, more isolated dark ear-coverts patch and whitish rump contrasting with darker back. Arctic Tern paler with much longer tail and long streamers. Non-br Black Tern darker above with dark patch on neck-sides. Compare also Common Tern.

White-winged Tern *Chlidonias leucopterus* `LC`
Vagrant (**Bahamas, Puerto Rico, St Croix, Guadeloupe, Barbados**).

L 23–27 cm; **WS** 58–67 cm. Monotypic. Bahamas (Great Inagua, Jun 1980; Cat I, spring 1999), Puerto Rico (Jan–Jun 2001), St Croix (Sept 1987, Sept 2013), Guadeloupe (ad, Sept 2015) and Barbados (Oct 1888, three, Oct–Nov 1996, Oct 2008, Oct 2010, Sept 2013, Dec 2015). Wet fields, inland lakes and other freshwater bodies, but vagrants could appear at just about any wetland. Uses contact-dipping, hover-dipping and hawks flying insects; does not plunge-dive. **ID** Small tern, with short, slightly notched tail, rather thin bill and relatively short broad wings. **Ad br** is black overall, with slightly paler back and contrasting white upperwing-coverts, undertail-coverts, rump and tail. In flight, striking pale grey upperwing with white coverts and blackish outer primaries, and pale grey underwing with contrasting black coverts. Bill blackish red and legs red. **Ad non-br** is pale grey with white underparts. Head white with blackish rear crown and isolated ear-coverts ('headphones'), underwing pale with slightly darker coverts, and upperwing has narrow dark leading edge. Bill black. Moulting ad has white-mottled head and underparts, a white rump and tail. **Juv** is like ad non-br, but has blackish-brown back and greyish wings, white rump and collar. **Voice** A loud "krek" or "kreek" when excited or alarmed. **SS** Unmistakable in breeding plumage. Non-br differs from Whiskered Tern in having dark restricted to hindcrown, finer bill and more isolated dark ear-coverts patch. Non-br Black Tern has slightly longer bill, more deeply forked tail, grey rump, dark mark on neck-sides, dark legs and darker wings, especially upper forewing. **AN** White-winged Black Tern.

Black Tern *Chlidonias niger* `LC`
Two subspecies placed in separate subspecies groups, of which both occur in region.

Eurasian Black Tern *Chlidonias (niger) niger*
Vagrant (**Barbados**).

L 23–28 cm; **WS** 57–65 cm. One subspecies in group. Juv, Aug 1998 (plus ad, perhaps this race, simultaneously but at different locality). Only other W Hemisphere record in NE Brazil, Sept 1986. **ID** Small, very dark tern with slightly forked short tail, black eyes and bill. **Ad br** is separable from American Black Tern probably only in comparative views by less blackish (slightly greyer) head and underparts, and when perched might be seen to have shorter legs. **Ad non-br** differs from American Black in lacking pale streaks on dark crown, having less extensive dark breast-side patches, and is overall larger and paler. **Juv** is generally as ad non-br, but further differs from American race in having whiter flanks and rump. **Voice** Not known to differ from American Black Tern. **SS** See also White-winged Tern.

American Black Tern *Chlidonias (niger) surinamensis*
Fairly common to very rare passage migrant (**throughout**).

L 23–28 cm; **WS** 57–65 cm. One subspecies in group. Often inland, at fresh and brackish ponds and rice fields, but also shallow coastal wetlands or even at sea on migration. Flight erratic and fluttery, pausing to pick small prey from surface, but rarely plunge-dives. Fairly common Jamaica and Puerto Rico; uncommon Cayman and Virgin Is; rare Cuba, Hispaniola, most of Bahamas and Barbados; very rare or vagrant elsewhere. Mar–Nov (mainly autumn). **ID** Small, very dark tern with slightly forked short tail, black eyes and bill, and somewhat White-winged Tern-like jizz in ad non-br plumages. **Ad non-br** has white forehead and hindneck, with blackish ear patch merging into dark cap. Upperwing and tail dark grey. Large dark patches on breast-sides. Underwing-coverts pale grey. **Ad br** has jet-black head, neck, breast and belly, and white vent. Some show patchy black-and-white pattern during moult. **1st-y** is like ad non-br but underparts vary from all white to patchy black and white. **Juv** is like ad non-br but extensively washed mid-brown above and sides dusky. Upperwing shows darker shoulder bar. Some show contrasting paler rump. **Voice** Relatively quiet compared to *Sterna* terns, but may utter high-pitched short reedy calls. **SS** The marsh tern most likely to be observed. Compare vagrant Whiskered and White-winged Terns. **TN** Has been treated as separate species from Eurasian Black Tern based on plumage, moult and structural differences.

Whiskered Tern
hybrida

adult non-breeding
adult breeding
adult non-breeding
juvenile

White-winged Tern

adult non-breeding
adult breeding
adult non-breeding
juvenile

Eurasian Black Tern
(Black Tern)

adult non-breeding
juvenile
adult non-breeding

American Black Tern
(Black Tern)

adult non-breeding
adult breeding
juvenile
adult non-breeding

145

Roseate Tern *Sterna dougallii* `LC`
Uncommon and local summer visitor throughout (**throughout**). **L** 33–43 cm; **WS** 72–80 cm. Two subspecies, of which nominate occurs in region. More maritime and pelagic than Common Tern, and flight shallower and faster. Feeds along tide-rips, in estuaries and several km offshore. Plunge-dives from greater heights and submerges more deeply than Common Tern. Breeds on barren, sandy or rocky islets. Mostly Mar–Oct, occasionally in winter (especially in S). **ID** Very pale with deeply forked tail and long tail-streamers. **Ad non-br** has very pale grey mantle and upperwing, black mask and nape, mottled black and white forecrown, and all-black bill. In flight, very faint dusky bar on lesser coverts, whitish secondaries and little dark on primaries (very faint on underwing). At rest, primaries show white line on inner edge and outer rectrices project far beyond tail (unless broken). **Ad br** has complete black cap, variable orange-red on bill (sometimes all red) and shows no trace of carpal bar. **Juv** has heavily black-scalloped upperparts and appears dark-capped. **1st-w** is like ad, but has more distinct carpal bar and darker secondaries. **Voice** Vocal at nest. Frequent contact calls include "pink" and "kr-rik" notes, higher-pitched and less harsh than Common Tern. **SS** Very similar to Common Tern, but has shorter wings and longer tail-streamers, which at rest project well beyond wingtips and in flight wings look set forward relative to total length. Much paler overall, especially wings, which show less black and no dark wedge on central primaries. Also, no dark trailing edge to underside of outerwing and all-white tail-streamers without dark edges. Additionally, in good light, ad br usually shows variable pink cast to breast, and legs are brighter red-orange. Non-br/imm show almost no carpal bar and juv has black legs. Bill black in all plumages, with variable red at base in ad breeding. See rarer Arctic and Forster's Terns.

Common Tern *Sterna hirundo* `LC`
Uncommon to rare resident, passage migrant and winter visitor (**throughout**). **L** 32–39 cm; **WS** 72–83 cm. Four subspecies, of which nominate occurs in region. Coastal lagoons, estuaries, rocky or sandy beaches, harbours, etc. Occasionally inland at rice fields and other wetlands. Also sometimes > 15 km offshore, where may feed in dense flocks over sea. Flight buoyant with deep wingbeats. Breeds in colonies on small islets. Very uncommon, local and erratic breeder Bahamas to Dominica; uncommon to rare visitor elsewhere. Year-round, but perhaps mostly May–Oct. **ID** Mid-sized, elegant tern with rather long, deeply forked tail and long bill. At rest, tail-streamers do not reach beyond wingtip. Mostly mid-grey upperparts and pale grey to whitish underparts. **Ad br** has complete black cap and usually black-tipped red bill. Legs reddish to orange. In flight, variable blackish wedge on upperside of central primaries, dark trailing edge to underside of primaries and narrow dark outer edges to streamers. **Ad non-br** has white lores and forehead, blackish bill and dull dark reddish legs. All non-br/imm plumages exhibit conspicuous dark carpal bar at wingbend. **Juv** is like ad non-br but tail shorter (no streamers), grey upperparts barred dark grey or brown; bill either blackish, has black maxilla and orange mandible, or black with orange base. **1st-w** is intermediate between juv and ad non-br. **Voice** A disyllabic, downslurred "kreee-arrr" or sharp "chip" in flight. Loud and raucous at colonies, where calls include "kierri-kierri-kierri". All calls lower-pitched and harsher than those of Arctic Tern. **SS** Separation from Roseate Tern covered therein; present species is probably routinely confused with latter in region. Compare rarer Arctic and Forster's Terns.

Arctic Tern *Sterna paradisaea* `LC`
Rare visitor or vagrant (**GA**, **LA**, **O**). **L** 33–36 cm; **WS** 76–85 cm. Monotypic. Pelagic away from breeding grounds, and status consequently perhaps under-estimated. Feeds by hovering and plunge-diving. Records (late Apr–Nov): Bahamas (Abaco, New Providence), Cuba (twice), Puerto Rico, Virgin Is, Guadeloupe (regular off E coast Apr–May) and Barbados (once). **ID** Mid-sized, relatively dark tern with very long tail-streamers. **Ad non-br** has blackish cap with large white forehead, blackish bill and legs, rather uniform grey body (including underparts) with narrow white cheeks, white rump and tail. In flight, outer primaries have inner webs grey and narrow black line at tips, and paler secondaries on underwing produce translucent 'window' across primaries. **Ad br** has complete black cap, red bill and legs, and long tail-streamers extending beyond wings at rest. **Juv/1st-w** has dark mantle and wings (including greater coverts), broad white area on secondaries and inner primaries, and rather diffuse greyish carpal bar; outertail feathers dark grey. **Voice** Usually quiet, but gives clear "pi-pi-pi-pi-pi-" and grating "krrri-errrr", higher-pitched and squeakier than Common Tern. **SS** From Common Tern by slightly slimmer build, rounder head, shorter bill, neck and legs, and longer tail-streamers, which at rest project slightly beyond wingtips, and in flight wings look set forward relative to total length. Mantle and wings darker mid-grey, contrasting more with white tail and rump. More uniform wings lack black wedge on primaries and flight feathers look translucent. Underwing has thin black line on trailing edge of primaries. Additionally, ad br has entirely blood-red bill and mid-grey breast and belly, contrasting with white cheeks, rump and tail. Non-br/imm show less contrasting carpal bar on wingbend. In juv, secondaries whitish, not grey like Common. Roseate Tern much paler overall, with black or mostly black bill in all plumages. Compare Forster's Tern.

Forster's Tern *Sterna forsteri* `LC`
Uncommon or rare to very rare winter visitor (**GA**, **O**), vagrant (**LA**). **L** 33–36 cm; **WS** 73–82 cm. Monotypic. Mainly nearshore seas and coastal lagoons. Slow wingbeats, feeding mainly by plunge-diving or surface-dipping. Most records Bahamas, Cuba, Hispaniola and Cayman Is; very rare Puerto Rico and Virgin Is; mainly Nov–Apr, but from mid Aug. **ID** Rather pale tern with deeply forked tail, white wingtips and very long tail-streamers. Very similar to Common Tern, especially in breeding plumage, but has slightly shorter wings, slightly longer bill and legs, and longer tail-streamers project beyond wingtips at rest. **Ad br** is also slightly paler, especially above, with silvery-white primaries, no black and slight contrast with rest of upperwing. Pale grey tail has thin white outer edges to streamers and white rump contrasts slightly with pale grey mantle and tail. Bill more orange than red. **Ad non-br** has bold black ear patch and dusky nape, instead of black mid to rear crown. **Imm** (like ad non-br) has carpal bar indistinct or absent. **Juv** has head pattern like ad non-br and thus different from juv Common. **Voice** Varied vocalizations (unlikely to be heard in region) include a hoarse "kyarr". All calls lower-pitched and more nasal than Common Tern's. **SS** In flight, all-white upperside to primaries always distinctive. Roseate and Common Terns have slightly shorter legs and bill, and Roseate has much faster wingbeats. Arctic Tern obviously darker, with shorter bill and legs, and translucent flight feathers.

adult breeding

adult non-breeding

juvenile

adult breeding with all-red bill

adult breeding

Roseate Tern
dougallii

juvenile

adult breeding

adult non-breeding

adult non-breeding

adult breeding

Common Tern
hirundo

adult breeding

adult non-breeding

adult breeding

Arctic Tern

juvenile

adult breeding

adult non-breeding

adult non-breeding

adult breeding

Forster's Tern

147

Sandwich Tern *Thalasseus sandvicensis* `LC`
Two subspecies placed in separate subspecies groups, of which one occurs in region.

Cabot's Tern *Thalasseus (sandvicensis) acuflavidus*
Fairly common to rare resident (**throughout**).
L 36–46 cm; **WS** 86–105 cm. One subspecies in group. Sandy or rocky beaches, bays, estuaries and harbours. Perches on buoys and piers, and feeds by plunge-diving, usually from greater height than other terns. Nests in colonies on small islands. Mainly Feb–Sept (breeds Mar–Jul), with rarer winter records (e.g. in Cuba). Commoner Bahamas, Greater Antilles and Virgin Is, with Puerto Rico and Virgin Is colonized recently (Hayes 2004); breeds S to Grenadines, but uncommon to vagrant in most of Lesser Antilles. **ID** Medium-sized pale tern with black shaggy crest, long slender bill, long pointed wings and moderately long, forked tail. **Ad br** has underparts, rump and tail white, mantle and wings very pale grey. Cap and crest solid black. Bill usually black tipped yellow (but yellow tip can be hard to see or absent), or sometimes mostly yellow or patchy yellow and black (*eurygnathus* 'Cayenne Tern'). Legs black (rarely yellow). **Ad non-br** is similar but forecrown white and mid-crown flecked black. **Juv** has reduced crest, black bill, less forked tail tipped black, and some dark grey and brownish mottling on upperparts. Legs dull blackish to yellowish. **Voice** Very vocal at colonies and typically gives loud, grating "kerrick" in flight. **SS** Gull-billed Tern looks heavier, has shorter less pointed wings, less forked tail and shorter, stouter, all-black bill. It prefers shallower water and has different feeding behaviour, often inland. Compare Roseate and Common Terns. **TN** Race *acuflavidus* (Cabot's Tern) may be closer to Elegant Tern (*T. elegans*) than nominate *sandvicensis*. Form *eurygnathus* ('Cayenne Tern') once considered a separate species but now as synonym of *acuflavidus*.

Royal Tern *Thalasseus maximus* `LC`
Common to fairly common resident but local and erratic breeder (**throughout**).
L 45–51 cm; **WS** 100–135 cm. Two subspecies, of which nominate occurs in region. Coastal lagoons, bays, estuaries, harbours and fishing ports. Follows vessels, often settles on buoys and piers. Gregarious, often with other terns. Plunge-dives. Flight buoyant with slow, steady wingbeats. Nests colonially on sandy barrier beaches and coral islands. **ID** Large tern with orange-red bill and moderately forked tail. In flight, all plumages have mostly white underwing with rather pale primaries. **Ad br** has head white with solid black cap and crest. Mantle and wings very pale grey with darker outer primaries. Legs black. **Ad non-br** has white forecrown with black confined to rear crown and nape. **Juv** is like ad non-br but less black on crown and no crest. Wings and mantle mottled brownish. Bill yellowish and feet dull yellowish to blackish. **Voice** Very vocal, especially at colonies: a characteristic loud, rolling "keer-reet". Alarm calls include a loud, hard "keet keet". All vocalizations higher-pitched than Caspian Tern. **SS** Caspian Tern larger and stockier, with heavier, coral-red bill. Non-br has full white crown streaked black (instead of white forecrown) and underwing has broad dark tips. Other terns with orange or red bills are much smaller. **TN** Genetic data emphasize that extralimital race *albididorsalis* (W Africa) is rather distinct from nominate and instead clusters with Old World Lesser Crested *Thalasseus bengalensis* and Greater Crested Terns *T. bergii* (Collinson *et al.* 2017).

STERCORARIIDAE
Skuas
7 extant species, 4 in region + 1 vagrant

Long-tailed Jaeger *Stercorarius longicaudus* `LC`
Uncommon (**Guadeloupe**) to very rare passage migrant or vagrant (**throughout**).
L 48–53 cm (including central tail-streamers of up to 22 cm); **WS** 105–117 cm. Two subspecies, of which at least one (unknown) occurs in region. Pelagic. Most records Mar–Jun and Aug–Oct, but very poorly known in region. Flight fairly graceful and almost tern-like. **ID** Small, slim and lightly built with long wings and short bill. Often stalls to settle on, or snatch items from, surface. **Ad non-br** has variable brown cap, whitish underparts washed pale yellowish on face that is often spotted or streaked darker, variable grey-brown breast-band, grey-brown scalloping or barring on flanks and pale brown upperparts. In flight, all-dark underwing typically with single pale shaft to outermost primary; upperwing has white shafts restricted to two outermost primaries and strong grey-brown barring on tail-coverts. Central rectrices short and usually pointed but can be blunt (like juv). **Ad br** has two colour morphs. **Pale morph** has very long, pointed central rectrices, dark cap, brown upperparts and whitish underparts with yellowish-buff wash to face and dusky-grey rear belly. In flight, dark flight feathers contrast with paler brown coverts; primaries as ad non-br. **Dark morph** (known only from Greenland) has dusky underparts. **Juv** has three colour morphs. **Pale morph** has whitish to pale yellowish head, underparts whitish, some with grey breast-band, variably scalloped grey-brown on flanks, and cold grey-brown upperparts with pale fringes. In flight white shafts typically restricted to 2–3 primaries (occasionally four, very rarely five), with heavily barred white and brown underwing- and tail-coverts. Central tail projection short and blunt. **Intermediate morph** has cold yellowish-grey collar separating grey face and body from paler central belly. **Dark morph** has dark body, fading rearwards on some. Barred white and brown underwing-coverts on some, others have largely dark underwing but tail-coverts usually strongly barred white and brown. **3rd-y** resembles ad but may retain some barring on wing-coverts, tail-coverts and dusky body plumage. **2nd-y** has variably barred underwing- and tail-coverts, and dirtier body. Sometimes indistinguishable from ad by 2nd-s. **1st-y** is like juv but new feathers lack pale fringes, central tail projection can be longer and more pointed. **Voice** Mewing and yelping "kreck", "kliu" and "kuep". **SS** Smaller, slighter, narrower-winged and shorter-billed with lighter, more buoyant flight than congeners, and much longer, pointed central tail in ad br and shorter, blunter central tail in juv. Arctic Jaeger usually shows more extensive white in primaries and when folded usually have pale or rusty tips, often but not always absent in Long-tailed, whilst pale and intermediate morph juv typically warmer-coloured. **AN** Long-tailed Skua.

'Cayenne Tern' first-winter

adult breeding

**Cabot's Tern
(Sandwich Tern)**

adult non-breeding

juvenile

adult non-breeding

Royal Tern
maximus

adult breeding

adult breeding pale morph

adult breeding dark morph

Long-tailed Jaeger
longicaudus

juvenile intermediate morph

Arctic Jaeger / Parasitic Jaeger   LC
Stercorarius parasiticus
Uncommon passage migrant and rare winter visitor (**throughout**). **L** 41–46 cm; **WS** 110–125 cm. Monotypic. Open seas. Aggressive, often chases other seabirds in dashing, acrobatic flight. In contrast to rest of region, considered common off Guadeloupe. Most records Mar–Jun and Sept–Nov, but sometimes earlier in autumn and in winter. Study by Rypdal (2018) suggested N European breeders occur throughout region, despite fact species has not been recorded in many territories. **ID** Medium-sized with pointed central tail projection. **Ad non-br** has two colour morphs. **Pale morph** has brown cap, pale yellowish face often spotted or streaked darker, whitish underparts with pale grey-brown breast-band, pale brown scallops or bars on flanks and brown upperparts. In flight, dark underwing is darkest on lesser coverts and axillaries, with white shafts to primary bases, brown-barred whitish undertail-coverts and often has similarly patterned uppertail-coverts. Central rectrices short and pointed. **Dark morph** is largely dark, sometimes with paler or rusty fringes, often with darker barring on flanks and variable paler barring on tail-coverts. **Ad br** has three colour morphs. **Pale morph** has long, pointed central tail, dark cap, brown upperparts, and whitish underparts with yellowish-buff face and variable grey-brown breast-band. In flight primary pattern as ad non-br. **Dark morph** is uniformly dark, sometimes with dark yellow shaft-streaks on neck-sides. **Intermediate morph** varies; paler individuals resemble pale-morph ad but have dark spots or bars on underparts. Dark individuals recall dark-morph ad but are paler, grey-brown, below and have paler face contrasting with dark cap. **Juv** has three colour morphs. **Pale morph** varies; head and body pale, belly sometimes almost whitish and upperparts have pale fringes. In flight buffish or whitish underwing-coverts, axillaries and tail-coverts with variable brown bars. **Dark morph** also varies; dark body fades rearwards on some. In flight largely dark underwing and tail-coverts, some with slightly paler bars. **Intermediate morph** is often rather uniform with rusty-brown underparts and rusty collar. In flight, brown base colour to dark-barred underwing-coverts, axillaries and tail-coverts reduces contrast. **3rd-y** recalls ad but can retain signs of immaturity, generally little barring on wing- and tail-coverts, or dusky body. **2nd-y** resembles ad but has variable bars on underwing- and tail-coverts. **1st-y** resembles juv but gradually loses rusty tinge, replaced feathers lack pale fringes, and central tail projection can be longer and more pointed. **Voice** Unlikely to be heard. A yelping "kyeew!..kyeew!..", sometimes longer and bisyllabic, a short "kiuk!" and a fast series of "kek-kek-kek-kek-kek" notes. **SS** See Long-tailed Jaeger. Pomarine Jaeger typically larger and always more heavily built with deep chest and slower, steadier flight on relaxed, measured wingbeats and fewer jerky, erratic movements. Ad br has central rectrices 'spoon-shaped' at tips with twisted shafts, whilst in juv they are very short and blunt. Juv almost never shows strongly contrasting pale head or belly, has bicoloured bill longer than congeners, and pale morph has white bases to underside of primaries and median coverts creating 'double patch'. **AN** Arctic Skua.

Pomarine Jaeger *Stercorarius pomarinus*   LC
Uncommon passage migrant and winter visitor (**throughout**). **L** 46–51 cm (including central tail-streamers of up to 11 cm); **WS** 125–138 cm. Monotypic. Open seas. Most records Oct–Jun, mainly off Bahamas, Cuba, Hispaniola and Lesser Antilles (where locally common in winter and spring, e.g. off Dominica and Guadeloupe). **ID** Fairly large with broad-based wings and central tail 'spoon-shaped' at tips, with twisted shafts in ad br. **Ad non-br** has two colour morphs. **Pale morph** has dark brown cap, whitish underparts with pale yellowish face, pale brown breast-band, and often extensive pale brown scalloping or barring on flanks. Dark underwing, darkest on lesser coverts and axillaries, with white shafts to primary bases and strong brown barring on white tail-coverts. Central rectrices short and blunt, often not projecting much. **Dark morph** has variable paler barring on tail-coverts and flanks, and pale fringes. **Ad br** has two colour morphs. **Pale morph** has twisted central rectrices with blunt 'spoon-shaped' tips, dark cap, brown upperparts and whitish underparts with yellowish-buff face, variable grey-brown breast-band. Pale orange-pink bill with dark tip. In flight primaries as ad non-br. **Dark morph** is uniformly dark, sometimes with dark yellow shaft-streaks on neck-sides. **Intermediate morph** varies; resembles dark-morph ad but usually paler, and dark cap contrasts with paler neck-sides. **Juv** has three colour morphs. **Pale morph** has head pale buff or greyish brown, underparts pale brownish white or sandy brown, and pale or rusty fringes above. Buffish or whitish underwing-coverts, axillaries and tail-coverts with variable brown barring. **Intermediate morph** has head and neck greyish brown without paler collar, sometimes darker contrasting with greyish-brown underparts. In flight pale ground colour to darker-barred underwing-coverts, axillaries and tail-coverts creates strong contrast. **Dark morph** has dark body, underwing-coverts variably barred and undertail-coverts usually distinctly, albeit perhaps narrowly, barred. **3rd-y** resembles ad but can retain some barring on wing-coverts, tail-coverts or dusky body. **2nd-y** has variable barring on underwing-coverts and tail-coverts. **1st-y** resembles juv but has paler neck and new feathers lack pale fringes. **Voice** Somewhat modulated nasal "nyAheheh" and short barks, but probably unlikely to be heard. **SS** See Arctic and Long-tailed Jaegers. **AN** Pomarine Skua.

Great Skua *Catharacta skua*   LC
Vagrant (**Guadeloupe**, **Martinique**). **L** 50–58 cm; **WS** 125–140 cm. Monotypic. Open seas. Perhaps regular albeit rare visitor, but just one documented record, NW of Guadeloupe, Mar 2016 (eBird). Most likely in boreal winter, but recent sight records mainly Apr–Jun (e.g. *AB* 37: 917, *NAB* 61: 526), plus others in Feb and Oct. Off Barbados an unidentified *Catharacta* Nov 1960 might have been Great Skua based on date (however, both Great and South Polar Skuas could occur in any month). **ID** Large, bulky skua with heavy bill and powerful, gull-like flight; soars more than jaegers. **Ad** is largely brown with coarse yellow-brown flecking. In flight, large white primary patches above and below. **Juv** has plainer reddish-brown plumage, less obvious white wing patches, and head often distinctly darker than body. **Voice** Generally silent at sea. **SS** Pale to intermediate morphs of South Polar Skua (most likely to be seen in austral winter, rather than boreal winter as in present species) display strong contrast between underwing and flanks. Dark morph has less distinct golden streaks and proportionately thinner bill, but perhaps best separated using moult pattern. In South Polar Skua, primary moult rapid (completed in c. 2 months), often showing several renewing primaries simultaneously.

adult breeding pale morph

juvenile pale morph

adult breeding dark morph

adult breeding pale morph

Arctic Jaeger / Parasitic Jaeger

adult breeding intermediate morph

adult breeding pale morph

Pomarine Jaeger

adult breeding dark morph

juvenile pale morph

adult breeding pale morph

adult

Great Skua

151

South Polar Skua *Catharacta maccormicki* LC
Rare to uncommon visitor (**LA**), vagrant (**Cuba**, **Puerto Rico**). **L** 50–55 cm; **WS** 130–140 cm. Monotypic. Pelagic. Probably regular, especially in boreal summer: first confirmed record, off Îles des Saintes, Guadeloupe, May 1967, originally ascribed to extralimital Brown Skua *C. antarctica lonnbergi* (Hudson 1968), but subsequently to present species (Devillers 1977); records Nov–Jun (probably Jul/Aug), mainly Mar–May and principally from extreme S of region, but one off Aruba (Luksenburg & Sangster 2013) and study using data-loggers (Kopp *et al.* 2011) suggest species could occur throughout basin. **ID** Large, broad-bodied and broad-winged with deep bill; very similar (especially dark morph) to Great Skua. **Ad** has three colour morphs. **Pale morph** has creamy to greyish-brown head, underparts and sometimes upper mantle, often with warm brown tinge when fresh, becoming whitish with wear and paler fringes can create an indistinctly scaled appearance, typically paler at bill base and sometimes darker around eye. Brown above, sometimes with pale shaft-streaks on scapulars and cold pale fringes to smaller wing-coverts. Pale head and underparts contrast with very dark axillaries and underwing-coverts, but also with brown back, rump and tail. Large white patch at base of primaries and can have pale line at edge of underwing-coverts during moult. **Dark morph** is fairly uniform blackish brown, with golden streaks on neck-sides the only warm coloration; some have pale forehead. **Intermediate morph** typically has dark hood, often with pale blaze at base of bill, paler brown nape, mantle and underparts, and dark brown upperparts often with pale shaft-streaks on scapulars. **Juv** resembles intermediate morph ad but has greyer head and underparts, lacks streaks on nape and neck-sides, and pale-headed birds can show dark around eye; often has pale bill with darker tip. **Voice** Calls include short duck-like quacks and a drawn-out modulated hoarse "hyeheheheh", but unlikely to be heard. **SS** See Great Skua.

ALCIDAE
Auks
24 extant species, 1 vagrant in region

Little Auk / Dovekie *Alle alle* LC
Vagrant (**Bahamas**, **Cuba**).
L 17–20 cm; **WS** 34–48 cm. Two subspecies, of which nominate has occurred in region. Pelagic in winter; inshore following storms. Quick, whirring wingbeats. Records Oct–May (mostly Nov–Mar); none in Cuba since 1980s. **ID** The only auk to have occurred in region. Small and compact but comparatively large head; short, stubby black bill. **Ad non-br** has white to upper breast and ear-coverts. **Ad br** has glossy black head, neck and upperparts, with browner face and upper breast. Wings black with white-streaked scapulars and white-tipped secondaries. Underwing appears all dark in flight. **Juv** is like ad br, but generally paler and browner. **Voice** Silent at sea. **SS** None.

TYTONIDAE
Barn-owls
16 extant species, 2 in region

Common Barn-owl / Barn Owl *Tyto alba* LC
Twenty-eight subspecies divided into eight subspecies groups, of which one occurs in region.

American Barn-owl *Tyto (alba) furcata*
Locally common (**GA**) to uncommon resident (**O**).
L 29–44 cm; **WS** 90–100 cm. Eight subspecies in group, of which three (*pratincola*[A] Bahamas and Hispaniola, *furcata*[B] Cuba, Jamaica and Cayman Is, *niveicauda*[C] on I of Pines) occur in region. Open areas at all elevations; often nests and roosts in buildings. Largely nocturnal, rarely hunts at dusk. North American migrants reach Greater Antilles, at least, in winter. Has recently bred and seems well established N & W Puerto Rico (race?). Reportedly introduced, race unknown, in 1980s to San Andrés (McNish 2011). **ID** Pale, long-legged owl with distinctive heart-shaped face. **Ad** (*furcata*) is highly variable. Typically orange-buff upperparts streaked greyish, white facial disc, buffish-white underparts usually with fine dark spots on breast and flanks. Whitish secondaries form pale area on closed wing; underwing mainly white. Sexes similar, but ♀ often slightly darker. **Juv** is similar to ad, or more heavily spotted. **GV** Considerable variation renders separating races difficult: *pratincola* (perhaps visits Cuba in winter) typically has whitish-buff underparts but otherwise very similar; *niveicauda* averages whiter and paler than *furcata*. **Voice** Harsh screeches and hisses. **SS** See Ashy-faced and Lesser Antilles Barn-owls. **TN** Recently considered to merit species status based on genetic evidence (Uva *et al.* 2018). Race *niveicauda* often subsumed in *furcata*.

Ashy-faced Owl *Tyto glaucops* LC
Three extant subspecies divided into two subspecies groups, of which both occur in region.

Ashy-faced Owl *Tyto (glaucops) glaucops*
Locally fairly common resident (**Hispaniola**).
L 27–33 cm. One extant subspecies in group. Endemic. Open woodland and scattered trees (including urban areas) but generally prefers more heavily forested areas than American Barn-owl; to 2000 m. Nocturnal, never observed hunting in daylight, but sometimes active in twilight at dusk/dawn. Current status in Haiti poorly known. **ID** Distinctive *Tyto* with silvery-grey facial disc. **Ad** has dark greyish-brown upperparts vermiculated pale brown and dusky, with buff or tawny mottling on wings, and tawny underparts with some transverse barring. Light and dark morphs, plus intermediates. **Juv** has darker face, closely resembling juv American Barn-owl. **Voice** A hissing cry, prefaced by series of higher-pitched ratchety clicks and a c. 2–3-second screech, "criiisssssh". **SS** Grey facial disc distinctive in ad. Juv smaller than young American Barn-owl, but plumage otherwise similar. See (allopatric) Lesser Antilles Barn-owl. **TN** See Lesser Antilles Barn-owl. **AN** Ashy-faced Barn-owl.

Lesser Antilles Barn-owl *Tyto (glaucops) insularis*
Fairly common resident (**LA**).
L 27–33 cm. Two subspecies in group, of which both (*nigrescens*[A] Dominica, *insularis*[B] St Vincent, Bequia, Union, Carriacou and Grenada) occur in region. Endemic. Nocturnal. Open woodland, scattered trees, farmland and caves. Behaviour poorly known, but probably similar to other barn-owls. Two records St Martin, Jul 2002, perhaps involved *nigrescens* but other barn-owls not eliminated (*NAB* 56: 497). **ID** Quite different from conspecific Ashy-faced Owl. **Ad** (*insularis*) differs from Ashy-faced by vinaceous-brown facial disc, blackish-grey upperparts sparsely spotted or streaked white, darker wings mottled and edged tawny, cinnamon-buff underparts mottled dusky with some small white spots, and marginally smaller size. **Juv** is similar to other young barn-owls. **GV** Race *nigrescens* almost lacks white spotting above and is less mottled below. **Voice** Little-known. Piercing scream and clicking notes

adult
pale morph

adult
pale morph

South Polar Skua

adult
dark morph

adult
non-breeding

adult
breeding

Little Auk / Dovekie
alle

American Barn-owl
(Common Barn-owl / Barn Owl)
furcata

Ashy-faced Owl

Lesser Antilles Barn-owl
(Ashy-faced Owl)

nigrescens

insularis

153

recorded on Dominica; other vocalizations probably similar to other *Tyto*. **SS** American Barn-owl larger with whitish facial disc and underparts. **TN** Sometimes treated as separate species *T. insularis* (König *et al.* 2008) or, more frequently, as races of Common Barn-owl.

STRIGIDAE
Typical Owls
220 extant species, 7 in region + 2 vagrants

Cuban Pygmy-owl *Glaucidium siju*   LC
Fairly common resident (**Cuba**).

L 16–17 cm. Three subspecies, all of which (nominate[A] most of mainland Cuba, and many offshore cays, *turquinense*[B] Pico Turquino, SE Cuba, *vittatum*[C] I of Pines) occur in region. Endemic. Semi-deciduous, evergreen and swamp woodland, pine forest, second growth, open areas with small palms (*Thrinax*), large gardens with many trees, and groves, to 1500 m. Partly diurnal and most frequently detected by vocalizations. **ID** Small owl with round head, indistinct facial disc and no ear-tufts. Occipital 'face' on nape unique in region. The most frequently observed owl in Cuba. **Ad typical morph** has greyish-brown head with white spots on crown, nape and head-sides, whitish eyebrows; mantle barred. Central breast generally whitish, bordered by densely barred upper breast-sides and buff-spotted flanks. **Ad rufous morph** is rufous-brown overall. **Juv** has unspotted crown. **GV** Race *turquinense* said to be darker with fewer spots and streaks on upperparts; *vittatum* larger with more distinctly barred upperparts. **Voice** Whistle-like "jiu" at short (c. 4-second) intervals, or repeated "jiu, jiu, jiu, jiu", starting softly and increasing in frequency and tone. **SS** Larger Bare-legged Screech-owl has long, bare tarsi, brown eyes and lacks 'false eyes' on nape, and much less likely to be seen in open by day. **TN** Race *turquinense* only rather doubtfully distinct, and true range and validity of *vittatum* also needs renewed scrutiny (Kirkconnell *et al.* in press).

Burrowing Owl *Athene cunicularia*   LC
Fairly common or common resident (**Bahamas**, **Hispaniola**), local (**Cuba**).

L 19–26 cm. Twenty-two subspecies, of which three (*floridana*[A] Bahamas, *guantanamensis*[B] SE Cuba, *troglodytes*[C] Hispaniola including I Gonâve and I Beata) occur in region. Two others (*amaura* Antigua and Nevis, *guadeloupensis* Guadeloupe) are extinct. Open terrain including scrub, abandoned fields, sandy pine savannas and golf courses. Active at all hours, often perching on ground or posts. Bobs when approached. Flight low and fast, direct and undulating. Race *floridana* reaches Cuba as vagrant or rare winter visitor (Kirkconnell *et al.* in press). **ID** Small, easily recognized owl with long, sparsely feathered legs, flat crown and lemon-yellow eyes. **Ad** has broad buff-white eyebrows and malar stripe; upperparts brown, with crown, back and scapulars spotted pale buff. Underparts whitish with broad brown bars. ♀ is more heavily barred below. **Juv** is brown above, streaked or spotted buff-white, with pale band on wings; chest dark brown. **GV** Races vary mainly in biometrics, depth of coloration, and strength and extent of markings, but much variation. **Voice** ♂ gives "coo cooooo", and 5–8 high-pitched and raspy "chéh" notes; ♀ a series of downslurred notes or warbles. **SS** Compare Bare-legged Screech-owl (endemic to Cuba), which is strictly nocturnal (unless disturbed at nest-hole) and confined to wooded habitats. The only small owl resident in Bahamas and on Hispaniola. **TN** Taxonomy of Cuban birds problematic: all originally assigned to *floridana*, prior to description of *guantanamensis*, but lack of specimens prevents designation of local populations on I of Pines and in W Cuba to race (Kirkconnell *et al.* in press).

Northern Saw-whet Owl *Aegolius acadicus*   LC
Vagrant (**Bahamas**).

L 18–21.5 cm. Two subspecies, of which nominate occurs in region. Abaco, sometime winter 1992/93, died in captivity (*NAB* 67: 533). Wooded areas, usually roosts in dense foliage. **ID** Comparatively small, distinctive, resolutely nocturnal owl. **Ad** has greyish-brown to warm brown upperparts, partially spotted and streaked white. Underparts white, broadly striped or blotched rufous-brown. Irides yellow to golden-yellow. **Juv** lacks any white above, except on wings and tail, has unmarked brown breast and buff belly, and black-brown facial disc with conspicuous Y-shaped white marking between and above eyes. **Voice** Unlikely to be heard, although sharp alarm calls may be given in winter. **SS** Quite unlike any other small owl in region: distinctively patterned facial disc, different, somewhat square head shape, bold scapular spots and rufous-streaked underparts.

Stygian Owl *Asio stygius*   LC
Fairly common to uncommon (**Cuba**) or rare resident (**Hispaniola**).

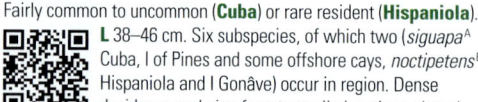

L 38–46 cm. Six subspecies, of which two (*siguapa*[A] Cuba, I of Pines and some offshore cays, *noctipetens*[B] Hispaniola and I Gonâve) occur in region. Dense deciduous and pine forests at all elevations, though much less common in lowlands of Dominican Republic and current status in Haiti unknown. Nocturnal. Day-roost usually in dense foliage, often near trunk. In alarm, becomes very slim, with ear-tufts erected vertically. Slow wingbeats sometimes interspersed by long glides. **ID** Large, very dark owl with prominent ear-tufts and yellow eyes. **Ad** (*siguapa*) has blackish facial disc, blackish-brown upperparts spotted and barred buff, dirty buff underparts heavily streaked and barred brown. **Juv** has sooty-black upperparts, more or less mottled white or buff; underparts buffish, heavily spotted and streaked sooty black. **GV** Race *noctipetens* perhaps averages even darker, with fewer and more restricted pale markings above. **Voice** Often only vocal when breeding. ♂ gives single deep, emphatic "whuoh" with downward inflection, repeated at intervals of 6–10 seconds, ♀ a short, cat-like "miah". Scratchy "whag-whag-whag" given by both sexes when excited. **SS** Compare Long-eared and Short-eared Owls. **TN** Race *noctipetens* often synonymized with *siguapa* (e.g. König *et al.* 2008).

Northern Long-eared Owl *Asio otus*   LC
Vagrant (**Cuba**).

L ♂ 35–38 cm; ♀ 37–40 cm; **WS** 90–100 cm. Four subspecies, of which one (*wilsonianus*) occurs in region. Oct 1932 (Bond 1956). Deciduous woodland, thickets and conifers. Nocturnal and crepuscular, but migrants can arrive on coast by day. Roosts against tree or in dense thicket. **ID** Mid-sized owl with long, narrow ear-tufts. **Ad** has yellow eyes, rufous facial disc with narrow dark border, mottled dark grey-brown and buff upperparts, orange-buff primary bases and pale buff underparts with dark central streaks (crossed by transverse bars) that typically extend over entire belly. **Juv** has facial disc, wings and tail as ad, rest of plumage greyish barred buff. **Voice** Generally silent in winter. **SS** Compare Short-eared Owl. Stygian Owl larger, darker, with more uniform blackish facial disc. **AN** Long-eared Owl.

Cuban Pygmy-owl
siju

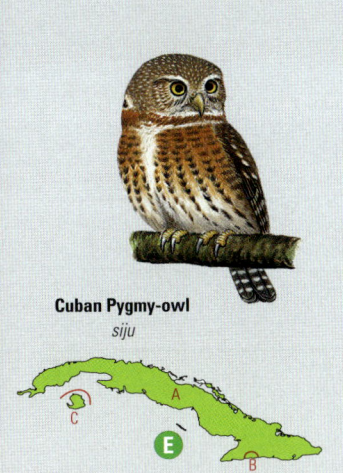

Burrowing Owl
troglodytes

Northern Saw-whet Owl
acadicus

Stygian Owl
siguapa

Northern Long-eared Owl
wilsonianus

155

Short-eared Owl *Asio flammeus* `LC`
Eleven subspecies divided into two subspecies groups, of which one occurs in region.

Common Short-eared Owl *Asio (flammeus) flammeus*
Locally common (**Cuba**, **Hispaniola**) to uncommon (**Puerto Rico**) or rare and irregular resident (**Cayman Is**), vagrant (**LA**). **L** 34–42 cm; **WS** 95–110 cm. Ten subspecies in group, of which four or five (nominate vagrant Grand Turk and St Barthélemy, *cubensis*[A] Cuba including several offshore cays, *domingensis*[B] Hispaniola, *portoricensis*[C] Puerto Rico, perhaps *pallidicaudus* vagrant Barbados) occur in region. Vagrants: specimen (of nominate race) and sight record, Grand Turk, Jan 1961 (Schwartz & Klinikowski 1963), St John, St Thomas, Tortola and St Croix, in Virgin Is (presumably *portoricensis*), St Barthélemy (assigned to nominate) and Barbados, Mar 2006 (either *portoricensis* or NE South American *pallidicaudus*, Buckley *et al.* 2009). Another c. 6 records from Bahamas, in at least Feb to Jun, on the Biminis, Exumas and Long I (e.g. *NAB* 61: 654, 64: 339, 67: 533), unascribed to race but perhaps *cubensis*. Revision required of claimed records of nominate, as both pre-date description of *cubensis* and knowledge that Cuban birds have invaded N to S Florida (Hoffman *et al.* 1999). Open areas in lowlands, especially rice fields and citrus plantations in Cuba, but also short-grass pastures and savannas elsewhere. Largely diurnal, most active at dawn and dusk. Stiff 'bouncing' wingbeats, quarters ground with frequent hovers and stalls; recalls a harrier. **ID** Mid-sized owl with large, round facial disc and very small (often-concealed) ear-tufts. **Ad** (*cubensis*) has bright yellow eyes, grey-white facial disc with black smudges around eyes, brown-and-buff upperparts, and pale buff underparts with dense breast streaking. In flight, long, broad wings buffish white on underside with broad blackish bands on primaries and carpal crescent. ♀ is browner above, buffier below and more heavily streaked. **Juv** has crown and rump dark brown, facial disc brown-black with feathers tipped buff, mantle incompletely barred, and underparts warm buff. **GV** Complex and potentially unresolved. Race *domingensis* overall paler both above and below, with less saturated underparts and possibly fewer streaks; *portoricensis* possibly indistinguishable from latter (Hoffman *et al.* 1999, König *et al.* 2008), but perhaps warmer brown, especially above (Garrido 2007). Race *pallidicaudus* (not always recognized) is plainly very similar to both *domingensis* and *portoricensis*, but is apparently longer-winged and perhaps shorter-tailed, arguably with rustier underparts. Vs. Greater Antillean birds, nominate race longer-winged and shorter-legged, upper back mostly tawny with dark brown central stripe, upper-tail-coverts tawny yellow, and underparts less buffy with streaking extending to tibial feathering. **Voice** Generally silent, ♂ gives 13–16 "hoo-hoo-hoo…" notes. Occasionally a barking call when flushed. **SS** Northern Long-eared Owl has longer, more conspicuous ear-tufts, warmer, rufous facial disc and belly streaking with transverse bars; in flight, richer, more orange-buff primary bases on upperwing, more evenly barred primaries on underside, with less solid dark bands, a less strongly barred tail and more extensively streaked underparts. **TN** Race *domingensis* formerly considered a separate species, including *portoricensis* (subsequently also *cubensis*). Cayman population (seems very rare and possibly erratic) never assigned to race and no specimens known (Bradley 2000); perhaps sustained, or periodically recolonized, from Cuba? Floridan population established via immigration from Cuba (Hoffman *et al.* 1999) and Cuban birds might also reach Hispaniola (Latta *et al.* 2006).

Jamaican Owl *Pseudoscops grammicus* `LC`
Fairly common resident (**Jamaica**).

L 27–33 cm. Monotypic. Endemic. Open woodland, forest edge, occasionally parks, plantations and large gardens. Commonest in lowlands, but to at least 600 m. Strictly nocturnal, repeatedly uses same day-roost for months. **ID** Mid-sized, warm-coloured owl with conspicuous ear-tufts and very dark hazel-brown eyes. **Ad** has rather well-marked rufous facial disc edged black and white, finely vermiculated upperparts, and dark brown shaft-streaks on underparts. Tail short, irregularly barred dark brown. **Juv** is paler above, with back pale greyish brown, rest of plumage light dull cinnamon-buff. **Voice** High, quavering hoot and guttural growl, "whow" or "tu-whoo", repeated every few seconds. Upslurred, wailing "kwe-eeh" given by fledged young. **SS** Risk of confusion with American Barn-owl, the only other owl in Jamaica, negligible.

Puerto Rican Screech-owl *Megascops nudipes* `LC`
Common resident (**Puerto Rico**), very rare or perhaps extinct (**Virgin Is**).

L 20–23 cm. Two subspecies, both of which (nominate[A] Puerto Rico, *newtoni*[B] Virgin Is, considered possibly extinct) occur in region. Endemic. Dense woodland, shade-coffee plantations and caves, to 900 m. Nocturnal; most vocal early morning and evening. Sometimes mentioned for Vieques, off SE Puerto Rico, but evidence for former presence very weak (Gemmill 2015). Most recently reported from Virgin Is, on St John, Jan 2015 (*NAB* 69: 307). **ID** Dimorphic screech-owl with mostly bare tarsi, rounded head without ear-tufts and reddish-brown eyes. **Ad brown morph** has dark facial disc, white eyebrows and chin. Dark brown upperparts with pale spots and irregular dark bars, scapulars with indistinct white spots. Underparts paler with dark vermiculations and shaft-streaks. **Ad rufous morph** (scarce) is reddish brown with plainer upperparts. **Juv** is nearly uniform olive-brown above, head and neck paler brown, underparts distinctly barred dusky brown and pale fulvous. **GV** Race *newtoni* has brown plumage rufous-tinged, and slightly less patterned underparts. **Voice** Primary song a deep, guttural, toad-like trill c. 3–5 seconds long. Other song a shorter trill (c. 2 seconds) with slight rise in middle, then drops and fades. Also cackling and cooing calls. **SS** Only small owl in Puerto Rico.

Bare-legged Screech-owl *Margarobyas lawrencii* `LC`
Fairly common resident (**Cuba**).

L 20–23 cm. Monotypic. Endemic. Deciduous and open woodland with small-stature palms (*Thrinax*) and in caves and on cliffs in or adjacent to woodland, throughout mainland Cuba and I of Pines, also some cays off N coast; to at least 1250 m. Nocturnal. Strongly declining at least locally. **ID** Small owl with proportionately large head (no ear-tufts) and long, bare tarsi. **Ad** has creamy-white facial disc with broad dark rim and long white eyebrows. Upperparts brown with white spots, wings barred; relatively long tail, outer rectrices with thin whitish bars. Underparts whitish with dark shaft-streaks. **Juv** is less spotted above and tail feathers usually plain brown (can occur in ad). **Voice** ♂ song a low, accelerating "cu-cu-cu-cucucuk" with bouncing-ball rhythm, rising slightly at end, also repeated "cu co, cu co, cu co...". ♀ utters harsh shriek, "hui hui hui hui ...". **SS** Burrowing Owl (sympatric but never syntopic) larger with feathered tarsi, yellow eyes, and barred (not streaked) underparts. See Cuban Pygmy-owl. **AN** Cuban Screech-owl, Bare-legged Owl.

Common Short-eared Owl
(Short-eared Owl)
domingensis

Jamaican Owl

brown morph

rufous morph

nudipes

newtoni

Puerto Rican Screech-owl

Bare-legged Screech-owl

157

CATHARTIDAE
New World Vultures
7 extant species, 1 vagrant in region + 1 introduced

Turkey Vulture *Cathartes aura* `LC`
Introduced, common and widespread (**GA**) to rare and local resident (**N Bahamas**), vagrant (**Virgin Is**).

L 62–81 cm; **WS** 160–182 cm. Four subspecies, of which nominate occurs in region. Open areas throughout, including open forest, scrubland, cane fields, pastures and towns. Continually rocks sideways, producing distinctive flight pattern with two-toned wings held above horizontal in strong dihedral and rarely flapping. Said to have been introduced to Jamaica in 17th century, spreading to Cuba, and introduced to Puerto Rico c. 1880 (although latter is sometimes claimed to have been colonized naturally); similarly, presence on Hispaniola either due to introduction (c. 1900 or as early as 1770!) or natural colonization. Probably formerly resident Cayman Is, but currently only an annual visitor. Some immigrants from North America probably reach region in winter. **ID** Dark vulture, with long tail and small head. **Ad** is mostly brownish black with bare dull pink to red head and neck. Bill whitish. Wings show pale grey feathers contrasting markedly with dark coverts. **Juv** is duller brown, with dull pinkish-grey head and dull blackish bill. **Voice** Largely silent. **SS** Rare American Black Vulture is all black (pale, not red head), has shorter tail and different flight pattern (rather flat profile, no rocking and more flapping), with white wingtips (instead of pale flight feathers).

American Black Vulture *Coragyps atratus* `LC`
Rare non-breeding visitor (**Cuba**), vagrant (**Bahamas**, **Jamaica**, **Puerto Rico**, **Grenada**).

L 56–74 cm; **WS** 133–160 cm. Three subspecies, of which probably two (nominate Bahamas and Greater Antilles, *brasiliensis* Grenada) occur in region. Open lowlands. Soars on mostly flat wings, alternated with occasional quick flaps, and does not rock sideways (unlike Turkey Vulture). More than ten records Cuba, but only one (involving 15 individuals) documented, virtually year-round; one record Puerto Rico, Dec 2018. Occasional records Grenada presumably involve race *brasiliensis* from South America. Exceptional report of c. 100 from Barbados (Bond 1965) not accepted. **ID** All-black vulture with small bare head and short tail. In flight, conspicuous whitish patch on underside of wingtip. **Ad** is dull black, with only slight iridescence on wings and tail. **Juv** is almost identical, but shows fewer folds in bare neck skin. **GV** Race *brasiliensis* smaller than nominate with pale area on underwing more extensive. **Voice** Largely silent. **SS** See much commoner and more widespread Turkey Vulture. Cuban Black Hawk has pale tail-band. **AN** Black Vulture.

PANDIONIDAE
Osprey
1 extant species, 1 in region

Osprey *Pandion haliaetus* `LC`
Four subspecies divided into two subspecies groups, of which one occurs in region.

Western Osprey *Pandion (haliaetus) haliaetus*
Common to uncommon resident, passage migrant and winter visitor (**throughout**).

L 53–61 cm; **WS** 127–174 cm. Three subspecies in group, of which at least two (*ridgwayi* resident Bahamas and Cuba, *carolinensis* migrant and winter visitor throughout) occur in region. Nominate race (Old World) potential vagrant; possible record, St Lucia, Oct–Nov 1988 (Keith 1997). Calm fresh, brackish or saltwater bodies. Plunges feet first and submerges in water to catch fish. Flaps more than glides. Nests along lakes, seashores, marshes and rivers. Race *carolinensis* mainly a common non-breeding visitor to Bahamas, Greater Antilles, Virgin and Cayman Is, uncommon Lesser Antilles, mostly Sept–Apr (Jul–May); has occasionally bred Cuba. Fairly common but local breeding resident in Bahamas and Cuba (*ridgwayi*). **ID** Large, with long, narrow wings and distinctive flight silhouette of wings bent at carpal, forming broad 'M'. **Ad** ♂ (*carolinensis*) has blackish-brown upperparts and broad eyestripe, and white head and underparts. Some show a few dark streaks on breast. In flight, heavily barred underwing and tail, and conspicuous blackish patch at wingbend, contrast with mainly white underparts and wing-coverts. **Ad** ♀ is slightly larger, with dark-streaked breast and crown. **Imm** is similar to ad ♀, but slightly paler above, with buff-edged feathers, giving mottled look. **GV** Race *ridgwayi* smaller with paler head and only trace of eyestripe; nominate differs most noticeably in having brown breast-band and much broader eyestripe. **Voice** Relatively silent away from nest, but noisy near it. Commonest call a single or repeated high-pitched whistle "cheek" or falling "piu-piu-piu-piu". **SS** Flight silhouette distinctive. Only raptor that plunges and submerges in water.

ACCIPITRIDAE
Hawks and Eagles
248 extant species, 13 in region + 4 vagrants

Hook-billed Kite *Chondrohierax uncinatus* `LC`
Very rare resident (**Grenada**).

L 38–43 cm; **WS** 76–83 cm. Two subspecies, of which one (*mirus*) occurs in region. Deciduous, evergreen and montane rainforests; also dry (xeric) scrub. Feeds almost exclusively on tree-snails. Race *mirus* endemic to Grenada, estimated to number just 15–30 birds in 1987, but 50–75 in 2000–2006. Threats include habitat loss (including by hurricanes) and introduced snails, which are probably too large for present species to tackle (Thorstrom & McQueen 2008).
ID Mid-sized raptor with large, deeply hooked bill. In flight, head projects and long and broad, butterfly-shaped wings look heavily barred blackish and white. Rather long tail and short yellow legs. Conspicuous orangey supraloral spot, pale eyes and greenish-yellow facial skin give parrot-like look. Unlike continental populations (nominate), no dark morph. **Ad** ♂ is dark grey above and usually grey barred whitish below. Tail has two broad pale grey to whitish bands. **Ad** ♀ is dark rufous-brown above with rusty nuchal collar and barred reddish brown below. **Imm** is dark brown above, with white cheeks and nuchal collar. Underparts, thighs and tail whitish variably barred dark brown. Tail has more bands, and eyes dark. **Voice** Gives 2–3 whistled notes, also shrill scream. **SS** Bill shape, orangey supraloral spot and, in flight, longish tail, projecting head and long broad, heavily barred wings distinguish all plumages in limited range. Additionally, pale eyes distinguish ad. See commoner, shorter-tailed Broad-winged Hawk. Cuban Kite allopatric. **TN** Previously considered conspecific with Cuban Kite. **AN** Grenada Hook-billed Kite.

juvenile

adult

adult

Turkey Vulture
aura

American Black Vulture
atratus

♂

♂

ridgwayi

carolinensis

Western Osprey
(Osprey)

♂

♀

♂

♂

♀

Hook-billed Kite
mirus

159

Cuban Kite *Chondrohierax wilsonii* `CR`
Extremely rare, almost extinct, resident (**Cuba**).
L 38–43 cm; **WS** 76–83 cm. Monotypic. Endemic. Lowland gallery forest (below 500 m) with tall trees. Very scarce, elusive and little-known. Feeds on snails. Now confined to far E Cuba (last reports 2004 and 2009). **ID** Mid-sized, forest raptor with massive hooked bill. Similar to race *mirus* of Hook-billed Kite. **Ad** differs by heavier, all-yellow (not dark) bill, yellow-green (not whitish) eyes and less variable plumage; both sexes show barred hindcollar. **Imm** has mostly blackish-brown upperparts, and white underparts and hindneck. Eyes dark. **Voice** Whistles similar to Hook-billed Kite (GMK pers. obs.). **SS** Very scarce. Huge, heavily hooked yellow bill, orangey supraloral spot, longish tail and heavily barred underwing unique in range, but see much commoner Broad-winged Hawk. **TN** See Hook-billed Kite. **AN** Cuban Hook-billed Kite.

Swallow-tailed Kite *Elanoides forficatus* `LC`
Uncommon to rare passage migrant (**Bahamas**, **Cuba**, **Cayman Is**), vagrant elsewhere.
L 52–66 cm; **WS** 119–136 cm. Two subspecies, of which nominate occurs in region. Usually near coasts, over fields and open woodland; alone, occasionally flocks of up to 20, rarely more, migrating together. Flight supremely effortless and graceful. Mostly Aug–Oct (Jul–Dec), rarely Feb–Apr (once Jun, Bahamas). Apparently increasing in region, with records in rest of Greater Antilles, Guadeloupe and Martinique. **ID** Elegant and distinctive raptor with long, deeply forked tail and narrow pointed wings. **Ad** has white head and underparts, black back, wings and tail. From below, black flight feathers contrast with white coverts. **Imm** has tail less deeply forked, black areas slightly browner, and sometimes head and upper breast indistinctly streaked brownish. **Voice** High, shrill, whistled "klee-klee-klee" or "peet-peet-peet". **SS** Unmistakable.

Western Marsh-harrier / Eurasian Marsh-harrier *Circus aeruginosus* `LC`
Vagrant (**Puerto Rico**, **Guadeloupe**, **Barbados**).
L 43–54 cm; **WS** 115–145 cm. Two subspecies, of which nominate occurs in the region. Habits and habitat like Northern Harrier. Puerto Rico, ♀, Jan–Mar 2004, imm, Jan–Feb 2006 (Merkord *et al.* 2006); Guadeloupe, imm, Nov 2002–Apr 2003 (Levesque & Malglaive 2004), ♀-type, Oct 2015–Mar 2016 (*NAB* 70: 129, 242); ♀-type, Barbados, Nov 2015–Jan 2016 (*NAB* 70: 242). **ID** Bulky harrier with broad, rounded wings and rather variable plumage. **Ad** ♂ has pale head, mainly brown upperparts, streaked underparts and solid dark brown belly. Tricoloured upperwing with brown coverts, silver-grey secondaries and primary-coverts, and black primaries. **Ad** ♀ is dark chocolate-brown with creamy crown, throat and shoulders. In flight, underwing has small pale flash on primaries and creamy leading edge to inner wing; indistinct cream breast-band. Rare melanistic morph all dark in both sexes. **Imm** ♂ acquires grey in plumage with age. **Juv** is like ♀ but lacks pale shoulder, often completely dark chocolate except creamy crown and throat. **Voice** Unlikely to be heard. **SS** Northern Harrier (commoner in Puerto Rico, vagrant further S) shares habitat and similar behaviour, but smaller and lighter built. Additionally, ad ♂ paler and largely grey above, and ad ♀/imm streakier below, without contrasting pale crown and throat. **TN** Previously considered conspecific with extralimital Eastern Marsh-harrier *C. spilonotus* (E Asia).

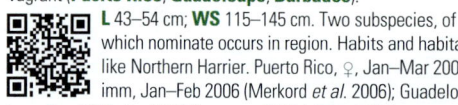

Northern Harrier *Circus hudsonius* `LC`
Uncommon to rare winter visitor and passage migrant (**GA**, **O**), vagrant (**LA**).
L 41–50 cm; **WS** 97–122 cm. Monotypic. Marshes, swamps, open savannas and rice fields. Flies low, with continuous flapping, alternated with distinctive tilting glides on wings held well above horizontal and head down, searching for prey. Mostly recorded Bahamas and Cuba, Aug–Apr. Vagrant as far S as St Lucia and Barbados. **ID** Medium-sized, typical harrier with owl-like facial disc, long wings, long, narrow, banded tail, long legs and prominent white rump. **Ad** ♂ has mostly pale grey upperparts and whitish underparts with reddish-brown spots on breast and flanks, and brown-washed breast-band. In flight, mostly white underwing contrasts with black tips to primaries and trailing edge, bolder on secondaries. **Ad** ♀ is mostly brown above, with narrow pale supercilium. Underparts buffier, heavily streaked brown. Pale brownish underwing has flight feathers barred brown and poorly defined dark trailing edge to primaries. **Imm** recalls ♀ but generally darker, has dark brown crown, rufous-orange underparts and somewhat variable (usually only breast) streaking. Underwing darker. **Voice** Usually silent. **SS** See vagrant Western Marsh-harrier. In Cuba shares habitat with Snail Kite, which lacks owl-like facial disc, has broad white tail base (instead of multi-banded tail) and flight more flapping on level wings, rarely gliding (and never on upraised wings). **TN** Previously considered conspecific with extralimital Hen Harrier *C. cyaneus* (Old World).

Sharp-shinned Hawk *Accipiter striatus* `LC`
Ten subspecies divided into four subspecies groups, of which one occurs in region.

Sharp-shinned Hawk *Accipiter (striatus) striatus*
Local and rare breeding resident (**GA**), casual visitor or vagrant (**O**). **L** 23–35 cm; **WS** 42–68 cm. Seven subspecies in group, of which four (*velox* winter visitor Bahamas, Cuba, Jamaica, Virgin Is, at least, *fringilloides*[A] resident Cuba, nominate[B] resident Hispaniola, *venator*[C] resident Puerto Rico) occur in region. Mature forest, mostly in montane areas, to 1050 m in Cuba and 1465 m on Hispaniola. Largely absent from coasts, except perhaps in Cuba (and North American migrants). Unobtrusive and often hard to find; furthermore, all endemic races increasingly threatened. Flight direct and swift with 3–4 fast wingbeats followed by short glide. Winter visitors occur Oct–Apr (probably throughout Greater Antilles). Vagrant Virgin Is (St John, Nov 1987; Guana, Oct 1998), perhaps St Kitts & Nevis, and San Andrés. **ID** Highly variable small forest *Accipiter*, with short rounded wings, small head and long, narrow, squar-tipped tail. **Ad** (*velox*) is dark steel-blue above and whitish variably barred rufous to tawny below. Tail has several pale and dark bands. Eyes orange-red to red, cere and legs yellow. ♀ is obviously larger than ♂. **Imm** is brown above, buffy below heavily streaked dark brown. **GV** Nominate race smaller, with tawny wash on head-sides, less conspicuously barred tail and narrower, more regular barring below; *fringilloides* similar, but head-sides cinnamon and thighs barred greyish brown; *venator* has dark upperparts, rufous head-sides, flanks and thighs, clear white belly and more distinct dark tail-bands than others. **Voice** High-pitched, sharp "kew-kew-kew-kew". **SS** Typical *Accipiter* with short rounded wings, long narrow tail and flap-flap-glide flight pattern. In Cuba, Cooper's Hawk (only hypothetical) larger, with longer, more rounded tail, larger head and, in ad, stronger contrast between crown and back; larger and more robust resident Gundlach's Hawk distinguished using similar characters.

Cuban Kite

Swallow-tailed Kite
forficatus

Western Marsh-harrier / Eurasian Marsh-harrier
aeruginosus

Northern Harrier

velox adult
striatus adult
striatus adult
striatus juvenile

Sharp-shinned Hawk

161

Cooper's Hawk *Accipiter cooperii* LC
Hypothetical (**Cuba**).

L 37–47 cm; **WS** 64–87 cm. Monotypic. More or less wooded habitats. Flight swift; 3–4 fast wingbeats and a short glide. Garrido (1985) mentioned without details two specimens he considered to be of this species, but we have not seen any unequivocally of this species from Cuba (GMK pers. obs.). Four seen in extreme W Cuba, late Aug–Sept 2007. Possible, Cayo Cantiles (Garrido & Schwartz 1969). **ID** Medium-sized, long-tailed forest hawk, with rounded wings, orange to red eyes, and greenish-yellow to orange-yellow cere and legs. All plumages very similar to *velox* Sharp-shinned Hawk. Differs by larger size, proportionately larger head, longer more rounded tail, slightly shorter wings. **Ad** has slate-black crown, paler nape and blue-grey underparts. ♀ much larger than ♂. **Juv** has brownish upperparts and brown-streaked and barred (on flanks) whitish underparts, yellow eyes, and yellow-green to yellow cere and feet. **Voice** Unlikely to be heard. **SS** Resident but rare Gundlach's Hawk has unbarred greyish chest and cheeks (greyer in E Cuba). See commoner and smaller Sharp-shinned Hawk.

Gundlach's Hawk *Accipiter gundlachi* EN
Rare resident (**Cuba**).

L 40–46 cm; **WS** 74–84 cm. Two subspecies, of which both (nominate[A] W & C Cuba, *wileyi*[B] E Cuba) occur in region. Endemic. Forest and its borders to 1100 m, also swamps and mangroves. Feeds mostly on birds, pursued at speed. Other habits little-known. **ID** Robust, long-tailed forest hawk, with orange to red eyes and yellow legs. All plumages very similar to *velox* Sharp-shinned Hawk, but much larger and heavier, with proportionately larger head, longer rounder tail and slightly shorter wings. **Ad** has unbarred (or scarcely barred) greyish throat, cheeks and sometimes upper breast, occasionally flecked reddish, becoming brick-red below (very narrowly barred white), and slaty-black crown contrasts with paler nape and blue-grey upperparts. Underwing-coverts brick reddish to orange, also spotted white. Cere greyish. ♀ is much larger than ♂. **Juv** is dark brown above, with brown to blackish streaking below, spotting on underwing, greenish-yellow irides and paler feet than ad. **GV** Race *wileyi* (E Cuba) very similar but ad slightly paler above and even greyer on cheeks and breast; juv has longer, darker streaks below and heavier markings on thighs. **Voice** Loud, harsh, cackling "kek-kek-kek-kek-kek-kek...." similar to Cooper's Hawk. **SS** See commoner and much smaller Sharp-shinned Hawk, and hypothetical Cooper's Hawk, which has browner, more boldy barred underparts. **TN** Validity of race *wileyi* needs corroboration. Garrido (1985) postulated that Gundlach's and Cooper's Hawks might be conspecific, and molecular work has confirmed their very close relationship (Breman *et al.* 2013).

Bald Eagle *Haliaeetus leucocephalus* LC
Vagrant (**Bahamas, Cuba, Puerto Rico, Virgin Is**).

L 70–96 cm; **WS** 180–237 cm. Two subspecies, of which one (presumably nominate) occurs in region. Generally prefers areas fringing water, e.g. coasts, estuaries, riparian habitats and lakes, usually in lowlands. Glides and soars, usually on flat wings. Abaco, Grand Bahama and Exumas, Bahamas, Dec 2000–Feb 2001 (*NAB* 55: 237), two, Mar 2002 (*NAB* 56: 373), Feb 2003 (eBird), Feb 2011 (*NAB* 65: 357), Nov 2012 (*NAB* 67: 174), Oct 2018, 2nd-y, Mar 2019 (eBird); five in W Cuba, imm, Nov 1997, ad, Feb 1998, ad, Dec 2002, 2nd-y, Dec 2002, imm, Mar 2004 (Suárez *et al.* 2005); Puerto Rico, Oct 1975; St John (Virgin Is), Feb 1977 (Raffaele 1989). **ID** Huge raptor with massive bill, proportionately short neck and long tail. **Ad** is dark brown with white head and tail, yellow bill and legs. Underwing dark brown. Eyes pale yellow.
♀ differs only in being larger. **Juv** varies with age (achieves ad plumage after fifth moult), mostly brown with pale mottling on underparts, whitish axillaries and underwing-coverts, and whitish centres to tail feathers. Tail longer than ad. Eyes and bill dark. Intermediate plumages like juv but progressively more extensive white on body, underwing and undertail. **Voice** A surprisingly weak, rather fast-paced series, "kah-kah-hah", but probably unlikely to be heard. **SS** Rare. Much larger than any other raptor.

Black Kite *Milvus migrans* LC
Seven subspecies divided into three subspecies groups, of which one occurs in region.

Black Kite *Milvus (migrans) migrans*
Vagrant (**Bahamas, Virgin Is, LA**).

L 44–66 cm; **WS** 120–153 cm. Four subspecies in group, of which nominate occurs in region. In native range uses wide range of habitats. Flight rather slow and usually fairly low, but agile and manoeuvrable. Glides and soars. Great Inagua, Bahamas, May 2018 (eBird); British Virgin Is, Oct 1999 (*NAB* 54: 110); Guadeloupe, Oct–Nov 2008 at least (*NAB* 63: 173) and Nov 2015–Mar 2016 (eBird); Dominica, Apr 1999 (*NAB* 57: 132); Barbados, Nov 2008 (Buckley *et al.* 2009). **ID** Long- and narrow-winged raptor with long, slightly forked tail and yellow bare parts. **Ad** is dark reddish brown, with finely streaked paler head. Upperwing has slightly paler brown central panel. In flight, underwing has darker primaries with slightly paler bases. Tail greyish brown. Eyes yellowish and bill black. **Juv** is generally paler and more heavily marked, with more contrasting pattern. Moderate white streaking on underparts and pale feather tips on mantle and coverts. Underwing-coverts cinnamon-brown. Tail faintly barred. **Voice** Unlikely to be heard. **SS** Very long wings, overall brown plumage and unbarred, slightly forked tail distinctive. Compare Western Marsh-harrier.

Mississippi Kite *Ictinia mississippiensis* LC
Passage migrant (**Cuba**), vagrant (**Cayman Is**, rest of **GA, San Andrés**).

L 31–38 cm; **WS** 75–83 cm. Monotypic. Over open woodland, savannas, semi-arid habitats and towns. Feeds largely on insects caught in flight, with much gliding, banking and wheeling. Sometimes in small, even large, groups, with virtually regular records in Cuba (all since 1999) in Sept–Dec and Mar/Apr (Kirkconnell *et al.* in press). Also: Cayman Is, two, Oct 2002 (*NAB* 57: 132), juv, Sept 2017 (eBird); six, Jamaica, Mar 2004 (*Cotinga* 23: 80, 90); juv, Dominican Republic, Feb 2013 (Hayes & Thorstrom 2014); ad, Puerto Rico, Feb 2016 (eBird); San Andrés, Oct 2017 (eBird). **ID** Slender raptor with very long, pointed wings. **Ad** has mid-grey body, darker wings and tail, and paler grey head. Upperwing shows some rufous at base of primaries and silvery-white secondaries. Eyes red, cere grey and legs orange. ♀ is slightly larger, with less contrasting head and less rufous in primaries. **Juv** has brown upperparts, creamy underparts heavily streaked rufous-brown, usually with conspicuous pale bars (sometimes incomplete) on undertail. Eyes brown. **Subad** resembles ad but has dark secondaries and juv tail feathers with pale bars on underside. **Voice** Mostly silent on migration. **SS** Falcon-like. Peregrine Falcon has similar shape, but larger with different plumage, flight pattern and behaviour. Compare rare ♂ Northern Harrier, which flies low to ground, with slow flapping and glides.

Cooper's Hawk

Gundlach's Hawk
gundlachi

Bald Eagle
leucocephalus

Black Kite
migrans

Mississippi Kite

163

Snail Kite *Rostrhamus sociabilis* LC
Fairly common resident (**Cuba**).

L 43–48 cm; **WS** 99–115 cm. Three subspecies, of which one (*plumbeus*) occurs in region. Freshwater marshes, reservoirs, canals and rice fields. Pairs or small groups. Flies low on slow, steady wingbeats, rarely gliding, head held down. Diet freshwater apple snails. Most numerous in W Cuba, but reported from six offshore cays and one site on I of Pines (Kirkconnell *et al.* in press). **ID** Mid-sized raptor with conspicuous white-based tail and well-hooked black bill. In flight, longish tail looks square-ended and long wings rounded. **Ad** ♂ has cere, facial skin and legs orange-red, slightly duller in non-br season, eyes red. Slaty black with contrasting white base to tail and undertail-coverts. **Ad** ♀ is slightly larger, dusky brown above, with whitish eyestripe. Underparts and underwing white to buff, variably but often heavily streaked and mottled dusky. **Imm** is similar to ad ♀, but bare parts duller orange and eyes dull brownish. **Voice** A distinctive, raspy, ratchet-like "ka-ka-ka-ka…" constantly repeated. **SS** Northern Harrier has owl-like facial disc, banded tail, longer and narrower wings, and glides with wings upraised. **TN** Cuban population sometimes separated as race *levis*. **AN** Everglade Kite.

Common Black Hawk *Buteogallus anthracinus* LC
Five subspecies divided into two subspecies groups, of which one occurs in region.

Common Black Hawk
Buteogallus (anthracinus) anthracinus
Uncommon resident (**St Vincent and Grenadines**, **Grenada**), vagrant elsewhere.

L 50–56 cm; **WS** 106–128 cm. Two subspecies in group, of which nominate occurs in region. Mainly montane forests. Regularly soars but perches rather unobtrusively. Vagrants can appear on coasts or in other habitats. **ID** Large, stocky hawk with broad wings and relatively short tail and legs. **Ad** is black with single broad white tail-band and narrow white tip. Cere and legs yellow. In flight from below, small white patches at base of primaries. ♀ is larger than ♂. **Juv/Imm** is blackish brown above with prominent whitish-buff supercilium. Underparts buffy, heavily streaked and blotched dusky. Tail has numerous narrow dark and pale bands. **Voice** Series of 10–15 shrill whistles, becoming higher-pitched and louder, and finally accelerating, often given in flight. **SS** Size, broad wings and short tail distinctive. Overall black plumage with single white tail-band in ad unique in range. Imm Broad-winged Hawk recalls imm, but much smaller and lacks prominent eyebrow. No overlap with Cuban Black Hawk. **TN** Previously considered conspecific with Cuban Black Hawk. Birds in S Lesser Antilles sometimes awarded separate race *cancrivorus*.

Cuban Black Hawk *Buteogallus gundlachii* NT
Fairly common resident (**Cuba**), vagrant (**Cayman Is**).

L 43–52 cm; **WS** c. 115 cm. Monotypic. Endemic. Largely coastal: cays, mangroves, forests and open areas near swamps and beaches. Soars frequently. Feeds heavily on crabs. Range includes I of Pines and many offshore cays, but not SE mainland. Vagrant Grand Cayman, Feb 2016–Jan 2017 (eBird). **ID** Fairly large with broad wings and relatively short tail and legs. Very similar to allopatric Common Black Hawk (considered conspecific until recently). **Ad** differs by overall browner plumage with greater number of, and more conspicuous, pale feather tips, ill-defined whitish malar and, in flight, much larger and more conspicuous white underwing patch. Cere, lores and legs orange-yellow to yellow. **Juv/Imm** is white below heavily mottled brown, with mottled, rather than obviously barred, thighs, unmarked white underwing patch (has dark markings in Common) and dark tail-bands are straight, narrow and parallel (vs. marginally broader and oblique in Common). **Voice** Very different from Common Black Hawk: repeated series of three (rarely four) short, nasal whistles, with emphasis on second, often rendered "ba-TIS-ta". **SS** Rather large size, broad wings and short tail distinctive. Overall brownish-black plumage with single white tail-band and large white underwing patch of ad unique in Cuba. Imm Broad-winged Hawk recalls imm, but is obviously smaller and lacks prominent eyebrow. **TN** See Common Black Hawk.

White-tailed Hawk *Geranoaetus albicaudatus* LC
Hypothetical (**St Vincent**).

L 44–60 cm; **WS** 118–143 cm. Three subspecies, of which one (*colonus*) might occur in region. Morph not stated, Jun 1964 (Gochfeld *et al.* 1973). Soars high; occasionally hovers. **ID** Large, bulky, long-winged, long-legged and variable hawk, with two morphs. **Ad pale morph** has mostly blackish-grey upperparts and head-sides, and white underparts, affording hooded appearance. Throat white or blackish grey. Shoulders chestnut. White rump and tail, latter with broad black subterminal band. ♀ is slightly larger than ♂. **Ad dark morph** (rare) is dark slate-grey, usually with some chestnut in wings. Thighs often barred rufous and white. Tail as pale morph. **Imm** is usually blackish with creamy patch on breast and sometimes on cheeks. Can have chestnut on shoulders. Tail greyish with multiple, faint darker bands. **Voice** Unlikely to be heard. **SS** Noticeable white tail and rump of ad (both morphs) distinctive. Imm larger and darker than most raptors in region.

Ridgway's Hawk *Buteo ridgwayi* CR
Very rare resident (**Dominican Republic**).

L 35–40 cm; **WS** 69–81 cm. Monotypic. Endemic. At least formerly in undisturbed rainforest, subtropical dry and moist forests, pine and limestone karst forest, occasionally secondary forest and farmland, but no longer found in all these. Lowlands and foothills, but to 2000 m. Formerly widespread on Hispaniola, including Haitian islands of Gonâve, the Cayemites and Île-à-Vache, and Dominican islands of Beata and Alto Velo, but lost from > 96% of original range during 20th century. Now endemic to Dominican Republic, where in recent decades definitely known only from Los Haitises National Park and nearby Samaná Peninsula in NE. Probably fewer than 100 pairs. Record (Apr 1984) from Culebra I, off NE Puerto Rico, was perhaps a Red-shouldered Hawk (not definitely recorded, but claimed in Bahamas, Mar 2001). Soars like typical *Buteo*, with broad wings and fanned tail, but flaps rather rapid, alternated with glides, in *Accipiter* fashion. **ID** Mid-sized, forest *Buteo* with longish legs. **Ad** ♂ has upperparts dark brownish grey, with grey head and neck, and rusty shoulders. Underparts rufous barred white, thighs reddish brown and throat paler. Tail grey with 3–4 narrow white bars and white tip. Legs yellow. **Ad** ♀ is slightly larger and browner, with less conspicuous rufous shoulders, paler breast with more barring and more heavily barred tail. **Imm** is greyish brown above with no rufous shoulders, and buffy white with brown streaking below.
Voice Shrill, squealing, whistled "kleeah" (mainly aggression), "weeup" (display and food passes) and a whistle-squeal.
SS Red-tailed Hawk much larger and ad has red tail. Broad-winged Hawk (vagrant on Hispaniola) has broader, bolder tail-bands (especially terminal). Compare vagrant Swainson's Hawk.

Snail Kite
plumbeus

Common Black Hawk
anthracinus

adult

juvenile

Cuban Black Hawk

adult

juvenile

dark morph

White-tailed Hawk
colonus

pale morph

adult

Ridgway's Hawk

immature

165

Red-shouldered Hawk *Buteo lineatus* LC
Hypothetical (**Bahamas**).

L 38–47 cm; **WS** 90–114 cm. Five subspecies, of which one (presumably either nominate or *extimus*) might occur in region. Sight, Grand Bahama, Mar 2001 (*NAB* 55: 371). See also Ridgway's Hawk. **ID** Medium-sized and slender-bodied, tan to rufous woodland raptor, with bold black-and-white pattern on wings and tail. Flight recalls Broad-winged Hawk. **Ad** has reddish shoulders and underwing-coverts, extensive pale spotting below and some dark streaks on breast. ♀ is larger and tends to have more obvious underparts markings. **Imm** displays less obvious wing and tail patterns, is streaked below, and mainly brown above with paler uppertail-coverts. **GV** Races differ in size and general coloration; *extimus* smallest and palest rufous below with greyish head.
Voice Commonest vocalization a scream "kee-aaah" or similar, with latter part drawn-out and ascending, usually repeated several times.
SS Pale crescent-shaped panels in wings diagnostic, but perched juv difficult to separate from same-age and similarly shaped Broad-winged Hawk, which has shorter legs, pale brown tail with dark brown bars (reverse in Red-shouldered) and lacks pale spots on secondaries.

Broad-winged Hawk *Buteo platypterus* LC
Common to rare resident (parts of **GA**, **LA**), winter visitor (**Cuba**), vagrant (**Bahamas**, **Jamaica**, **Hispaniola**, **Guadeloupe**, **Barbados**).

L 32–42 cm; **WS** 74–96 cm. Six subspecies, all of which (nominate frequent passage migrant and winter visitor Cuba, vagrant elsewhere Greater and Lesser? Antilles, *cubanensis*[A] Cuba, *brunnescens*[B] Puerto Rico, *insulicola*[C] Antigua, *rivierei*[D] Dominica, Martinique and St Lucia, *antillarum*[E] St Vincent and the Grenadines, and Grenada) occur in region. All types of forest to at least 1200 m (Cuba), preferring dense broadleaf, less frequently open woodland and farmland. On Antigua also towns. Soars often, tail usually broadly spread and occasional slow, deep wingbeats. Often nests near wet or mesic areas with good canopy cover. Fairly common resident Cuba (formerly more numerous), Antigua and Dominica S to Grenada, rare, very local and declining Puerto Rico. Flocks of migrants typically number < 20 individuals, mainly moving through W Cuba, where nominate recorded late Jul–early May. **ID** Typical, mid-sized *Buteo* with broad wings and rather short, broad tail. **Ad** is mostly brown above and buffy white densely barred and spotted rufous-brown below. In flight looks rather pale, with pale brownish underwing conspicuously rimmed black. Usually fanned tail has several blackish and white bands, that closest to tip broader. **Imm** has underparts streaked brown and tail has more, thinner and fainter bands. **GV** Race *cubanensis* smaller and ad more streaked than barred below, more like imm nominate; *brunnescens* darker with blacker streaking; *insulicola* smallest race and much paler overall, with whitish throat; *antillarum* also smaller than nominate, with more rufous underparts and wing linings; *rivierei* is second-smallest race but darker than *insulicola*, with streaked throat, more rufous breast and rufous-toned wing linings. Nominate has rare dark morph, which is blackish throughout (unclear if ever recorded in region). **Voice** Loud, high-pitched, piercing squeal "keeeeeeeee"; ♂ higher-pitched than ♀. **SS** Compare larger Red-tailed Hawk. In Dominican Republic see rare Ridgway's Hawk, and on Grenada, rare, longer-tailed and larger-billed Hook-billed Kite. Also vagrant Short-tailed and Swainson's Hawks.

Short-tailed Hawk *Buteo brachyurus* LC
Hypothetical (**Cuba**).

L 37–46 cm; **WS** 83–103 cm. Two subspecies, of which one (*fuliginosus*) claimed in region. Soars like typical *Buteo*, with broad wings and fanned tail. Two pale morph, Sept 2007 (Rodríguez Santana 2010) and three (morph/s unknown), Sept 2016 (eBird), in extreme W Cuba. Could occur, especially given recent discovery that comparatively significant numbers of raptors migrate between Florida and Yucatán Peninsulas via W Cuba, but proof required. **ID** Mid-sized, compact, polymorphic *Buteo*. **Ad pale morph** has upperparts and head brownish black, white throat, narrow forehead and underparts, giving hooded appearance. Rufous patch between hindneck and breast-sides. In flight, whitish wing linings contrast with dark-barred flight feathers. Tail greyish brown with several dark bands and whitish tip. Cere and legs yellow. **Ad dark morph** (rare) is blackish overall, with tail as pale morph. Underwing has pale flight feathers and black linings. **Imm pale morph** usually as ad but more or less streaked and spotted brown below and on head-sides, with more tail-bands. **Imm dark morph** as ad but underparts mottled white.
Voice Generally silent. **SS** Hooded appearance of ad pale morph distinctive. Much commoner Broad-winged Hawk has broader, more obvious subterminal tail-band and different underwing pattern. Juv of latter is more streaked below and has paler underwing. Pale morph similar to rare Swainson's Hawk, which is larger and has breast-band. Compare much commoner and larger Red-tailed Hawk.

Swainson's Hawk *Buteo swainsoni* LC
Vagrant (**Bahamas**, **Cuba**, **Hispaniola**).

L 48–56 cm; **WS** 117–137 cm. Monotypic. Liable to be seen over more or less open areas. Flies with wings slightly upturned in almost Turkey Vulture-like fashion. Records (some documented): Bahamas (Abaco), pale morph, Jan–Mar 2012 (*NAB* 66: 360, 568); Cuba, pale morph, Apr 2014, pale morph, Oct 2014, two different juv, Nov 2017, plus total of 31 claimed moving through W Cuba, Aug–Sept 2007 (Rodríguez-Santana 2010, Kirkconnell *et al.* in press); Dominican Republic, Apr 1996 (Bradshaw *et al.* 1997). Perhaps Jamaica, Oct 1974 (Bond 1976). Several records Trinidad & Tobago (Hayes 2001, Kenefick *et al.* 2011) suggest it could occur in Lesser Antilles.
ID Large, variable *Buteo* with relatively long pointed wings and longish tail. **Ad pale morph** has dark brown upperparts. Prominent rufous chest-band (dark at long range) contrasts with white throat and rest of underparts. In flight, white wing linings contrast with dusky flight feathers. Greyish tail has numerous indistinct darker bands and black subterminal band. **Ad dark morph** is mostly dark brown with tail as pale morph. This morph and intermediates undocumented (but reported) in region. **Imm pale morph** usually has head, upper back and underparts whitish, streaked and mottled brown. **Voice** Unlikely to be heard. **SS** Rufous chest-band of ad pale morph distinctive. Slenderer shape, with longer more pointed wings, distinguish all plumages from other *Buteo*. Larger Red-tailed Hawk lacks breast-band and has streaked belly. Smaller ad Broad-winged Hawk has more prominent tail-bands and paler underwing with black rim. Pale-morph Short-tailed Hawk (hypothetical) has similar underwing pattern, but smaller and lacks breast-band.

Red-shouldered Hawk

extimus

lineatus

Broad-winged Hawk

platypterus adult

cubanensis adult

cubanensis juvenile

platypterus adult

Short-tailed Hawk
fuliginosus

dark morph

pale morph

Swainson's Hawk

dark morph

intermediate

pale morph

167

Red-tailed Hawk *Buteo jamaicensis* `LC`
Fifteen subspecies divided into two subspecies groups, of which one occurs in region.

Red-tailed Hawk *Buteo (jamaicensis) jamaicensis*
Common to rare resident (**Bahamas, GA, N LA**) vagrant (rest of **LA, Cayman Is**).
L 45–65 cm; **WS** 107–141 cm. Fourteen subspecies in group, of which two (*solitudinis*[A] Bahamas and Cuba, nominate[B] Jamaica, Hispaniola and Puerto Rico to N Lesser Antilles) occur in region. Fairly catholic: open country, coasts, farmland, woodland, even towns, to at least 1700 m in Cuba and 2440 m on Hispaniola. Most often seen soaring with occasional slow, deep wingbeats. Common resident on larger islands of N Bahamas, Greater Antilles, Virgin Is, St Barthélemy, Saba, St Kitts & Nevis; rare St Eustatius; vagrant elsewhere, e.g. Cayman Is and St Lucia. **ID** Large, robust-bodied, variable *Buteo* with broad, rounded wings. **Ad** is mostly dark brown above and buffy to whitish below usually with contrasting dark belly-band. Tail reddish with narrow dark subterminal band. Broad, rounded wings show dark wrist patch and leading edge. **Imm** has tail faintly barred greyish brown and more heavily streaked underparts. **GV** Race *solitudinis* slightly larger than nominate.
Voice Raspy descending "keeeeer". **SS** Larger than other regularly occurring *Buteo*. Unbarred reddish tail should distinguish ad from other raptors. Apart from size, imm Broad-winged Hawk has broad tail-bands and flaps more in flight.

TROGONIDAE
Trogons
43 extant species, 2 in region

Cuban Trogon *Priotelus temnurus* `LC`
Fairly common resident (**Cuba**).
L 23–25 cm. Two subspecies, of which both (nominate[A] mainland Cuba, plus Palma, Caguanes, Guajaba and Sabinal cays, *vescus*[B] I of Pines) occur in region. Endemic. Reasonably common and widespread on main island, rare N cays, uncommon I of Pines. Wooded habitats, including pines, but primarily dense dry and shady woodlands, to 1300 m. Perches distinctively upright; flight undulating. **ID** Only trogon in Cuba, with spiky-looking tail. **Ad** has iridescent blue crown and green back, white throat and breast, plus red belly. Wings and broadly flared tail patterned in blue, black, green and white.
Juv differs only in being duller. **GV** Race *vescus* very slightly smaller (no overlap in wing length of ♀♀, but published sample sizes tiny).

Voice Song has typical trogon pattern, albeit softer and more burry than most, offering a frequently heard background sound in Cuban forested landscapes: "toco-toco-tocoro-tocoro…". **SS** Unmistakable. **TN** Race *vescus* probably needs re-evaluation (Kirkconnell *et al.* in press). Suggestion that birds on cays off N Cuba could merit subspecific treatment (Collar 2001) very unlikely.

Hispaniolan Trogon *Temnotrogon roseigaster* `LC`
Fairly common but local resident (**Hispaniola**).
L 27–30 cm. Monotypic. Endemic. Forests, including mature pine and deciduous broadleaf, principally at c. 350–3000 m, but exceptionally to sea level. Still relatively common in appropriate, undisturbed habitat. **ID** Typically patterned New World trogon. **Ad** ♂ has yellow bill, blackish mask with orange eye, green crown and upperparts, distinct white bars on wing panel and regular white notches on outer webs of primaries. Chin to upper belly grey, rest of underparts red; uppertail green-blue, undertail mostly black with very broad white tips. **Ad** ♀ lacks white notches on primaries, has wing panel unbarred grey-green, less vivid red belly. **Juv** is apparently undescribed. **Voice** Song "kuh kwao", "kuh kwao kwao", "toca-loro" or "cock-craow", repeated several times; calls include puppy-like barks if disturbed, a low rattle and whimpering whistles. **SS** None; the only trogon on Hispaniola.

MEROPIDAE
Bee-eaters
31 extant species, 1 vagrant in region

European Bee-eater *Merops apiaster* `LC`
Vagrant (**St Lucia**).
L 28 cm (with streamers, up to 2.5 cm more). Monotypic. Ad, Feb 2014, was first for New World (Anon. 2015, *NAB* 68: 291). In native range occupies open habitats. Often hunts high with hirundine-like flight pattern. **ID** Only bee-eater recorded in region. **Ad** ♂ has chestnut cap and mantle, yellow throat with black gorget, greenish-blue underparts; green tail, with streamers; primaries and tertials green to green-blue, but rest of upperwing mahogany, dusky band on trailing edge. **Ad** ♀ tends to have greener scapulars and lower back, less intensely green lesser coverts, rufous median and greater coverts edged green, and yellow throat and turquoise breast and belly paler. **Juv** is also paler, with ad chestnut and rufous parts suppressed by green. **Voice** Quite vocal. Typical call a distinctive, mellow, liquid and burry "prreee" or "prruup", frequently repeated. **SS** Multicoloured plumage unmistakable.

adult

Red-tailed Hawk
jamaicensis

juvenile

♂ ♀

Cuban Trogon
temnurus

Hispaniolan Trogon

♂ ♀

European Bee-eater

169

TODIDAE
Todies
5 extant species, 5 in region

Cuban Tody *Todus multicolor* LC
Common resident (**Cuba**).
L c. 10–11 cm. Monotypic. Endemic. Widespread in xeric, mesic, wet and riparian woodland, thickets, deciduous and semi-deciduous forest, pines and secondary vegetation, to c. 1300 m, throughout mainland Cuba and most larger cays off N coast. Perches still for long periods, thus remarkably unobtrusive, although tends to bob head up and down. **ID** Striking, little, mainly vivid green-and-grey bird. **Ad** is bright green above, with yellow lores, prominent sky-blue patches below ear-coverts (deeper blue on I of Pines) and on carpal. Below, much pink on flanks, whitish belly and yellow undertail-coverts. Blue carpal more prominent in ♂, but of little use in field. **Juv** has short bill, dull green back, grey or pink bib, pale grey underparts and brown or slate-coloured eyes (white in ad). **Voice** Highly distinctive: a hard, rapid chattering and comparatively loud "tot-tot-tot", like brief burst of machine-gun fire; also mechanical wing noise in flight. **SS** The only tody in Cuba, and unmistakable in all but briefest view. **TN** Race *exilis* (described from NE Cuba) not accepted.

Narrow-billed Tody *Todus angustirostris* LC
Common resident (**Hispaniola**).
L c. 11 cm. Monotypic. Endemic. Primarily in dense, wet montane forest, as well as dry lower montane forest, pines and shade-coffee plantations, to 3000 m (mainly at 900–2400 m). Overlaps locally with Broad-billed Tody in lowlands and lower montane forest. **ID** The more highland-dwelling of the two todies on Hispaniola. **Ad** is brilliant green above, with a few grey feathers on lower cheeks, red throat, whitish breast and belly, pink flanks, yellow vent. Iris whitish; maxilla black, mandible red usually with blackish tip. **Juv** has shorter bill, dull green upperparts, grey or pink bib, much grey on breast, and lacks colourful pink and yellow patches of ad. **Voice** Frequent is a repeated "chick-kweee" or "chip-chee", with accent on second syllable, and a chattering "chippy-chippy-chippy-chip" that drops in pitch but not in tone. **SS** Broad-billed Tody, but differs in whitish (not yellowish) belly, narrower bill with mandible usually blackish distally, and iris whitish (not slate-coloured); behaviour generally more active and, especially, note different voices. **TN** Populations N & S of Neiba Valley/Cul de Sac Plain differ quite markedly in genetics (Overton & Rhoads 2004).

Broad-billed Tody *Todus subulatus* LC
Common resident (**Hispaniola, including Gonâve I**).
L c. 11–12 cm. Monotypic. Endemic. Primarily arid and semi-arid, lowland scrub-forest and subdesert, but locally in lower montane forest, second growth, shade-coffee plantations, pines and humid ravines, to 1700 m. **ID** Generally at lower elevations and in drier habitats than Narrow-billed Tody, but with significant overlap. Largest tody. **Ad** has bright green upperparts, with conspicuous yellow-green lores and supercilia, a few grey feathers below ear-coverts, red throat, and greyish-yellow underparts with pink flanks. Iris slate; bill broad, maxilla black, mandible red. **Juv** has shorter bill, duller green back, grey or pink bib, much grey on breast, and lacks pink and yellow patches of ad. **Voice** A near-monotonous, complaining "terp, terp, terp...", i.e. single-noted, vs. double note of Narrow-billed Tody. Mechanical wing noise, "burrrrrrr", frequently heard in flight. **SS** Can only be confused with Narrow-billed Tody, with which it overlaps to some extent in both altitude and habitat; voice is best means of distinguishing them.

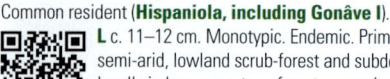

Jamaican Tody *Todus todus* LC
Fairly common resident (**Jamaica**).
L 9–10.8 cm. Monotypic. Endemic. Widespread in wet, mesic and dry forest, including mangroves; most abundant in hills and mountains, but rare above 1500 m. Like other todies perches unobtrusively at low levels. **ID** Typical tody with bright green upperparts. **Ad** has small blue-grey subauricular patch, red throat, yellowish-green breast-sides, greenish-yellow belly, with sparse pink on flanks. Iris white to slate, maxilla black, mandible red. ♀ has less prominent blue-grey face patch. **Juv** has shorter bill, duller green back, grey or pink bib, much grey on greenish-washed breast and lacks blue-grey and yellow patches of ad. **Voice** Generally silent, but during Dec–Jul gives a loud "beep" and rapid, guttural throat-rattling "frrrup" calls in territorial displays, plus loud hissing "cheep" in alarm. Like other todies, wings make diagnostic buzzing noise in flight. **SS** Unmistakable given a reasonable view. **TN** Birds trapped at 1200 m significantly heavier and longer-winged than those at 650 m; perhaps subspecifically distinct.

Puerto Rican Tody *Todus mexicanus* LC
Fairly common resident (**Puerto Rico**).
L c. 11 cm. Monotypic. Endemic. Range of habitats from rainforest to arid scrub and dense thickets, shade-coffee plantations, mesic forest and karst landscapes, often near streams, in lowlands to hills. Behaviour as other todies. **ID** Typical tody with bright grass-green upperparts. The only tody in Puerto Rico. **Ad** has small blue carpal patch (inconspicuous in field), a few grey feathers on lower cheeks, red throat, whitish breast (often streaked or washed grey), yellow belly and flanks. Iris slate; maxilla black, mandible red. ♀ has carpal patch duller, iris white. **Juv** has shorter bill, pale grey bib, yellowish belly; over time, bib becomes pink, then red, belly whitens, and bright yellow flanks develop. **Voice** Most frequently heard is characteristic loud, nasal "beep" or "bee-beep"; also an insect-like "pree". Wing rattles in flight mainly given during courtship or territory defence. **SS** Wholly unmistakable.

Cuban Tody

Narrow-billed Tody

Broad-billed Tody

Jamaican Tody

♂ variant
♂
♀

Puerto Rican Tody
♂ ♀

171

ALCEDINIDAE
Kingfishers
119 extant species, 2 in region

Ringed Kingfisher *Megaceryle torquata* LC
Fairly common (**Dominica**) to scarce resident (**Guadeloupe**), vagrant (**Grenada**, **Barbados**).

L 38–42 cm; **WS** 61–74 cm. Three subspecies, of which one or two (nominate probable vagrant, *stictipennis* resident) occur in region. Rivers, lakes, reservoirs, but possible at almost any wetland, especially vagrants (presumably from South America). Flies high with floppy, almost tern-like wingbeats. Single detailed report from Grenada—♀, May 1998—strongly suggestive of nominate race (Smith & Smith 1999a), which was predicted to occur occasionally (Bond 1936). Hypothetical Puerto Rico (1960, 1995) and Montserrat, presumably attributable to wandering *stictipennis*. No recent records Martinique, where formerly said to be common. **ID** Largest kingfisher in Americas. **Ad** ♂ has bluish-grey head and upperparts, small white loral spot, broad white collar and deep rufous underparts. In flight, black primaries and white underwing-coverts. **Ad** ♀ has blue-grey band on upper breast separated from rufous belly by narrow white one; deep rufous underwing- and undertail-coverts. **Juv** is like ♀, but has streaky upperparts, paler underparts, rufous-washed grey breast, underwing-coverts partly white. **GV** Nominate (probable vagrant) lacks white spots on inner webs of secondaries, whereas white spots extend from inner to outer webs in resident endemic race *stictipennis*. **Voice** Loud, rattling "klek-klek-klek-klek…" in alarm. Also single "keck", frequently given in high flight. **SS** Both sexes of Belted Kingfisher extensively white below.

Belted Kingfisher *Megaceryle alcyon* LC
Fairly common non-breeding visitor (**throughout**).

L 28–33 cm; **WS** 49–57 cm. Monotypic. Calm water, both fresh and saline. To almost 1100 m in Dominican Republic. Mostly Aug–May, but observed all months, although summers only occasionally even in N of region. **ID** Large kingfisher with heavy, pale-based bill. **Ad** ♂ has bluish-grey head, breast-band and upperparts, white loral spot, collar, underparts and underwing-coverts. **Ad** ♀ has rufous breast-band, sides and flanks to upper breast. **Juv** is like ad ♀ but has rufous wash to grey breast-band, strongest in juv ♀. **Voice** Commonest call a loud, harsh rattle, "kekity-kek-kek-kek-tk-ticky-kek". **SS** Compare larger and (in region) range-restricted Ringed Kingfisher.

Green Kingfisher *Chloroceryle americana* LC
Hypothetical (**St Kitts & Nevis**).

L 18.5–20.5 cm; **WS** 28–33 cm. Five subspecies, of which one (race unknown) might occur in region. Favours streams, lakes and rivers, but vagrants could appear at any wetland habitat. Reports from Dominican Republic (c. 1986) and Bahamas (Eleuthera, spring 1989) not accepted. St Kitts, on unknown date in 2003, accepted locally, but no documentation. **ID** Mid-sized, distinctive kingfisher. **Ad** ♂ has dark green upperparts with bronzy reflections on crown, white collar, bright rufous breast, white belly, and green bars on flanks and breast-sides. White spots on primaries, secondaries and wing-coverts form indistinct lines. **Ad** ♀ has green (not rufous) breast-band. **Juv** is like ♀, but duller, less bronzy, with buff spots on crown and wing-coverts. **Voice** Dry rasping, often followed by sputter, "dzeew dzeew kuk-kuk-kuk dzeew". Also clicking "tick tick tick" in alarm. **SS** No other kingfisher with dark green upperparts has occurred in region.

RAMPHASTIDAE
Toucans
50 extant species, 1 introduced in region

Channel-billed Toucan *Ramphastos vitellinus* VU
Introduced resident (**Grenada**).

L 46–56 cm. Monotypic. Present in and around Grand Etang Forest Reserve; first reported Aug 2001, but apparently released in 1989 (Frost & Massiah 2003). Forested areas. **ID** Unique, large, mainly black bill. **Ad** is black above and on belly, with bright red tail-coverts, white-sided orange-centred bib, broad red breast-band, blue facial skin. **Imm** is duller, with muted reds, orange and yellow, face paler. **Voice** Song of noisy and clear elements mixed variously, disyllabic "kee-ark" to "kerrrk" to "keee", singly or in series, c. 30–50/minute. **SS** None. Despite large size, more likely to be heard than seen, so knowledge of voice important to aid detection.

PICIDAE
Woodpeckers
254 extant species, 12 in region + 1 vagrant

Antillean Piculet *Nesoctites micromegas* LC
Locally fairly common resident (**Hispaniola**).

L 14–16 cm. Two subspecies, of which both (nominate[A] mainland Hispaniola, *abbotti*[B] Gonâve I) occur in region. Endemic. Wide range of dense wooded habitats, from pines to mangroves, rarely plantations and orchards, but unquestionably most numerous in mixed desert scrub, thorn-forest in semi-arid regions and drier broadleaf forest, to 1770 m, but most regular at 400–800 m. **ID** Largest piculet and only one in West Indies. **Ad** ♂ has olive-green upperparts and pale underparts heavily spotted and streaked; yellow crown with red central patch; sharply pointed, short, stout bill. **Ad** ♀ is larger and lacks red on crown. **Juv** is duller, both sexes with crown as ad ♀ but yellow less bright, and belly obscurely barred (not streaked); may briefly become bare-crowned, but soon acquires ad feathers. **GV** Race *abbotti* paler with greyer upperparts, sometimes has white spots extending to mantle-sides, plainer white throat, less heavy streaks below, and both sexes have less yellow on crown. **Voice** Single mechanical "pit" and "pew" in alarm; loud, rapid, musical whistling "kuk-ki-ki-ki-ke-ku-kuk" between pair members and in territorial encounters; short series of weak "wiii" notes, and continuous noisy "yeh-yeh-yeh-yeh" during fights. **SS** Unmistakable, although beware capacity to fool inexperienced observers into thinking they are not watching a woodpecker.

Ringed Kingfisher
torquata

Belted Kingfisher

Green Kingfisher
americana

Channel-billed Toucan

abbotti   *micromegas*

Antillean Piculet

173

Ivory-billed Woodpecker *Campephilus principalis* CR
Two subspecies placed in separate subspecies groups, of which one occurs in region.

Cuban Ivory-billed Woodpecker
Campephilus (principalis) bairdii
Extremely rare resident, perhaps extinct (**Cuba**).
L 48–53 cm. One subspecies in group. Endemic. Undisturbed mixed woodland and tall pines (*Pinus cubensis*) with abundant dead trees and some large palms in mountains and hills; probably confined to extreme E Cuba since at least mid-20th century. Historical reports from lowlands perhaps doubtful. Last unambiguous sighting Mar 1988, with fewer than a handful of undocumented sightings, and rumours of its persistence, since. **ID** Very large, distinctive woodpecker. **Ad** ♂ is mainly glossy black, with large red crest, white neck-stripe, white tramlines on mantle-sides and largely white flight feathers. Chisel-shaped bill pale, ivory-coloured. **Ad** ♀ is slightly smaller with longer crest sometimes slightly upcurved, and no red on head. **Juv** is browner than ad, crest shorter, white tips to all primaries but less white on underwing, head pattern as ad ♀, but ♂ gradually acquires red. **Voice** Common note in alarm "kent" or "hant", timbre like clarinet or toy trumpet, repeated up to six times; instrumental signal of single or double raps. **SS** No other large, black-and-white woodpecker shares same range. Undulating flight distinctive vs. distantly seen Cuban Crow. **TN** Cuban race sometimes considered possibly a separate species; molecular analysis (Fleischer *et al.* 2006) suggests it is (or was) genetically distinct from nominate (in continental USA).

Fernandina's Flicker *Colaptes fernandinae* VU
Rare and local resident (**Cuba**), vagrant (**Bahamas**).
L 30–34 cm. Monotypic. Endemic. Semi-deciduous, evergreen and swamp woodland, and open areas, especially with abundant palms (usually *Coccothrinax*, *Sabal parviflora*, *Roystonea*), to mid elevations. Apparently in inexorable decline, and population now fewer than 1000 individuals. Vagrant: Grand Bahama, Oct 1964, after a hurricane (Bond 1965). **ID** Large and very attractive woodpecker. **Ad** ♂ is mainly yellowish brown closely and quite heavily barred black throughout, except on cinnamon-coloured face and lightly streaked crown, which occasionally shows some red on nape; obvious black malar. **Ad** ♀ differs in having black malar heavily streaked white, and never any hint of red on nape. **Juv** is duller and browner, with less barring above and broader markings below. **Voice** Loud "pic" series, slower and lower-pitched than Cuban Flicker; also typical "wicka" series, loud, nasal "ch-ch-ch" when breeding. **SS** Only liable to confusion with Cuban Flicker, but latter has grey crown, black patch on breast and black spots elsewhere on unbarred underparts.

Yellow-shafted Flicker / Northern Flicker LC
Colaptes auratus
Four subspecies divided into two subspecies groups, of which both occur in region.

Yellow-shafted Flicker *Colaptes (auratus) auratus*
Vagrant (**Bahamas**).
L c. 30–35 cm. Two subpecies in group, of which one (probably nominate) occurs in region. Grand Bahama, Dec 2005–Jan 2006 (*NAB* 60: 301). **ID** Much like Cuban Flicker in plumage and structure.

Ad/juv differs in shorter tail, less olive upperparts, less heavily barred rump and tail, less obvious spots on underparts and breast crescent, and less obviously yellow underside to wings. **Voice** Not known to differ from Cuban Flicker. **SS** See Cuban Flicker.
TN Usually treated as conspecific with extralimital Gilded *C. cafer* (SW USA and NW Mexico) and Guatemalan Flickers *C. mexicanoides* (S Mexico to NC Nicaragua), under name Northern Flicker, due to high level of interbreeding between taxa at their geographical boundaries.

Cuban Flicker *Colaptes (auratus) chrysocaulosus*
Fairly common but local resident (**Cuba**, **Cayman Is**).
L c. 33 cm. Two subspecies in group, of which both (*chrysocaulosus*[A] mainland Cuba and larger cays off N coast, *gundlachi*[B] Grand Cayman) occur in region. Endemic. Semi-deciduous and evergreen woodland, second growth, pines, coastal vegetation adjacent to forest and mangroves (on smaller islands), to at least 1250 m (Cuba). Race *gundlachi* has perhaps wandered to Little Cayman and Cayman Brac (undocumented eBird reports). **ID** Classic *Colaptes*, with strikingly yellow underwing and undertail in flight. **Ad** ♂ has brown upperparts barred black, white belly with neat black spots, red crescent on nape, mostly grey head, and broad black breast crescent and malar. **Ad** ♀ is slightly smaller, lacks black malar and breast crescent is slightly less deep. **Juv** is often more greyish on head, crown sometimes barred, usually with black loral patch, much broader dark bars above, more black on tail-coverts, smaller breast patch and larger spots; ♂ usually has dull red on crown and less distinct malar, ♀ head even less marked. **GV** Race *gundlachi* smaller and shorter-tailed, with smaller red nape patch. **Voice** Common call a descending "peah" or "klee-yer"; also "wicka" and variants during aggressive encounters or courtship; soft "wee-tew" when approaching mate, or "wa-wa-wa" in flight. **SS** See Fernandina's Flicker (Cuba) and compare West Indian Woodpecker (no black malar and breast patch) and much smaller Yellow-bellied Sapsucker (large white wing blaze and different head pattern). **TN** Race *chrysocaulosus* sometimes mooted as potential split, but seems insufficiently well differentiated morphologically, and a molecular study found hardly any differences from rest of superspecies complex (Manthey *et al.* 2017).

Yellow-bellied Sapsucker *Sphyrapicus varius* LC
Fairly common winter visitor (**O**, **GA**), vagrant (**LA**).
L c. 19–21 cm. Monotypic. Forest, open woodland and semi-open areas, occasionally coastal palm groves and suburbs, to 1300 m (Cuba). Mainly Oct–Apr, but regular from Aug/Sept until early May in N of region (exceptionally late Jun on Hispaniola). Most frequent Bahamas (where also exceptionally seen in summer; Buden 1987) and Cuba, becoming less common over rest of Greater Antilles, Cayman and Virgin Is (e.g. Bond 1957) and San Andrés, and only vagrant Lesser Antilles, on St Martin, St Barthélemy, St Kitts, Guadeloupe and Dominica. **ID** Distinctive woodpecker that drills series of horizontal holes in live trees. **Ad** ♂ has red crown plus throat, striking black-and-white facial pattern, black breast-band, mainly yellowish underparts, and black upperparts heavily marked white, including large blaze on wing. **Ad** ♀ has white chin and throat, usually a paler red crown (sometimes mixed black), or patchy red on forehead, occasionally top of head all black or with a few buff spots.
Voice Only vocalization likely to be heard a soft mew in alarm, louder and hoarser with increasing excitement. **SS** Unlikely to be confused; despite overlap with several, generally similar-sized species in Greater Antilles, its head and upperparts pattern are quite unique, especially large white wing patch (also well visible in flight).

Cuban Ivory-billed Woodpecker
(Ivory-billed Woodpecker)

Fernandina's Flicker

Yellow-shafted Flicker
(Yellow-shafted Flicker / Northern Flicker)
auratus

gundlachi

chrysocaulosus

Cuban Flicker
(Yellow-shafted Flicker / Northern Flicker)

Yellow-bellied Sapsucker

not to scale

variant

175

Cuban Green Woodpecker *Xiphidiopicus percussus* LC
Fairly common resident (**Cuba**).

L c. 21–25 cm. Two subspecies, of which both (nominate[A] mainland Cuba and N cays, *insulaepinorum*[B] I of Pines and S cays) occur in region. Endemic. Semi-deciduous, evergreen and riparian woodland, second growth, palms, taller swamp woodland, sandy coast vegetation and mangroves, to at least 1300 m. Curiously absent from Sierra Maestra, in SE Cuba. Claims from Hispaniola and Jamaica lack foundation. **ID** Beautiful and uniquely plumaged woodpecker. **Ad** ♂ has olive-green upperparts, barred wings, mostly white head, unmarked pale yellow belly, red lower throat and upper breast, and red forehead to nape. **Ad** ♀ is smaller and shorter-billed, has most of crown black with very fine white streaks, and more extensive barring below. **Juv** is duller and more barred above and below, sometimes with hint of red on mantle, fully barred tail, throat more brownish black, and smaller belly patch often orange or orange-red; both sexes have more black on forecrown, ♀ soon acquires white-streaked black crown. **GV** Race *insulaepinorum* averages smaller and paler, with smaller red throat, paler yellow belly and fully barred tail. **Voice** Short harsh "jorr" or "gwurr" in short series; also higher "eh-eh-eh"; mewing or squealing "ta-há", like call of Yellow-bellied Sapsucker. **SS** None; highly distinctive. **TN** Described races *monticola* (E Cuba) and *cocoensis* (Cayo Coco and nearby cays) treated as synonyms of nominate, and *gloriae* (Cayo Cantiles) and *marthae* (Cayo Caballones) as synonyms of *insulaepinorum*.

Guadeloupe Woodpecker *Melanerpes herminieri* NT
Fairly common resident (**Guadeloupe**).

L c. 24–29 cm. Monotypic. Endemic. All available forest types within range, but prefers humid semi-deciduous and evergreen forest, with highest densities in evergreen second growth, then swamp forest and rainforest. To c. 1000 m, but commonest below 700 m. One record on Antigua, c. 80 km N of Guadeloupe, Sept 1975 (Holland & Williams 1978); probably blown there by a tropical storm. **ID** Rather small, distinctive woodpecker that appears all black in field. **Ad** has head and upperparts black glossed blue, sooty-black underparts with dull red tips to feathers of throat to belly, and blackish-brown flanks and undertail-coverts. Long bill somewhat pointed, up to 20% shorter in ♀. **Juv** has duller, less glossy, browner plumage, and underparts suffused dull orange-red. **Voice** Main call a single or repeated "kwa", ♀ higher-pitched. Also variable "wa-wa-wa" or "kakakakaka" calls when excited, and loud "ch-arrgh" in series of 3–8 notes, apparently in contact. Drums in relatively slow rolls. **SS** No other woodpecker on Guadeloupe (in any case, its plumage is highly distinctive and unusual).

Puerto Rican Woodpecker *Melanerpes portoricensis* LC
Common resident (**Puerto Rico**).

L 23–27 cm. Monotypic. Endemic. Like most Caribbean congenerics, relatively widespread in wooded areas, from mangroves and plantations to montane forest and suburban gardens. Sometimes small flocks in non-breeding season. Also Vieques, where rare. Specimens (ANSP) labelled St Thomas generally considered to reflect labelling error (Wetmore 1927); we treat reports for St Croix, both historical (Newton & Newton 1859) and recent (*fide* L. Yntema), with circumspection. **ID** Very distinctive *Melanerpes*. **Ad** ♂ is glossy blue-black from crown to tail, except white rump. Throat and breast red, continuing narrowly onto belly; forehead and lores white, and rest of underparts pale buffish brown. **Ad** ♀ is smaller and shorter-billed, has chin, throat and often moustachial largely brown, red of belly usually narrower and sometimes orange-yellow; occasionally much like ♂. **Juv** is less glossy, with red below less extensive and more orange; ♂ usually has a few red feather tips on crown. **Voice** Wide variety of calls, most commonly "wek, wek, wek-wek-wek-wek-wek" or similar, increasing in volume and speed; also a rolling "gurrr-gurrr", "kuk" notes like domestic hen, and "mew" notes. Drums weakly and infrequently. **SS** None.

Hispaniolan Woodpecker *Melanerpes striatus* LC
Common resident (**Hispaniola**).

L c. 20–24 cm. Monotypic. Endemic. Ubiquitous in wooded areas, from mangroves, coastal scrub and semi-arid country in lowlands to humid forest in mountains, including pine woodland, and around settlements; most numerous in cultivated areas with trees and palms, especially in hills. To 2400 m. Nests in loose colonies. Also on Beata I. **ID** Easily identified. **Ad** ♂ has strikingly blackish and greenish-yellow barred upperparts, black tail and red rump, dark buffy-olive underparts, greyish face and mainly red crown and nape. **Ad** ♀ is smaller and less bulky, with shorter bill and black crown often white-speckled at sides. **Juv** has red of nape and tail-coverts more orangey and less extensive, both sexes have white-spotted black crown with few red tips. **Voice** Highly variable: long series of up to 23 notes in long-distance communication, several connected "waa" notes during encounters, with an aggressive "wup", more defensive "ta" and "ta-a" calls; also 3–5 distinct notes combined in short "bdddt". **SS** Larger than only other woodpeckers on Hispaniola, resident Antillean Piculet and winter visitor Yellow-bellied Sapsucker. Longer-billed than both with very different plumage, especially prominently barred upperparts. Lacks white wing patch of the sapsucker.

Jamaican Woodpecker *Melanerpes radiolatus* LC
Common resident (**Jamaica**).

L c. 24–26 cm. Monotypic. Endemic. All manner of wooded areas and elevations, including mangroves, copses, citrus groves, plantations, gardens, dry and wet limestone forest and lower montane rainforest. Highest densities in secondary mesophytic forest. **ID** Fairly typical *Melanerpes*. **Ad** ♂ has mainly blackish upperparts and tail, narrowly barred white, mainly red crown and nape, whitish face and dull tan-coloured underparts, with some red on vent and barring on rear flanks and undertail-coverts. **Ad** ♀ is less bulky than ♂, has crown grey to dark grey, sometimes buff-tinged or partially black. **Juv** is duller than ad, greyer below with yellower belly patch, more diffuse flank bars, brown eyes (red in ad), both sexes with red on crown, but less in ♀. **Voice** Most frequently a loud "kaaa", sometimes repeated 2–3 times; single "kao" in mild alarm; very loud "kaaaah" apparently in advertisement; "wee-cha weecha" in intraspecific encounters; also "krirr, krirr" and more intimate "whirr-whirr" when breeding. Loud drumming by both sexes. **SS** Only other woodpecker in Jamaica is uncommon winter visitor Yellow-bellied Sapsucker, which is smaller with quite different upperparts and head pattern.

Cuban Green Woodpecker

percussus

insulaepinorum

Guadeloupe Woodpecker

Puerto Rican Woodpecker

Hispaniolan Woodpecker

Jamaican Woodpecker

177

West Indian Woodpecker *Melanerpes superciliaris* LC

Five subspecies divided into two subspecies groups, of which both occur in region.

West Indian Woodpecker
Melanerpes (superciliaris) superciliaris
Fairly common (**Cuba**) to uncommon resident (**Bahamas**). **L** c. 27–32 cm. Four subspecies in group, all of which (*nyeanus*[A] Grand Bahama and San Salvador in N & E Bahamas, *blakei*[B] Great Abaco in N Bahamas, nominate[C] Cuba and many offshore cays, *murceus*[D] I of Pines, Cayo Largo and Cayo Real) occur in region. Endemic. Wooded habitats, including semi-deciduous, evergreen, swamp, tropical karstic, riparian and pine forests, palms, coastal vegetation and scattered trees, in mainland Cuba to at least 1000 m. Race *murceus* perhaps extinct on Cayo Largo and Cayo Real; nominate has perhaps wandered to Cayo Largo del Sur (Kirkconnell *et al.* in press). Race *nyeanus* uncommon and probably extirpated on Grand Bahama, while no more than 160 pairs were estimated on San Salvador in the 1970s (Miller *et al.* 2018). **ID** Generally the most ubiquitous woodpecker in its range. **Ad** ♂ has upperparts broadly barred black and white, red crown to hindneck, and buffish-cinnamon to brownish-grey underparts, except red vent. **Ad** ♀ is slightly shorter-billed and less bulky, with whitish forehead and crown, blackish rear crown joining black behind eye. **Juv** is less contrastingly patterned, often with red tinge above, larger but more diffuse red area below, and both sexes have red crown, on ♀ admixed black. **GV** Race *murceus* resembles nominate, but smaller in bill, wing and, particularly, tail measurements; *nyeanus* much smaller but rather variable, with only small amount of black behind eye or none, markings on uppertail-coverts more bar-like and underparts slightly greenish-tinged; *blakei* averages larger and darker than *nyeanus*, with pale bars above narrower and often tinged greenish buff on mantle, greyer and darker face and underparts, and more black around and behind eye. **Voice** Typically a loud "krrru" or repeated "krrruu-krrru-krru-krru"; also "waa", "key-ou", and continuous "ke-ke-ke-ke-ke". **SS** No other sympatric woodpecker shows same combination of boldly black-and-white barred upperparts and mainly red crown. **TN** Grand Bahama population sometimes separated as race *bahamensis*, and in Cuban S cays, proposed races *sanfelipensis* (Cayo Real, in Cayos de San Felipe) and *florentinoi* (Cayo Largo) considered inseparable, and both perhaps extinct.

Cayman Woodpecker
Melanerpes (superciliaris) caymanensis
Fairly common resident (**Grand Cayman**). **L** c. 27–32 cm. One subspecies in group. Endemic. Found in most forest types, including mangroves and palms. Has wandered to Little Cayman, Mar 1997 (Bradley 2000). **ID** Distinctive woodpecker. **Ad** differs from conspecific West Indian Woodpecker in having dark barring above much narrower, pale bars on back often strongly buffish, more evenly barred tail with white reaching outer webs of central pair of feathers, less regular uppertail-coverts markings, no black around and behind eye, and red of nasal tufts much paler and restricted. ♀ has greyish (not black) hindcrown. **Juv** is probably inseparable from same-age West Indian Woodpecker. **Voice** No known differences from West Indian Woodpecker. **SS** Compare larger Cuban Flicker (resident) and smaller Yellow-bellied Sapsucker (winter visitor), the only other woodpeckers on Grand Cayman.

Red-bellied Woodpecker *Melanerpes carolinus* LC
Vagrant (**Bahamas**).

L c. 24 cm. Monotypic. ♀, Grand Bahama, Feb 1999 (*NAB* 53: 215), assigned to S Florida race *perplexus* (not recognized here). **ID** Typical *Melanerpes* with predominantly pale underparts, some colour on face and white-and-black-barred upperparts. **Ad** ♂ has pinkish to reddish-orange lower forehead, bright red upper forehead to hindneck and uppermost mantle. Variable orange or pinkish tinge on lores to cheeks and chin. Breast tinged olive, buffish or pink, central belly pale red or pink. **Ad** ♀ is slightly smaller, with grey upper forehead and crown (occasionally small red central patch), usually less extensive reddish on cheeks, less reddish below, and often smaller or paler belly patch. **Juv** has barring above less contrasting than ad, and is darker below with variable streaking on breast and usually more extensive but diffusely barred lower underparts, belly patch often indistinct; ♂ has smaller and paler red crown patch than ad, crown mostly grey with black bars (occasionally, all blackish), ♀ with even less red on nape but small patch of red on crown. **Voice** Very vocal; calls include vibrato "churr", singly or in series of up to four, in contact, also highly variable single "chip" or "chup" notes. **SS** Compare West Indian Woodpecker.

Hairy Woodpecker *Leuconotopicus villosus* LC

Seventeen subspecies divided into six subspecies groups, of which one occurs in region.

Bahamas Hairy Woodpecker
Leuconotopicus (villosus) maynardi
Fairly common resident (**Bahamas**), vagrant (**Puerto Rico**). **L** c. 18–20 cm. Two subspecies in group, of which both (*piger*[A] Grand Bahama, Abaco and Mores, in N Bahamas, *maynardi*[B] Andros and New Providence, in C Bahamas) occur in region. Primarily in pines for breeding, but will forage in other woodland types. Usually in pairs. Vagrants: Providenciales, Apr 1986 (Caicos Is) and Mona I, Mar 1974 (Puerto Rico). **ID** Largely black-and-white woodpecker. **Ad** ♂ (*piger*) has mostly black upperparts, mainly white face and underparts; most distinctive is red hindcrown. **Ad** ♀ lacks red on head. **Juv** has black areas browner, underparts somewhat darker with streaks or bars, some bars on outer tail, and ♂ has orangey or red patch on crown, usually much smaller in ♀. **GV** Race *maynardi* has unstreaked back, plain white outer tail, and paler unmarked underparts with buff-tinged breast. **Voice** Loud high-pitched "keek" and "keek kit-kit-kit-kit-kit-kt" series. Both sexes drum, producing short rolls that slow towards end. **SS** Only possible confusion is with winter visitor Yellow-bellied Sapsucker, but it has quite different head and upperparts patterns. **TN** Relationships of Bahamian races to rest of Hairy Woodpecker complex require study. Often placed in genus *Dryobates*.

superciliaris

blakei

West Indian Woodpecker

Cayman Woodpecker
(West Indian Woodpecker)

Red-bellied Woodpecker

Bahamas Hairy Woodpecker
(Hairy Woodpecker)
maynardi

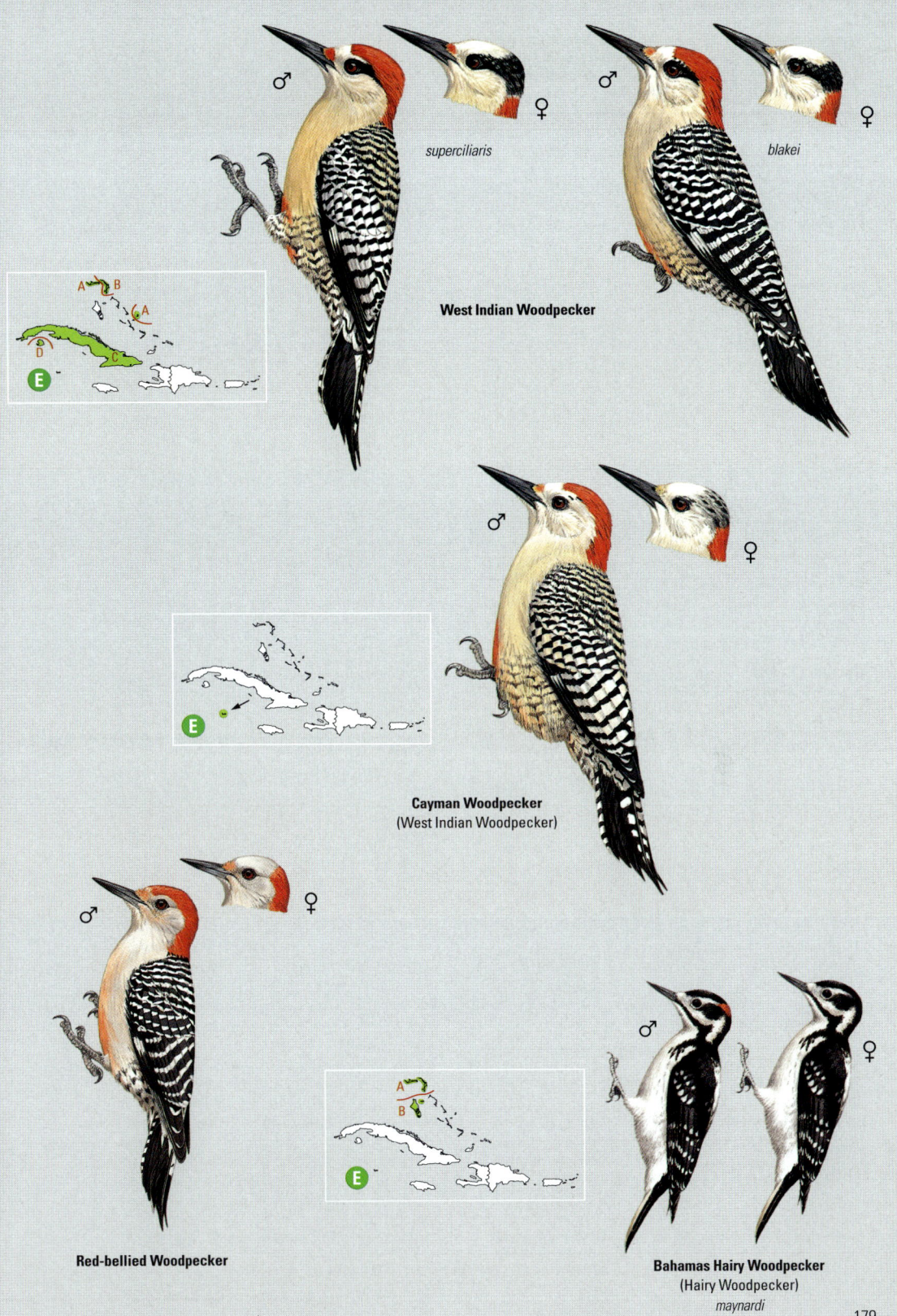

FALCONIDAE
Falcons and Caracaras
64 extant species, 4 in region + 3 vagrants

Crested Caracara *Caracara cheriway* `LC`
Uncommon and local resident (**Cuba**), vagrant (**Jamaica**).
L 49–58 cm; **WS** 107–130 cm. Monotypic. Semi-arid open country: palm savannas, grasslands, cut-over areas, mangroves and pastures, mostly in lowlands and on offshore cays. Highly opportunistic and a scavenger. Flies rather low on steady wingbeats and perches conspicuously on fences or exposed branches. **ID** Imposing appearance, with large bill, bushy crest and long neck, tail and legs. **Ad** is mostly brownish black with white throat and neck. Crown and crest black. Breast and upper back whitish barred black. Reddish facial skin and tail white, finely barred black with broad black tip. Legs orange to yellow. In flight, white patches near wingtips contrast with blackish underwing. **Imm** is duller and browner overall, with buffy underparts streaked brownish. Legs pinkish grey to yellowish. **Voice** Usually silent. When agitated, harsh metallic-sounding rattles "ca-ca-ca-ca". **SS** Nothing similar in region. **TN** Formerly treated as conspecific with extralimital Southern Caracara *C. plancus* (South America S of Amazon). **AN** Northern Crested Caracara.

Common Kestrel / Eurasian Kestrel `LC`
Falco tinnunculus
Twelve subspecies divided into two subspecies groups, of which one occurs in region.

Common Kestrel *Falco (tinnunculus) tinnunculus*
Vagrant (**Guadeloupe**, **Martinique**).
L 27–35 cm; **WS** 57–79 cm. Ten subspecies in group, of which at least nominate occurs in region. Ad ♀, Martinique, Dec 1959 (Pinchon & Vaurie 1961); ♀, Désirade I, Guadeloupe, Apr 2009 (Levesque *et al.* 2012). Habitat and habits similar to American Kestrel's, but hovers more readily and for longer. **ID** Mid-sized falcon with long narrow-based wings and long tail. Similar to American Kestrel, but larger and has single (not double) black moustachial. **Ad** ♂ has grey head (not just blue-grey crown) with finely black-streaked crown, giving hooded appearance. Back spotted (rather than barred) black. Inner half of upperwing concolorous with back; outer blackish. Tail grey with broad black subterminal band. Cere and legs yellow. **Ad** ♀ is slightly larger, generally paler, more uniform and more heavily spotted above, with less contrasting plumage. Head brownish and tail brownish grey with multiple black bands. **Imm** is like ad ♀, with broader streaks below and slightly paler cere and legs. **Voice** Liable to be silent in region. **SS** Compare commoner and smaller Merlin, also American Kestrel.

American Kestrel *Falco sparverius* `LC`
Common to rare resident (virtually **throughout**) and winter visitor (**Cuba, O**).
L 21–31 cm; **WS** 51–61 cm. Seventeen subspecies, of which at least four (*sparverioides*[A] Bahamas, Cuba and Jamaica; *dominicensis*[B] Hispaniola, *caribaearum*[C] Puerto Rico, Virgin Is and Lesser Antilles, nominate migrant S to Cayman Is) occur in region. Varied dry, open or semi-open habitats, including grasslands, palm savannas, pastures, coasts, scrub, towns, parks and forest edges. At all elevations but less numerous in highlands; to 1300 m in Cuba and 2840 m on Hispaniola. Perches on bare branches, fences and at roadsides, often bobbing tail. Fast wingbeats and often hovers briefly, but less frequently and for much shorter periods than Common Kestrel. Nests in cavities. Generally common resident Bahamas S to St Lucia and Grenada; rare elsewhere. Inter-island movements probable, as birds resembling *sparverioides* have occurred off Hispaniola, and *dominicensis* reported S Bahamas and N Cuba. Migrants (nominate race) fairly common, mainly Aug–May, mostly Cayman Is, Bahamas and Cuba. Vagrants to Barbados unidentified (either *caribaearum*, or *isabellinus* from South America), and race *paulus* from USA suspected to occasionally reach Bahamas. **ID** Highly variable but distinctive small falcon, with long tail and two bold black bars on white face. **Ad** has rufous upperparts barred black, bluish-slate crown, sometimes with orangey central patch. Underparts vary from creamy to pale rufous with black spots and blotches on lower breast and flanks. ♂ has rufous tail with broad, black terminal band and blue-grey upperwing; ♀ has generally more barred upperparts, rufous tail with multiple dark bands, and rufous, barred black upperwing. **Imm** is similar to respective ad but has paler and less obvious pattern, and boldly streaked breast. **GV** Distinctive *sparverioides* (Bahamas, Cuba and Jamaica) slightly smaller, has narrower cheek-stripes and occurs in two morphs: a light morph, paler overall than nominate, with largely white underparts, and dark morph, much darker overall, with dark grey upperparts and rufous underparts; *dominicensis* paler than nominate (much like pale-morph *sparverioides*) with deep rufous breast and some flank spots; *caribaearum* smaller than others, richly coloured with heavy dark underpart markings. ♂ *isabellinus* (potential vagrant) usually has large rufous crown patch, barred scapulars and spots below. Race *paulus* is comparatively small with relatively large head and rather variable plumage, but ad ♂ tends to be very sparsely marked below, with spots confined to sides and flanks. **Voice** High-pitched "killi-killi-killi". **SS** Distinctive, but see slightly larger Merlin. In E Lesser Antilles, see vagrant Common Kestrel.

Merlin *Falco columbarius* `LC`
Fairly common to rare winter visitor and passage migrant (**throughout**).
L 24–32 cm; **WS** 53–73 cm. Nine subspecies, of which nominate occurs in region. Open to semi-open areas, from coastal lagoons and other wetlands in lowlands where shorebirds abound, to c. 1200 m in Cuba and 1465 m on Hispaniola. Also cities, grassland and open woodland. Passage migrant throughout in Oct and Mar: common Bahamas, less so Greater Antilles, Virgin and Cayman Is, and generally rare Lesser Antilles. Some winter throughout until Mar (extremes late Jul–early May), but largely absent rest of year. **ID** Fairly small, compact falcon with pointed wings, narrow tail and ill-defined dark 'sideburns'. **Ad** ♂ has uniform dark bluish-grey upperparts, whitish forehead and weak eyebrow. Underparts whitish to creamy, heavily streaked brown. Tail has 3–4 pale greyish bands. Cere and legs yellow. **Ad** ♀ is larger, with brown upperparts. **Juv** is like ad ♀ but usually has greenish-yellow to pale yellow cere and legs. **Voice** Rapid, accelerating series of strident "kee-kee-kee-kee-ki-kikiki" calls, but rarely heard. **SS** Recalls streaky imm Peregrine Falcon, which is much larger, with broader, more pronounced 'sideburns'. In addition, ad of latter darker above. Slightly smaller American Kestrel has rufous back and bold facial pattern. In flight, could be mistaken for dark Rock Dove, but Merlin's flight is faster, with quicker, steadier wingbeats.

not to scale

adult

Crested Caracara

♂ ♀

Common Kestrel
(Common Kestrel / Eurasian Kestrel)
tinnunculus

dominicensis

♂ ♀

American Kestrel

♂ *caribaearum* ♀

sparverioides

light morph ♂

dark morph ♂

dark morph ♀

♂ *sparverius* ♀

♂ ♀

Merlin
columbarius

181

Bat Falcon *Falco rufigularis* LC
Vagrant (**Barbados**, **Grenada**).

L 23–30 cm; **WS** ♂ 51–58 cm, ♀ 65–67 cm. Three subspecies, of which one (presumably nominate) occurs in region. Grenada, two very vague records mentioned in (Devas 1954, considered doubtful by Bond 1972); Barbados, Mar 2018 and Mar 2019 (eBird). Usually in forested areas and at edges, occasionally urban areas. Perches on high exposed branches, antennas, etc. Often active at dusk. **ID** Rather small, compact and dark falcon with long, pointed wings and narrow tail. Upperparts and head blackish, giving hooded appearance. **Ad** has throat and upper breast whitish to cinnamon, reaching neck-sides. Broad 'vest' black, finely barred white. Lower underparts dark rufous. Tail blackish with several fine whitish to greyish bars, and white or buff tip. Cere, eye-ring and legs bright yellow. **Imm** has buffier throat and slightly paler and duller underparts. Legs paler yellow. **Voice** Lone birds generally quiet. **SS** Aplomado Falcon (also vagrant) is larger and paler, with prominent head-band. In flight, could be mistaken for slightly smaller White-collared Swift.

Aplomado Falcon *Falco femoralis* LC
Vagrant (**Puerto Rico**).

L ♂ 35–38 cm, ♀ 43–45 cm; **WS** 76–102 cm. Three subspecies, of which nominate occurs in region. Laguna Cartagena, Jan 2008 (Mathys 2011). Pastures, farmland, dry scrub and other open areas, usually in lowlands. **ID** Medium-sized falcon with pointed wings and long, narrow tail. **Ad** has upperparts slate, and broad whitish head-band encircling crown. Cheeks, head-sides and throat whitish to creamy, with black 'sideburn' below eye. Sides and narrow breast-band black, finely barred white. Chest whitish with tawny flush, sparsely streaked black. Lower underparts pale rufous. Tail blackish with several white bands and tip. In flight, conspicuous white trailing edge to wings. **Imm** is duller overall, browner above, buffier and streakier below. Breast-band incomplete or absent. **Voice** Unlikely to be heard. **SS** Long tail, whitish head-band and dark 'vest' distinctive. See also vagrant Bat Falcon.

Peregrine Falcon *Falco peregrinus* LC
Uncommon to rare and local winter visitor and passage migrant (**throughout**), has bred (**Cuba**, **Dominica**).

L 35–51 cm; **WS** 79–114 cm. Nineteen subspecies, of which two (*anatum* widespread and commoner, *tundrius* apparently much rarer) occur in region. Mainly over offshore cays and and coastal wetlands with many shorebirds or waterfowl. Inland wetlands, again if there are waterbirds to hunt. Perches on high bare branches, towers, antennas, buildings and church steeples. Fast, powerful pigeon-like flight. To at least 1250 m in Cuba and 1440 m on Hispaniola. Mainly Aug–May, with peak passage Oct/Nov and Mar, but recorded in very small numbers year-round, with at least two breeding records (ascribed to *anatum*) in Cuba and one in Dominica. **ID** Large, stocky falcon with pointed wings and narrow tail. **Ad** (*anatum*) has upperparts and tail mostly bluish slate. Crown, nape and broad 'sideburns' black, giving distinctive 'helmeted' appearance. Sides of neck and underparts whitish to buffy, densely barred and spotted dusky. ♀ is much larger than ♂. **Imm** is dark brown above, also with 'helmeted' look, and creamy, heavily streaked brown below. **GV** Race *tundrius* smaller and much paler, with narrower dark moustachial and large pale cheeks. Ad has unmarked or faintly marked white breast, and juv a broad dark eyestripe and cream-buff forehead and supercilium. **Voice** Mostly silent in region. **SS** Streaky imm recalls Merlin, which is much smaller and more lightly built. **AN** Peregrine.

CACATUIDAE
Cockatoos
21 extant species, 1 introduced in region

White Cockatoo *Cacatua alba* EN
Introduced, doubtfully established rare resident (**Puerto Rico**).

L c. 46 cm. Monotypic. Native to N Moluccas (Indonesia). Restricted to urban areas in Bayamón and Guaynabo municipalities; first reported mid 1990s (Pérez-Rivera 1998), but perhaps still dependent on feeders. **ID** The only white cockatoo more or less established in region (see Similar species). **Ad** is all white with underwing and undertail washed yellow. Broad erectile crest, spread sideways to give impression of massive head. ♂ has black eyes, ♀ red-brown. **Juv** has dark grey eyes. **Voice** Typical loud, cockatoo screeching, and nasal monosyllabic or bisyllabic "keh" or "keeh-ah". **SS** Tanimbar Corella *C. goffiniana* (not yet established Puerto Rico) smaller with much smaller crest and reddish-pink lores. Sulphur-crested Cockatoo *C. galerita* (also present Puerto Rico and New Providence, Bahamas) larger with prominent yellow crest. **AN** White-crested Cockatoo, Umbrella Cockatoo.

PSITTACIDAE
Parrots
375 extant species, 12 in region + 11 introduced

Monk Parakeet *Myiopsitta monachus* LC
Introduced and locally common resident (**Puerto Rico**, **Cayman Is**).

L 28–29 cm. Three subspecies, of which one (unknown) occurs in region. Native to SE South America. In West Indies, only in coastal palm groves and gardens on Puerto Rico (probably introduced c. 1950s) and George Town and elsewhere on Grand Cayman (c. 1987), though latter population has been largely eradicated recently. Occasionally wanders to Virgin Is (from Puerto Rico?). Records in Bahamas, initially Eleuthera in 1980, subsequently also Grand Bahama. Nests conspicuously, often on human structures. **ID** Fairly large, distinctive parakeet with long, pointed green tail and orange bill. **Ad** is bright green above, with whitish-grey face, throat and breast, faintly yellowish belly, and deep blue flight feathers. **Imm** has forehead tinged green. **Voice** Commonest call a drawn-out nasal grating note, typically repeated in long series, e.g. "rrreh-rrreh-rrreh…". Also an upslurred less grating "krree" and subdued conversational calls. **SS** Greyish-white throat and breast distinctive.

White-winged Parakeet *Brotogeris versicolurus* LC
Introduced, locally common resident (**Puerto Rico**).

L 22–23 cm. Monotypic. Native to Amazon Basin. In Puerto Rico (where introduced c. 1950s) lowland wooded and urban areas. Noisy and conspicuous, in flocks of up to 1000 individuals (Lever 2005); one of most successful avian introductions to region. **ID** Rather slim-bodied parakeet with distinctive wing pattern. **Ad** is green with slightly bluish face. In flight, dark green primaries, large whitish triangular wing patches and yellow secondary-coverts. **Juv** is similar. **Voice** Common calls a shrill "chra" or bisyllabic "chra-chra". Also fast chattering "cra-cra-cra-cra-cra". Large flocks produce near-continuous cacophony. **SS** No other parakeet has distinctive tricoloured wing pattern. **AN** Canary-winged Parakeet.

adult

adult

Bat Falcon
rufigularis

Aplomado Falcon
femoralis

tundrius
adult

anatum juvenile

adult

Peregrine Falcon

White-winged Parakeet

♂ ♀

Monk Parakeet
monachus

White Cockatoo

183

Black-billed Amazon *Amazona agilis* VU
Fairly common resident (**Jamaica**).

L 25–26 cm. Monotypic. Endemic. Wet limestone forest, generally at 100–1400 m, in Cockpit Country E to Mt Diablo and E slopes of John Crow Mts. Readily identified in flight by fast, fluttering wingbeats. Usually in small flocks; roosts communally, often with Yellow-billed Amazon (especially) and Jamaican Parakeet. **ID** The smaller and scarcer of Jamaica's two *Amazona*. **Ad** is all green, with red flecks on forehead, dark spot on ear-coverts, soft blue primaries, red bases to outertail feathers and yellowish-green undertail-coverts. ♂ has primary-coverts red, ♀ partially or all green. **Juv** is like ♀, with all-green primary-coverts. **Voice** Varied screeches, most very nasal. When perched, growling "rrak" or "muh-weep", with richer and more varied tones than Yellow-billed Amazon; in flight "tuh-tuk", higher-pitched than Yellow-billed. **SS** All-green plumage and (in ♂) red primary-coverts distinguish it from Yellow-billed Amazon, which has pinkish throat, blue forecrown, yellow bill and generally yellower-green plumage. Compare Jamaican Parakeet. **AN** Black-billed Parrot.

White-fronted Amazon *Amazona albifrons* LC
Introduced, rare and local resident (**Puerto Rico**).

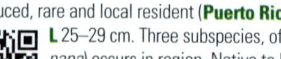

L 25–29 cm. Three subspecies, of which one (perhaps *nana*) occurs in region. Native to Middle America. Confined to environs of Mayagüez where originally bred in nestboxes at Univ. of Puerto Rico (present since at least 2001, but numbers dwindling). Typically in pairs or small flocks. **ID** Small *Amazona* with red eye patch and white forehead. Secondaries blue, no speculum. **Ad** ♂ has red primary-coverts, obvious in flight. **Ad** ♀ has green primary-coverts and therefore appears more uniform in flight. White forehead smaller than ♂. **Imm** is like ♀ but duller. **Voice** Varied screeches and other raucous calls, also a harsh grating chattering. In flight, mostly harsh "crek-crek" or "kyak-yak-yak" calls. **SS** Hispaniolan Amazon lacks red eye patch and primary-coverts. **AN** White-fronted Parrot.

Yellow-billed Amazon *Amazona collaria* VU
Fairly common resident (**Jamaica**).

L 28–31 cm. Monotypic. Endemic. Mid-elevation wet limestone forest to 1200 m, occasionally at sea level and in semi-wooded areas and cultivation. Outside breeding season, gathers in (sometimes large) roosts with other native parrots. Slower wingbeats than Black-billed Amazon, more typical of other *Amazona*. **ID** The larger and more widespread of Jamaica's amazons. **Ad** has mostly yellowish bill, white forehead and eye-ring, bluish forecrown, pinkish throat and upper breast. Blue flight feathers, and maroon-based tail. **Juv** as ad. **Voice** Varied screeches and mostly loud low-pitched squawks when perched. Flight call diagnostic, rolling squawks followed by high-pitched upslurred squeal, "krra-ah-eeeeh!", lower-pitched than Black-billed Amazon, with more protracted last syllable. **SS** See Black-billed Amazon. **AN** Yellow-billed Parrot.

Cuban Amazon *Amazona leucocephala* NT
Common to locally fairly common resident (**Cuba**, **Bahamas**, **Cayman Is**).

L 28–33 cm. Four subspecies, all of which (nominate[A] Cuba and I of Pines, *bahamensis*[B] Great Abaco and Great Inagua, Bahamas, *caymanensis*[C] Grand Cayman, *hesterna*[D] Cayman Brac and formerly Little Cayman) occur in region. Endemic. Locally common Cuba, especially in remoter and better-protected areas, and I of Pines. Formerly more widespread Bahamas, where stronghold now Abaco, with recent records (provenance unknown) on New Providence. Fairly common Grand Cayman and Cayman Brac. Wide range of habitats including limestone forest, dry mixed broadleaf woodland, savannas with pines and palms, mangroves, plantations and gardens, to at least 700 m in Cuba. **ID** The only *Amazona* in range; considerable insular variation. **Ad** is green with prominent dark scaling, white on forehead and around eye, pinkish-red cheeks and throat, and purplish belly patch. Flight feathers blue. **Imm** is less strongly scaled, with reduced purple on belly. **GV** Race *caymanensis* has white restricted to forehead and mid-crown, little or no pink on throat or purple on belly; *hesterna* smaller than nominate, with larger belly patch; and *bahamensis* frequently lacks belly patch, whereas white of crown is more extensive below and behind eye. **Voice** Varied screeching calls, most rather nasal. In flight, a disyllabic, donkey-like braying. **SS** None; however, a population of **Yellow-naped Amazon** *A. auropalliata* did become briefly established on Grand Cayman in 1990s (see illustration on p. 187). **TN** Birds of I of Pines formerly recognized as *palmarum*, but only very doubtfully distinct. In Bahamas, race *bahamensis* extinct on Acklins I, and extant populations proposed as races *abacoensis* (Great Abaco) and *inaguaensis* (Great Inagua); separately, Russello *et al.* (2010) suggested all three forms are phylogenetic species. **AN** Rose-throated Parrot, Cuban Parrot.

Hispaniolan Amazon *Amazona ventralis* VU
Locally common resident (**Hispaniola**).

L 28–31 cm. Monotypic. Endemic. Patchily distributed Haiti, Dominican Republic and, at least historically, on satellite islands of Gonâve, Beata and Saona, favouring lowland palm savannas and more humid montane evergreen forest, to 1500 m. Numbers much reduced during 20th century, now largely confined to protected areas. Introduced Puerto Rico, St Croix and St Thomas, but probably not established (Raffaele 1989, Lever 2005). **ID** Similar to Cuban Amazon. **Ad** has prominent white forehead and lores bordered by blue suffusion, blackish ear-coverts, reddish spot on chin, and variable maroon patch on belly. Vivid blue flight feathers. **Imm** lacks blue on face and cheeks, and has paler belly patch. **Voice** Varied screeches and calls, mostly rather high-pitched. Also chattering "chachachachacha". In flight, typically bisyllabic, including a barking "wi-chah" or higher-pitched reedy "wi-chih". **SS** Only *Amazona* on Hispaniola. In Puerto Rico, no other parrot has dark ear-coverts or maroon belly patch, but see White-fronted Amazon. **AN** Hispaniolan Parrot.

White-fronted Amazon
nana

♀

♂

Black-billed Amazon

E

Yellow-billed Amazon

E

caymanensis

bahamensis

leucocephala

Cuban Amazon

Hispaniolan Amazon

E

E

185

Puerto Rican Amazon *Amazona vittata* `CR`
Very rare resident (**Puerto Rico**).
L 29–30 cm. Two subspecies, of which both occurred but only nominate extant. Endemic. Moist montane forest at 200–600 m. Once abundant, population now much reduced; was confined to Luquillo Forest, in NE, but second population established using released birds in Río Abajo State Forest, in W of island. Full impact of Hurricanes Irma and Maria, in 2017, remains to be seen. **ID** Rather plain-looking *Amazona* with broad white eye-ring. **Ad** appears rather scaled; red forehead, bright blue remiges and red-based outertail feathers. **Juv** is similar. **GV** Extinct race *gracilipes* (Culebra I, off NE Puerto Rico) was smaller. **Voice** Varied, rather high-pitched screeches and calls, including distinctive bugling flight call, "kar…kar". **SS** Compare head patterns of (introduced) Red-crowned, Yellow-crowned and Orange-winged Amazons. Hispaniolan Amazon (introduced but rare in Puerto Rico) has white forehead, dark ear-coverts and maroon belly. **AN** Puerto Rican Parrot.

Red-crowned Amazon *Amazona viridigenalis* `EN`
Introduced, locally uncommon resident (**Puerto Rico**).
L 30–35 cm. Monotypic. Native to NE Mexico. In Puerto Rico, favours lowland forest and scrub near coast, especially in SE. **ID** Stocky parrot. **Ad** has prominent red lores and forecrown, bluish-grey patch behind eye, bright green cheeks and blackish-scaled nape. In flight, red speculum, dark blue primaries and broad yellowish tips to outer rectrices. ♀ has less red on head. **Imm** is like ♀ but duller. **Voice** In flight, a diagnostic, shrill, upslurred squeal typically followed by several lower-pitched grating barks "cleeuw! … crreh-crreh-crreh"; less raucous and rasping than other parrots. **SS** Puerto Rican Amazon has red restricted to forehead, no blue behind eye, no red speculum, and is confined to montane forest. **AN** Red-crowned Parrot, Green-cheeked Amazon.

Red-necked Amazon *Amazona arausiaca* `VU`
Locally fairly common resident (**Dominica**).
L 33–40 cm. Monotypic. Endemic. Favours montane rainforest, chiefly at 300–800 m, sometimes in more open cultivated and coastal areas. Noisy and gregarious. **ID** The smaller of Dominica's two *Amazona*. **Ad** is primarily green with violet-blue face, red forecrown, red-and-yellow speculum, and blackish-tipped primaries. **Juv** is duller with less blue on face and little or no red on neck. **Voice** Disyllabic, drawn-out "rrr-eee", higher-pitched than Imperial Amazon. Also varied screeches and rolling calls, sounding rather thin and parakeet-like. **SS** Imperial Amazon—the only other *Amazona* on Dominica—unmistakable if seen well, by extensive purple body plumage and being larger; also typically at higher altitudes than present species, and more elusive. **AN** Red-necked Parrot.

St Lucia Amazon *Amazona versicolor* `VU`
Uncommon resident (**St Lucia**).
L 42–46 cm. Monotypic. Endemic. Primarily in moist tropical forest in montane interior of C & S, at 500–900 m; occasionally forages in adjacent second growth and cultivated areas. **ID** Unmistakable, large *Amazona*, the only parrot in St Lucia. **Ad** has striking blue forehead and face, scaled green on hindcrown, nape and mantle. Red patch on foreneck and upper breast, with green-and-maroon mottling on lower breast, and greenish-yellow vent. In flight, red speculum and dark blue primaries. **Juv** is duller with reduced blue on face, and less red on foreneck. **Voice** Wide variety of screeches, high-pitched squeals and rolling barks, over very wide frequency range. Flight call a repeated, grating "rrrek". **SS** None. **AN** St Lucia Parrot.

Imperial Amazon *Amazona imperialis* `EN`
Uncommon resident (**Dominica**).
L 45–51 cm. Monotypic. Endemic. Moist forest in mountain valleys, chiefly at 600–1300 m. Forages quietly in canopy. Small population (250–350 individuals prior to Hurricane Maria in 2017) between Morne Diablotin area, Northern and Central Forest Reserves, and recently re-established population in Morne Trois Pitons National Park. Numbers severely impacted by hurricanes, in past sometimes being reduced to just c. 50 birds. **ID** Largest and very distinctive *Amazona*. **Ad** has brown face and cheeks, scaled purple nape and underparts, and dull green wings with red carpal, purple speculum and blackish-blue primaries. Tail reddish brown with greenish-blue tip. **Imm** has dull rufous face, and green nape and neck. **Voice** In flight, distinctive metallic trumpeting, "eeeee-er" that descends at end. Perched, a variety of screeches, squeals and rolling barks. **SS** See Red-necked Amazon, the only sympatric congener. **AN** Imperial Parrot.

Orange-winged Amazon *Amazona amazonica* `LC`
Introduced, locally common to uncommon resident (**Puerto Rico, LA**).
L 31–32 cm. Monotypic. Native of South America. In Puerto Rico, where probably introduced in 1960s, local but occurs in several areas; commoner Martinique, where fairly widespread in C of island; in Barbados poorly known but first noted 1950s. Also present St Croix, Guadeloupe (not established) and Grenada, with isolated occurrences on St Thomas (Virgin Is) and Anguilla. Favours second growth and urban areas with ornamental trees. **ID** Relatively distinctive *Amazona*. **Ad** has yellow forecrown and throat divided by blue lores and broad diffuse superciliary. Body and wings green, latter with orange-red patch and blue primaries. Tail green with yellow tips; lateral feathers tinged orange-red with dark green central bar. **Imm** is indistinguishable from ad. **Voice** Screeches and rolling calls. In flight, a diagnostic, repeated, upslurred, shrieking "scrreeeh!" or squealing "queek!", higher-pitched and less raucous than most *Amazona*. **SS** In Puerto Rico compare slightly larger and also introduced (but not established) **Yellow-headed** *A. oratrix* and **Yellow-crowned Amazons** *A. ochrocephala*, neither of which have blue eyebrow and usually have red shoulders. **AN** Orange-winged Parrot.

Puerto Rican Amazon
vittata

Red-crowned Amazon

Red-necked Amazon

St Lucia Amazon

Imperial Amazon

Orange-winged Amazon

Yellow-headed Amazon

Yellow-crowned Amazon

Yellow-naped Amazon

187

St Vincent Amazon *Amazona guildingii* `VU`
Uncommon resident (**St Vincent**).

L 40–46 cm. Monotypic. Endemic. Mature moist forest, with some preference for elevations below 700 m, where larger trees and thus more nest-sites exist, occasionally wandering to cultivated areas. Recorded most frequently in upper reaches of Buccament, Cumberland and Wallilibou Valleys. **ID** Large, strikingly plumaged *Amazona* with two colour morphs and many intermediates. **Ad** has distinct whitish face and yellow-tipped blue tail. **Yellow-brown morph** is commoner and itself variable; largely golden-bronze with pale bluish patch behind eye, yellow greater underwing-coverts and underside to remiges, black primaries with yellow bases, and deep blue secondaries with orange bases. **Green morph** tends to lack most golden coloration, with dull green upperparts, more extensive blue on head (usually forming indistinct collar); greater underwing-coverts and underside of remiges green. **Imm** is like respective ad morph, but duller. **Voice** Varied screeches and rolling barks. Flight call a repeated, grating "rrrek" or loud un-parrot-like "quaw…quaw…quaw". **SS** No other parrots in St Vincent. **AN** St Vincent Parrot.

Green-rumped Parrotlet *Forpus passerinus* `LC`
Introduced, common resident (**Jamaica**).

L 12–13 cm. Five subspecies, of which nominate occurs in region. Native of NE South America. Rather common and widespread in Jamaica (first noted 1918), but probably always rare Barbados (where introduction dates from early 1900s and died out there in early 1960s. Introduced unsuccessfully Martinique. Favours second growth and relatively open areas in lowlands and hills. **ID** Tiny parrot, smaller than House Sparrow. **Ad** is essentially all green, brightest on rump. ♂ has bluish-tinged rump and violet-blue band on wings. ♀ lacks blue, but has yellower forehead. **Imm** is like ad. **Voice** Call a passerine-like bisyllabic "chidit", or sometimes longer "chididididit". Groups in flight call continuously, producing a relatively loud tinkling twittering. **SS** None.

Jamaican Parakeet *Eupsittula nana* `NT`
Common resident (**Jamaica**), introduced (**Dominican Republic**).

L 22–26 cm. Monotypic. Endemic. Common and widespread Jamaica, especially in mid-elevation limestone forest to 1500 m, but also scrub, plantations and even gardens; rarely visits arid S forests. Apparently introduced Dominican Republic (sometime between 1970s and early 1990s) where mainly present in SW and increasing; favours deciduous woodland and pines, principally in foothills. A few records Puerto Rico, first Apr 1993 (Salguero-Faría & Roig-Bachs 2000). **ID** Rather small with long, pointed tail. **Ad** is green, yellower on ear-coverts and rump, with broad white orbital skin, olive-brown throat and underparts, and blue-fringed flight feathers. **Imm** is similar to ad. **Voice** Flight call a screeching "creek" or "clack", often doubled or tripled. Perched calls similar, but interspersed by chattering notes. Less raucous than sympatric (in Dominican Republic) Hispaniolan Parakeet. In Jamaica, calls might be confused, if bird not seen, with Jamaican Woodpecker. **SS** The only parakeet in Jamaica. In Dominican Republic, Hispaniolan Parakeet has yellowish-green (not olive) underparts and distinct red shoulder patch. **TN** Often treated as conspecific with extralimital Aztec Parakeet *E. astec* (Middle America). **AN** Olive-throated Parakeet.

Orange-fronted Parakeet *Eupsittula canicularis* `LC`
Introduced, uncommon and local resident (**Puerto Rico**).

L 23–25 cm. Three subspecies, of which nominate occurs in region. Native to W Middle America. In Puerto Rico, woodland and urban areas with ornamental trees, primarily in E lowlands. **ID** Small, pale-billed parakeet. **Ad** is dull green with yellowish orbital skin, orange-peach forehead, dull blue mid-crown, pale olive-brown throat and breast, and blue primaries. **Imm** has reduced orange-peach on forehead. **Voice** Flight calls range from high-pitched shrill "creeh creeh" to lower-pitched harsh grating screeches. When perched, similar calls interspersed by chattering notes. **SS** Brown-throated Parakeet (no longer extant in Puerto Rico) superficially similar, but note head patterns. **AN** Orange-fronted Conure.

Brown-throated Parakeet *Eupsittula pertinax* `LC`
Introduced, locally common resident (**LA**, **O**).

L 23–28 cm. Fourteen subspecies, of which at least three (*aeruginosa*[A] San Andrés, nominate[B] St Thomas, *xanthogenia*[C] Saba) occur in region. Native to S Central & South America. Fairly common St Thomas (particularly E), in hillside thickets and woodland, having been introduced c. 1860s from Curaçao (Lever 2005). Uncommon Tortola and Saba (where birds closely recall *xanthogenia*). Reported Dominica, St John, Jost van Dyke, Guadeloupe, Martinique and, recently, San Andrés (where seems fairly common, and birds apparently correspond to *aeruginosa*; Donegan & Huertas 2011). Formerly on Puerto Rico (last reported 1982) and nearby Culebra and Vieques (not since 1976). Usually in small groups. **ID** Rather large, black-billed parakeet. **Ad** has rusty-orange face, grading to deep yellow on ear-coverts, blue mid-crown, dull olive breast, and greenish-yellow underparts with small orange belly patch. Flight feathers and tail edged and tipped dullish blue. **Imm** is duller and largely lacks yellow. **GV** Race *xanthogenia* has orange-yellow extending to crown and nape, without blue on former; *aeruginosa* similar to latter but has blue on crown and buff forehead. **Voice** Rather noisy. Flight calls include high-pitched screeching and harsh grating "scraart scraart" cries, rapidly repeated. Also shorter, bisyllabic "tchrit tchrit" and "cherr cheedit". **SS** Dark bill and orange face distinctive throughout. **AN** Brown-throated Conure.

188

green morph

yellow-brown morph

St Vincent Amazon

not to scale

Green-rumped Parrotlet
passerinus

Jamaican Parakeet

Orange-fronted Parakeet
canicularis

aeruginosa

pertinax

xanthogenia

Brown-throated Parakeet

189

Blue-and-yellow Macaw *Ara ararauna* [LC]
Introduced, doubtfully established resident (**Puerto Rico**).
L 81–86 cm. Monotypic. Native to South America. In Puerto Rico, most records around San Juan and Guaynabo; usually pairs or small groups with **Red-and-green Macaws** *A. chloropterus*. Present since at least 1990s. Both of these species and extralimital Scarlet Macaw *A. macao* (Central and South America) are said to have been introduced in Cuba recently (Navarro Pacheco 2018), however, without details. **ID** Large, very distinctive macaw. **Ad** has dull green crown vs. otherwise blue upperparts, bare white face crossed by narrow dark lines, and golden-yellow underparts, including underwing-coverts. **Imm** is similar to ad. **Voice** Loud raucous cries including harsh "raaa" in flight; generally rather guttural with wavering quality, sounding somewhat more nasal and fuller than Red-and-green Macaw. **SS** Likely to be confused only with even larger Red-and-green Macaw, but only at long range or against light. Latter species known from small number of escapees in Puerto Rico (and apparently Cuba). **AN** Blue-and-gold Macaw.

Red-masked Parakeet *Psittacara erythrogenys* [NT]
Introduced, locally common resident (**Puerto Rico**).
L 29–33 cm. Monotypic. Native to W Ecuador and NW Peru. Released Grand Cayman c. 1992 and breeding first noted 1995 but apparently no longer established, with first records in Puerto Rico also in 1992, and was locally common in N of island prior to the passage of Hurricane Maria in Sept 2017. **ID** Mid-sized, pale-billed parakeet. **Ad** is green with white orbital ring, and red head, shoulders, lower thighs and outermost underwing-coverts. **Imm** has many fewer red feathers. **Voice** Squeaky notes and screeches, also braying phrases, with distinctly nasal quality. Vocal, especially in flight, groups typically maintaining a continuous loud screeching chatter. **SS** None; the only parakeet in range with extensive red on head. **AN** Red-masked Conure.

Cuban Parakeet *Psittacara euops* [VU]
Uncommon and local resident (**Cuba**).
L 24–27 cm. Monotypic. Endemic. Savannas, notably those rich in palms, cultivation with groves of trees, and wooded edges; apparently requires proximity of larger tracts of forest; to 400 m. Declining; now confined to remote or better-protected regions (including around Zapata Peninsula, Trinidad Mts, Sierra de Najasa, and E mountains). Formerly I of Pines. Typically in small flocks of up to 30. **ID** Only native parakeet in Cuba; large pale bill. **Ad** is rather uniform green with white orbital ring, scattered red flecks on head and underparts, and red leading edge to wing and smaller underwing-coverts. **Juv** has fewer red markings. **Voice** Flight call a loud and characteristic "crick-crick-crick". Perched, emits piercing "cree-eek" and "cree-cree", higher-pitched than other parakeets and very different from Cuban Amazon. **SS** Rose-ringed Parakeet is only other parakeet to have been recorded in Cuba. **AN** Cuban Conure.

Hispaniolan Parakeet *Psittacara chloropterus* [VU]
Locally common resident (**Hispaniola**).
L 30–33 cm. Two subspecies, of which both (nominate Hispaniola, *maugei* Mona I, off Puerto Rico) occur in region, but second-named is extinct (not seen since 1892). Endemic. Now mostly seen in pine zone, shade-coffee plantations and other undisturbed or less modified habitats, to 1800 m (mainly above 900 m). In Dominican Republic, largely confined to remoter areas, e.g. Sierra de Bahoruco, Sierra Neiba, Cordillera Central, and lowlands around Lago Enriquillo, but also common in Santo Domingo. In Haiti, stronghold is Massif de la Hotte. Reported I Beata. Flocks of > 50 now unusual, but larger numbers occasionally at communal roosts. Extinct introduction in SW Puerto Rico; one pair on Guadeloupe. **ID** Large, rather uniform green parakeet. **Ad** has white orbital ring, yellowish-green underparts, red leading edge to wing and outermost underwing-coverts contrasting with otherwise dull yellowish underwing and undertail. **Imm** has less or no red in wing. **GV** Extinct race *maugei* duller below, with more red in greater underwing-coverts. **Voice** Screeching flight and perched calls, e.g. a rather grating "kree". **SS** Jamaican Parakeet (Dominican Republic alone) is sole sympatric parakeet, lacks red on wing, has olive underparts and blue on remiges. **TN** Olson (2015) advocated recognizing extinct race *maugei* at species level. **AN** Hispaniolan Conure.

Rose-ringed Parakeet *Psittacula krameri* [LC]
Introduced, fairly common to rare resident (**GA**, **LA**).
L 37–43 cm. Four subspecies, but that in region is unknown. Native to Sahelian Africa and S Asia. Viable feral population Barbados (first noted 1960s), common in wooded areas around Bridgetown; uncommon Martinique. Also present Jamaica (since 2004), Puerto Rico (first report 1979), St Martin (Brown & Collier 2005a), and bred Grand Cayman in early 1990s. Single report from Cuba and a few recent ones from Guadeloupe. **ID** Large yellowish-green parakeet, with long tail, red maxilla and black mandible. **Ad** ♂ has thin black loral line (narrowly meeting on forehead), black chin extending to form fine collar, which narrows on hindneck, and lilac nape. **Ad** ♀ has head all green (no black collar and lores) and shorter tail. **Juv** as ♀ but duller, with even shorter tail. **Voice** Sharp, high-pitched "chi, chi, chi..." and "chew-ee", usually given in flight; also melodious trills. **SS** None; compare other introduced parakeets. **AN** Ring-necked Parakeet.

Red-and-green Macaw
Ara chloropterus

Red-masked Parakeet

Blue-and-yellow Macaw

not to scale

Cuban Parakeet

Hispaniolan Parakeet
chloropterus

Rose-ringed Parakeet
krameri

♂ ♀

191

TITYRIDAE
Tityras and allies
49 extant species, 1 in region

Jamaican Becard *Pachyramphus niger* LC
Fairly common resident (**Jamaica**).

L 18 cm. Monotypic. Endemic. Tall, mostly open, forest and edges at low to mid elevations; also more closed woodland, pastures with scattered trees and large gardens. Bulky globular nest of plant materials; often conspicuous, can be useful clue to species' presence. **ID** Sturdy-bodied, short-tailed flycatcher-like bird. **Ad** ♂ is all black, except white mark at base of wing (visible in flight). **Ad** ♀ has deep brown crown, reddish-brown upperparts, cinnamon cheeks and throat to upper breast, and pale grey underparts. **Juv** is similar to ♀. **Voice** Song two "queeck" notes followed by a melodic, syncopated phrase that rises before falling over last two syllables; "Co-ome and tell me what you hee-ear". **SS** ♀-type birds should prove unmistakable. ♂'s head and bill shape, and habits (the becard typically perches prominently, and hovers and sally-gleans for prey) distinguish it from all-black icterids, especially much rarer and less conspicuous Jamaican Blackbird.

TYRANNIDAE
Tyrant-flycatchers
449 extant species, 32 in region + 6 vagrants

Yellow-bellied Elaenia *Elaenia flavogaster* LC
Fairly common resident (**St Vincent and Grenadines**, **Grenada**), vagrant (**Martinique**).

L 16–17 cm. Four subspecies, of which nominate occurs in region. Lowland forest edge, open woodland, coastal scrub and gardens; absent from well-forested areas. Alone or in pairs. Conspicuous, perches in open and calls frequently, appearing agitated or excited, with raised crest. One record Martinique (Sept 1997). **ID** Medium-large elaenia with conspicuous bushy crest. **Ad** has brownish-olive head and upperparts, and faint whitish eye-ring. Crest frequently erected and parted to reveal semi-concealed white crown patch; when flattened it protrudes slightly from nape. Two bold white wingbars. Belly and vent pale yellow. Rather short bill dull blackish with pinkish-based mandible. **Juv** is browner above, with buff wingbars and no coronal patch. **Voice** Call a rough, wheezing "breeer". Song, frequently given as duet by pairs, repeated hiccuping series of "whik-breeer" phrases. Dawn song slightly more musical rhythmic series of repeated phrases. **SS** Caribbean Elaenia duller with smaller crest and less yellow on belly. Larger *Myiarchus* flycatchers have darker head and upperparts, and longer, slightly hooked bill.

Caribbean Elaenia *Elaenia martinica* LC
Seven subspecies divided into two subspecies groups, both of which occur in region.

Chinchorro Elaenia *Elaenia (martinica) cinerescens*
Common (**Cayman Is**) to rare resident (**San Andrés and Providencia**).

L 16–18 cm. Four subspecies in group, of which two (*caymanensis*[A] Cayman Is, *cinerescens*[B] San Andrés and Providencia) occur in region. Near-endemic. Dry scrub and woodland, and coastal habitats including mangroves. Habits not known to differ from Caribbean Elaenia. **ID** Very similar to all races of Caribbean Elaenia, especially *riisii*. **Ad** (*caymanensis*) is generally more uniformly coloured and paler above than latter (but differences probably only detectable in specimens in series). **GV** Race *cinerescens* has somewhat darker, slightly browner upperparts. **Voice** Similar to Caribbean Elaenia. **SS** Bushy crest and bold wingbars distinctive within limited range.

Caribbean Elaenia *Elaenia (martinica) martinica*
Common and widespread resident (**Puerto Rico**, **LA**).

L 16–18 cm. Three subspecies in group, all of which (nominate[A] Lesser Antilles from Saba and St Eustatius S to Grenada, *riisii*[B] Puerto Rico and satellites, Virgin Is, Anguilla, St Martin, St Bartholomew, Antigua and Barbuda, *barbadensis*[C] Barbados) occur in region. Near-endemic. Common and almost ubiquitous on most islands; colonized mainland Puerto Rico in early 1960s, where has been subject to major population fluctuations. In woodland, scrub, dry to moist forests, and edges, primarily in lowlands, but also in mountains on volcanic islands. Alone or in pairs, usually in canopy and often on semi-exposed perches, but not very conspicuous or noisy. Often silent for long periods. **ID** Medium-large elaenia with bushy crest, very similar to Yellow-bellied Elaenia (slight overlap). **Ad** has upperparts dull olive to brownish olive, with pure white coronal patch (often invisible) and very thin whitish eye-ring, dusky wings with two whitish bars, and yellowish to whitish fringes to remiges. Throat sooty grey, becoming whitish or dull yellowish on belly and vent. **GV** Subtle: race *riisii* smaller and paler than nominate, whereas *barbadensis* is larger than nominate and somewhat darker below. **Voice** A piping, whistled "whi-eeep" or slightly burry "pe-weeer". Dawn song a loud clear burry "pee-wee-reereeree". All vocalizations more musical and less harsh than Yellow-bellied Elaenia. **SS** In S Lesser Antilles, Yellow-bellied Elaenia has slightly more spiky crest, more yellow on belly and different voice. No overlap with other *Elaenia*. Puerto Rican and Lesser Antillean Flycatchers are obviously larger and lack wingbars, and have far more robust and longer bills.

Hispaniolan Elaenia *Elaenia cherriei* LC
Locally common resident (**Hispaniola**).

L 14.5–16 cm. Monotypic. Endemic. Primarily montane pine forest, but also broadleaf woodland, open country with scattered trees and even disturbed areas, at 500–2500 m. Forages, often in pairs, from near ground to canopy. Sallies to snatch prey from surface of leaves or branches. Regularly joins mixed-species flocks in pines. **ID** Smaller elaenia, very similar to allopatric Large Jamaican Elaenia (which see). **Ad** has largely greyish-white underparts, with little or no yellowish on breast and belly, two wingbars and white, semi-concealed, coronal patch. Mainly black mandible usually with pinkish base. **Juv** lacks crown patch. **Voice** Call a short emphatic "wheep!", repeated regularly. Presumed song a fast descending series of c. 15 harsh notes: "pwee-chi-chi-chiup, see-ere, chewit-chewit" or a more mellow "whee-ee-ee-ee-ee" at dawn. **SS** Only elaenia in range. Hispaniolan Pewee has slightly darker underparts and lacks wingbars. Stolid Flycatcher obviously larger with distinctly yellow underparts and is mainly found in lowlands. **TN** Most authorities treat as conspecific with Large Jamaican Elaenia (as Greater Antillean Elaenia), but split on molecular evidence.

Jamaican Becard

Yellow-bellied Elaenia
flavogaster

Chinchorro Elaenia
(Caribbean Elaenia)
cinerescens

Caribbean Elaenia
martinica

Hispaniolan Elaenia

Large Jamaican Elaenia *Elaenia fallax* LC
Locally common resident (**Jamaica**).

L 15 cm. Monotypic. Endemic. Humid forest, edges, thickets and open country with scattered trees, mostly at 500–2000 m in breeding season, when apparently confined to Port Royal Mts, Blue Mts and St Andrew; island-wide in winter (Nov–Mar), but very inconspicuous then. Forages in pairs, but will join mixed-species flocks, from low to canopy. Sallies to glean insects from leaves and twigs, and hover-gleans for fruits. **ID** Rather small elaenia with faint eyestripe and eye-ring, rounded head (with indistinct crest) and small blackish bill with pink mandible tipped black. **Ad** has head and upperparts olive-brown with whitish, usually concealed, coronal patch. Wings dusky with two bold whitish wingbars and tertial fringes. Underparts pale grey, washed yellow, with faintly streaked neck and breast. **Juv** lacks coronal patch. **Voice** Call a single "chewit" and rather long downslurred plaintive whistled "wheeeeer". Song not known to differ from Hispaniolan Elaenia. **SS** Only *Elaenia* in Jamaica. Small Jamaican Elaenia lacks wingbars, with more distinctive facial markings and yellow coronal patch. **TN** See Hispaniolan Elaenia. **AN** Greater Antillean Elaenia (when lumped with Hispaniolan Elaenia).

Small Jamaican Elaenia *Myiopagis cotta* LC
Locally common resident (**Jamaica**).

L 12–13 cm. Monotypic. Endemic. Most frequent in wet forest at mid elevations, but occurs in open woodland, second growth, shade-coffee plantations and dry forest from sea level to 2000 m. Forages from understorey to canopy, typically 4–8 m above ground. Sallies from perch and gleans from vegetation in flight. **ID** Smaller flycatcher with tiny black bill and no wingbars. **Ad** has greenish-olive upperparts, slightly darker crown with bright orange-yellow coronal patch (rarely exposed). Whitish supercilium and grizzled face. Flight feathers edged pale yellow. Throat whitish, breast pale yellow and rest of underparts brighter yellow. **Juv** lacks coronal patch and has pale grey underparts, yellower on belly. **Voice** Rapid, high-pitched "ti-si-si-sip" or "si-sip", last note lower. **SS** See Large Jamaican Elaenia. **AN** Jamaican Elaenia.

Great Kiskadee *Pitangus sulphuratus* LC
Vagrant (**Barbados**).

L 20.5–23.5 cm. Ten subspecies, of which one (probably *trinitatis*) occurs in region. Photographed, Jan–Feb 2016 (*NAB* 70: 242). Perches conspicuously, often in open. Hawks insects from perch, hovers to glean fruits; often drops to ground. **ID** Large colourful tyrant-flycatcher with powerful hooked black bill. **Ad** has black crown with semi-concealed yellow coronal patch (often visible). Head-sides black with broad white supercilium from forehead to nape. Upperparts olive-brown to rich brown with extensive rufous on wings and tail. White throat contrasts with bright yellow rest of underparts. **Juv** lacks yellow on crown. **Voice** Quite noisy, with loud varied calls, commonest an emphatic "kiss-ka-dee". **SS** Extremely rare in region, but unmistakable.

Tropical Kingbird *Tyrannus melancholicus* LC
Rare and irregular visitor (**Grenada**), vagrant (**Cuba**, **Cayman Is**).

L 18.4–24 cm. Three subspecies, of which one (*satrapa*) occurs in region. On Grenada, semi-arid scrub with scattered trees and other open areas, often near coasts; records all months and has bred, but sometimes none for several years. Vagrant Cuba (c. 5 records) and Grand Cayman (Oct 1995). Also several records, first Oct 2009, of either Tropical or extralimital **Couch's Kingbird** *T. couchii* (S Texas to Belize) in Bahamas (none heard to vocalize). Multiple claims San Andrés and Providencia, most or all perhaps Grey Kingbirds, but a now lost specimen (FMNH 43096) attributed to Couch's was collected 1886/87 (Yojanan Lobo-y-Henriques 2014); May 2017 photo from St Thomas is misidentified Grey Kingbird. Usually alone, perches obviously and feeds primarily by aerial hawking. **ID** Fairly large, rather colourful tyrant-flycatcher with yellow rear underparts. **Ad** has grey head with semi-concealed orange crown patch. Dusky lores and ear-coverts create faint mask. Back greyish olive and rather long, notched tail dusky brown. Whitish throat and bright yellow breast to vent, with indistinct yellow-olive breast-band. Fairly stout, large, black bill. **Juv** is slightly duller overall, with paler head, whiter throat and coronal patch absent or greatly reduced. **Voice** A series of repeated sharp trills "tip-tri-tip-tri-tip-tri....", similar but softer, less emphatic and often more rapid and twittering than Grey Kingbird. **SS** Compare Western Kingbird (also vagrant to Cuba) and widespread Grey Kingbird. Formerly conspecific Couch's Kingbird (no definite records) separated by smaller bill on average, slightly greener (less grey) back, less deeply notched tail, but mainly by voice, a plain and nasal "pit", "kip" or similar, singly or in slow series, and a more distinctive shrill yet buzzy and nasal "breeeer".

Cassin's Kingbird *Tyrannus vociferans* LC
Vagrant (**Cuba**).

L 20.5–23 cm. Two subspecies, of which one (presumably nominate) occurs in region. Photographed, Guanahacabibes Peninsula, Oct 2017 (Castellón Maure *et al.* 2017). Habits like those of other kingbirds. **ID** Medium-large kingbird. **Ad** is similar to same-age Tropical Kingbird but differs by much shorter bill, greyer back, and grey (rather than yellow, washed olive) breast. Relatively short dark brown tail is blunt-ended, rather than notched, with narrow buffy tips. Orange-red crown patch usually concealed. **Juv** is duller and slightly browner above than ad, with bold buff fringes to wing-coverts and duller underparts. **Voice** A burry "ch-beehr" with emphasis on second syllable. **SS** Very rare Western Kingbird has paler grey head, back and breast, contrasting dark wings, and white-edged black outertail feathers; vocalizations differ.

Western Kingbird *Tyrannus verticalis* LC
Rare passage migrant/winter visitor (**N Bahamas**), vagrant (**Cuba**, **Puerto Rico**, **Swan Is**).

L 19.5–24.1 cm. Monotypic. Dry open country and coastal areas, Oct–May, mainly Apr–May and Aug–Nov, with six records in Cuba, a recent one in Puerto Rico in Apr, and three on Swan Is, Nov 1958 (Bond 1959). In Bahamas, reported S to Eleuthera. Habits similar to Grey Kingbird. **ID** Fairly large tyrant-flycatcher similar to Tropical Kingbird. **Ad** differs by shorter bill and generally paler and greyer plumage, with pale grey head to breast and hindneck. Dark grey lores and upper ear-coverts create slight mask; orange-red coronal patch (usually invisible). Rest of upperparts greyish olive with darker wings. Dark tail conspicuously white-edged. Rear underparts yellow. **Juv** lacks crown patch, has more olive upperparts, buffy-fringed wing-coverts, brownish-tinged breast and paler yellow belly. **Voice** Commonest call a sharp, emphatic "whit", "bec" or "bek"; often gives short bursts of querulous and bickering chatter. **SS** Compare much commoner Grey Kingbird, and equally scarce or rarer Tropical (also vagrant to Cuba) and Cassin's Kingbirds (one record).

Small Jamaican Elaenia

Great Kiskadee
trinitatis

Large Jamaican Elaenia

Cassin's Kingbird
vociferans

Tropical Kingbird
satrapa

Couch's Kingbird

Western Kingbird

195

Eastern Kingbird *Tyrannus tyrannus* LC
Uncommon to rare passage migrant (**Cuba**, **O**), vagrant (elsewhere). **L** 19–23 cm. Monotypic. Semi-open woodland, borders, farmland and other semi-open habitats in lowlands. Never in forest interior (unlike resident Loggerhead Kingbird). Feeds mainly by aerial hawking. Perches in open. Commoner Sept–Oct (Jul–early Nov), sometimes in large numbers (e.g. San Andrés, Cayman Is), less frequent late Mar–early May. Exceptionally rare Jamaica and Puerto Rico. Occasional winter reports, e.g. Bahamas (Dec 1996 and 1999), W Cuba (Jan 2016) and Cayman Brac (Jan 1985). **ID** Medium-large, distinctively two-toned kingbird, with relatively short black bill. **Ad** has crown and head-sides black, with semi-concealed orange-red (rarely yellow) crown patch. Can appear slightly crested. Above dark blackish grey with two indistinct whitish wingbars and edges to wing feathers. Rather square black tail with obvious white tip. Underparts pure white. **Juv** is much duller and paler, with slight brownish-cinnamon edges to nape, wings and rump feathers, no crown patch and little or no white on tail tip. **Voice** Usually silent on migration. **SS** Larger Loggerhead Kingbird has heavier bill, paler grey back and less white on tail. The Cuban endemic Giant Kingbird is also larger with much heavier bill and no white on tail. Compare less contrastingly patterned Grey Kingbird.

Grey Kingbird *Tyrannus dominicensis* LC
Common resident (**Hispaniola** to **LA**) and breeding visitor (elsewhere). **L** 21–25 cm. Two subspecies, both of which (*vorax*[A] Lesser Antilles, nominate[B] elsewhere) occur in region. Conspicuous in open and semi-open areas, including dry forest borders, cactus scrub, open woodland, cleared areas, outskirts of towns, mangroves and agricultural areas, but not dense forest, to c. 1100 m, but usually below 500 m. Assumes high, conspicuous perches and feeds primarily by aerial hawking. Most breeders in Bahamas, Cuba, Jamaica and Cayman Is absent late Nov(Dec)–(Feb)Mar, but apparently increasing numbers winter there (especially Bahamas and Cuba). Largely resident elsewhere, but some move locally during this period too. Migration patterns not fully understood. **ID** Fairly large, mostly grey tyrannid with quite heavy black bill. **Ad** has mostly grey head and upperparts, usually invisible yellow-orange crown patch, and distinct blackish mask. Faint whitish wingbars. Slightly notched tail dusky brown to blackish. Throat white, breast greyish-tinged. **Juv** is browner above, with brownish or cinnamon fringes to wing-coverts, rump and tail, and no crown patch. **GV** Race *vorax* larger and slightly darker, with bigger bill. **Voice** Main call a loud, emphatic "pi-tirr-ri" or "pit-piteerri-ri-ree", reminiscent of Tropical Kingbird but harsher. Song a more complex, musical, chattered version of call. Also a harsh "peet" and "burr". **SS** Eastern, Loggerhead and Giant Kingbirds have mostly black (not grey) head. Further, Giant (only Cuba) has heavier bill, and Eastern and Loggerhead show greater contrast between dark upperparts and white underparts. Compare Tropical Kingbird, Northern Mockingbird and rare Fork-tailed Flycatcher. **TN** Race *vorax* only weakly differentiated and species probably best treated as monotypic.

Loggerhead Kingbird *Tyrannus caudifasciatus* LC
Seven subspecies divided into three subspecies groups, of which all three occur in region.

Loggerhead Kingbird
Tyrannus (caudifasciatus) caudifasciatus
Common and widespread resident (**Bahamas**, **Cuba**, **Cayman Is**, **Jamaica**).

L 23–26 cm. Four subspecies in group, all of which (*bahamensis*[A] N Bahamas, nominate[B] mainland Cuba, I of Pines and many cays, *caymanensis*[C] Cayman Is, *jamaicensis*[D] Jamaica) occur in region. Endemic. Prefers more heavily wooded habitats than Grey Kingbird: semi-deciduous and evergreen forests, second growth, pine and riparian forests, to mid elevations in Cuba. In some areas also gardens, settlements and open woodland in Nov–Mar, when Grey Kingbird is largely absent from relevant range. Perches in open and forages by sallying at low to mid levels. **ID** Fairly large and distinctively two-toned, with heavy black bill. **Ad** has blackish head, with yellow to pale orange but rarely visible coronal patch. Above dark grey, with conspicuous whitish edges to wing-coverts. Square-ended tail has buff-white tips. Below mostly white, with variable yellow wash to belly. **Juv** has brownish wingbars and no crown patch. **GV** Race *bahamensis* rather distinctive by browner head, rusty rump, yellow-washed underparts and narrower buff tail tip; *caymanensis* is more yellow below than all races except *bahamensis* and has a slightly paler head than nominate; *jamaicensis* has always yellow crown patch, almost no yellow below and broad white tail tip. **Voice** Variable, most frequent call a loud, rolling and sputtering "teeerrp" or rising "pit-pit-pit-pit-pit-tirr-ri-ri-reee" chatter. Also bubbling sounds. **SS** Eastern Kingbird is smaller, with less robust bill, blacker back and broader white tail tip. In Cuba, much rarer Giant Kingbird has heavier bill and no white in tail, among other differences. Grey Kingbird has paler, greyer crown and upperparts, and is largely absent from same islands in Nov–Mar (note this is not true in relation to Hispaniolan and Puerto Rican Kingbirds). **TN** Proposed I of Pines race *flavescens* synonymized with nominate.

Hispaniolan Kingbird *Tyrannus (caudifasciatus) gabbii*
Widespread but uncommon and local resident (**Hispaniola**). **L** 23–26 cm. One subspecies in group. Endemic. Dry to humid forests (both broadleaf and pine), shade-coffee plantations and other semi-open areas, to 2000 m, but commoner at mid to upper elevations. Rare in lowland cactus scrub. Essentially replaces Grey Kingbird in densely forested areas and at higher elevations. Habits similar to Loggerhead Kingbird. **ID** Similar to allopatric Loggerhead Kingbird. **Ad** differs by average slightly larger size, dark brown (rather than dark grey) back and tail, offering little contrast with brownish-black head. Wings and tail edged pale rufous, latter without pale tip. Whitish underparts lightly washed grey on breast. **Juv** lacks yellow coronal patch. **Voice** Dawn song a rolling trill, "br-r-r-r-r-r-r", usually ending in several loud notes, longer than that of other Loggerhead Kingbirds. Rarely vocalizes at other times. **SS** Stolid Flycatcher has much smaller bill and yellowish-washed underparts. Grey Kingbird has paler crown and grey upperparts, contrasting with black mask. Compare migrant Eastern Kingbird (not yet recorded on Hispaniola, but could occur). **TN** Hispaniolan and Puerto Rican Kingbirds have each been suggested to warrant species status (Garrido *et al.* 2009), but differences thought insufficient.

Puerto Rican Kingbird *Tyrannus (caudifasciatus) taylori*
Fairly common resident (**Puerto Rico**, including **Vieques I**). **L** 23–26 cm. One subspecies in group. Endemic. Habitat and habits similar to Loggerhead Kingbird. **ID** Similar to allopatric Loggerhead and Hispaniolan Kingbirds. **Ad** shows relatively little contrast between head and upperparts, large yellow crown patch but no white tail tip (as Hispaniolan), slight rusty tinge to fringes of wing feathers, and greyish underparts tinged brown on neck-sides. **Juv** lacks yellow coronal patch. **Voice** Similar to Loggerhead and Hispaniolan Kingbirds, but song comparatively short like *jamaicensis* Loggerhead. **SS** Compare Grey Kingbird, Puerto Rican Flycatcher and smaller migrant Eastern Kingbird. **TN** See Hispaniolan Kingbird.

Eastern Kingbird

Grey Kingbird
dominicensis

jamaicensis

bahamensis

caudifasciatus

Loggerhead Kingbird

Hispaniolan Kingbird
(Loggerhead Kingbird)

Puerto Rican Kingbird
(Loggerhead Kingbird)

Giant Kingbird *Tyrannus cubensis* `EN`
Rare and local resident (**Cuba**).

L 23–26 cm. Monotypic. Endemic. Pine forest, and open riparian woodland, semi-deciduous and evergreen, but in some areas is specialized on woodland/open country ecotone, often around *Ceiba pentandra* trees; to c. 1100 m. Perches exposed on large trees. Hawks flying insects. Formerly also Great Inagua and Caicos Is, where never known to breed but now extinct. Jan 2019 photo claim from Dominican Republic insufficient for documentation. **ID** Large kingbird; very big, heavy, black bill, with fairly arched culmen. **Ad** has blackish crown and nape, with semi-concealed orange coronal patch. Head rather rounded, not crested. Upperparts scarcely contrast with head; narrow but conspicuous whitish edges to wing feathers. Slightly notched dusky tail, usually with very inconspicuous pale tip. Below white, sometimes washed pale greyish on upper breast. **Juv** lacks coronal patch. **Voice** Song loud, declaiming "tooe-tooe-tooee-tooee-toee", very different to Loggerhead Kingbird. Also unusual-sounding antiphonal duets. Call a whistled "pee-puurr". **SS** Only kingbird with such a massive bill. Much commoner Loggerhead Kingbird appears smaller, with less deep-based bill, even darker head, and more white in tail. Compare migrant Grey and Eastern Kingbirds; note that former can also appear to have an unusually large bill.

Scissor-tailed Flycatcher *Tyrannus forficatus* `LC`
Very rare passage migrant (**GA**, **O**).

L 19–38 cm including tail-streamers (c. 15 cm without them). Monotypic. Records Bahamas (Grand Bahama, Abaco, San Salvador, Great Inagua), Cuba, Cayman Is, Hispaniola and Puerto Rico (only vagrant on last two-named); late Oct–Dec (rarely Jan–Mar). Open to semi-open scrub, grassland or savanna with scattered trees, forest edges, even suburban areas, mostly in lowlands. Habits like Fork-tailed Flycatcher. **ID** Elegant, long-tailed flycatcher, similar to Fork-tailed Flycatcher. **Ad** has pale grey head and back (rather than black cap and dark grey back). Variably pink-tinged back and rump. Wing feathers show clear whitish edges. Below, tinged grey on breast and pink-orange or salmon rearwards. Considerable salmon wash to underwing-coverts and reddish patch on axillaries, most obvious in flight but sometimes visible when perched. Tail-streamers show some white or pinkish white at base. ♀ is duller overall, with paler pink areas and shorter tail than ♂. **Juv** as ad ♀ but even paler, with duller and more greyish-brown upperparts, and indistinct pale pinkish wash below. **Voice** Common calls a sharp, dry and harsh "pik", or "kek", repeated "kee-kee" or "ka-leep", more nasal "bik err". **SS** Long tail plus pink or salmon tones of ad unique. Juv similar to juv Fork-tailed Flycatcher, which has dark cap, and they are unlikely to overlap temporally in West Indies.

Fork-tailed Flycatcher *Tyrannus savana* `LC`
Rare and irregular visitor (**Grenada**, **Grenadines**, **Barbados**), vagrant (elsewhere).

L ♂ 37–40.5 cm, ♀ 28–30 cm, both including tail-streamers. Four subspecies, of which at least one (probably nominate) occurs in region. Open terrain, especially pastures or savannas with trees and bushes, but rarely in forested areas. Usually perches low, atop bushes, small trees, fences or wires. Mostly Jun–Oct, with peak Jul/Aug on Grenada (sometimes in large numbers), mainly Apr–May and Aug–Sept on Barbados, perhaps a mix of non- and failed breeders and overshooting austral migrants; exceptional N to S Bahamas (Jun–Sept), Cayman Is (Oct–Nov), Cuba (Feb 1952 and Nov), Jamaica (Mar 1991) and Puerto Rico (Jun–Oct). **ID** Distinctive, two-toned flycatcher with extremely long black tail (in ad). All plumages show deeply forked tail. **Ad** ♂ has black cap, with large, usually concealed yellow crown patch. Head rounded (not crested). Above grey, with blackish-brown wing feathers edged whitish. Below white. Tail in br season has very long (20–29 cm) white-edged outer feathers, shorter in moult. **Ad** ♀/**imm** are duller, with shorter streamers (14–16 cm). **Juv** as ad ♀, with even shorter tail and browner upperparts. **Voice** A thin, low-pitched, weak, rather creaky-sounding "tic", "jek" or "jiit" call. **SS** Ad unmistakable. Juv and imm distinguished by deeply forked tail. Compare especially young or moulting individuals with Eastern Kingbird, which has darker back, white tail tip and more crested appearance. Compare Grey Kingbird and Scissor-tailed Flycatcher.

Sad Flycatcher *Myiarchus barbirostris* `LC`
Common and widespread resident (**Jamaica**).

L 16.5–17 cm. Monotypic. Endemic. Humid forest and woodland to 2000 m, but less frequent in semi-arid lowlands and fairly open forest at higher elevations, rarely in mangroves. Sallies from perches 3–9 m above ground to snatch prey from leaves, often returning to same perch. **ID** Smallest of Jamaican *Myiarchus* with dark crown, rather broad pinkish-black to blackish bill, and little or no rufous in tail. **Ad** has slightly crested smoky olive-brown head. Rest of upperparts largely olive-brown, with faint whitish wingbars. Throat and upper breast pale greyish to whitish. Rest of underparts lemon-yellow (usually extending to lower breast). **Juv** has yellow confined to vent. **Voice** Emphatic "pip, pip-pip". Sometimes "pip-pip-pireee", rising at end. Also single "huit" and whistled note at dawn. **SS** Rufous-tailed Flycatcher much larger with extensive rufous in wings and tail. Flatter-headed Stolid Flycatcher larger and whiter on throat and breast, with prominent white wingbars and edges to flight feathers, and more rufous in tail. Compare much more different Large Jamaican and Caribbean Elaenias.

Giant Kingbird

Scissor-tailed Flycatcher

Fork-tailed Flycatcher
savana

Sad Flycatcher

199

Rufous-tailed Flycatcher *Myiarchus validus* `LC`
Fairly common resident (**Jamaica**).

L 24 cm. Monotypic. Endemic. Primarily in moist forest, but also dry scrub and second growth, at 300–2000 m (seldom lower or higher), but commonest at mid elevations. Sallies for prey from perch in dense vegetation below canopy. **ID** Large *Myiarchus* with much rufous in wings and tail, and blackish bill usually with obvious pinkish base. **Ad** has dark olive-brown upperparts, and smoky-brown crest. Wings appear mainly rufous at rest and most tail feathers largely rufous. Throat and breast dark grey, vent yellow, but no strong demarcation. **Juv** has more whitish underparts, is greyer above, with less obvious fringes to wing feathers and more orange at base of mandible. **Voice** Distinctive loud, piercing, increasingly fast-paced and descending "pree-ee-ee-eeee", some notes rather squeaky-sounding and the whole rather stuttering, given frequently in territory defence. Occasional "chi-chi-chiup" and disyllabic "wick-up" or clicking note. **SS** Largest *Myiarchus* in Jamaica (and entire region). Nominate Stolid Flycatcher lacks rufous in wings or tail and has prominent white edges to wing feathers. Sad Flycatcher much smaller and more yellowish below, with rounder, less bushy crown.

La Sagra's Flycatcher *Myiarchus sagrae* `LC`
Common to uncommon resident (**Bahamas**, **Cuba**, **Cayman Is**).

L 17–19 cm. Two subspecies, both of which (*lucaysiensis*[A] Bahamas, most numerous on Abaco, Andros, Grand Bahama, Green Cay, Inagua and New Providence, nominate[B] Cuba, I of Pines and many offshore cays, and Grand Cayman) occur in region. Endemic. All manner of woodland, including pines, dense thickets and mangroves, to mid elevations. Absent Turks & Caicos (single undocumented record, Feb 2015). Typically forages in understorey, mainly by snatching prey during hovering flight. Perches less vertically than many congeners. **ID** Fairly small, rather dull *Myiarchus* with pale underparts and usually all-black bill. **Ad** has olive-brown upperparts, slightly crested crown (often appears flat-headed) and two inconspicuous whitish wingbars. Throat to upper breast pale grey, below whitish to pale yellow, becoming slightly yellower on vent and lower flanks. Primaries have thin rufescent edges. **Juv** has slightly more rufous on wing and tail feathers. **GV** Race *lucaysiensis* larger and browner above, rufous inner webs to outer tail feathers produce distinctly rufous underside to tail (much-reduced or no, and paler, rufous in nominate), and has slightly different voice. **Voice** Calls include a short, plaintive "huit" and more prolonged whistles. Also bisyllabic "tra-hee". Song of nominate a "weeet-ze-weer", whilst that of race *lucaysiensis* is said to be a modified "huit" and rolling "brrr-r-r", but small number of available recordings suggest no definite differences between them. **SS** Migrant Great Crested Flycatcher has more contrasting plumage, bright yellow abdomen and more vertical posture. No other *Myiarchus* shares same range. See smaller Caribbean Elaenia in Cayman Is.

Stolid Flycatcher *Myiarchus stolidus* `LC`
Common resident (**Jamaica**, **Hispaniola**).

L 20 cm. Two subspecies, both of which (nominate[A] Jamaica, *dominicensis*[B] Hispaniola and satellites) occur in region. Endemic. Lowland forest, arid woodland, scrub and mangroves; on Hispaniola, also pine woodland, to at least 1800 m. Feeds by hover-gleaning. **ID** Mid-sized *Myiarchus*, with conspicuous white fringes on wing feathers, like allopatric La Sagra's Flycatcher. **Ad** differs from latter by virtue of its two prominent white wingbars, primaries narrowly edged rufous and rest of flight feathers broadly fringed whitish. Inner webs of all but outer tail feathers fringed rufous (producing largely dull grey undertail with rufous restricted to centre). Lower underparts yellower, variably demarcated from grey breast. **Juv** lacks rufous on rectrices. **GV** Race *dominicensis* has more obvious rufous primary panel, and rufous on inner webs of all tail feathers (producing largely rufous undertail), and throat and breast perhaps slightly darker. **Voice** Commonest call an ascending whistle that falls at end: "wheeeeur". Dawn song a variable rolling "whee-ee-ee, swee-ip, bzzrt" and ascending whistles, including elements similar to La Sagra's Flycatcher (requires further study, see TN). **SS** Migrant Great Crested Flycatcher (rare) slightly larger, has more contrasting underparts with bright yellow abdomen, and more vertical posture. In Jamaica, see larger Rufous-tailed and smaller Sad Flycatchers. **TN** Joseph *et al.* (2004) showed Jamaican Stolid Flycatcher is more closely related to Cuban La Sagra's Flycatcher than it is to Hispaniolan Stolid Flycatcher. Vocal data incomplete, but also suggest nominate *sagrae* is closer to Jamaican *stolidus* than to either Bahaman *lucaysiensis* or Hispaniolan *dominicensis* (GMK).

Puerto Rican Flycatcher *Myiarchus antillarum* `LC`
Common (**Puerto Rico**) to uncommon or rare resident (**Virgin Is**).

L 18.5–20 cm. Monotypic. Endemic. Deciduous and humid forest, arid lowland scrub, mangroves, and coffee and citrus plantations, below 800 m. Inconspicuous and inactive, often noticed only by voice. Takes insects on wing, from perches at low to mid levels. Present on Puerto Rico's satellites, and in Virgin Is from St Thomas N & E to Necker I and Anegada. **ID** Mid-sized, rather dull *Myiarchus* with whitish lower underparts, very similar to allopatric La Sagra's Flycatcher. **Ad** shows faint brownish wash on lower breast-sides, very pale rear underparts with hardly any or no yellow, fairly prominent crest (never looks flat-headed) and no or almost no rufous on undertail. **Juv** is similar to ad. **Voice** Distinctive plaintive whistle, "whee", sometimes descending at end. Dawn song like Stolid Flycatcher, but terminal element unmodulated, e.g. "whee-wickup-wheer". Disyllabic "wickup" forms central element of dawn song, and also given at other times. Also an emphatic "huit, huit" and rolling "pee-r-r-r". **SS** Migrant Great Crested Flycatcher (very rare) has more contrasting underparts, with bright yellow belly. Compare Caribbean Elaenia.

Lesser Antillean Flycatcher *Myiarchus oberi* `LC`
Common to rare resident (**LA**).

L 19–22 cm. Four subspecies, all of which (nominate[A] Guadeloupe and Dominica, *berlepschii*[B] St Kitts, Nevis, Antigua and Barbuda, *sclateri*[C] Martinique, *sanctaeluciae*[D] St Lucia) occur in region. Endemic. Primarily edges of dense woodland and plantations at or above 100 m (rarely to 900 m), but on St Lucia mainly in humid upland forest. Infrequently in second growth or scrub lower down. Generally common, but uncommon on Guadeloupe. Perches vertically for long periods, peering about in search of prey, which is primarily seized by hover-gleaning. **ID** Medium-large *Myiarchus* with obvious rufous in wings and tail. **Ad** has slightly crested head, dark brownish-olive upperparts, whitish to buffish fringes to wing-coverts forming wingbars, and prominent rufous-edged flight feathers. Tail dark brown with conspicuous rufous inner webs. Throat and breast grey, slightly paler on throat. Rest of underparts yellow, washed greenish on flanks and not sharply demarcated from grey of breast. Bill black. **Juv** has even brighter fringes to wing and tail feathers. **GV** Race *sanctaeluciae* larger than nominate, *berlepschii* smaller than nominate with paler rufous in tail, perhaps with paler yellow belly. Race *sclateri* obviously smallest, overall darker upperparts, with no rufous in tail and less in primaries, and wingbars almost obsolete. **Voice** Loud, plaintive whistle "peeu-wheeet". Also short whistles

Rufous-tailed Flycatcher

La Sagra's Flycatcher
sagrae

dominicensis

stolidus

Stolid Flycatcher

Puerto Rican Flycatcher

Lesser Antillean Flycatcher
oberi

201

"oo-ee, oo-ee" or "e-oo-ee". Dawn song like Puerto Rican Flycatcher, but differs in lower-frequency whistled components. Any differences between races require study and elucidation. **SS** Only *Myiarchus* in range. Compare smaller, generally paler and shorter-billed Caribbean Elaenia.

Great Crested Flycatcher *Myiarchus crinitus*   LC
Very rare passage migrant (**Cuba**, **San Andrés**), vagrant (elsewhere).

L 18–21.5 cm. Monotypic. Wooded habitats, including dry to humid forest and edges; Sept–Nov and Feb–May (Cuba). Uncommon transient on San Andrés (especially Oct). Vagrants N Bahamas (e.g. Andros, Abaco, Eleuthera), Cayman Is (Sept and Oct), Hispaniola (Nov and Apr) and Puerto Rico ('winter' and Dec). **ID** Medium-large, brightly marked *Myiarchus* with large blackish bill and extensively pinkish-based mandible. **Ad** is olive-brown above with darker crown, rufous-edged primaries and two whitish wingbars. Inner webs of rectrices almost entirely rufous, producing extensively rufous undertail. Throat and breast grey, sharply demarcated from bright yellow belly and vent. **Juv** is duller than ad. **Voice** Diagnostic call a loud, harsh, rising whistled "wheeep". **SS** Sharply demarcated underparts distinctive. In Cuba and Bahamas, compare much duller La Sagra's Flycatcher. On Hispaniola see Stolid Flycatcher, and in Puerto Rico slightly smaller Puerto Rican Flycatcher. See vagrant Brown-crested Flycatcher.

Brown-crested Flycatcher *Myiarchus tyrannulus*   LC
Seven subspecies divided into four subspecies groups, of which one occurs in region.

Brown-crested Flycatcher
Myiarchus (tyrannulus) magister/cooperi
Vagrant (**Bahamas**).

L 20–23 cm. Seven subspecies in four groups, of which one (presumably either *magister* or *cooperi*) occurs in region. Grand Bahama, Feb 2007 (sound-recorded, *NAB* 61: 346), Apr 2010 (*NAB* 64: 513). Latter undocumented. Vagrants likely to occur in any lightly wooded area. **ID** Large *Myiarchus* with bushy crest or peaked nape, rufous inner webs to rectrices and two broad whitish wingbars. **Ad** (*cooperi*) is greyish olive above, with browner crown, primaries edged rufous-brown, narrow white edge to outer webs of inner secondaries, outer webs of tertials and larger wing-coverts broadly edged whitish. Throat and breast pale grey. Lower underparts yellow. Undertail shows conspicuous rufous centre reaching tip. Bill blackish, often with paler base to mandible. **GV** Race *magister* larger with relatively long body and large bill, also whiter throat and brighter yellow below. **Voice** Characteristic call a sharp "huit", sometimes mixed (especially when excited) with burry "breer" or "burrt" notes in sputtering phrases. **SS** Similar size to Great Crested Flycatcher, which has more extensively rufous undertail, darker grey chin to upper breast and face (latter contrasting less with dark crown), stronger yellow belly, dark olive breast-sides. Additionally, in fresh plumage Great Crested has conspicuously broader white margin to innermost tertial (which character becomes steadily less useful with wear). Vs. La Sagra's Flycatcher, larger, heavier-billed, more colourful, especially below, and more rufous tail. Compared to other congenerics, including some resident in region (e.g. Grenada, Lesser Antillean and Stolid Flycatchers), present species most reliably identified by voice. **TN** Both *magister* and *cooperi* recognized as subspecies groups under taxonomy followed herein.

Grenada Flycatcher *Myiarchus nugator*   LC
Common resident (**St Vincent and Grenadines**, **Grenada**).

L 20 cm. Monotypic. Humid forest and edges, second growth and open areas around settlements, especially near palms, at all elevations. Behaviour like other *Myiarchus*. **ID** Medium-large *Myiarchus*, very similar to nominate race of allopatric Lesser Antillean Flycatcher. **Ad** differs by slightly larger, heavier black bill, usually with pink-based mandible, slightly bushier crest, and greyish-brown upperparts (lacking obvious olive tones). **Juv** is presumably paler than ad. **Voice** Loud "quip" or harsh "queuk". Vocabulary lacks prolonged plaintive whistles. **SS** Only *Myiarchus* in range. Compare smaller, generally paler and shorter-billed Caribbean Elaenia. **TN** Has been treated as race of Brown-crested Flycatcher, from which may differ only in voice (Lanyon 1967, Sari & Parker 2012).

Common Vermilion Flycatcher *Pyrocephalus rubinus*   LC
Ten subspecies divided into two subspecies groups, of which one occurs in region.

Common Vermilion Flycatcher
Pyrocephalus (rubinus) obscurus
Vagrant (**Cuba**).

L 13–14 cm. Nine subspecies in group, of which one (unknown) occurs in region; speculated to be *blatteus* (presumably on geographic grounds). ♀-type, Guanahacabibes Peninsula, Oct 2016 (Kirkconnell *et al.* in press). Usually in scrub, savannas, farmland, coastal habitats and other open areas in lowlands. Not shy, perches conspicuously on bare branches, fences, wires and other exposed places. **ID** Small, distinctive flycatcher. **Ad** ♂ has bright crimson crown and underparts, dark brown upperparts and broad mask. **Ad** ♀ is greyer or browner above. Breast usually whitish, streaked dusky, and pinkish to peach-coloured rear underparts. Supercilium and forehead whitish. **Juv** as ad ♀ but much duller with browner, streakier upperparts and whitish underparts heavily streaked brownish. Belly sometimes tinged yellowish (but no red or pinkish tones). **Voice** Call a high, sharp "peent", "pisk" or "peep". **SS** Unmistakable. **AN** Vermilion Flycatcher.

Euler's Flycatcher *Lathrotriccus euleri*   LC
Five subspecies divided into two subspecies groups, of which one occurs in region.

Lawrence's Flycatcher *Lathrotriccus (euleri) flaviventris*
Probably extinct resident (**Grenada**).

L 12.7–13.5 cm. Three subspecies in group, of which one (*flaviventris*) occurs in region. Humid montane forests of Grand Etang to c. 450 m, but no definite reports since 1950 and must now be extremely rare, if not extirpated. Alone or in pairs, rarely with mixed-species flocks. Perches unobtrusively in shady understorey. Sallies to foliage, returning to same perch or a new one. **ID** Fairly drab, medium-small flycatcher with rather flat bill and pinkish mandible. **Ad** has thin whitish eye-ring and indistinct supraloral-stripe. Head rather rounded (not conical or crested). Upperparts mid olive-brown, with warmer rump. Wings dusky with two buff/cinnamon wingbars. Below yellowish with dull greyish-olive breast-band and greyish-white throat. Legs blackish brown. **Juv** is undescribed. **Voice** Unknown. South American races give a plaintive, murmuring, descending "pee, de-dee-dee-dee-dee", with emphasis on first note. **SS** Little if any overlap with pewees and *Empidonax* flycatchers,

Great Crested Flycatcher

Brown-crested Flycatcher
magister
cooperi

Common Vermilion Flycatcher
saturatus

Grenada Flycatcher

Lawrence's Flycatcher
(Euler's Flycatcher)
flaviventris

which are more olive above, and have whitish (not buffy) wingbars and different voice. Additionally, pewees have more conical head and no eye-ring.

Eastern Phoebe *Sayornis phoebe* LC
Rare winter visitor and passage migrant (**Bahamas**, **Cuba**), vagrant (**Cayman Is**).

L 12.4–16.8 cm. Monotypic. Woodland and edges, mostly in lowlands, often near fresh water. Forages at low to mid levels, mostly via aerial sallies, and typically flicks tail downwards while perched. Present Sept–Feb, with most Cuban and Bahamas records during last decade. **ID** Dull-coloured flycatcher with black bill, longish dark tail and no eye-ring. **Ad** has dark greyish-olive upperparts with darker crown. Wings dusky with whitish feather edges, sometimes forming pale panel but not wingbars. Below whitish (with pale yellow wash in fresh plumage in Sept–Oct). Legs black. **Juv** is browner above, with two buff wingbars and cinnamon rump, but unlikely to be seen in region as plumage quickly moulted. **Voice** Common call a sharp "chip" or "piik". **SS** Wood-pewees have paler bill and distinct whitish wingbars, and do not pump tail. *Empidonax* often have obvious eye-ring and wingbars (juv of present species has buff—not whitish—wingbars and cinnamon rump, and is very unlikely to be seen in West Indies). La Sagra's Flycatcher has larger bill, pale wingbars, different voice, and does not dip tail.

Acadian Flycatcher *Empidonax virescens* LC
Uncommon (**San Andrés**) to rare passage migrant (**Bahamas**, **Cuba**), vagrant (**Cayman Is**, **Puerto Rico**, **Barbados**).

L 13–14.5 cm. Monotypic. Open woodland, edges, second growth, clearings and even gardens, in lowlands. Perches upright and occasionally jerks tail upwards. Forages at all levels, using less exposed perches than pewees. Records early Sept–early Nov and late Feb–May. Vagrants Cayman Is (Sept/Oct 1994), Puerto Rico (Oct 2005, Dec 2006) and Barbados (Nov 1961). **ID** Drab, mid-sized *Empidonax* characterized by large broad bill (averaging largest of genus), with black maxilla and pale pinkish-yellow mandible. Forehead flat, sloping up to peak on rear crown. Narrow but sharply defined yellowish eye-ring surrounds prominent dark eye. **Ad** has olive-green upperparts and dark wings with white to yellowish or buffy wingbars and fringes to tertials. Relatively long wings extend well beyond uppertail-coverts at rest; primary projection longest of genus. Below whitish, washed pale olive on breast and pale yellow on belly/vent. Legs grey (often black in congeners). **Juv** has bright buff wingbars. **Voice** Common call a short, explosive "peek!" or "fweet", usually at well-spaced intervals, but seldom heard in region. **SS** Very similar to other *Empidonax*, but note broad bill, uniform greenish upperparts, long wings and grey legs. Yellow-bellied Flycatcher is yellower below, including throat, with more prominent eye-ring, whereas Willow Flycatcher has less conspicuous eye-ring and browner upperparts. Best separated by voice, but all species are usually quiet on passage. Wood-pewees lack eye-ring and usually perch more exposed. Eastern Phoebe lacks wingbars and eye-ring.

Yellow-bellied Flycatcher *Empidonax flaviventris* LC
Very rare passage migrant (**Cuba**), vagrant (**Bahamas**, **San Andrés**).

L 12.5–15 cm. Monotypic. Dry to humid woodland in lowlands. Records only in autumn, early Sept to mid Oct; occurrence in spring mentioned (e.g. by Raffaele *et al.* 1998) but in error. Claimed twice in Bahamas (Oct 1976, Oct 2007) and photographed on San Andrés (Oct 2018). Occurrence in Jamaica (Raffaele *et al.* 1998) refers to Jamaican Vireo misidentified in hand (Levy in prep.). Perches upright, unobtrusively and occasionally jerks tail up. **ID** Drab, mid-sized *Empidonax*, very similar to fractionally larger Acadian Flycatcher. Appears large-headed, with rounded crown and short tail. **Ad** differs subtly by yellower underparts including throat, marginally smaller bill and shorter wings, with bright white wingbars; primary projection shorter than Acadian. Tail also slightly shorter and mandible entirely pale orange. Legs dark grey. **1st-w** has slightly buffy-tinged wingbars and less yellowish underparts. **Voice** Calls a soft but high-pitched, vaguely disyllabic, liquid "beeyip" or upslurred disyllabic "tur-eee", the latter recalling Eastern Wood-pewee. **SS** Willow and Alder Flycatchers have whiter throat, browner upperparts and lack conspicuous eye-ring. Acadian (more frequent in region) has peaked rear crown, longer primary projection and different call. Note 1st-w Acadian and some other *Empidonax* have variable yellow wash below. Wood-pewees (Eastern especially is much commoner) lack eye-ring, have longer primary projection and assume more exposed perches. Compare Least Flycatcher and Eastern Phoebe.

Willow Flycatcher *Empidonax traillii* LC
Very rare passage migrant (**Cuba**, **San Andrés**), vagrant (**Jamaica**, **Puerto Rico**).

L 13.3–17 cm. Four subspecies, of which at least one (unknown, possibly nominate) occurs in region. Most records mid Sept–early Nov, others late Mar and May (Pacheco 2012, Kirkconnell *et al.* in press). Wooded areas, edges of wetlands, thickets and gardens, in lowlands. Perches upright and occasionally flicks tail downwards. **ID** Large *Empidonax*. **Ad** differs from Acadian Flycatcher in subtly browner upperparts, inconspicuous eye-ring and shorter primary projection. Obviously whitish throat with contrasting dull brownish-olive wash on breast, darkest at sides, and darker head. Pale lores. Greener above in fresh plumage (autumn), but usually less so than Acadian. Legs blackish. **1st-w** is browner above, yellower below, with broader yellowish-buff wingbars. **Voice** Common call a liquid upslurred "whuit" or "whit!" (most similar to Least Flycatcher). **SS** See Alder Flycatcher. Yellow-bellied Flycatcher is obviously yellower below (especially throat) with more prominent eye-ring. Eye-ring also less distinct than in smaller Least Flycatcher. Eastern and Western Wood-pewees have slightly longer bill and primary projection, different voice and usually perch in more exposed sites.

Alder Flycatcher *Empidonax alnorum* LC
Vagrant (**Cuba**, **San Andrés**).

L 13–17 cm. Monotypic. Known solely from several birds trapped on San Andrés (Pacheco 2012); two specimens W Cuba, Oct 1966 and 1967 (Bond 1968), one of which has some characters matching Willow Flycatcher (with which this species was, until 1970s, treated as conspecific), whereas the other appears identifiable as the present species based on detailed biometrics (Navarro Pacheco 2018, Kirkconnell *et al.* in press). Habits as Willow Flycatcher. **ID** Virtually identical to Willow Flycatcher and most individuals inseparable unless vocalizing or handled. **Ad/1st-w** differs very subtly by slightly greener upperparts (vs. W races of Willow), longer tail, bolder wingbars and more conspicuous eye-ring (but nominate Willow can have an eye-ring and wingbars of equal strength, and upperparts of identical shade). **Voice** Common call a short, liquid, frog-like "pic", or "peep" (recalling some *Myiarchus* flycatchers). **SS** Willow Flycatcher almost identical and only safely distinguished by voice or potentially by detailed in-hand examination (even then, not all are safely identifiable). Compare other, more common *Empidonax* and wood-pewees.

Eastern Phoebe

Acadian Flycatcher

Yellow-bellied Flycatcher

Willow Flycatcher
traillii

Alder Flycatcher

205

Least Flycatcher *Empidonax minimus* `LC`
Vagrant (**Cayman Is**, **San Andrés**).

L 12.5–14 cm. Monotypic. Grand Cayman, Mar 1904 (NHMUK 1904.8.17.311), Apr 1971 (unconfirmed), Feb 1972 (Bradley 2000), Sept 2010, Oct 2018 (eBird); San Andrés, trapped, Oct 2005 (Salaman *et al.* 2008). Unconfirmed records Cuba (Kirkconnell *et al.* in press). Prefers woodland edge and semi-open country. Often very active at lower and mid-levels, flicks both wings and tail, and frequently dips and shivers tail. **ID** Small, compact, large-headed, narrow-tailed *Empidonax* with short medium-width bill. **Ad** has bold white eye-ring forming teardrop behind eye, brownish-grey upperparts with white wingbars and fringes, greyish-washed breast, and short wings (unusually short primary projection). **Juv** is browner above, whiter below and wingbars are tinged buff. **Voice** A sharp liquid "whit", repeated rapidly (recalls Willow Flycatcher). **SS** Other *Empidonax* slightly larger. Note behaviour. Wood-pewees also slightly larger, lack eye-ring and have longer primary projection.

Western Wood-pewee *Contopus sordidulus* `LC`
Very rare passage migrant (**Cuba**), vagrant (**Jamaica**, **San Andrés**).

L 14–16 cm. Four subspecies, of which one (probably *saturatus*) occurs in region. Few records, most W & C Cuba in Sept–Oct and late Mar–late Apr. Habitat and habits similar to Eastern Wood-pewee. **ID** Very similar to marginally longer-tailed Eastern Wood-pewee and accurate identification must ideally rely on careful evaluation of multiple features, preferably including voice. **Ad** differs subtly by slightly more uniformly darker and browner plumage, especially head-sides (highly influenced by ambient light) and duskier below. Upper wingbar weaker (both wingbars equally prominent in Eastern). Mandible usually dusky with pale orange base (rather than entirely pale orange), but overlap especially in young birds. When perched, may hold tail more in line with back than Eastern, which tends to hold tail more vertically, thus at slight angle to line of back. **1st-w** is virtually identical to 1st-w Eastern. **Voice** Commonest call a harsh, burry, "dzree", very different to Eastern Wood-pewee, but similar "pee" also given. Song, not often heard from migrants, a burry first note and short second one, "brhee—it", often followed by quick trisyllabic "ti-a-lit". **SS** See Eastern Wood-pewee for differences from region's resident pewees and migrant *Empidonax*.

Eastern Wood-pewee *Contopus virens* `LC`
Uncommon to fairly common passage migrant (**W Caribbean**), vagrant (elsewhere).

L 13.5–15 cm. Monotypic. Most frequent Aug–Nov, less so Mar–May and rare Oct–Nov (Bahamas, Cuba, San Andrés). Wooded habitats, including second growth, thickets and gardens, mostly in lowlands. Perches upright, rather conspicuously, on dead branches or other exposed perches, usually fairly high. **ID** Rather drab, with peaked crown, black bill with pale orange mandible tipped dark. **Ad** has greyish-brown to greyish-olive upperparts, and usually no eye-ring. Two whitish wingbars and fringes to tertials. Whitish throat and underparts washed dark grey on sides and breast, sometimes forming faint breast-band. Rear underparts often tinged pale yellow. **Juv** has browner upperparts, broader buffy wingbars, slightly darker underparts and usually partly dark mandible. **Voice** Most common a plaintive whistle "pee-o-wee", with second note lower (regularly given by migrants). Also a series of "pwik" calls suggestive of Alder Flycatcher (and other *Empidonax*). **SS** See much rarer Western Wood-pewee for differences. Resident pewees more colourful below or have an obvious eye-ring. *Empidonax* usually show an obvious eye-ring, shorter primary projection, and consistently perch less conspicuously; nor do they systematically return to same perch.

Cuban Pewee *Contopus caribaeus* `LC`
Common resident (**N Bahamas**, **Cuba**).

L 15–16.5 cm. Four subspecies, all of which (nominate[A] in Cuba, I of Pines and many cays, *bahamensis*[B] Grand Bahama, Great Abaco, New Providence, Eleuthera, Cat and Andros, *morenoi*[C] Canarreos archipelago, *nerlyi*[D] Jardines de la Reina archipelago and nearby islands, off S Cuba) occur in region. Endemic. Broadleaf and pine forests, edges, swamps, mangroves, brushy scrub and thickets, to 1800 m in Cuba. Perches low, sallying to capture airborne insects (sometimes with loud snap of bill), often returning to same perch; flicks tail on landing. **ID** Distinctive by virtue of conspicuous white crescent behind eye and tufted crest, often held erect. **Ad** has olive-grey to olive-brown upperparts, slightly darker on crown. Dusky-brown wings with two faint wingbars. Throat and breast beige-grey, washed olive on sides. Belly and vent more buffy yellow. Black bill broad and flat, with orange-yellow mandible. **Juv** has broader, buffy-white wingbars and paler mandible. **GV** Race *bahamensis* duller and greyer than nominate, with less contrast between crown and back, and paler below, with only slight yellow tinge on belly; *morenoi/nerlyi* very similar and intermediate between nominate and *bahamensis*, with mainly buffish underparts. **Voice** Rather vocal, throughout day. Song a prolonged descending "weeeooooo". Call a repeated "weet" or doubled feeble "vee-vee" reminiscent of La Sagra's Flycatcher. **SS** Wood-pewees and Eastern Phoebe duller and lack eye-ring. La Sagra's Flycatcher obviously larger, with longer black bill. **TN** Complex racial and inter-racial variation. Race *florentinoi* (Cayo Anclitas, off S Cuba) possibly valid, but here included in *nerlyi*, whereas *sanfelipensis* (Cayos San Felipe, W of I of Pines) here regarded as synonym of nominate. Race *morenoi* reported by some authors to occur at one locality on mainland Cuba.

Jamaican Pewee *Contopus pallidus* `LC`
Fairly common resident (**Jamaica**).

L 15 cm. Monotypic. Endemic. Foothills and montane forest, less often at edges, in openings below canopy, at mid to high elevations. Behaviour as Cuban and Hispaniolan Pewees. **ID** Rather nondescript dark pewee. **Ad** has generally dark plumage, no eye-ring or crescent behind eyes, round head with almost no crest, short bill and slightly notched tail. **Juv** has greyer underparts, paler breast, paler mandible and more obvious cinnamon wingbars. **Voice** Call a plaintive "pee", rarely "pee-wee". Song a rising then falling "oéeoh". Dawn song two alternating phrases, "paléet, weeléah". **SS** Migrant wood-pewees much rarer (both vagrants in Jamaica), have peaked (not rounded) crown, less yellow or buff below, and different voices. Sad Flycatcher has longer black bill and crested head. Other *Myiarchus* much larger. Rarer *Empidonax* usually show an eye-ring.

Least Flycatcher

Western Wood-pewee
saturatus

Eastern Wood-pewee

caribaeus

bahamensis

nerlyi

Cuban Pewee

Jamaican Pewee

207

Hispaniolan Pewee *Contopus hispaniolensis* LC
Common and widespread resident (**Hispaniola**), vagrant (**Bahamas**, **Puerto Rico**).

L 15–16 cm. Two subspecies, both of which (nominate[A] mainland Hispaniola, *tacitus*[B] Gonâve I, off W Haiti) occur in region. Endemic. Dry and humid woodland, including pine and broadleaf, edges, shade-coffee plantations and orchards, to at least 2000 m, but commonest in montane pine forest. Not shy, sallies for airborne insects, frequently returns to same perch, flicking tail on landing. Joins mixed-species flocks in pines. Vagrant Mona I (Puerto Rico) and Providenciales (Turks & Caicos). **ID** Similar to nominate race of allopatric Cuban Pewee. **Ad** has slightly greyer upperparts and rounder head, with less peaked crown, and no eye-ring or crescent behind eye. Underparts have olive, yellow or brown wash. Bill relatively long and broad, blackish with pinkish to orangey base to mandible. **Juv** has pale feather fringes to crown, back and wing-coverts. **GV** Race *tacitus* said to be greyer above and paler buff on rear underparts (Wetmore & Swales 1931). **Voice** Call a strong, mournful "purr, pip-pip-pip-pip". Dawn song a loud, rapid series of paired notes rising in pitch, "shurr, pet-pit, pit-pit, peet-peet". **SS** Only resident pewee on Hispaniola, where Eastern Wood-pewee is only vagrant. Hispaniolan Elaenia paler with prominent wingbars. Stolid Flycatcher much larger and browner, with obvious yellow on underparts.

Lesser Antillean Pewee *Contopus latirostris* LC
Three subspecies divided into three subspecies groups, all three occur in region.

Puerto Rican Pewee *Contopus (latirostris) blancoi*
Fairly common but local resident (**Puerto Rico**).

L 15 cm. One subspecies in group. Endemic. Mainly in montane moist forest and shade-coffee plantations, less frequently in drier forest, mangroves and scrub in lowlands; mainly W & C of island. Forages from low perches to ground, usually in open areas below canopy. Perches erect and sallies for flying insects. **ID** Small pewee with black, rather broad and flat bill, and pinkish-based mandible. **Ad** has dark olive-brown head and upperparts, blackish-brown wings and tail, and no wingbars. Throat and breast dull ochre tinged greyish olive, buffier on belly and vent. **Juv** has pale feather fringes above, broad cinnamon to brown wingbars, and slightly paler and greyer throat and breast. **Voice** Sweet, high-pitched trill, rising up scale (like water filling a glass). Dawn song a repetitive trill. Call a repeated "peet-peet-peet". **SS** Only pewee resident in Puerto Rico. Compare Hispaniolan Pewee (vagrant to Puerto Rico) which has duller underparts. Puerto Rican Flycatcher larger with longer bill and paler, duller underparts. Caribbean Elaenia paler and duller below, with prominent wingbars. **TN** The three subspecies groups are sometimes treated as species.

Lesser Antillean Pewee *Contopus (latirostris) brunneicapillus*
Fairly common but local resident (**Guadeloupe**, **Dominica**, **Martinique**).

L 15 cm. One subspecies in group. Endemic. Habitat and habits similar to Puerto Rican Pewee. Multiple reports from St Kitts & Nevis, but none documented. **ID** Very similar to Puerto Rican Pewee. **Ad** differs in its generally paler underparts, with pale yellow-brown to whitish throat, shading to ochre-buff on breast to vent. **Juv** is undescribed. **Voice** Very few recordings, but day song apparently double-noted, the first note slightly longer and more level, the second shorter, more buzzy and declining, the whole sounding less squeaky than conspecific St Lucia Pewee. **SS** Only pewee in range. Lesser Antillean Flycatcher is larger with rufous in wings and yellower belly. Caribbean Elaenia paler and duller below, with bold wingbars. **TN** See Puerto Rican Pewee.

St Lucia Pewee *Contopus (latirostris) latirostris*
Fairly common resident (**St Lucia**).

L 15 cm. One subspecies in group. Endemic. Mostly in moist and humid montane forests at higher and mid altitudes; seldom in drier habitats near sea level. Habits as Puerto Rican Pewee. **ID** Small distinctive pewee. **Ad** has rufous-cinnamon throat and entire underparts, plus narrow rufous rump patch, and maxilla usually all yellow-orange. **Juv** has pale feather fringes above, and broad cinnamon wingbars, perhaps paler and greyer throat and breast. **Voice** Song a short, somewhat squeaky-sounding trill, usually starting louder and fading. Call rather sharp "peet" notes. **SS** Only pewee in range. See Puerto Rican Pewee for distinctions from Lesser Antillean Flycatcher and Caribbean Elaenia. **TN** See Puerto Rican Pewee.

VIREONIDAE
Vireos
64 extant species, 17 in region + 1 vagrant

Philadelphia Vireo *Vireo philadelphicus* LC
Rare passage migrant (**GA**, **O**), vagrant (**LA**).

L 11.5–13 cm. Monotypic. Records Oct–Nov and Feb–Apr, mainly Bahamas and Cuba, becoming increasingly rare to E & S. Perhaps occasionally winters. Lightly wooded areas and coastal vegetation, but could appear anywhere with trees. Mainly in canopy, where can join mixed-species flocks. **ID** Smallish, short-billed, warbler-like vireo with gentle facial expression. Grey crown, broad white supercilium, dark grey eyestripe extending to bill base, and yellowish underparts (very variable). **Ad** has darker crown, dull greyish-olive nape to upperparts, pale yellow throat, stronger on breast, yellowish-white belly and vent, greyish-yellow flanks. Darker primary-coverts contrast with paler rest of wing. **1st-w** is generally drabber, with brownish-washed upperparts and more distinct pale edges to wing-coverts. **Voice** Commonest call a nasal "queek". **SS** Warbling Vireo, even rarer in region, has greener crown, weaker face pattern, supercilium being narrower and diffuse, pale lores; throat white, breast whiter. Bulkier Red-eyed Vireo has stronger face pattern due to whiter, more contrasting supercilium outlined black, crown bluish grey, eyes red. Recalls Tennessee Warbler, which has much slenderer, sharply pointed bill.

Hispaniolan Pewee
hispaniolensis

St Lucia Pewee
(Lesser Antillean Pewee)

Lesser Antillean Pewee

Puerto Rican Pewee
(Lesser Antillean Pewee)

Philadelphia Vireo

Warbling Vireo *Vireo gilvus* **LC**

Four subspecies divided into two subspecies groups, of which one occurs in region.

Eastern Warbling Vireo *Vireo (gilvus) gilvus*
Rare passage migrant to vagrant (**Bahamas**, **GA**).
L 12.5–14 cm. One subspecies in group. Records (mainly mainland Cuba and Bahamas, on Abaco and New Providence) late Sept–mid Nov and Apr, with midwinter records in Dec (Bahamas) and early Jan (Bahamas, Cuba). Very rare E & S of Cuba. Woods and gardens. Often with mixed-species flocks. Extralimital Western Warbling Vireo *V.* (*g.*) *swainsoni* rarely eliminated for records in West Indies. **ID** Small, 'washed-out' vireo with blank facial expression. Very similar to Philadelphia Vireo. Pale grey upperparts contrast little with crown, hint of dark eyestripe and diffuse greyish-white supercilium. **Ad** has forehead to nape mid-grey, pale lores, mid-grey upperparts faintly tinged greenish, especially at rear, uniformly coloured wing, off-white throat, yellowish-tinged breast and flanks, whitish lower belly and vent tinged yellowish at sides; iris dark brown, legs blue-grey. **1st-w** has brown upperparts, cinnamon-buff tips to greater coverts and white underparts. **Voice** Commonest call a harsh upslurred scolding "shreee". **SS** Face pattern in Red-eyed and Black-whiskered Vireos more marked. Philadelphia Vireo has shorter bill and tail, different facial expression with darker and more prominent eyestripe reaching bill base, contrasting dark primary-coverts, more yellowish-tinged breast and greener upperparts.

Yellow-green Vireo *Vireo flavoviridis* **LC**
Rare to fairly common passage migrant (**San Andrés and Providencia**).
L 14–15 cm. Five subspecies, of which one (presumably nominate) occurs in region. Recorded spring (Feb–Mar) and autumn (Oct–Nov), apparently more frequently during former. Can be expected elsewhere in W Caribbean basin. Dry woodland. **ID** Robust vireo similar to Red-eyed and Black-whiskered Vireos. **Ad** has crown mid-grey, narrow and often weak blackish-grey upper border to supercilium, lores dark grey, ear-coverts greenish grey, nape and upperparts dull greenish with yellow-green patch on folded wing. Throat and central breast whitish, neck-sides and flanks greyish yellow, becoming brighter yellow rearwards (especially in autumn); iris dull red, bill greyish, legs blue-grey or slate-blue. **1st-w** is brownish-tinged above, with indistinct edges to larger wing-coverts, iris brown. **Voice** Song a persistent series of usually disyllabic phrases, occasionally trisyllabic or monosyllabic, interspersed by longer pauses, "viree, viree, fee, vireo" etc. Generally simpler, more monotonous and with shorter phrases than Red-eyed Vireo. Calls a nasal mew, high nasal "chaaa" in scolding, and rasping alarm. **SS** Larger Black-whiskered Vireo has longer, heavier bill and usually obvious black malar, and lacks obviously yellow underparts, which feature also helps distinguish Red-eyed Vireo (migrant). The latter is obviously less greenish above, has a marginally shorter bill and more boldly delineated supercilium.

Yucatan Vireo *Vireo magister* **LC**
Common resident (**Grand Cayman**).
L 15 cm. Four subspecies, of which one (*caymanensis* Grand Cayman) occurs in region. Elsewhere occurs only in SE Mexico and on small islets off Belize and Honduras. Occurs in all wooded areas on island, but can be secretive and rather quiet; forages at all heights. Recent decline across W third of Grand Cayman. **ID** Large dull vireo, with heavy bill. **Ad** has whitish supercilium, prominent dark eyestripe, dusky ear-coverts and greyish crown. Above dull olive, flight feathers with narrow pale olive edges. Throat and breast dull buffy white, sides and flanks greyish, belly and vent buffy white; iris brownish, grey bill with paler base, legs bluish or blue-grey. **Juv** is undescribed. **Voice** Song a series of 2–3-syllable phrases, "sweet brid-get", separated by pauses (similar to Black-whiskered Vireo). Calls: a soft, dry chatter (similar to Yellow-green Vireo, a potential vagrant to Cayman), low, frequently repeated "bik-bik", and loud "wik". **SS** Both Red-eyed (migrant on Grand Cayman) and Black-whiskered Vireos (summer visitor) have stronger face patterns and redder eyes; furthermore, former has whiter supercilium narrowly outlined black, and grey crown contrasting with olive upperparts; Black-whiskered has thin dusky malar and greater contrast between pale supercilium and grey crown. Compare Philadelphia and Warbling Vireos, latter not yet recorded on Cayman and former very rare; also Vitelline, Swainson's and Tennessee Warblers.

Red-eyed Vireo *Vireo olivaceus* **LC**
Ten subspecies divided into two subspecies groups, of which one occurs in region.

Red-eyed Vireo *Vireo (olivaceus) olivaceus*
Fairly common to uncommon passage migrant (**throughout**).
L 14–15.5 cm. One subspecies in group. Generally common on passage through N & W of region (albeit curiously few reports from Hispaniola), but less common in Virgin Is and Lesser Antilles (with records wholly lacking for many islands); mainly late Aug–mid Nov, mid Feb–late Apr (rarely into May). Wooded habitats, including dry and humid areas, open woodland, scrub, shade-coffee plantations and gardens. Mostly in canopy, where deliberately gleans foliage and sometimes sallies for prey. Often joins mixed-species flocks. **ID** Robust vireo with grey crown, long whitish supercilium outlined in black, and olive-green upperparts. **Ad** has supercilium extending above ear-coverts, dull grey lores, greenish-grey ear-coverts; flight feathers and wing-coverts edged olive-green. Throat to belly greyish white, with greenish-yellow tinge on lower flanks and vent; iris deep red; legs blue-grey. **1st-w** has brown or grey-brown iris, upperparts tinged brownish, wing-coverts with indistinctly paler tips. **Voice** Not very vocal in region, but song (occasionally heard in spring) of short, abrupt phrases, separated by brief pauses, "here-I-am, up-here, see-me?, see-mee?", ascending and descending in pitch. Call a mewing "myah" (similar to Yellow-green and Black-whiskered Vireos). **SS** Ad Black-whiskered Vireo has narrow yet obvious blackish malar, which can be much weaker or even lacking in 1st-w; then concentrate on creamier-white and obviously dark-bordered supercilium and more contrasting crown (vs. upperparts) of present species. On Grand Cayman, compare Yucatan Vireo.

Eastern Warbling Vireo
(Warbling Vireo)

Yellow-green Vireo
flavoviridis

Yucatan Vireo
caymanensis

Red-eyed Vireo

211

Black-whiskered Vireo *Vireo altiloquus* **LC**
Common resident (**O**, **LA**), summer visitor (**Bahamas**, **GA**) and passage migrant (**throughout**).

L 15–16.5 cm. Six subspecies, of which five (nominate[A] Jamaica E to Puerto Rico, *barbatulus*[B] Bahamas, Cuba, Little Cayman and Cayman Brac, *grandior*[C] Providencia, *canescens*[D] San Andrés, *barbadensis*[E] St Croix S to Barbados) occur in region. Resident in S (Lesser Antilles and small islands in W Caribbean), but mainly summer visitor to Bahamas, Cayman Is and Greater Antilles, present Feb–Nov; however, at least in Cuba and Hispaniola, increasing numbers overwinter (or migrants arrive from further N, e.g. Florida). Scale of passage probably under-appreciated, e.g. nominate collected Grand Cayman (Bradley 2000). Most forest types, including semi-arid coastal scrub, mangroves, broadleaf forest and shade-coffee plantations, but usually shuns pines; to at least 1700 m (typically much lower). Mainly in canopy. **ID** Grey-brown crown with long slightly buffy supercilium and variable malar. **Ad** has lores and eyestripe dull greyish brown, ear-coverts and area below eye buff-brown, narrow blackish malar, nape and back more greenish brown than crown, flight feathers edged dull greenish brown, throat and breast grey-white tinged yellowish, belly whitish, tinged yellow, stronger on lower belly and vent; iris red or reddish brown. **Juv** is browner than ad, with hint of pale wingbar and brown iris. **GV** Race *barbatulus* generally smaller, with slightly shorter bill, duller grey crown, whiter supercilium, dull olive back and rump tinged greyish; *grandior* has heavier bill, narrower malar, yellowish-olive flanks, pale yellow vent; *canescens* has long bill but is duller, greyish olive above and on sides, with flanks and vent whitish; *barbadensis* like *barbatulus* but bill and tarsi longer, crown greyish and only fore-supercilium whitish. **Voice** Monotonous but variable, short, loud and melodious series of 2–3 syllables, separated by pauses (each lasting c. 1 second); given day-round (but not in winter). Calls include a thin, mewing "tsit" and nasal "yeea", very similar to analagous vocalizations of Red-eyed and Yellow-green Vireos. Migratory races have larger repertoires. **SS** Most similar is smaller-billed Red-eyed Vireo, but it lacks dusky malar (individually variable in Black-whiskered), has whiter underparts, bolder grey crown, and whiter supercilia and underparts. See Philadelphia, Warbling (both very uncommon in region) and Yucatan Vireos (nests Grand Cayman, where present species does not breed), all of which have weaker face patterns and lack malar stripe.

Yellow-throated Vireo *Vireo flavifrons* **LC**
Uncommon to rare passage migrant and winter visitor (**throughout**).

L 13–15 cm. Monotypic. Generally uncommon across N of region (Bahamas, Cayman Is and Greater Antilles), becoming steadily rarer S & E (basically vagrant in Lesser Antilles, except Guadeloupe); present early Aug–early May (mainly Sept–Mar). Scrub, thickets and woodland, usually alone (but occasionally in mixed-species flocks), foraging at all heights, but typically rather secretive and deliberate. **ID** Bold yellow 'spectacles', bright yellow throat and chest, white belly and bold white wingbars. **Ad** has crown, neck-sides, ear-coverts and upper back deep olive-green, lower back and shoulders dull grey, supraloral and eye-ring lemon-yellow; flight feathers narrowly edged grey-white, two conspicuous white wingbars, throat to breast bright lemon-yellow, belly and undertail-coverts white, greyer on rear flanks; iris brown; legs bluish grey. **1st-w** is washed brownish above, throat pale buffy yellow. **Voice** Song (rarely heard in region) a wheezy series of short, varied notes with brief pauses "chee-wee, chee-woo, u-wee…"; also harsh scolding "chi-chur" and quiet "werr". **SS** White-eyed Vireo has obviously pale eyes, mostly white underparts, yellow confined to flanks, and duller upperparts. Blue-headed Vireo (only superficially similar) has underparts mostly white, bluish-grey head, white 'spectacles', yellowish wingbars. Cuban, Flat-billed and Thick-billed Vireos all lack combination of yellow 'spectacles' and contrasting yellow-and-white underparts. On Hispaniola, compare also Pine Warbler.

Blue-headed Vireo *Vireo solitarius* **LC**
Rare passage migrant and winter visitor (**Bahamas**, **Cuba**), vagrant (**Cayman Is**, **Jamaica**, **Hispaniola**).

L 13–15 cm. Two subspecies, of which one (presumably nominate) occurs in region. Recorded Sept–Apr. Only vagrant in Jamaica (*Gosse Bird Club Broadsh*. 22: 25), Cayman Is (Oct 2002, Nov 2014) and Dominican Republic (Feb 2005). Low, dense vegetation with abundant tangles, exceptionally mangroves. Solitary, rarely with mixed-species flocks. Forages deliberately, gleaning and picking prey from leaves. **ID** Stocky, large-headed and -billed vireo. **Ad** has prominent white supraloral and eye-ring, slate-grey or blue-grey head, sometimes suffused slightly greenish, dark greenish-grey upperparts; flight feathers narrowly edged greenish yellow, with two prominent whitish or yellowish wingbars, rectrices edged off-white. Most underparts white, slightly yellower breast, sides to flanks deeper yellow suffused grey; iris dark brown; legs bluish grey. **1st-w** is browner and drabber than ad. **Voice** Calls include low "teewee", longer trills; also short, staccato series of "ti" notes and some scolds. **SS** No other vireo combines blue-grey head, white 'spectacles' and yellowish-white wingbars; White-eyed Vireo has some pale grey in head-sides and nape, and pattern of underparts similar, but crown olive-grey, 'spectacles' yellow, eyes white.

Blue Mountain Vireo *Vireo osburni* **NT**
Local and uncommon resident (**Jamaica**).

L 12.5–15 cm. Monotypic. Endemic. Humid and moist montane forest, borders and shade-coffee plantations; mainly 500–2000 m. Mostly alone, secretive in dense foliage, rarely with mixed-species flocks. **ID** Robust, grey-bodied, large-billed vireo without wingbars or face markings. **Ad** has forehead to nape dull grey-brown, upperparts tinged olive-brown, with greyish-green edges to flight feathers; lores and ear-coverts dull grey-brown, chin and throat off-white tinged yellowish, breast more yellowish-mottled greyish brown, flanks brighter yellow, vent dull yellowish white; iris reddish brown, dark heavy bill, grey legs. **Juv** has yellow below restricted to rearmost parts. **Voice** Song a deliberate trilling whistle, slightly descending in pitch at end; alarm a harsh, descending "burr". **SS** Other duller, less well-marked vireos that migrate through Jamaica have long pale supercilia and dark eyestripes, also slighter and slenderer bills. Smaller Jamaican Vireo much yellower, with bold white wingbars and white eyes. Rather duller migrant parulids generally very rare in Jamaica and tend to favour lower storeys of forest, and always have much weaker bills.

altiloquus

grandior

barbatulus

Black-whiskered Vireo

Yellow-throated Vireo

Blue-headed Vireo
solitarius

Blue Mountain Vireo

213

Jamaican Vireo *Vireo modestus* `LC`
Fairly common resident (**Jamaica**).

L 12–13 cm. Monotypic. Endemic. All manner of bushy habitats from drier lowlands to more humid forest edges and thickets in mountains. Forages at various heights, usually gleaning from leaves, sometimes hawking airborne prey. Regularly flicks tail. **ID** Round-crowned, small-bodied vireo with bold white wingbars, pale eyes and flesh-coloured mandible. **Ad** has diffuse pale supercilium, yellower supraloral and pale ear-coverts, dull grey-green upperparts; flight feathers with greenish-yellow edges, rectrices with diffuse greenish edges to outer webs. Throat whitish, breast dull yellowish grey, darker at sides, belly brighter yellow-grey; iris pale blue-grey to white, legs grey. **Juv** has head greyer, yellow below reduced to narrow belly patch, greyish-brown iris. **Voice** Fast, short and high-pitched phrases, each repeated for several minutes before switching to new phrase: "sewi-sewi", "twee-weet-weet-wuu", etc. Calls include a loud scolding. **SS** White-eyed Vireo (a rare winter visitor) shares pale eyes and white wingbars, but has bold yellow 'spectacles', grey nuchal collar and yellow confined to sides. Other migrant vireos have longer and bolder supercilia, and lack bold wingbars.

Flat-billed Vireo *Vireo nanus* `LC`
Uncommon and local resident (**Hispaniola**).

L 12–13 cm. Monotypic. Endemic. Mostly in semi-arid scrub and thickets, but also humid, mesic or hill forest, to 1200 m. Range includes Î de la Gonâve and, at least formerly, I Saona. Deliberate and agile, mostly low down, sometimes to ground or higher. **ID** Small vireo with yellowish underparts, pale eyes, tiny yellowish 'spectacles' and two narrow whitish wingbars. **Ad** has dark lores, pale yellowish supraloral and broken eye-ring, dull greenish-grey crown and upperparts, and flight feathers with narrow grey-green edges, outer rectrices with whitish tips. Throat to breast dull greyish yellow, below brighter (sometimes yellow confined to central belly), becoming greyish yellow on flanks; iris whitish grey; bill wide, flattened and triangular, dark grey and pinkish; legs bluish grey. **Juv** lacks blackish on lores. **Voice** Chattering, high-pitched series of pleasing "weet" notes; also more whistled series of clear "wii" notes. Calls include a harsh scold. **SS** White-eyed Vireo has bold yellow 'spectacles', much whiter underparts, bolder wingbars and yellowish-edged flight feathers. Yellow-throated Vireo shares white wingbars, but is much yellower on face to breast. Black-whiskered and other rarer migrant vireos lack wingbars, have darker eyes, bold whitish supercilia and dark eyestripes.

Bell's Vireo *Vireo bellii* `LC`
Four subspecies divided into two subspecies groups, of which one occurs in region.

Bell's Vireo *Vireo (bellii) bellii*
Vagrant (**Bahamas**).

L 11.5–12.5 cm. Four subspecies, of which presumably nominate occurs in region. Trapped and photographed, San Salvador, Jan 2012 (*NAB* 66: 360). **ID** Small, very active, short-billed but comparatively long-tailed vireo. **Ad** has crown and nape dull greyish brown, more greenish back and rump. Eye-ring and supraloral off-white, ear-coverts brownish grey, two variably prominent, narrow wingbars, chin and throat whitish, breast yellowish white, belly and vent yellowish white; bill blackish, paler at base and on mandible, legs dark greyish blue to black. **Juv** has brownish-washed upperparts, whiter underparts, relatively duller wingbars. **Voice** Most likely to be heard are a three-note alarm, an aggressive "zip-zip-zip", and a "chee". **SS** From similar Warbling Vireo by smaller and slimmer build, white wingbar and more spectacled appearance. Can recall Ruby-crowned Kinglet, but note thicker bluish legs and stout bill.

Puerto Rican Vireo *Vireo latimeri* `LC`
Fairly common resident (**Puerto Rico**).

L 12–13 cm. Monotypic. Endemic. Main island of Puerto Rico, with unconfirmed record on Vieques I (Gemmill 2015). Arid coastal scrub, moist limestone and humid montane forest, borders, and shade-coffee plantations. Usually forages in low strata; meticulous and deliberate, regularly joins mixed-species flocks in some areas, but not others. **ID** Small vireo with short whitish supercilium and broken eye-ring. **Ad** has dull grey-green crown and nape, brighter upperparts, and yellow-green fringes to flight feathers; tail feathers with olivaceous edges. Throat whitish grey, contrasting with yellowish-grey breast, yellower belly, flanks duller and central belly whiter; iris reddish brown, pale mandible, legs bluish grey. **Juv** has two faint wingbars, whiter underparts, more yellow-olive on flanks and vent. **Voice** Melodious, whistled series of 3–4 notes, repeated for several minutes, then changes to new phrase. Calls: a rattling, hoarse mewing and repeated "tup". **SS** Black-whiskered Vireo shares reddish eyes and nearly unmarked wings, but pale supercilium is long and broad, crown and eyestripe dusky grey, with narrow dusky malar.

Mangrove Vireo *Vireo pallens* `LC`
Ten subspecies divided into three subspecies groups, of which one occurs in region.

Providence Vireo *Vireo (pallens) approximans*
Common resident (**Providencia**).

L 11–12 cm. One subspecies in group. Endemic. Principally in dry scrub and woodland, typically at slightly higher elevations on island, and rare or absent in mangroves. Single undocumented claim from San Andres (*NAB* 67: 358). **ID** Chunky, neat little vireo, with comparatively large, broad-based bill. **Ad** has supraloral buffy or lemon-buff, upperparts dull greenish brown, flight feathers narrowly edged yellowish green, tertials edged whitish and two prominent white wingbars, tail dull dark grey, outer webs edged greenish. Chin whitish yellow, below tinged yellow or buff; iris brownish straw, yellow or dirty white (very variable), bill dark brown or grey-brown with paler mandible, legs grey-blue. **Juv** has face pattern less distinct. **Voice** Song described as a brief chatter (Bond 1961), a series of fast downstrokes (but slower and over smaller frequency range vs. San Andres and other Mangrove Vireos). Harsh alarm calls with longer notes and shorter inter-note intervals vs. similar taxa. **SS** Most similar to allopatric Thick-billed Vireo; only other vireos recorded on Providencia are larger with longer bills and quite different face patterns. San Andres Vireo has much greyer crown, nape and neck-sides. **TN** Treated as separate species by both Gill & Wright (2006) and Donegan *et al.* (2015); in contrast, many authorities consider *approximans* a race of Thick-billed Vireo.

Jamaican Vireo

Flat-billed Vireo

Bell's Vireo
bellii

Puerto Rican Vireo

Providence Vireo
(Mangrove Vireo)

215

San Andres Vireo *Vireo caribaeus* VU
Common resident (**San Andrés**).

L 12.5 cm. Monotypic. Endemic. Dry scrub, sparse dry woodland with cocoa plantations, spring creepers and low open canopy, and denser, more humid woodland with higher canopy and some epiphytes, typically at slightly higher elevations, being rare or absent in mangroves. **ID** Solid, round-crowned, compact vireo. **Ad** has partial eye-ring and supraloral yellowish, upperparts dull greenish grey, becoming more greenish on lower back/rump, flight feathers edged greenish yellow, tertials edged whitish, with off-white wingbars. Tail dull dark grey-brown, edged dull greenish. Chin and throat off-white, suffused yellow on breast and belly, flanks deeper and stronger yellowish green; iris grey or grey-brown, bill dark horn, legs grey or leaden. **Juv** is undescribed. **Voice** Three song types: simplest involves rapid repetition of single syllable 2–20 times, producing chatter song similar to Mangrove Vireo (which see); second a serial repetition, 1–15 times, of two syllables, similar to Jamaican Vireo song; third comprises 3–4 syllables, uttered randomly. Calls include single contact note while foraging, a "chee-chee" in agitation, and raspy buzz in aggression. **SS** The only small vireo resident on San Andrés, although Mangrove Vireo (which see) has been claimed there, and White-eyed Vireo is a winter visitor (white eyes and supraloral).

White-eyed Vireo *Vireo griseus* LC
Seven subspecies divided into two subspecies groups, of which one occurs in region.

White-eyed Vireo *Vireo (griseus) griseus*
Uncommon to rare winter visitor and passage migrant (**GA**, **O**), vagrant (**Guadeloupe**).

L 11–13 cm. Six subspecies in group, of which two (*noveboracensis*, nominate) occur in region. Mostly in Bahamas and Cuba, becoming progressively rarer further E through Greater Antilles; mainly present Oct–Apr, but arrives from early Sept and recorded to May on San Andrés. Scrub, coastal thickets, mangroves and open, brushy woodland. Usually seen alone; rather secretive in dense undergrowth. **ID** Small-bodied vireo with white eyes and yellow 'spectacles'. **Ad** has bright yellow supraloral and eye-ring, olive-green upperparts with greyer neck-sides and nape, flight feathers and tail edged greenish yellow, and two broad whitish-yellow wingbars. Below greyish white, becoming yellow on flanks; iris pale bluish grey to white, legs bluish grey. **1st-w** has duller 'spectacles' and iris greyish brown. **GV** Race *noveboracensis* (commoner in region) has brighter yellow flanks. **Voice** Song (rarely heard in region) comprises 3–7 loud, slurred syllables, often with sharp chip note at beginning and end (similar to, but faster and more emphatic than, Thick-billed Vireo). Calls include a nasal churr, soft "pick" and mewing notes. **SS** Only other vireo with white eyes, white wingbars and 'spectacles' is Flat-billed Vireo (Hispaniola) but its 'spectacles' are whitish, with pale tips to outer rectrices, narrower and fainter wingbars, and more extensively pale yellow underparts. Yellow-throated Vireo has bright yellow throat to lower breast, brighter olive upperparts, and dark eyes. Thick-billed Vireo also more yellowish below, with darker bulging eyes.

Thick-billed Vireo *Vireo crassirostris* LC
Fairly common to uncommon and local resident (**Bahamas**, **Cuba**, **Cayman Is**, **Hispaniola**).

L 13–14 cm. Five subspecies, of which all (nominate[A] Bahamas, *stalagmium*[B] Caicos Is, *tortugae*[C] Î de la Tortue, off Haiti, *cubensis*[D] mainly Cayo Paredón Grande, off N Cuba, *alleni*[E] Cayman Is) occur in region. Endemic; only casual visitor to S Florida (USA). Brush, thickets, wooded edges, dry broadleaf forest, second growth and mangroves. Probably most numerous in parts of Bahamas, where regularly uses more xeric habitats. Apparently extirpated as breeder on Little Cayman; only occasional on other cays off N Cuba, where habitat rapidly being encroached by burgeoning tourist infrastructure. Sluggish and keeps low, in dense vegetation. **ID** Stocky and large-billed, with pale amber eyes, yellow supraloral, two white wingbars. **Ad** has broad lemon-yellow supraloral, paler and broken eye-ring (lower half almost white), blackish-grey lores, dull greenish-grey upperparts with flight feathers narrowly fringed greenish yellow, and two whitish wingbars. Throat dull yellowish, brighter on breast and belly, greyish-yellow flanks; iris pale brown, bill stout and heavy, legs grey. **Juv** lacks blackish on lores, has underparts olive-yellow, wingbars fainter. **GV** Race *stalagmium* has shorter wings and tail, less deep bill, generally more yellow on vent; *tortugae* distinctly buff-washed underparts; *cubensis* brownish-grey tinge on neck and upper back; *alleni* similar to last, but has brighter yellow lores. **Voice** Song, mostly heard Dec–Jul, an emphatic series often ending in sharp "chik", similar to but slower and less emphatic than White-eyed Vireo. Calls include low "turrr", buzzy "shhh" and nasal "enk". **SS** Confusion unlikely. White-eyed Vireo has yellowish 'spectacles', white eyes, yellow confined to flanks; and Yellow-throated Vireo has darker eyes, lemon-yellow 'spectacles' and chin to lower breast, contrasting white belly and bolder wingbars; neither is resident. No overlap with Flat-billed Vireo.

Cuban Vireo *Vireo gundlachii* LC
Common resident (**Cuba**).

L 13 cm. Four subspecies, of which all (*gundlachii*[A] most of Cuba, I of Pines and several N cays, *sanfelipensis*[B] Cayo Real, *magnus*[C] Cayo Cantiles, *orientalis*[D] SE Cuba but perhaps also many larger cays) occur in region. Endemic; vagrant to S Florida. Mostly lowlands, but all elevations, in dense scrub, thickets and most forest types (except pines). Singles or pairs forage deliberately, rather low above ground; often with small mixed-species flocks, especially with Yellow-headed Warbler (where they overlap). **ID** Stocky, rather large-billed vireo. **Ad** has yellow to creamy-white lores, eye-ring and small post-ocular spot, dull olive-grey upperparts with pale yellowish-fringed flight feathers and two narrow pale wingbars. Throat and breast dull yellowish, paler on belly, greyer on sides and vent; dark brown iris, pale mandible, legs lead-grey. **Juv** is generally duller than ad. **GV** Race *orientalis* generally greyest and paler yellow below; *magnus* larger, longer-winged and longer-tailed, less olivaceous above and paler yellow below; *sanfelipensis* whiter below than nominate, especially chin and throat. However, none might merit recognition (Buden & Olson 1989, Kirkconnell *et al.* in press). **Voice** Song a loud whistling, high-pitched, variable "wee-chivee, wee-chivee", or "chuee-chuee" and other variants. Also a rapid, descending series of "chi" notes, a scolding "kik" and soft rattle. **SS** Yellow-throated and White-eyed Vireos (neither resident) have bolder wingbars; White-eyed has pale eyes and mostly whitish underparts; and Yellow-throated has bicoloured underparts, bolder eye-ring and more olive upperparts.

San Andres Vireo

White-eyed Vireo
griseus

alleni

crassirostris

Thick-billed Vireo

Cuban Vireo
gundlachii

217

LANIIDAE
Shrikes
33 extant species, 1 vagrant in region

Loggerhead Shrike *Lanius ludovicianus* `NT`
Vagrant (**Bahamas**).
L 18–22 cm. Seven subspecies, of which one (*migrans*) occurs in region. Specimen (NHMUK 1897.5.31.4), Andros, Dec 1896 (Sharpe 1897, Bond 1951); sight, Grand Bahama, Oct 1964 (Bond 1965) and Great Exuma, Oct 1966 or 1967 (Bond 1968). **ID** Grey-and-black shrike with mid-length tail. **Ad** has black mask, sometimes with very narrow whitish supercilium, grey upperparts, white-tipped scapulars, black wings with broad white tips to tertials and secondaries, white bar on primaries, and black tail with all but central feathers tipped white. Below whitish, flanks sometimes washed pale grey, occasionally some faint vermiculations on breast. Bill black, in non-br season often has pale base to mandible; legs blackish. ♀ often has lores duller, upperparts less bluish, sometimes more greyish below, legs browner. **Juv** is generally pale brownish grey with fine dark vermiculations above and below, mask narrower and browner, buffish wingbar, smaller white primary patch and pale base to mandible. **Voice** Repeated harsh screeching "tscheeer-tscheeer…" and sharp "bzeeek-bzeeek" perhaps most likely to be heard. **SS** No other shrike has occurred in region.

CORVIDAE
Crows
130 extant species, 4 in region + 2 vagrants

Cuban Crow *Corvus nasicus* `LC`
Fairly common resident (**Cuba, Bahamas**).
L 40–42 cm. Monotypic. Endemic. Present in open and semi-deciduous, evergreen and swamp woodland, second growth, and scattered trees and palm groves near forest, principally in lowlands, on mainland Cuba (where most widespread and numerous in E third), I of Pines and four cays off NC coast, as well as on Turks & Caicos (mainly Providenciales, North Caicos, Middle Caicos). Vagrant North Andros, Jan 2015 (*NAB* 69: 313). **ID** All-black crow with upturned nasal bristles that do not cover nostrils, and relatively long bill. **Ad** is deep black with slight blue-purple gloss, and patch of bare skin behind eye and at base of bill. **Juv** is slightly duller than ad. **Voice** Vocalizations quite musical; rather less nasal and abrupt than those of Cuban Palm Crow. Ringing, high-pitched "aaaaauh" rising in inflection, liquid bubbling, trilling and chattering parrot-like calls, raven-like croaks and varied guttural chattering. **SS** From rarer and more geographically restricted Cuban Palm Crow by vocalizations, longer and less stout bill, longer wings and slower, heavier wingbeats in flight. At rest, Cuban Crow appears shorter-tailed because of its longer wings. Nasal bristles are longer in latter species, covering nostrils, although very good views will be required to establish this feature. Cuban Crow perhaps descends more regularly to ground, where gait more walking than hopping.

Jamaican Crow *Corvus jamaicensis* `LC`
Locally common resident (**Jamaica**).
L 35–38 cm. Monotypic. Endemic. Range perhaps expanding, in wet limestone forests, montane pastures, wooded hills and open parkland, including landscapes with settlements and agriculture, at all elevations, but principally at low altitudes during dry season. **ID** Smallest and dullest Caribbean crow; body plumage loose and soft, nostrils sometimes fully exposed. **Ad** has black head, becoming more dull grey or sooty black over rest of plumage, with some violet iridescence on wings, tail and neck; iris grey-brown or reddish brown (indicative of age or sex), bare suborbital patch evident on most individuals. **Juv** is undescribed. **Voice** Remarkably variable, especially jabbering outbursts, which are frequent and include mix of chuckling, chattering and gobbling, strung together in garbled melody, sometimes likened to a turkey being strangled; also known to croak like raven and to give harsh "craa craa". **SS** The only crow on Jamaica.

White-necked Crow *Corvus leucognaphalus* `VU`
Locally common to uncommon resident (**Hispaniola**).
L 42–46 cm. Monotypic. Endemic. Most closely associated with mature pine and broadleaf forests in mountainous and hilly, inaccessible country to 2650 m (rare above 1500 m); also coastal mangrove and cactus forests. Recorded on several major satellite islands. Has declined and range retracted due to excessive hunting; formerly also Puerto Rico, but extirpated by early 1960s. **ID** Largest Caribbean crow, with long tapered bill, nasal bristles upturned and sweeping over culmen, intense plush-like black feathers at base of culmen between bill and eye. **Ad** is black glossed violet or blue; iris red-brown (yellow also reported, possibly age-specific), bare skin around bill base and below eye. **Juv** is slightly duller than ad. **Voice** Varied, raven-like vocalizations, including guttural "culik-calow-calow", deep "wallough", and varied high-pitched musical, gurgling, bubbling and laughing notes, e.g. "klook". **SS** Compare Hispaniolan Palm Crow (the only sympatric corvid).

Fish Crow *Corvus ossifragus* `LC`
Vagrant (**Bahamas**).
L 36–41 cm. Monotypic. Multiple records: one photographed, Grand Bahama, Feb 1997–Jul 1998, with presumably same Abaco, Aug 1997 (*FN* 51: 810, 933; 52: 133, 508), one Abaco, Aug 2012 (eBird) and two different individuals, Oct–Nov 2015 (*NAB* 70: 130); up to two Grand Bahama, Oct 2017–Feb 2019 at least (eBird). Abundant in neighbouring Florida. **ID** Larger crow with proportionately rather long tail but short bill. **Ad** is all black with bluish-violet and green gloss on head, upper- and underparts; iris dark brown. **Juv** is duller and fluffier in appearance, with grey-blue iris. **Imm** is mostly black, with brown eyes, feathers of wing and tail often brown and contrasting with body. **Voice** Nine different calls, of which short, nasal "awwr" in aggression, shorter and higher-pitched "ewh" and distinctive disyllabic nasal "uh-uh" in contact, perhaps most likely to be heard. **SS** Compare Cuban and House Crows, the only other largely black corvids likely to appear in N Bahamas.

Loggerhead Shrike
migrans

Cuban Crow

Jamaican Crow

White-necked Crow

Fish Crow

219

Palm Crow *Corvus palmarum* `LC`
Two subspecies placed in separate subspecies groups, of which both occur in region.

Cuban Palm Crow *Corvus (palmarum) minutus*
Rare and local resident (**Cuba**).
L 34–38 cm. One subspecies in group. Endemic. Lowland cultivation with tall, scattered palms in Camagüey, Sancti Spíritus, Cienfuegos, Villa Clara and Pinar del Río provinces, but very few localities outside first-named. Formerly more common in W & C Cuba. Less regularly observed on ground than Cuban Crow. **ID** Small, rather tame crow with short, stout, sharp-pointed bill and well-developed nasal bristles covering nostrils. **Ad** has coarse black plumage with some purple-blue iridescence. **Juv** is somewhat duller than ad. **Voice** A somewhat sheep-like "caw", sharp, high-pitched "craa" and deep, harsh, burring sounds like a frog, all of which help to distinguish it from Cuban Crow. **SS** Only sympatric crow is Cuban Crow (which see); they are regularly seen together in Sierra de Najasa, C Cuba. **TN** Cuban and Hispaniolan Palm Crows are sometimes treated as two species based on morphology, behaviour and vocalizations (Garrido *et al*. 1997a), and their eggs also differ (pale blue with well-defined markings in present species vs. pale green with diffuse markings; P. Regalado). A molecular study found that, while they are each other's closest relatives, genetically they differ as much as many traditionally recognized species (Jønsson *et al*. 2012).

Hispaniolan Palm Crow *Corvus (palmarum) palmarum*
Locally common resident (**Hispaniola**).
L 34–38 cm. One subspecies in group. Endemic. Mainly in pine forest but also dry scrub and humid broadleaf forest, to 3000 m (principally at 1300–1900 m) in Dominican Republic, and in similar habitats to sea level in Haiti. **ID** Very similar to Cuban Palm Crow. **Ad** is all black with purplish-blue sheen on mantle and wing-coverts in fresh plumage; dull brownish black in worn plumage. **Juv** is somewhat duller than ad. **Voice** Vocalizations less variable and more distinctly nasal than in White-necked Crow: a harsh and usually complaining "aar" or "cao cao", typically in pairs or short series. Often flicks tail downwards when calling. **SS** Only sympatric corvid is White-necked Crow, which is larger and distinctly bulkier-bodied, with longer wings, reddish eyes (vs. brown) and different vocalizations (arguably best distinction). Present species also has steadier flapping flight than latter. **TN** See Cuban Palm Crow.

House Crow *Corvus splendens* `LC`
Vagrant (**Cuba**, **Barbados**).
L 40–43 cm. Four subspecies, of which that in region is unknown. Two presumably ship-assisted records: Barbados, May–Aug 2008 (Buckley *et al*. 2009); Cayo Guillermo, Cuba, at least Nov 2007–Apr 2011 (Kirkconnell *et al*. in press). Native to S & SE Asia, but has spread via ship-assisted passage and deliberate releases virtually throughout Indian Ocean and Red Sea coastal regions, and has been seen, principally near ports, in many parts of world. **ID** Relatively small, slim-bodied crow. **Ad** has broad dull greyish collar extending to nape, upper mantle and breast; glossy black wings and tail. Bill fairly slender. **Juv** is similar but duller and less glossy. **Voice** Very noisy. Flat, toneless "kaaan-kaaan" is most frequently heard call. **SS** Greyish collar and head shape distinctive within region.

ALAUDIDAE
Larks
92 extant species, 1 vagrant in region

Horned Lark *Eremophila alpestris* `LC`
Twenty-eight subspecies divided into six subspecies groups, of which one occurs in region.

American Horned Lark *Eremophila (alpestris) alpestris*
Vagrant (**Bahamas**).
L 14–17 cm. Thirteen subspecies in group, of which one (probably nominate) occurs in region. Photographed, New Providence, Jan 2012 (*NAB* 66: 360); photographed, Cat I, Dec 2014 (*NAB* 69: 307–308; eBird). Most likely to be observed on open bare ground near coasts, especially beaches. **ID** The only lark to have occurred in region. **Ad** ♂ has yellow forehead, supercilium, throat and ear-coverts, contrasting blackish central crown with elongated lateral feathers, blackish lores and broad band from eye to cheeks, and black breast-band. Above warm rufous-brown tinged pinkish, tail blackish, central feathers paler and greyish, outer ones edged white. Underparts below breast-band whitish, washed pinkish rufous at sides. In non-br plumage head pattern partly obscured by pale fringes and 'horns' often invisible. ♀ is slightly duller, with narrower black bands on head and breast than ♂. **Voice** Usual flight call "eeh" or "ééh-ti", or liquid "tur-reep"; also an occasional harsh "tsrr". **SS** Unmistakable given reasonable views.

HIRUNDINIDAE
Swallows and Martins
89 extant species, 11 in region + 3 vagrants

Northern House Martin *Delichon urbicum* `LC`
Vagrant (**Guadeloupe**, **Barbados**).
L 13–14 cm. Two subspecies, of which one (possibly nominate) occurs in region. Four records: Barbados, eight, Oct–Nov 1999, singles, Nov 2000 and Jun 2002 (Buckley *et al*. 2009); Guadeloupe, Aug 2006 (*NAB* 61: 161). Flight rather slow, with long glides and gentle arcing movements. **ID** Black-and-white swallow with forked tail and white rump. **Ad br** ♂ has glossy blackish-blue crown and back, white rump, black wings and tail, latter moderately forked. Below white. **Ad non-br** ♂ has brown-grey mottling on rump, face-sides, throat and flanks. **Ad** ♀ is greyer below. **Juv** is duller and browner, with less blue gloss on upperparts and shorter tail. **Voice** Dry twitter, sometimes with more emphatic and drawn-out "chierr" notes; also short or double-noted trills. **SS** Other swallows lack white rump and underparts, and forked black tail. **AN** Common House Martin.

Cuban Palm Crow
(Palm Crow)

Hispaniolan Palm Crow
(Palm Crow)

House Crow
splendens

♀ / ♂ non-breeding

♂ breeding

American Horned Lark
(Horned Lark)
alpestris

Northern House Martin
urbicum

221

Cliff Swallow *Petrochelidon pyrrhonota* `LC`
Uncommon to rare passage migrant (**throughout**) and winter visitor (**C & S LA**).

L 13–15 cm. Four subspecies, of which one (presumably nominate, less likely *tachina*) occur in region. Usually small numbers (rarely up to c. 100) on passage, often with other hirundines; Aug–Dec, (late Feb) Mar–May (early Jun). Tiny numbers overwinter in far S of region, e.g. Guadeloupe, Barbados (Buckley *et al*. 2009), but most winter in E South America, especially Brazil. Open areas near water, over grassland and coasts, sometimes towns; flight usually fast, low over water. **ID** Small and stocky, with reddish-brown throat, buffy collar and cinnamon-buff rump; tail square-ended. **Ad** has white to buffy forehead, deep blue crown, glossy deep blue mantle and back, mantle streaked whitish, pale cinnamon-rufous to cinnamon-buff rump, blackish-brown wings and tail. Head-sides and throat chestnut, patch of black on throat; breast and sides grey-brown, below whitish, underwing-coverts grey-brown. **1st-w** is duller and browner, ear-coverts dark brown, throat and forehead vary in colour and pattern, pale feather fringes above. **GV** Races differ in size and coloration; *tachina* smaller than nominate, forehead cinnamon to pale buff. **Voice** Calls a low, soft and rolled "werr" and thin "churr". **SS** See Cave Swallow. Barn Swallow has long deeply forked tail; if tail-streamers missing, still shows white tail-band, no pale rump or nuchal collar, and throat is uniform rufous or chestnut. **AN** American Cliff Swallow.

Cave Swallow *Petrochelidon fulva* `LC`
Six subspecies divided into three subspecies groups, of which one occurs in region.

Caribbean Cave Swallow *Petrochelidon (fulva) fulva*
Fairly common resident (**GA**), rare passage migrant or vagrant (**LA**, **O**). L 12–14 cm. Four subspecies in group, of which all (*cavicola*[A] Cuba, I of Pines and many satellite cays, *poeciloma*[B] Jamaica, nominate[C] Hispaniola including large satellites, *puertoricensis*[D] Puerto Rico) occur in region. Near-endemic; since 1987, race *cavicola* has bred in Florida (SE USA) and presumably this race also at one site on Andros (Bahamas). Prefers open areas, cliffs with caves and man-made structures (barns, wooden roofs, viaducts and bridges), to c. 1640 m in Hispaniola. Believed resident in Greater Antilles, but numbers in Cuba and on Hispaniola decline Sept–Feb, when some or all might leave (being potentially replaced by Floridian migrants), with concomitant evidence of migration through Cayman Is and to some extent elsewhere in region S to St Vincent and San Andrés (race or races unknown), but no evidence of where these winter, despite recent record in Venezuela (Escola *et al*. 2011) and a few on Guadeloupe. Typically in flocks (occasionally up to 500 birds), flying low over fields and water, with other swallows, perching on utility wires. **ID** Structure and plumage very similar to Cliff Swallow. **Ad** has dark rufous-buff rump and forehead, pale reddish-brown ear-coverts, throat, breast and sides, glossy blue-black crown, mantle and back, latter streaked whitish, blackish-brown wings and almost square-ended tail. Belly whitish, longer undertail-coverts brownish centrally; underwing-coverts pale brownish. **Juv** is browner and less glossy above, with greyish (not white) tramlines, and duller face and throat. **GV** Race *cavicola* has deeper blue crown and broader white streaks above; *poeciloma* has paler forehead, more extensive cinnamon below, rustier undertail-coverts; *puertoricensis* has more intense cinnamon or chestnut coloration, dark rusty wash on undertail-coverts. **Voice** Song starts with squeaks, then a warble and finally series of two-tone notes; calls include a "che", short "weet", "cheweet", and series of "che" notes; higher, sweeter and more single-syllabled than Cliff Swallow. **SS** Distinguished from rarer Cliff Swallow by darker, less contrasting but marginally more extensive forehead patch (without pale 'headlights' of Cliff), paler throat, nape and head-sides, giving more capped appearance; much overlap in rump colour, although Cave averages darker (Hough 2000). In comparative views or with considerable experience, Cave Swallow appears more compact and stocky, and more bull-necked or neckless than Cliff Swallow. Barn Swallow without long tail-streamers still has white band in tail, deep rufous-chestnut throat outlined black, no nuchal collar. **TN** Taxonomy basically follows Garrido *et al*. (1999).

Barn Swallow *Hirundo rustica* `LC`
Eight subspecies divided into two subspecies groups, of which one occurs in region.

American Barn Swallow *Hirundo (rustica) erythrogaster*
Common passage migrant and local winter visitor (**throughout**). L 17–19 cm. One subspecies in group. Passage mainly Aug–Nov, Feb–Jun, with small to large numbers locally in Dec–Jan (e.g. Hispaniola, S Lesser Antilles). Over fields, swamps, marshes, coasts and towns, to c. 1100 m, in small low-flying flocks; flight rather fast and can appear almost directionless. Eurasian Barn Swallow *H. r. rustica* potential vagrant to S Lesser Antillies. **ID** Elegant with long tail-streamers, white tail-band and underparts, but cinnamon throat and blue breast-sides. **Ad** has forehead rufous-chestnut, crown and upperparts glossy steel-blue, wings and tail black, white patches on most rectrices, outer feathers greatly elongated, throat bright rufous-chestnut, broad steel-blue breast-sides, and rest of underparts, and underwing-coverts, buffy white or pinkish buff to orangey (brightest on vent). ♀ is less glossy than ♂ with shorter tail-streamers. **Juv** is duller, red parts paler, tail shorter, has pale creamy tips to greater coverts and can show greyish-brown band on lower breast. **GV** Eurasian Barn Swallow can show complete blue breast-band (including young birds), has longer tail-streamers and never shows such deeply saturated underparts, but palest American can overlap and young Eurasian can be buff-coloured (Jiguet & Zucca 2005). **Voice** Thin, unmusical "chit" calls, sometimes in short series, running into stuttering, and chirping contact calls. **SS** See Cave and Cliff Swallows, both of which have short, notched tails without white band, and have coloured rump and nuchal collar (latter can be hard to see), among other features.

Collared Sand Martin / Bank Swallow `LC`
Riparia riparia
Uncommon to rare passage migrant (**throughout**) and winter visitor (**S LA**).

L 12–13 cm. Five subspecies, of which nominate occurs in region. Mainly Aug–Nov and Mar–Jun, but records also Dec–Feb, and overwinters regularly in small numbers in S Lesser Antilles. Open areas near water and coasts, towns. Associates with other migrant swallows in loose flocks, often low over water; flight fast and direct. **ID** Small, with greyish-brown upperparts, underparts white with greyish-brown breast-band. **Ad** has remiges and rectrices darker brown, tail shallowly forked, breast-band sometimes extends to central belly and underwing-coverts dark brown. **Juv** has pale feather fringes above, rufous-buff wash on face, neck-sides and chin to breast. **Voice** Fast series of quick "chrrrt" notes. **SS** No other swallow in region has clear dusky breast-band, but juv Tree Swallow brownish above with faint dusky breast-sides, and Northern Rough-winged Swallow can show pale brownish-washed breast. **AN** Sand Martin.

Cliff Swallow
pyrrhonota

fulva

puertoricensis

Caribbean Cave Swallow
(Cave Swallow)

adult

adult   juvenile

American Barn Swallow
(Barn Swallow)

Collared Sand Martin / Bank Swallow
riparia

223

Tree Swallow *Tachycineta bicolor* `LC`
Common passage migrant and winter visitor (**GA**, **O**), mainly vagrant (**LA**). **L** 12–15 cm. Monotypic. Mainly present Sept–May, but handful of summer records Cuba. Open areas mostly in and around wetlands, including wet rice fields. Feeds low over water and fields, usually in lee of wind and mainly in loose flocks, often with other hirundines. **ID** Proportionately broad-winged with notched tail; blue upperparts, white underparts. **Ad** has blacker wings and tail, latter notched; grey-brown underwing-coverts. ♀ is duller, sometimes with brown forehead. In autumn, both sexes tend to appear greener, less blue, above. **Juv** has brown upperparts and grey-brown wash on breast-sides. **Voice** A liquid twittering. **SS** Other blue-and-white swallows are greener (Bahama Swallow) or golden-green (Golden Swallow) above, among other differences. Juv much like Collared Sand Martin, but averages larger, without distinct breast-band. Compare vagrant Violet-green Swallow and less numerous Northern Rough-winged Swallow.

Violet-green Swallow *Tachycineta thalassina* `LC`
Vagrant (**Puerto Rico**). **L** 12–13 cm. Two subspecies, of which nominate occurs in region. Photographed, Feb 2010 (*NAB* 64: 339). Often near water on migration. **ID** Attractive swallow with unique head pattern. **Ad** has forehead to back matt green, tinged violet on nape, rump and uppertail-coverts violet-blue, extensive white on sides of rump (can appear white-rumped), tertials edged white when fresh, tail short and shallowly forked. Underparts and head-sides white, underwing-coverts pale grey. ♀ is duller with browner head. **Juv** is browner. **Voice** Contact call "chee-chee". **SS** Differs from Tree Swallow in having iridescent blue-green rather than iridescent blue plumage, more extensive white on head-sides and white on rump-sides. Compare Golden Swallow (now confined to Hispaniola) and Bahama Swallow.

Golden Swallow *Tachycineta euchrysea* `VU`
Rare and local resident (**Hispaniola**). **L** 12–12.5 cm. Two subspecies, of which both (nominate Jamaica but now extinct, *sclateri* Hispaniola) occur in region. Endemic. Montane humid broadleaf and pine forests, and adjacent open fields, mainly at 750–2000 m, but possibly moves lower post-breeding, to sea level. Alone or in small groups, feeding mostly low above ground. Population small and declining on Hispaniola; not seen in Jamaica since early 1980s, where island-wide surveys in 1994–2012 (Graves 2014) and Jan–Mar 2015 (Proctor *et al.* 2017) failed to locate species, which is probably extirpated there. **ID** Small, bright bronzy to golden-green and white, with moderately forked tail. **Ad** ♂ (*sclateri*) is iridescent bronzy green on most upperparts; wings and tail dusky bronze-green, tail slightly forked, underparts white but underwing-coverts dusky with bronze-green margins. **Ad** ♀ is duller, with some grey-brown mottling on breast. **Juv** is even duller and browner on upperparts, with extensive grey-brown mottling on breast. **GV** Nominate (probably extinct) is more golden above, with shorter wings and shallower tail-fork. **Voice** Twittering "tchee-weet" heard during breeding season. **SS** Ad Tree Swallow slightly larger, with less forked tail and lacks bronzy/golden sheen on upperparts.

Bahama Swallow *Tachycineta cyaneoviridis* `EN`
Uncommon resident (**Bahamas**), vagrant (**Cuba**). **L** 13.5–15 cm. Monotypic. Endemic. Breeds Grand Bahama, Great Abaco and Andros, possibly still on New Providence. Pine forest, other woodland, marshes and coasts, plus clearings, urban parks and towns. Forages over open areas, in small flocks, at various heights. Post-breeding dispersal reaches S to N & E Cuba (where basically vagrant) and W to Florida (USA). Perhaps declining due to loss of nest-sites. **ID** Dark greenish and snowy white, with deeply forked tail. **Ad** has upperparts green, rump more bluish green, wings and tail blue, underparts and underwing-coverts white. ♀ is duller and less metallic green above, with shorter tail-fork. **Juv** is sooty brown above with green highlights (strongest in ♂), even less deeply (but still obviously) forked tail and dusky-washed (but still contrasting paler) underwing-coverts. **Voice** Metallic "chep", "chi-chep". **SS** Tree Swallow has blacker wings (including dark underwing-coverts), only slightly notched tail and, in ad, bluer upperparts.

White-winged Swallow *Tachycineta albiventer* `LC`
Vagrant (**Martinique**, **Grenada**). **L** 14 cm. Monotypic. Martinique, sight, Aug 1993 (Feldmann *et al.* 1999, Belfan & Conde 2016); Grenada, photo, Dec 2005 (*Cotinga* 26: 77, 94). Open waterbodies, including rivers, mangroves, lakes and reservoirs. In native range, singles or small loose flocks feed low over water or adjacent open areas. Flight generally rather fast and direct. **ID** Chunky-bodied swallow, with large white rump and markings on wings. **Ad** has cap and upperparts glossy blue-green, mostly blackish wings and black tail with shallow fork; underparts and underwing-coverts clean white. ♀ shows less white on wing than ♂. **Juv** is browner, with even less white on wing, greyer underparts. **Voice** Call a sweet-sounding "zreeet". **SS** Confusion unlikely; no other blue-and-white swallow shows extensive white on wings.

Tree Swallow

Violet-green Swallow
thalassina

euchrysea

sclateri

Golden Swallow
sclateri

Bahama Swallow

White-winged Swallow

225

Purple Martin *Progne subis* **LC**
Uncommon to rare passage migrant (**throughout**).
L 18–19 cm. Three subspecies, of which possibly all (nominate, *arboricola*, *hesperia*) occur in region. Recorded Aug–Nov (exceptionally Jul and Dec), Jan–Mar (rarely to May). Very few records outside Greater Antilles and Bahamas, but true scale of passage through region probably under-appreciated due in part to potential for confusion with both Cuban (especially) and Caribbean Martins. Open areas, including cultivation, marshes and towns. Singles or small flocks, often with other hirundines; flight involves much gliding and circling. **ID** Large, steel-blue ♂, ♀ has prominent pale collar. **Ad** ♂ steel-blue with concealed white tufts on sides; tail moderately forked. **Ad** ♀ is much duller, with grey forehead, pale nuchal collar, dusky-grey throat finely streaked, dusky-brown upper breast and sides scaled and streaked greyish white. **Juv** has grey-brown upperparts and throat, rest of underparts grey-white; acquires blue feathers with age, especially on head and underparts; wings and tail shorter. **GV** Races differ in size and ♀ plumage: *arboricola* largest, ♀ with whitish forehead and hindcollar, much paler below than nominate; *hesperia* (not definitely recorded) smallest with plumage as *arboricola*. **Voice** Calls include melodious, low, and semi-musical whistles, descending "cherr", and a vibrating gurgling, but rarely heard in region. **SS** ♂ inseparable in field from ♂ Cuban Martin; ♀ Cuban has dusky throat, breast and onto flanks, belly whiter, lacks pale nuchal collar. See Caribbean Martin.

Caribbean Martin *Progne dominicensis* **LC**
Fairly common and widespread summer visitor and passage migrant (**virtually throughout**).
L 18–20 cm. Monotypic. Near-endemic breeder; otherwise nests only on Tobago. Records N to Bahamas. Mostly recorded Feb–Oct, but exceptionally in winter N to Puerto Rico (Dec/Jan). Non-breeding grounds still largely unknown although a ♀ was tracked to E Brazil (Perlut *et al.* 2017); very rare on Barbados in winter and no records on Guadeloupe mid Nov– mid Jan. Semi-open areas often near water and coasts; to c. 1500 m. Forages alone or in small groups, high or low above ground. Flight alternates gliding and flapping. **ID** Large steel-blue martin similar to Purple and Caribbean Martins. **Ad** ♂ is mostly glossy steel-blue, wings and tail black, tail moderately forked; lower breast to undertail-coverts white, undertail-coverts spotted dark, underwing-coverts grey-brown. **Ad** ♀ is more grey-brown, with pale throat. **Juv** is much duller than ad ♀; 1st-y ♂ resembles ad ♀, but faintly glossed blue above. **Voice** Song is a gurgling, including short and high-pitched "twick-twick" notes and more melodious warbles. Calls include "zwot" in contact and clear "kweet" or "peak" in alarm. **SS** ♂ Purple and Cuban Martins are completely steel-blue, without any white below. ♀ Purple has whitish underparts streaked and scaled dusky, with pale nuchal collar. ♀ Cuban has duller upper breast extending more onto flanks, and tail is slightly less forked. In practice ♀♀ of these three species will require very good views to discriminate and in many cases certain identification will be impossible.

Cuban Martin *Progne cryptoleuca* **LC**
Fairly common summer visitor (**Cuba**).
L 18–19 cm. Monotypic. Endemic breeder. Present in Cuba late Jan–early Nov (with one winter record, either this species or Purple Martin), but bulk arrives mid Feb/early Mar and leaves by late Sept. Reports in Bahamas (Eleuthera, Mar), Jamaica (Feb), Puerto Rico (at least twice in Jan) and Guadeloupe (♂ paired with ♀ Caribbean Martin) but perhaps none absolutely proven, with unidentified martins as far afield as Barbados possibly this species. Open woodland with palms, lake edges, marshes with dead trees, and especially urban areas and cities. Alternates flaps and glides in leisurely flight; nests in loose colonies. Wintering areas unknown. **ID** Very similar to Purple Martin. **Ad** ♂ is glossy steel-blue, with concealed white band on lower belly; wings and tail mainly sooty black. **Ad** ♀ is duller, more sooty brown with blue feather fringes above, greenish-blue gloss on wings and tail, sooty-grey head- to breast-sides and underwing-coverts, rest of underparts white, with dark shafts on breast to undertail-coverts. **Juv** is much duller; 1st-y ♂ acquires some blue feathers below. **Voice** Vocalizations apparently identical to Purple Martin. **SS** ♂ not safely separated from Purple Martin, but tail of latter shorter and less forked, and lacks concealed white in centre of belly; ♀ paler than ♀ Cuban Martin, with dusky throat and breast shading into whiter central belly. ♂ Caribbean Martin has pure white lower breast/belly. See comments under latter.

Northern Rough-winged Swallow **LC**
Stelgidopteryx serripennis
Uncommon passage migrant and rare winter visitor (**GA**, **O**), vagrant elsewhere.
L 12.5–14.5 cm. Four subspecies, of which two (nominate, *psammochroa*) occur in region. Records late Jul–early May, but mainly Nov–Apr. Perhaps overlooked to some extent, but occasionally seen in large numbers. Most records presumably involve nominate, but two Cuban specimens identified as *psammochroa* (USNM). Open areas and wetlands. Flight fast and direct, with slow wingbeats, regularly low over water or ground. **ID** Broad-winged, tail short and square; nondescript brown-and-buff plumage. **Ad** has greyish-brown upperparts, rump paler, wings and tail darker; pale grey-brown chin to breast and flanks, throat buffier when fresh, rest of underparts dull white; underwing-coverts grey-brown. **Juv** has cinnamon wingbars. **GV** Race *psammochroa* paler, especially on crown and rump, than nominate. **Voice** Series of fast "brrrt" notes, also a gurgling trill. **SS** Other dull-coloured swallows include juv Tree Swallow, which has whiter throat and breast, with dusky brown limited to sides; and smaller Collared Sand Martin, with more forked tail and dull brown breast-band.

Purple Martin
subis

Caribbean Martin

Cuban Martin

Northern Rough-winged Swallow
serripennis

not to scale

SITTIDAE
Nuthatches
32 extant species, 1 in region

Bahama Nuthatch *Sitta insularis* `CR`
Extremely rare resident (**Bahamas**).
L c. 10.5 cm. Monotypic. Endemic. Confined to mature Caribbean pine (*Pinus caribaea*) forests on Grand Bahama, favouring pure stands over other forested habitats. Fairly common in late 1960s and 1970s, but numbers declined by > 95% between 1969 and 1993. Hurricanes Matthew and Irma, in Oct 2016 and Sept 2017, are believed to have further impacted the now tiny population; nuthatches were extremely difficult to find in spring 2018 (Sessa-Hawkins & Hermes 2019). Climbs up and down trunks, often working upside-down; either alone or in small groups. **ID** Small nuthatch; the only representative of its family in West Indies. **Ad** has brown crown with much darker mask, bluish-grey upperparts, white patch on hindneck and dull whitish underparts with buffy tones. **Juv** is apparently undescribed. **Voice** A short swallow-like, high-pitched warble combining both slightly squeaky notes and short trills apparently is the principal vocalization. **SS** None. **TN** Usually considered conspecific with extralimital Brown-headed Nuthatch *S. pusilla* (SE USA).

POLIOPTILIDAE
Gnatcatchers
15 extant species, 2 in region

Blue-grey Gnatcatcher *Polioptila caerulea* `LC`
Fairly common resident (**Bahamas**) or winter visitor (**Cuba**, **Cayman Is**), vagrant (**Hispaniola**).
L 10–12 cm. Nine subspecies, of which one or two (probably nominate[A] winters Cuba and Cayman Is, *caesiogaster*[B] resident Bahamas) occur in region. Residents in Bahamas in scrub and woodland, including pines. Migrants present Cuba (including I of Pines and many offshore cays) mid Aug to second week of May, in semi-deciduous and evergreen woodland, coastal thickets, swamp woodland, second growth, scrub and pine forest, to 1200 m. Uncommon to rare Cayman Is (also in mangroves) and vagrant Hispaniola (Jan 1987, Sept 2018). Nominate never definitely identified in Bahamas (or anywhere in region), but continental birds presumably also winter and migrate through the islands.
ID Sprightly, long-tailed, pale- and slim-bodied passerine, with fairly long, fine bill. **Ad br** ♂ (*caesiogaster*) has U-shaped black mark over bill to behind eye, complete white eye-ring, bluish-grey crown and upperparts, broad white edges to tertials, black tail with mostly white outer three rectrices, and off-white underparts. **Non-br** ♂ lacks black on head. **Ad** ♀ is like non-br ♂, but upperparts washed brownish. **Juv** resembles ♀. **GV** Nominate differs from *caesiogaster* only in measurements and in being paler below. **Voice** Thin, nasal contact calls, generally in short bursts of 2–6 notes, rendered "zeee", "spee" or "chay", faster and more intense in agitation. Song a repeated "spee-spuu-spee-speet". Also longer, complex, rambling jumbles of sharp "chip" notes, high-pitched whistles, mewing notes, trilled series of high sharp notes, and occasional crude mimicry of other birds, usually lasting > 10 seconds. **SS** No other gnatcatcher occurs in Bahamas; endemic Cuban Gnatcatcher has distinctive face pattern. **TN** Race *caesiogaster* sometimes synonymized with nominate (e.g. Buden 1987).

Cuban Gnatcatcher *Polioptila lembeyei* `LC`
Locally common resident (**Cuba**).
L 10–11 cm. Monotypic. Endemic. C & E mainland Cuba, plus several N cays, mainly in coastal, semi-arid or semi-xerophytic (cactus) scrub, below 100 m elevation. Commonest on S coast of Santiago de Cuba and Guantánamo provinces. **ID** One of the prettiest gnatcatchers. **Ad** ♂ has distinctive black crescent behind eye curving behind ear-coverts. Bluish-grey above, whitish on face and underparts; long black tail strongly graduated, with white outer feathers. **Ad** ♀ is similar but paler, and black crescent thinner. **Juv** is olive-grey above, with buffy flanks, creamy belly, and indistinct crescent. **Voice** Song a sustained, disorganized series of warbles, whistles and chattering notes, sometimes with mimicked calls of other species; louder, more varied and melodious than Blue-grey Gnatcatcher. Common call a buzzy "speeee", similar to Blue-grey; also a repeated "pip" or "pyip". **SS** Face pattern should immediately separate it from Blue-grey Gnatcatcher, and note latter's wider range in Cuba in winter.

TROGLODYTIDAE
Wrens
93 extant species, 2 in region + 1 vagrant

Zapata Wren *Ferminia cerverai* `EN`
Rare and very local resident (**Cuba**).
L 15.5–16 cm. Monotypic. Endemic. Only in Zapata Swamp, on S coast of Cuba, where restricted to sawgrass (*Cladium jamaicense*) and *Typha dominguensis* marshes, with scattered bushes (used as songposts). Probably fewer than 150 pairs. **ID** Distinctive. **Ad** has blackish-brown crown, inconspicuous eyestripe, dark brown upperparts with blackish bars across wing-coverts and flight feathers. Rectrices long, with diffuse fluffy ends, blackish brown with fine greyish-brown bars, Breast pale brown, flanks darker with prominent transverse blackish bars at rear. **Juv** is like ad, but has fine black speckles on throat and flanks, and less distinct barring. **Voice** ♂ song 4–7 clear gurgling whistles interspersed by harsher churrs, often lasting c. 1 minute or more; ♀ song, often in duet with ♂, simpler, resembling extended series of call notes, "achut-chut-chut-chut-churr". Calls include low harsh "chut chut", "churr-churr" etc. **SS** Distinctive bill shape and overall body structure; note tiny range. Extremely unlikely to be confused, but much darker and differently marked Marsh Wren (which see) has been claimed from Zapata Swamp.

Bahama Nuthatch

♀ / non-breeding ♂

♂ breeding

Blue-grey Gnatcatcher
caerulea

Cuban Gnatcatcher

Zapata Wren

House Wren *Troglodytes aedon* LC
Thirty-one subspecies divided into four subspecies groups, of which three occur in region.

Northern House Wren *Troglodytes (aedon) aedon*
Vagrant (**N Bahamas**, **Cuba**).
L 11.5–12.5 cm. Two subspecies in group, of which nominate occurs in region. Bahamas (mainly sight records): Bimini, Nov 1962 (Brudenell-Bruce 1975), Great Exuma, late Feb 1964 (note coincidence with sole Cuban record; Bond 1964, Buden 1992), one specimen plus one seen, South Bimini, Nov 1965 (Paulson 1966), New Providence, Nov 1967 (Bond 1968), Harbour I, Nov–Dec 1968 (Bond 1969), Grand Bahama, Nov 1997 (*FN* 52: 133), Feb 2001 (eBird), Abaco, Dec 2007, Apr 2009, Jan 2010, Nov 2012, Mar 2013, Oct 2014 (*NAB* 62: 320, 63: 521, 64: 339, 67: 175, 67: 534, 69: 170); Cuba: ad ♂ collected, La Habana province, Jan 1964 (Bond 1964). Claimed Jamaica, Aug 1988, but no documentation (Hood-Daniel 1988). In native range found in very wide range of habitats, including heavily disturbed or modified areas. Bold and curious. **ID** Familiar North American wren but vagrant to region. **Ad** has mid-brown upperparts, becoming more rufescent on tail, and paler, buffy-white or greyish-white underparts, a pale and rarely striking supercilium, plus finely black-barred wings, tail and rear flanks. **Juv** has dusky mottling on breast and less distinct barring on flanks. **Voice** Various low short calls, churrs and rattling notes, also high squeaky calls. Song (has been heard in region) a loud, confident bubbling, cascading series of complex phrases. **SS** Far more contrastingly plumaged Marsh Wren (unconfirmed in region) is tied to wet habitats and far more skulking. Larger Zapata Wren (highly restricted range) is also confined to wetlands and is structurally quite different.

Antillean House Wren *Troglodytes (aedon) martinicensis*
Fairly common to very local resident (**LA**).
L 11.5–12.5 cm. Six subspecies in group, all of which (*guadeloupensis*[A] Guadeloupe, *rufescens*[B] Dominica, *martinicensis*[C] Martinique, *mesoleucus*[D] St Lucia, *musicus*[E] St Vincent, *grenadensis*[F] Grenada) occur or did occur in region; those on Guadeloupe (last definitely seen 1973) and Martinique (not seen since 1886) are possibly or certainly extinct. Endemic. Of the extant races, those on Dominica and Grenada are still common and widespread in forest, but species is uncommon and now apparently confined to forest and scrub in lowlands on St Vincent, extremely rare and confined to NE coastal scrub on St Lucia, and only locally numerous but found in lowlands and mountains on Grenada. **ID** Unlike North American races, most Lesser Antillean birds apparently do not cock their tails. **Ad** (*rufescens*) is, vs. Northern House Wren, much more uniform over head, upperparts and underparts, being generally very warm chocolate- to buff-brown, with only indistinct supercilium, becoming brighter and paler rufous on rump, has closely dark-barred tail, narrowly barred wings and black-barred undertail-coverts. **Juv** has scalloped underparts. **GV** Race *mesoleucus* much paler and more rufescent above and white below; *musicus* larger than last-named with rust-coloured upperparts, obscurely dark-barred back and no ventral barring; *grenadensis* most similar to the two extinct races, but larger and paler than *guadeloupensis* and lacks any dark barring on back; *martinicensis* had dark brown upperparts narrowly barred black; and *guadeloupensis* was smaller and more rufescent above than previous race. **Voice** Bond (1965) considered songs of *musicus* and *mesoleucus* 'inferior' to the more beautiful warble of *rufescens*, and simultaneously reported that *grenadensis* is somewhat intermediate and therefore more like songs heard from races on Trinidad (*clarus*) and Tobago (*tobagensis*) of the Southern House Wren group. To our ear, *grenadensis* and *mesoleucus* sound very similar to Southern House Wrens in South America, whereas both *rufescens* and, especially, *musicus* are really rather different. **SS** Compare conspecific Southern House Wren, which has wandered to Barbados; no other wren occurs in this part of region. **TN** The six Lesser Antilles races possibly warrant species rank, or are sometimes treated together with the Southern House Wren group as a separate species. Molecular work should target relationships among the different taxa endemic to the Antilles. The House Wren complex is one of many examples of West Indies vertebrate taxa that urgently require additional taxonomic research that will in turn inform monitoring and conservation efforts.

Southern House Wren *Troglodytes (aedon) musculus*
Vagrant (**Barbados**).
L 11.5–12.5 cm. Twenty subspecies in group, of which one (perhaps *tobagensis*) occurs in region. Singing ad, photographed, Sept 1998–Feb 2000 (Buckley *et al.* 2009). **ID** A small wren which, unlike most taxa of Antillean House Wren, will readily accept human-modified habitats. **Ad** (*tobagensis*) has greyish-brown crown and nape, grey-white supercilium, and mid-brown upperparts with relatively broadly barred wings and tail; throat and breast, especially, are strikingly whitish, becoming buffier on vent. **Juv** shows some dark scalloping on flanks and rear underparts. **GV** Race *clarus* (Trinidad and continental South America), a potential vagrant, reportedly differs from *tobagensis* in its shorter and slighter bill and much less white underparts, but we have been unable to adequately check differences ourselves (and the literature is ambiguous on this issue). **Voice** Song is rich and varied, comprising rather fruity notes, e.g. "chewee chu chewee". Calls as Northern House Wren. **SS** The Barbados bird was distinguished from Antillean House Wrens by its lack of any capped effect, more obvious supercilium, greyish-brown upperparts and relatively stout bill (Buckley *et al.* 2009).

Sedge Wren *Cistothorus stellaris* LC
Vagrant (**Bahamas**).
L 10–12 cm. Monotypic. Photographed, Grand Bahama, Nov 2016 (Levesque *et al.* 2019). On migration and in winter uses weedy fields, agricultural areas, saltmarshes, fresh and brackish sedge meadows. **ID** Smallish wren. **Ad** has buff-white streaks on blackish crown and back, orange-buff shoulders, breast-band and flanks, and lower flanks with darker bars. Indistinct whitish-buff supercilium, wing-coverts buff-brown with some buff streaks, and whitish buff below. Bill dull blackish above, yellowish below. **Juv** resembles ad, but has reduced white streaking above and paler underparts. **Voice** Both sexes give variety of call notes in winter, including a rich "chip" that is often doubled. Song is presumably unlikely to be heard in region. **SS** Most similar to Eastern Marsh Wren, which could occur in similar habitats, but separated by streaked crown, smaller size, lack of bold white supercilium, and shorter bill. **TN** Often considered conspecific with extralimital Grass Wren *C. platensis* (NC Mexico discontinuously S to Cape Horn and Falkland Is); because *platensis* is the oldest available name, it is used for the composite species.

variant

Northern House Wren
(House Wren)
aedon

guadeloupensis

mesoleucus

rufescens

Antillean House Wren
(House Wren)

Southern House Wren
(House Wren)
tobagensis

Sedge Wren

Marsh Wren *Cistothorus palustris* [LC]

Sixteen subspecies divided into three subspecies groups, of which one occurs in region.

Eastern Marsh Wren *Cistothorus (palustris) palustris*
Hypothetical (**Cuba**).

L 11.5–12.5 cm. Eight subspecies in group, of which one (presumably nominate) has been claimed in region. One singing, Zapata Swamp, Aug 1993, lacks substantiative documentation (but was from an area where Zapata Wren does not occur). Vagrancy to West Indies possible, given that nominate race is migratory. In winter, in continental USA, occurs in both freshwater and brackish habitats. Forages low in marshy vegetation, right down to water level. **ID** Boldly patterned wren. **Ad** has black-brown crown and obvious pale grey supercilium. Central back dull blackish, contrastingly streaked whitish; rectrices medium brown, strongly barred darker brown, especially at sides. Underparts pale greyish, becoming buffy at sides, lower belly buff, becoming richer and warmer on flanks and vent. **Juv** is generally duller than ad, white streaks on back reduced or absent, barring on flight-feathers more diffuse, often has dull buff breast-band. **Voice** ♂ song a bubbling chatter (♀ does not sing). Most frequently heard call a sharp "tsuk", frequently repeated. **SS** Very dark cap, white-streaked mantle and warm, rusty, upperparts distinguish from other wrens in region. Note also behaviour.

STURNIDAE
Starlings
123 extant species, 1 introduced in region

Common Starling / European Starling [LC]
Sturnus vulgaris
Introduced, locally fairly common resident (**N Bahamas**, **Jamaica**), vagrant (rest of **GA**).

L 20–23 cm. Thirteen subspecies, of which one (presumably nominate) occurs in region. Well established in Jamaica (where initially introduced in 1903/04) and Bahamas (Abaco, first seen 1956), rare winter visitor to Grand Bahama (once a flock of 100) and Biminis, Cuba and Cayman Is in Oct–Apr (but only a vagrant on last two). Probably also only vagrant Puerto Rico (first seen 1973), Virgin Is (St Croix) and Barbados (flock of 30 in Jul 2006). Open habitats, including farms, coasts, city parks and other urban and rural areas, in lowlands. Forages in small to large groups. Flight direct, with wings distinctively swept back. Forms large communal roosts in some areas, e.g. Kingston (Jamaica). **ID** Short-tailed with mostly glossy black plumage and sharp, pointed bill. **Ad br** has metallic greenish and purple gloss, with usually a few indistinct paler spots above and below, and yellow bill. **Ad non-br** is heavily flecked buff to white, especially below, with blackish-brown bill. **Juv** is greyish brown, paler on underparts. Bill brownish. **Imm** during first complete moult shows features intermediate between juv and ad non-br. **Voice** A rich repertoire of whistles, squeaks, warbles, chirps and raspy notes. Often contains some mimicry of local species. **SS** Note distinctive shape of short tail and pointed bill, behaviour and direct flight. Grackles and cowbirds have longer tails and less pointed bills.

MIMIDAE
Mockingbirds and Thrashers
34 extant species, 9 in region + 1 vagrant

Grey Catbird *Dumetella carolinensis* [LC]
Common to rare passage migrant and winter visitor (**GA**, **O**), vagrant elsewhere.

L 21–24 cm. Monotypic. Thickets, dense undergrowth, scrub and second growth, to c. 1400 m, but commoner below 1000 m. Mostly mid Sept–May, being more or less common in Bahamas, Cayman Is, Cuba and Jamaica, but rare on Hispaniola and small islands in SW Caribbean, and only vagrant to Puerto Rico, Virgin Is and Guadeloupe. Moderately shy, more often heard than seen. Usually at low levels or on ground. **ID** Smaller, grey mimid that often cocks its long blackish tail upwards. **Ad** is almost entirely dark grey, with black cap and reddish-brown undertail-coverts. Wings short and rounded. Short bill is black. **Juv** is very similar to ad, but more brownish grey overall, with slightly mottled underparts and grey eyes. **Voice** Characteristic, downslurred, harsh cat-like mews. Also soft, low-pitched "quirt" or "turr". Alarm a loud, harsh chatter," chek-chek-chek" or "pert-pert-per". Song a rambling series of phrases, including mews, gurgles and squeaks, imitating other birds. **SS** Distinctive.

White-breasted Thrasher *Ramphocinclus brachyurus* [EN]
Two subspecies placed in separate subspecies groups, both occur in region.

White-breasted Thrasher
Ramphocinclus (brachyurus) brachyurus
Scarce and local resident (**Martinique**).

L 20–21 cm. One subspecies in group. Endemic. Dry woodland, valleys and ravines, and transition zone between coastal woodland and interior rainforest, but absent from latter. Confined to Caravelle Peninsula, in E of island, where population estimated at no more than 400 birds. Forages mainly on ground; often droops and flicks wings while tossing leaves aside. Breeds cooperatively. **ID** Distinctive bicoloured appearance, and rather long, slightly decurved black bill. **Ad** has dark sooty-brown upperparts and flanks, with darker lores to ear-coverts. Underparts white, with brownish undertail-coverts and mottled body-sides. Eyes usually dark reddish and legs blackish. **Juv** is brown below, becoming whiter with age. **Voice** Repeated short, mellow and varied phrases, each of several syllables. Call "chek-chek-chek". **SS** Bicoloured appearance unique on Martinique. **TN** The two subspecies groups are sometimes considered to be different species (DaCosta *et al.* 2019).

St Lucia Thrasher
Ramphocinclus (brachyurus) sanctaeluciae
Uncommon and local resident (**St Lucia**).

L 23–25 cm. One subspecies in group. Endemic. Habitat and habits similar to White-breasted Thrasher of Martinique, but population (once thought to be as small as 46 pairs) now known to be considerably larger (1100–2200 birds). Largely confined to drier forest in E of island, being apparently excluded from wetter and more humid woodland by presence of Brown Trembler (as on Martinique). **ID** Very similar to allopatric White-breasted Thrasher. **Ad** differs from latter by obviously larger size, darker upperparts, more immaculate white underparts, and size dimorphism (♂ larger than ♀). **Juv** is even more similar to same-age White-breasted Thrasher. **Voice** Song similar to White-breasted Thrasher. Call notes "tschhhhhhh". **SS** Very distinctive within range. **TN** See White-breasted Thrasher.

Eastern Marsh Wren
(Marsh Wren)
palustris

Common Starling / European Starling
vulgaris

♂ breeding
♀

Grey Catbird

White-breasted Thrasher

St Lucia Thrasher
(White-breasted Thrasher)

233

Scaly-breasted Thrasher *Allenia fusca* `LC`
Fairly common to rare resident (most of **LA**).

L 23 cm. Five subspecies, all of which (*hypenema*[A] from St Martin, Saba and St Barthélemy S to Guadeloupe, nominate[B] Dominica, Martinique, Grenada, *schwartzi*[C] St Lucia, *vincenti*[D] St Vincent, *atlantica*[E] possibly extinct Barbados) occur in region. Endemic. Moist to semi-arid habitats, from forest to semi-open woodland, mostly in lowlands. Also around settlements, but more dependent on wetter forest than Pearly-eyed Thrasher. Fairly common from Saba and St Barthélemy S to St Lucia, but rare and local on Grenada, vagrant on Grenadines, and very rare Barbuda and possibly extinct Barbados (where *atlantica* known from just two specimens and seven sight records, the last in 1994). Largely arboreal, forages in canopy on Dominica and Guadeloupe, but much lower on Montserrat and St Kitts. **ID** Large brownish mimid with pale yellow eyes and slightly decurved, rather short black bill, recalling a thrush. **Ad** has dark grey-brown upperparts with darker-centred feathers and a whitish wingbar. Underparts whitish heavily scaled greyish brown. Tail has large white tips. **Juv** has darker eyes. **GV** Race *hypenema* largest, paler than nominate; *schwartzi* intermediate in size with more white in tail; *vincenti* smallest, darker overall with narrow white margins to flank feathers; *atlantica* more reddish brown above, with less white in tail. **Voice** Series of varied notes and phrases, somewhat similar to Tropical Mockingbird, but softer, slower and less forceful. **SS** Compare larger, paler-billed Pearly-eyed Thrasher and paler, more uniform Tropical Mockingbird. **TN** Further study of racial variation required.

Pearly-eyed Thrasher *Margarops fuscatus* `LC`
Generally common resident (**Bahamas**, **Hispaniola**, **Puerto Rico**, most of **LA**).

L 28–30 cm. Four subspecies, three of which (nominate[A] Bahamas S from Abaco and New Providence, E Dominican Republic, Beata I, Puerto Rico and Mona I, Virgin Is E to Antigua, *densirostris*[B] Montserrat and Guadeloupe S to Martinique, *klinikowskii*[C] St Lucia) occur in region. Near-endemic; otherwise only on Bonaire (with single record on Curaçao), in Leeward Antilles. Dense scrub, thickets, forest and farmland at all elevations, from mangroves and palm groves to mountains; also urban and suburban areas. In process of colonizing Dominican Republic (first reported on mainland in 1984), with one report in N Haiti. On Dominica, commoner above 250 m in wetter forest; common on Guadeloupe. Vagrant Jamaica and Barbados. Largely arboreal, mainly at mid levels to canopy. Aggressive, competing with other birds for cavity nest-sites. **ID** Large mimid with conspicuous white eyes and large yellowish bill. **Ad** has upperparts brown with darker feather centres, giving faintly scaled appearance. Underparts dull white streaked brown, less marked on belly. Tail feathers have large white tips. **Juv** resembles ad. **GV** Race *densirostris* larger than nominate, darker above and more broadly streaked below; *klinikowskii* even larger, with breast markings more contrastingly dark brown and white, and more extensive white tips to tail. **Voice** Song a series of 1–3-syllable phrases, e.g. "pío-tareeu-tsee" with lengthy pauses between them. Often sings well into day and at night. Also raucous calls, including a harsh "chook-chook" and guttural "craw-craw". **SS** In Lesser Antilles, Scaly-breasted Thrasher is smaller and slightly darker, with shorter black bill, and single whitish wingbar. Compare slightly smaller mockingbirds (typically in more open habitats).

Brown Trembler *Cinclocerthia ruficauda* `LC`
Fairly common to rare resident (**LA**).

L 23–26 cm. Three subspecies, all of which (nominate[A] Dominica and Martinique, *tremula*[B] Saba, Barbuda, St Kitts, Nevis, Montserrat and Guadeloupe, *tenebrosa*[C] St Vincent and Grenada) occur in region. Endemic. Humid and wet forest, to c. 900 m. Also swamp forest, secondary and drier woodlands, occasionally plantations. On St Lucia reported only in dry forest and scrub, but presence (and whether mere visitor or resident) requires proof. Fairly common Saba, Guadeloupe and Dominica, uncommon St Kitts, Nevis, Montserrat and St Vincent, rare (perhaps erratic) Martinique (no recent records) and Grenada, and vagrant Antigua, Barbuda, St Eustatius and Virgin Is (St Thomas). Searches tangles, epiphytes, ferns, hollows and crotches on trunks and branches, also on ground. Often droops wings and cocks tail over back. **ID** Forest mimid with very long, slightly decurved black bill, conspicuous yellow eyes, and relatively short tail usually held cocked. **Ad** has rufous-brown upperparts with greyer head, and more reddish-brown rump, tail and wings. Throat pale greyish buff, greyish brown on breast, rest of underparts buffy brown. ♀ is slightly smaller than ♂, but has notably longer bill. **Juv** has faint dusky-grey spots on breast. **GV** Race *tremula* larger and longer-billed, with sexual dimorphism more marked (♂ obviously heavier but shorter-billed); *tenebrosa* intermediate in size, darker above, breast more strongly and extensively suffused greyish, and shows less sexual dimorphism in bill length. **Voice** Main song a series of loud and varied notes and phrases, harsh to melodious, at intervals of c. 2 seconds. Also harsh alarm call, "yeeak". **SS** Grey Trembler has greyer upperparts and whiter underparts. **TN** Formerly considered conspecific with Grey Trembler. Race *tremula* has occasionally been considered a separate species, and a recent genetic study recovered substantial phylogeographic structure within Brown Tremblers (DaCosta *et al.* 2019).

Grey Trembler *Cinclocerthia gutturalis* `LC`
Fairly common resident (**Martinique**, **St Lucia**).

L 23–26 cm. Two subspecies, both of which (nominate[A] Martinique, *macrorhyncha*[B] St Lucia) occur in region. Endemic. Mature moist and wet forest, usually at higher elevations; less common in second growth, drier scrub and open woodland in lowlands. Mainly arboreal, but feeds at all heights in vegetation, and also on ground. Often droops wings and cocks tail over back. **ID** Medium-large, greyish forest mimid with very long, slightly downcurved black bill, conspicuous pale yellow eyes and relatively short tail. **Ad** has warm grey-brown upperparts, with darker lores and ear-coverts, and duskier wings. Below mostly brownish grey, with flanks browner or more olivaceous. ♀ is slightly smaller, but has longer bill. **Juv** is browner, with mottled chest. **GV** Race *macrorhyncha* has paler, whitish to pale grey underparts, with vent and flanks pale cinnamon to brown, and sides tinged olive greyish. **Voice** Repeated notes and phrases, both harsh and melodic, and includes quavering whistles, absent from Brown Trembler's song. **SS** Brown Trembler has browner upperparts and darker, buffier or browner underparts; on St Lucia, it has been reported solely from xeric habitats, although confirmation of its presence there is required. **TN** See Brown Trembler.

fusca

vincenti

Scaly-breasted Thrasher

fuscatus

klinikowskii

Pearly-eyed Thrasher

tremula

ruficauda

tenebrosa

Brown Trembler

gutturalis

macrorhyncha

Grey Trembler

235

Northern Mockingbird *Mimus polyglottos*   LC
Common resident (**GA**, **O**, **Virgin Is**).

L 23–25 cm. Three subspecies, of which one (*orpheus* Bahamas S to Greater Antilles, Cayman Is and Virgin Is) occurs in region. Nominate also claimed to reach West Indies in winter, but we are unaware of confirmed records. Open and semi-open habitats, including semi-arid scrub, scattered bushes or trees, open mangroves, scrubby and disturbed vegetation, gardens, parks and other rural and urban areas, mostly in lowlands, rarely above 500 m. Conspicuous and opportunistic, forages mostly on ground, with tail often held cocked. Runs and hops on ground. **ID** Mid-sized mockingbird with white in wings and tail. **Ad** has whitish supercilium. Upperparts mostly dull brownish grey. Two narrow white wingbars and large white patch on primaries contrast with darker rest of wings, especially in flight. Underparts mostly greyish white. Long blackish-grey tail has white outer edges. Longish, slightly decurved bill blackish, eyes yellow and legs dusky. **Juv** has faint brownish breast spots, and paler legs. Eyes greyish. **GV** Nominate race differs in being darker and more saturated, above and below, vs. *orpheus*, and in being larger compared to some populations of latter. **Voice** Loud, conspicuous and persistent song is a sustained, disjointed delivery of harsh to mellow notes and short phrases, of great diversity. Often incorporates calls and songs of other species in its repertoire. Sings day and night. Alarm call a loud, explosive "chack". **SS** Bahama Mockingbird is larger without white in wings, overall browner above, and has streaked underparts. Grey Kingbird lacks white in wings and tail, and behaves quite differently. Compare allopatric Tropical Mockingbird. **TN** Birds from Haiti sometimes treated as race *dominicus*, and those in Bahamas as races *elegans* and *delinificus*.

Tropical Mockingbird *Mimus gilvus*   LC
Fairly common but somewhat local resident (**LA**, **San Andrés**).

L 23–25.5 cm. Ten subspecies, of which two (*antillarum*[A] Lesser Antilles, *magnirostris*[B] San Andrés) occur in region. Widespread in Lesser Antilles from Antigua S, but only vagrant on Barbados and St Kitts & Nevis. Dry scrub, agricultural areas and around habitation, mostly in lowlands. Conspicuous and opportunistic; often feeds on ground. Habits similar to Northern Mockingbird. **ID** Mid-sized mockingbird with long tail conspicuously tipped white; very similar to Northern Mockingbird. **Ad** (*antillarum*) differs in broad, blackish eyestripe and more conspicuous white supercilium. Wings lack prominent white on primaries and tail has no white edges, but has noticeable white tip. Eyes darker reddish brown (rather than yellow). **Juv** is browner than ad, with breast and flanks streaked dusky, and eyes dark brown. **GV** Race *magnirostris* larger and heavier-billed, with more contrasting dark wings and brighter yellow or orange iris. **Voice** Song a varied and long-continued sequence of diverse mellow to harsh notes and trills, with considerable repetition, similar to Northern Mockingbird but does not mimic other species. Alarm a loud, harsh "chek". Often sings at night. **SS** Grey Kingbird has much shorter tail and different behaviour. No known overlap with Northern or Bahama Mockingbirds. **TN** Race *magnirostris* sometimes treated as separate species.

Bahama Mockingbird *Mimus gundlachii*   LC
Fairly common (**Bahamas**) to rare and local resident (**Cuba**, **Jamaica**), vagrant (**Puerto Rico**).

L 28 cm. Two subspecies, of which both (nominate[A] Bahamas, cays off N Cuba, Turks & Caicos Is, *hillii*[B] S Jamaica) occur in region. Endemic (vagrant to Florida, USA). Coastal strand, semi-arid scrub, grassy areas with scattered trees, palmettos and plantations around habitation; lowlands. Confined to arid scrub-woodland on limestone in S Jamaica. Prefers denser, taller vegetation than Northern Mockingbird, and more arboreal than latter. Little overlap in Jamaica, where Northern is in more open areas. Declining on some cays off N Cuba due to development of touristic infrastructure and associated arrival of Northern Mockingbird. Where sympatric with Pearly-eyed Thrasher, occupies scrub and woodland away from settlements. One record Puerto Rico, Mar–Apr 2007 (*NAB* 61: 652, 654). **ID** Large brownish mockingbird with long tail. **Ad** has brown upperparts streaked dusky, with streaks more conspicuous on flanks. Dull blackish malar. Lacks prominent white patch on primaries and wingbars are duller than Northern Mockingbird. Broader, almost fan-shaped tail lacks white edges but has white tip (unlike Northern). Eyes yellow to orange. **Juv** is more densely marked below than ad. **GV** Race *hillii* more prominently streaked above, with larger white tips on tail (mainly inner webs). **Voice** Loud series of abrupt varied notes and phrases, with repetition, less varied and more musical than Northern Mockingbird, and does not mimic other species. **SS** Pearly-eyed Thrasher has darker upperparts, pale bill and white eyes. Slightly smaller Northern Mockingbird differs as mentioned above.

Brown Thrasher *Toxostoma rufum*   LC
Vagrant (**Bahamas**, **Cuba**).

L 29–30.5 cm. Two subspecies, of which at least one (presumably nominate) occurs in region. Dense shrubbery, well-vegetated residential and coastal areas. Records mainly Sept–Dec (once late Jun), mostly in Bahamas (especially Grand Bahama and Abaco, also Harbour I). Forages near-exclusively on ground. **ID** Large, distinctive mimid with long tail, long slightly curved blackish bill and long legs. **Ad** has mostly reddish-brown upperparts and tail, with two whitish wingbars. Throat white and moustachial dark brown. Buffy-white underparts boldly streaked dark brown. Conspicuous yellow-orange eyes. **Juv** has rufous upperparts mottled buffy, buff wingbars and grey eyes. **Voice** Varied calls include a typical, abrupt "kek" or "tchuck", also higher whistled or nasal and slurred "teea, teea" notes. Song unlikely to be heard in region. **SS** Long tail, rufous upperparts, streaky underparts and yellow eyes in combination are unique.

TURDIDAE
Thrushes
176 extant species, 18 in region + 1 vagrant

Eastern Bluebird *Sialia sialis*   LC
Vagrant (**Bahamas**, **Cuba**).

L 16.5–21 cm. Eight subspecies, of which nominate occurs in region. Records in Bahamas (Grand Bahama, Abaco, Harbour I) and W Cuba, late Oct–early Apr. Mentioned for Virgin Is, but details lacking. Fields with hedges and wooded edges. **ID** Unmistakable bright blue bird, smaller than many other thrushes. **Ad** ♂ has rich blue upperparts and head-sides, chestnut-orange throat to belly, white lower belly to vent. **Ad** ♀ is paler and duller, with brownish-grey upperparts and blue flight feathers, rump and tail. **Juv** is brown speckled buff above, always with trace of blue on wings and tail, greyish white mottled dark below. **Voice** Calls include a loud "tu-a-wee". **SS** Distinctive, although there is one record on Eleuthera (Bahamas) of a ♀ *Sialia* that was thought to be possibly a Mountain Bluebird *S. currucoides* (Connor & Loftin 1985).

Northern Mockingbird
orpheus

Tropical Mockingbird
antillarum

magnirostris

gundlachii

Brown Thrasher
rufum

hillii

Bahama Mockingbird

Eastern Bluebird
sialis

237

Cuban Solitaire *Myadestes elisabeth* NT
Common but local resident (**Cuba**). **L** 19–20.5 cm. Two subspecies, of which both (nominate[A] W & E Cuba, *retrusus*[B] I of Pines) occur in region. Endemic. Mainly in canopy of dense humid hill and montane semi-deciduous and pine forests, often near rocky outcrops (which it uses to project voice). In W virtually confined to Sierra de los Órganos, Sierra del Rosario and Sierra de la Güira, but is more widespread in E mainland. I of Pines population extirpated in 1930s. **ID** Nondescript, with long tail and short bill, resembling some Old World flycatchers. **Ad** has olive-brown upperparts and whitish underparts, whitish eye-ring, broad dark brown malar, white outer tail feathers, blackish maxilla and yellow mandible with black tip. **Juv** is flecked and streaked buff above, mottled brown on breast and flanks. **GV** Race *retrusus* was slightly greyer above, with pale buff supercilium, almost pure white below. **Voice** Song is high-pitched, flute-like, melodious and far-carrying, 'like rubbing a wet finger against the rim of a fine porcelain cup', similar to allopatric Rufous-throated Solitaire. Call a short whistle. **SS** Might recall La Sagra's Flycatcher, which has slight crest, lacks eye-ring and malar, and has cinnamon-brown tail.

Rufous-throated Solitaire *Myadestes genibarbis* LC
Fairly common resident (**Jamaica, Hispaniola, S LA**). **L** 19–20.5 cm. Six subspecies, of which all (*solitarius*[A] Jamaica, *montanus*[B] Hispaniola, *dominicanus*[C] Dominica, nominate[D] Martinique, *sanctaeluciae*[E] St Lucia, *sibilans*[F] St Vincent) occur in region. Endemic. Humid montane forest, pure broadleaf or mixed with pine, and scattered groves along streams, to 1800 m (Hispaniola), but generally less numerous below c. 400 m. Some evidence of post-breeding movements to lower elevations on some islands. Arboreal, often sallies from perch to catch airborne prey, but also gleans insects from leaves. **ID** Attractive thrush with slate-grey upperparts, chestnut throat and vent. **Ad** has white crescent below eye, whitish chin and dark malar, pale grey underparts, and white outertail feathers visible in flight. **Juv** has distinct orange-buff spots and streaks on upperparts, orange-buff and slaty scallops below, with plain orange-buff undertail-coverts. **GV** Race *solitarius* slightly larger than nominate, with longer tail, slightly brighter throat; *montanus* shorter-tailed and paler-throated, with ear-coverts barely marked; *dominicanus* has broader streaks on ear-coverts, darker, especially below, where also more extensively grey; *sanctaeluciae* has slightly paler rufous throat than last, with more white in tail and more orange on lower belly; *sibilans* has much darker upperparts and paler underparts, with thin black malar below ochraceous-rufous stripe, throat less clearly delimited from breast. **Voice** Song a 'hauntingly beautiful minor-key whistle', highly ventriloquial, comprising clear and semi-discordant flute-like whistles, sometimes "twuit, toi, tu-tu-tu-tu". Calls include single long "toot" like distant car horn. **SS** Unlikely to be confused, only solitaire in range.

Wood Thrush *Hylocichla mustelina* NT
Rare passage migrant and winter visitor (**Cuba, San Andrés**), vagrant (elsewhere). **L** 18–21.5 cm. Monotypic. Recorded mostly mid Sept–Nov and Mar–Apr, but also less frequently in Dec–Feb, and as late as May in N Bahamas. Vagrant to Cayman Is and throughout most of Greater Antilles, but apparently regular in Cuba and on San Andrés (Pacheco 2012). Deciduous woodland, large gardens and similar areas, foraging on ground. **ID** Slightly larger, bigger-headed and stockier than *Catharus* thrushes, none of which is so heavily spotted below or rusty above. **Ad** has rufous-brown crown, nape and neck-sides, becoming mid-brown on upperparts, with fairly bold pale eye-ring, and whitish underparts with bold blackish spots on breast to flanks. **1st-w** is similar to ad but has paler tips to greater coverts. **Voice** Call sharp short "pit-pit-pit" notes or low clucks. **SS** Ovenbird is smaller with cinnamon crown bordered black.

Swainson's Thrush *Catharus swainsoni* LC
Generally uncommon passage migrant (mainly **GA, O**). **L** 16–18 cm. Three subspecies, of which presumably all three (*incanus*, nominate, *appalachiensis*) occur in region (but when racial designation attempted, usually ascribed to nominate). Uncommon or locally common to rare migrant principally through W half of region, especially Bahamas, Cuba, San Andrés and apparently Virgin Is, mostly between mid Sept and late Nov and mid Mar to mid May. Recorded for first time in Puerto Rico in 2005 (Lewis 2007) but at least twice subsequently. Vagrant to a few of Lesser Antilles, e.g. Guadeloupe (Oct 2005). Open woodland and tree clumps with abundant leaf litter, also gardens. Forages on ground. **ID** Medium-sized thrush with conspicuous buff-yellow eye-ring, lores and throat. **Ad** has brownish-olive upperparts, and whitish belly spotted dark brown on breast and flanks. **1st-y** has pale-tipped greater coverts forming spotty wingbar. **GV** Race *incanus* greyer above than nominate, with more rufous tail, paler buff on head and duller-looking below, with darker and denser breast spotting; *appalachiensis* darker and more rufous than others. **Voice** Calls include mellow, liquid, upslurred sharp "whit". **SS** Veery, Grey-cheeked and Bicknell's Thrushes lack eye-ring, and present species tends to have buffier face and more prominent loral-stripe than any of these (or Hermit Thrush, which shows obvious contrast between tail/rump and rest of upperparts); W race of Veery can have upperparts very similar to present species, but has less heavily marked breast and clean white belly and flanks, also lacks eye-ring, and is relatively longer-tailed. Extralimital Russet-backed Thrush *C. ustulatus* (breeds W USA, winters Mexico S to Panama) has rufous-brown upperparts, narrower and less contrasting eye-ring, subtler spotting below and on average shorter tail, but would be exceptionally difficult to separate in the field. **TN** Almost universally considered conspecific with Russet-backed Thrush, from which differs only slightly in visible morphological characters (wing and tail length potentially useful for trapped birds) but more so in voice (especially song). *C. ustulatus sensu lato* has been rarely identified to race in West Indies, but given its more westerly migration route we suspect that Russet-backed will prove to be a vagrant, at most, to West Indies. **AN** Olive-backed Thrush.

Rufous-throated Solitaire

genibarbis

dominicanus

sibilans

Cuban Solitaire
elisabeth

Wood Thrush

swainsoni

incanus

Swainson's Thrush

239

Veery *Catharus fuscescens* **LC**
Rare passage migrant (**O**, **GA**).

L 17–18 cm. Four subspecies, all of which (*salicicola*, *subpallidus*, nominate, *fuliginosus*) could occur in region, and *salicicola* and nominate have been confirmed via specimens. Principally moves through W of region, N Bahamas, Cuba, Jamaica, Cayman Is and San Andrés, becoming steadily rarer further E, in S Bahamas and rest of Greater Antilles; mainly Aug–Oct and Apr–May, but several extraordinary winter records in Cuba and Jamaica (usually in S & E Amazonian Brazil and N Bolivia at this season). Open woodland with dense understorey, but also scrub and gardens; retiring, feeds on ground. **ID** Brightest and least-marked *Catharus*, with marginally longer tail than others. **Ad** has tawny-brown upperparts (brightest on tail), indistinct pale eye-ring visible only when close, off-white loral-stripe, very lightly brown-spotted creamy-buff throat, indistinct submoustachial, whitish underparts and slightly greyish flanks. **1st-w** has bright cinnamon-buff tips to greater coverts. **GV** Race *fuliginosus* slightly larger than nominate, with deeper brown upperparts, larger and sharper spots on breast; *salicicola* has more olive-brown upperparts, darker and more rounded breast spots; *subpallidus* drabbest, brown above with well-defined breast spots and marginally paler buff throat. **Voice** Calls include distinctive nasal, rough, downslurred "feuw" or "breeuh". **SS** More rufous above but less spotted on breast that any other *Catharus* (but bear in mind possibility of confusion with extralimital Russet-backed Thrush *C. usulatus*). See larger Wood Thrush.

Grey-cheeked Thrush *Catharus minimus* **LC**
Uncommon to rare and local passage migrant (**O**, **GA**), vagrant (**LA**).

L 17–18.5 cm. Two subspecies, of which both (*aliciae*, nominate) occur in region. Principally moves through W of region, with most records on Grand Bahama, mainland Cuba and San Andrés, Sept–Nov and Mar–May; very rare elsewhere, and only vagrant anywhere in Lesser Antilles. One Jun record in Jamaica. On ground in woodland with reasonably dense undergrowth. Requires careful separation from Bicknell's Thrush, and that the two were formerly treated as conspecific has undoubtedly inhibited our knowledge of their status in region. **ID** Largest and on average darkest-looking *Catharus*; does not regularly cock tail like slightly smaller Hermit Thrush. **Ad** has olive-tinged grey-brown upperparts, dull whitish underparts with slightly greyish face, well-defined blackish lateral throat-stripe, indistinct but variable pale eye-ring (can be quite obvious), dark brown spots on throat and breast, and brownish-olive wash to flanks **1st-w** has pale buff tips to greater coverts forming subtle wingbar, rather smudgier and less distinct spotting on breast and more obvious flesh base to mandible. **GV** Variation clinal and race *aliciae* differs only slightly from nominate. **Voice** Calls include variable, thin, weak, slightly burry or scratchy "pseer", "phreu" or "what?"; in contact (often at night on migration) a piercing "pweep" or "pe-i-i-ir" or "jee-er", similar to extralimital (but potential vagrant) Russet-backed Thrush *C. ustulatus* (see under Swainson's Thrush). **SS** Marginally smaller Bicknell's Thrush has warmer and plainer face (head-sides more streaked/spotted, buffier throat, very vague eye-ring), more chestnut-toned tail, greyer below, shorter primary projection and orange-yellowish (not pinkish) mandible. Swainson's Thrush (which see) has conspicuous eye-ring. Veery has finer spots on underparts and more rufous upperparts. Hermit Thrush has contrasting reddish-brown tail and note face pattern.

Bicknell's Thrush *Catharus bicknelli* **VU**
Rare and local winter visitor (**GA**), rare migrant (**Bahamas**).

L 16–17 cm. Monotypic. Breeds in NE USA and S Canada, and entire population winters in Hispaniola (mainly present Oct–Apr) and to lesser extent SE Cuba, Blue Mts of E Jamaica (no records since 1998) and Puerto Rico including Vieques I (McFarland *et al.* 2013, Gemmill 2015), in moist broadleaf forest or mixed with a few pines, rarely in pine-dominated forest, always with dense understorey, from sea level to 2250 m but mainly > 1000 m. Rare migrant, Oct–Nov and Apr–May, through Bahamas (Eleuthera) and W Cuba. Forages inconspicuously on ground. **ID** Small *Catharus*, very similar to Grey-cheeked Thrush (but any midwinter record in region more likely to be present species). **Ad** has dull olive-brown upperparts, chestnut-tinged tail, dull whitish underparts, with greyish-brown ear-coverts and lores, indistinct pale eye-ring, darkish brown spots on throat and breast, dark bill with extensive orange-yellow base to mandible. **1st-w** has pale buff tips to greater coverts forming subtle wingbar. **Voice** Calls include highly variable, harsh downslurred whistle, e.g. "beer", "peert" or "quee-a", or soft low "chook". Like all *Catharus* responds readily to playback of calls in winter, but does not regularly vocalize spontaneously. Infrequently gives subdued version of song in winter; very similar to Grey-cheeked, but higher-pitched, even more nasal and wiry, "chook-chook wee-o wee-o wee-o-tee-t-ter-ee". **SS** See Grey-cheeked Thrush. Among other differences, Veery has more finely spotted underparts, Swainson's Thrush has yellowish eye-ring and more buff-tinged face, and Eastern Hermit Thrush has reddish-brown tail, browner upperparts, more prominent eye-ring, paler legs and longer primary projection.

Hermit Thrush *Catharus guttatus* **LC**
Eight subspecies divided into three subspecies groups, of which one occurs in region.

Eastern Hermit Thrush *Catharus (guttatus) faxoni*
Rare winter visitor (**Bahamas**), vagrant (**Cuba**).

L 16–18 cm. Two subspecies in group, of which one (*faxoni*) occurs in region. Known virtually solely from N Bahamas (Andros, Eleuthera, Grand Bahama, New Providence), with most records involving trapped birds, late Oct–late Apr (e.g., Bond 1961, 1969, 1972, 1974, 1980); vagrant Cuba, on Cayo Coco, Dec 1995, Apr 2001 (Wallace *et al.* 1999, *Cotinga* 17: 85). Forest thickets, where forages on ground. **ID** On average the smallest *Catharus*, habitually cocks and gently lowers its relatively short tail, movements that can be shared by congenerics but rarely performed with such emphasis. **Ad** has rufous-tinged greyish-brown upperparts becoming warmer on rump and dull rufous-brown on tail (often visible from below), with rustier edges to wing-coverts, reasonably prominent dull white or pale buff eye-ring, distinct lateral-throat stripe and buff-mottled lores. Whitish below, washed buff on breast with rather bold dark brown spots (sometimes sparser and less prominent), greyer on flanks; bill dark, with pale base to mandible. **1st-w** has buff-tipped median and greater coverts. **Voice** Calls include a nasal "aaaank" (not unlike Grey Catbird) and a single or doubled low "chuck". **SS** Other *Catharus* generally lack contrastingly rufous tail, and combination of bold underparts spotting and reasonably obvious eye-ring also useful.

Veery
fuscescens

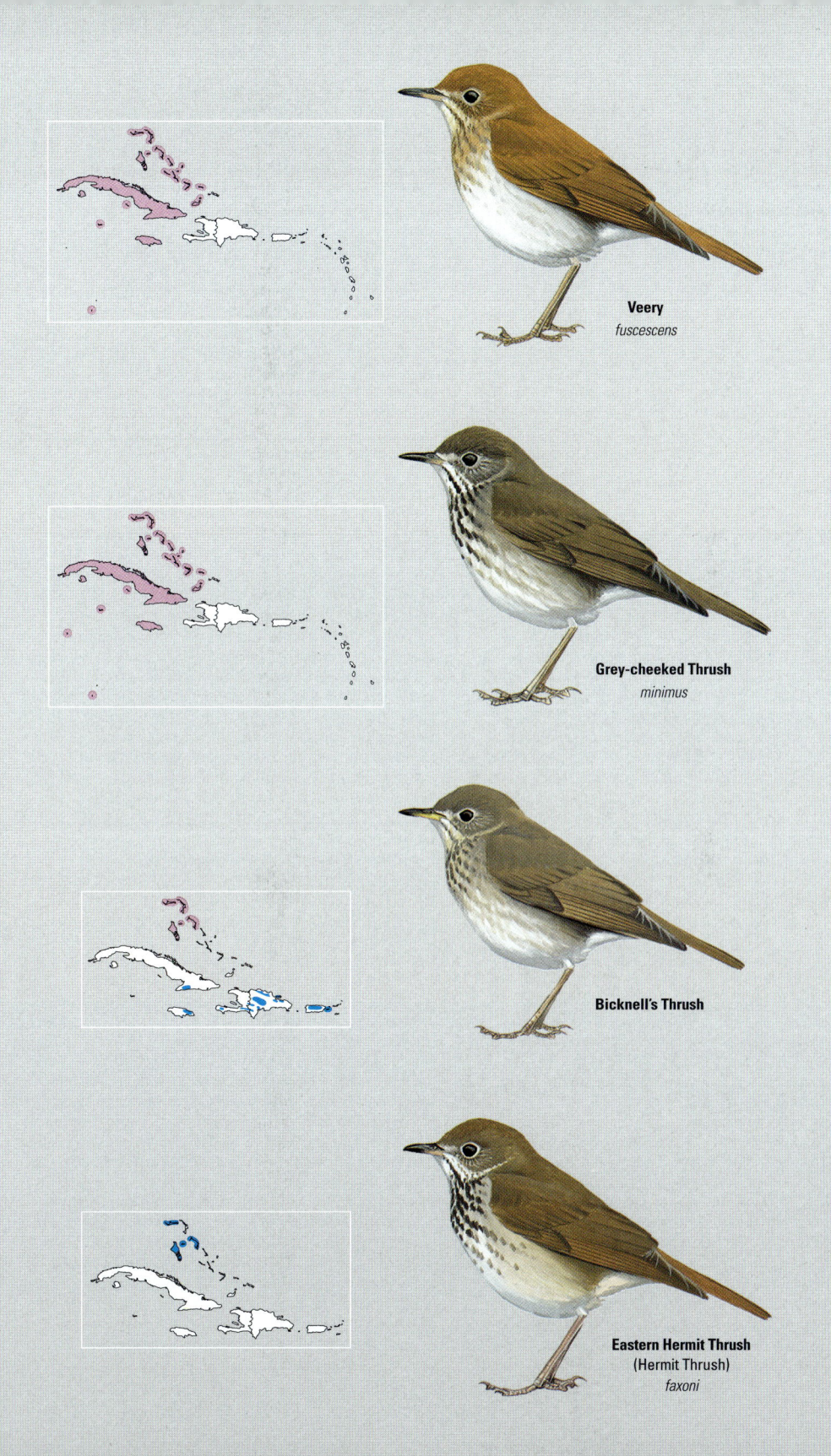

Grey-cheeked Thrush
minimus

Bicknell's Thrush

Eastern Hermit Thrush
(Hermit Thrush)
faxoni

241

Cocoa Thrush *Turdus fumigatus* `LC`
Five subspecies divided into two subspecies groups, of which one occurs in region.

Lesser Antillean Thrush *Turdus (fumigatus) personus*
Fairly common resident (**St Vincent**, **Grenada**).
L 21.5–24 cm. Two subspecies in group, of which both (*bondi*[A] St Vincent, *personus*[B] Grenada) occur in region. Endemic. Probably less common on Grenada than St Vincent. Humid forest, cacao plantations and cultivation with many trees, favouring higher elevations; forages on ground. **ID** Large, nondescript thrush. **Ad** (*bondi*) has rufous-brown upperparts and paler underparts, dark-streaked whitish-buff throat, whitish belly and vent. **Juv** has dark brown upperparts and buffier underparts. **GV** Race *personus* is more rufous above and buffish-brown below. **Voice** Series of short, loud, musical and usually slurred phrases with some notes repeated, gliding smoothly over narrow pitch range, e.g. "pree-er, churry, churry, o-ee-o, lulu, o-e-er, cheer-er, wu-e, wu-e, e-a-oeeo, te-a, te-a, e-o-to-e, cheer-o, o-ee, urr, wu-ee-er, toee-tu-tu, o-ee-o". Calls include "bak", warning "chat-shat-shat", harsh "kik-ik-ik-ik" alarm. **SS** Spectacled Thrush has conspicuous yellow eye-ring, yellowish bill and white belly.

Spectacled Thrush *Turdus nudigenis* `LC`
Common resident (**LA**).
L 23–24 cm. Two subspecies, of which nominate occurs in region. Fairly common to common Martinique, St Lucia (mainly dry lowland habitats), St Vincent and Grenadines, and Grenada (especially in lowlands on two last-named), and uncommon Guadeloupe and Dominica, where first reported as recently as 1997 and 1993, respectively (Evans & Jones 1997, Levesque & Jaffard 2002). Vagrant St Martin (*NAB* 60: 301) and Nevis (Biro & Ludlow 2015). Forest, dry and humid woodland, second growth, plantations and forest borders. Mainly arboreal, but occasionally on ground. **ID** Bulky thrush with broad yellow eye-ring. **Ad** has olive-brown upperparts, paler underparts with dark-streaked white throat, whitish belly and vent; yellow bill. **Juv** has upperparts like ad, but with narrow eye-ring, narrow buff streaks and orangey-spotted wingbars, dark brown-mottled underparts, whitish chin, and whitish belly to vent. **Voice** Series of short, quiet, monotonous but melodious phrases, more musical than many *Turdus* but more halting than Lesser Antillean Thrush, mixed with many high notes, e.g. "clee-er… weer-o…wureer, wureer…". Calls include distinctive querulous, nasal, cat-like "queeow" or sharply upslurred "cue-erree" or "miter-ee"; also "tak-tak-tak". **SS** See Lesser Antillean Thrush; Forest Thrush (generally of higher elevations and exclusively heavily forested areas) has scaled underparts. **AN** Bare-eyed Robin.

Forest Thrush *Turdus lherminieri* `VU`
Uncommon to rare resident (**LA**).
L 25–30 cm. Four subspecies, all of which (nominate[A] Guadeloupe, *dorotheae*[B] Montserrat, *dominicensis*[C] Dominica, *sanctaeluciae*[D] St Lucia) occur in region. Common on Guadeloupe, uncommon Montserrat and Dominica, probably very rare St Lucia (where formerly numerous but hunted heavily in 1800s). Endemic. All strata in humid primary and secondary forest and edges at middle and upper elevations; particularly numerous in swamp forest on Guadeloupe. On Montserrat favours mature mesic forests with dense canopy.
ID Large, brown, retiring thrush **Ad** has warm brown upperparts and head-sides, broad yellowish-red eye-ring, whitish underparts with buff-bordered dark brown scalloping, yellow iris, bill and legs; those on Basse-Terre are significantly larger than Grande-Terre. **Juv** is slightly paler, with vague thin pale streaks above, more mottled effect below. **GV** Race *dorotheae* has rufous-edged throat feathers, and longer, more pointed white centres to breast feathers; *dominicensis* smaller than nominate, darker upperparts and breast with much smaller 'scales', white belly, different bill pattern compared to nominate, and paler legs and eye-ring; *sanctaeluciae* also smaller than nominate, paler upperparts, spots on breast larger, buffish. **Voice** Song, given from concealed perch and mainly at night (especially with full moon), a musical cadence of clear notes, fairly loud and far-carrying. Differences exist in ♂ song and its timing on Dominica (race *dominicensis*). Calls include sharp "chuk" or "chuk-chuk". **SS** Compare Spectacled Thrush, which also occurs on Guadeloupe, Dominica and St Lucia. **TN** Arnoux (2012) identified three clades showing particularly well-defined morphological (and genetic) differentiation, on St Lucia, Dominica and Montserrat plus Guadeloupe.

White-eyed Thrush *Turdus jamaicensis* `LC`
Fairly common resident (**Jamaica**).
L 23–24 cm. Monotypic. Endemic. Montane wet forests and gullies, also shade-coffee plantations and other wooded areas, at all elevations; reportedly most numerous at mid-altitudes in Feb–Jul, but no definite evidence of elevational movements. Forages secretively in dense vegetation at all levels. **ID** Striking large-bodied thrush with piercing white eyes. **Ad** has slate-grey upperparts, pale grey-brown underparts, dark chestnut-brown head, white chin and throat streaked chestnut-brown, joining small white crescent on upper breast; white vent. **Juv** is like ad, but displays heavy breast streaking, and rufous streak-spots on wing-coverts, scapulars and mantle; irides grey. **Voice** Song a series of musical phrases (each one repeated 2–3 times), including whistled "hee-haw", resembling Northern Mockingbird but louder and less variable. Double-noted contact call; other calls harsh and shrill. **SS** White-chinned Thrush has darker throat, dark iris, orange bill and legs, and white wing patch.

La Selle Thrush *Turdus swalesi* `VU`
Uncommon and very local resident (**Hispaniola**).
L 26–27 cm. Two subspecies, both of which (nominate[A] S Haiti and SW Dominican Republic, *dodae*[B] C Dominican Republic) occur in region. Endemic. In Haiti uncommon and very local in Massif de la Selle, and in Dominican Republic (where discovered only in 1975) rare and local in Sierra de Bahoruco, Cordillera Central and Sierra de Neiba. Secretive inhabitant (most frequently seen at dawn and dusk) of dense shrubby understorey in montane broadleaf forest, occasionally pine forest with well-developed ground layer, at 1400–2100 m. Mostly forages on ground, but sings from treetops. **ID** Large-bodied thrush, superficially recalling more familiar American Robin, with conspicuous orange bill. **Ad** has blackish head and upperparts, white-streaked throat, slate-grey upper breast, deep rufous-chestnut mid-belly and flanks, blackish thighs, white lower belly, white-tipped blackish vent; yellow to reddish-orange eye-ring. **Juv** is undescribed. **GV** Race *dodae* has mantle, back and scapulars tinged grey-olive. **Voice** Song (mostly given early and late in day) a long-sustained, slow and deliberately spaced series of loud, low, fluty phrases, e.g. "tu-re-oo" and "cho-ho-ho", followed by a "zeek", but also described as 'bubbling, rollicking'. Calls include loud "wheury-wheury-wheury" in alarm, and various gurgling notes. **SS** Eastern Red-legged Thrush has paler upperparts without chestnut on belly, red (not black) legs and white in tail, among other differences; American Robin (an exceptional vagrant to Dominican Republic) is paler above with different underparts pattern.

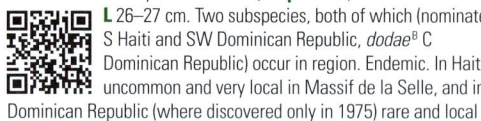

Lesser Antillean Thrush
(Cocoa Thrush)
personus

Spectacled Thrush
nudigenis

Forest Thrush
lherminieri

dominicensis

White-eyed Thrush

dodae

swalesi

La Selle Thrush

243

American Robin *Turdus migratorius* LC
Rare passage migrant (**Bahamas**, **Cuba**), vagrant elsewhere.

L 23–28 cm. Six subspecies, of which two (nominate, *achrusterus*) occur in region. Rare passage migrant mainly through N Bahamas and Cuba, Sept–early Dec and Mar–Apr, exceptionally Jul (*NAB* 60: 590), but only definite records of race *achrusterus* (from Cuba) were in late Dec and Jan, during unusually cold winters further N. Vagrant: Jamaica (*Gosse Bird Club Broadsh.* 20: 24), Dominican Republic (Nov 1985) and Puerto Rico (Mona I, Dec 1971; Aguadilla, Apr 2005). Open woodland, gardens and coastal scrub, foraging boldly on ground in open. Much less secretive than most resident thrushes, with exception of any of Red-legged Thrushes. **ID** Large, distinctive and familiar thrush (at least to North American observers). **Ad** has blackish head, broken white eye-ring, fleck above lores and black-streaked throat. Otherwise, dark grey above with white-tipped blackish tail. Orange-rufous underparts, white vent. Bill yellow. ♀ is usually duller, with white areas buffier. **1st-w** has pale-tipped greater coverts and buff-edged primary-coverts. **GV** Race *achrusterus* smaller, browner above, tawnier below, with reduced tail spots. **Voice** Calls include spirited "cluk-cluk", varying in pitch and volume with intensity of alarm, often preceded by one or more "sheek" notes. **SS** Similar-sized Northern Red-legged Thrush has red legs, blackish bill and grey breast; Western Red-legged Thrush has red bill and legs. On Hispaniola, compare La Selle Thrush.

White-chinned Thrush *Turdus aurantius* LC
Common resident (**Jamaica**).

L 24–26.5. Monotypic. Endemic. The larger and more common and widespread of the two thrushes endemic to Jamaica. Forest and woodland, citrus and banana plantations, adjacent pastures, well-treed cultivation and gardens, mainly at middle and high elevations, but regularly at sea level on N & SW coasts, especially. Mostly forages on ground, where regularly encountered at roadsides early and late in day. **ID** Conspicuous black-tipped orange bill and yellow-orange legs (brighter when breeding). **Ad** has glossy dark brownish-grey upperparts, shading to mid-grey below, palest on belly and flanks, with white wing patch (on inner greater coverts), tiny chin patch, mid-belly and tips to undertail-coverts. **Juv** is like ad but has browner crown to upper back, breast and flanks, white on chin absent or reduced, and duller bare parts. **Voice** Song, given late Feb–Aug, a musical lullaby-like "turé-too-too", repeated almost indefinitely. Calls include shrill whistling "p'liss, p'liss" and prolonged chicken-like clucking. **SS** See White-eyed Thrush.

Northern Red-legged Thrush *Turdus plumbeus* LC
Common and widespread resident (**N Bahamas**).

L 25–28 cm. Monotypic. Endemic (vagrant to Florida, USA). Resident Grand Bahama, Abaco, Andros, New Providence, Eleuthera and Cat I. Woodland at all elevations and of all types (semi-deciduous, pine), *Casuarina*, coppices, scrub, cactus brush, dense undergrowth, groves in cultivation, wooded gardens, favouring riparian habitats and vicinity of permanent water, especially in drier areas. Mostly forages on ground. Has wandered S to Great Inagua. **ID** Conspicuous orange-red eye-ring and legs. **Ad** has slate-grey upperparts, black throat, short white submoustachial and chin patch, black tail, mid-grey underparts; bill blackish. **Juv** has lightly mottled blackish upperparts, buffier with black spotting below, bill pale yellowish. **Voice** Song a melodious but laboured, monotonous series of phrases comprising 1–3 syllables, "chirruit, chirruit eeyu biyuyu pert, squeer squit, seeer cheweap, screeet chirri", recalling Pearly-eyed Thrasher. Calls include rapid high-pitched "weecha weecha weecha" or "cha-cha-cha", also weak sibilant "slee", and loud high "wiit-wiit" or "wet-wet" in alarm. **SS** No other thrush in range shares orange-red legs and eye-ring. **TN** Treated as conspecific with Western and Eastern Red-legged Thrushes by most authorities, under name Red-legged Thrush *T. plumbeus*.

Western Red-legged Thrush *Turdus rubripes* LC
Common and widespread resident (**Cuba**, **Cayman Is**).

L 25–28 cm. Three subspecies, all of which (nominate[A] W & C Cuba, I of Pines and many offshore cays, *schistaceus*[B] E Cuba, in Holguín, Santiago de Cuba and Guantánamo provinces, *coryi*[C] Cayman Brac) occur in region. Endemic. Various types of woodland at range of elevations (to c. 1975 m) including semi-deciduous forest, pines, scrub, wooded gardens and urban parks. Habits as Northern Red-legged Thrush. Formerly occurred on Swan Is, where ten specimens (considered indistinguishable from Cuban birds, but *coryi* undescribed at time) collected late 19th century, but not recorded since despite being searched for extensively in early 1900s (Ridgway 1888, Paynter 1956, Gallardo 2014). **ID** Very similar to allopatric Northern and Eastern Red-legged Thrushes. **Ad** has slate-grey upperparts, white chin and submoustachial, black throat, greyish breast, orange-buff belly, black tail; blackish to reddish-orange bill, orange-red eye-ring and legs. **Juv** has lightly mottled blackish upperparts, buffier with black-spotted underparts. **GV** Race *schistaceus* has reddish bill and greyish belly; *coryi* recalls nominate but smaller, with white malar. **Voice** Song a melodious but monotonous series of one- to three-syllable phrases, "chirruit, chirruit eeyu biyuyu pert, squeer squit, seeer cheweap, screeet chirri", with short pauses between phrases. Calls not known to differ from those of Northern Red-legged Thrush. **SS** Compare American Robin. **TN** See Northern Red-legged Thrush. Apparently extinct population on Swan Is originally described as race *eremita*, but subsequently synonymized.

Eastern Red-legged Thrush *Turdus ardosiaceus* LC
Common and widespread resident (**Hispaniola**, **Puerto Rico**, **Dominica**).

L 25–28 cm. Two subspecies, of which both (nominate[A] Hispaniola including Î de la Tortue, Î de la Gonâve and I Saona, and Puerto Rico, *albiventris*[B] Dominica) occur in region. Endemic. Various types of forests to c. 2440 m (on Hispaniola), from xeric and second-growth woodland to humid forest (can be abundant in moist broadleaf forest), pines, dry forest (rarely semi-desert scrub) and shade-coffee plantations. Has wandered to Vieques I, off SW Puerto Rico, Mar 1988, Sept 2000, Jan 2005 (Gemmill 2015), and perhaps also to St John (Virgin Is). Behaviour as Northern (and Western) Red-legged Thrushes. **ID** Similar to Northern and Western Red-legged Thrushes, but very distinctive in range. **Ad** has slate-grey upperparts, white chin and throat boldly striped black, black tail with white tips (obvious when flushed), greyish underparts, paler on belly. **Juv** lacks bold stripes on throat, bill pale yellowish. **GV** Nominate has coral-red bill, orange to yellowish legs; race *albiventris* similar but has shorter wing, longer legs, whiter belly, and yellow bill, eye-ring and legs. **Voice** Song a long, melodious, monotonous series of one- to three-syllable phrases, "chirruit, chirruit" or "pert-squeer" includes regular sharp "pit" note between phrases; mimics other birds and said to recall Pearly-eyed Thrasher, but more musical with shorter pauses between phrases. Calls include harsh "tsurrip, tsurrip tsurrip", low "weecha" and rapid, high-pitched "chu-week, chu-week, chu-week". **SS** Compare La Selle Thrush on Hispaniola. **TN** See Northern Red-legged Thrush.

American Robin
migratorius

White-chinned Thrush

Northern Red-legged Thrush

Western Red-legged Thrush
rubripes

albiventris

ardosiaceus

Eastern Red-legged Thrush

MUSCICAPIDAE
Old World Flycatchers and Chats
335 extant species, 1 vagrant in region

Northern Wheatear *Oenanthe oenanthe* LC
Vagrant (**Bahamas, Cuba, Puerto Rico, Guadeloupe, Barbados**). **L** 14.5–15.5 cm. Three subspecies, of which at least one (possibly *leucorhoa*) occurs in region. Bahamas: Eleuthera, Oct 1976 (Connor & Loftin 1985), Andros, Oct 1981 (Norton 1982); Cuba: Oct 1903 (Robinson 1905); Puerto Rico: two, Sept 1966 (Bond 1967), Sept 2011 and Sept 2016 (eBird); Guadeloupe: Oct 2012 (*NAB* 67: 175); Barbados: Dec 1955–Jan 1956, Oct–Dec 1994 (Buckley *et al.* 2009). Favours open areas, e.g. fields, littoral scrub and stony areas. Very active and perches rather erect. Feeds on ground, often flicks and fans tail. **ID** Slender, terrestrial bird, with much white on tail. **Ad non-br** has mostly pale greyish-brown upperparts with darker ear-coverts and wings, and narrow whitish supercilium. Prominent white rump and tail, latter with black central feathers and broad terminal band, conspicuous in flight. Underparts pale buff. **Ad br** ♂ (unlikely to occur) is pale grey from crown to back, with black wings. Black mask contrasts with supercilium. Tail as non-br. **Ad br** ♀ has browner back and crown, and less distinct face pattern. **1st-w** is like ad non-br, but some show characteristic moult limit in inner greater coverts (juv coverts slightly browner than ad-like ones) and extent of black on outer rectrices greater than in ad. **Voice** Calls include throaty "chack" or "tuc", and whistled "wheeet". **SS** Long tail and legs, extensive white on tail and rump, erect posture and terrestrial habits distinctive.

REGULIDAE
Kinglets and Firecrests
6 extant species, 1 in region

Ruby-crowned Kinglet *Regulus calendula* LC
Very rare winter visitor (**N Bahamas, Cuba**), vagrant (**Jamaica, Dominican Republic**). **L** 10–11 cm. Three subspecies, of which nominate occurs in region. Scrubby vegetation, woodland, second growth, mainly in lowlands but to c. 1450 m; recorded Oct–Apr. Hops and flits actively in search of insects, often flicking wings. **ID** Tiny short-tailed passerine with bold white eye-ring and thin blackish bill. **Ad** ♂ has olive-green to greyish upperparts with two white wingbars (upper can be barely visible, lower conspicuous) and obvious white eye-ring, interrupted at upper and lower edges, contrasting with beady black eye. Bright scarlet median crown-stripe rarely seen (except in aggression or excitement). Underparts dull greyish white. **Ad** ♀ lacks scarlet crown-stripe. **Juv** as ad ♀, but upperparts and wingbars tinged brownish and below more yellowish. **Voice** Short, dry "chet" notes, often strung together in prolonged chatter, but seldom heard in region. **SS** Prominent broken eye-ring and tiny size distinctive. Vireos are obviously larger, larger-headed and more heavily built, with stouter hooked bill. See also Tennessee and other migrant warblers.

DULIDAE
Palmchat
1 extant species, 1 in region

Palmchat *Dulus dominicus* LC
Very common resident (**Hispaniola**). **L** 18–20 cm. Monotypic. Endemic. One of commonest birds on Hispaniola (including Gonâve I and Saona I), being numerous everywhere up to 1500 m (more rarely to c. 1825 m). Primarily in royal palm (*Roystonea hispaniolana*) savannas, but has adapted extremely well to man-made habitats such as farms, parks, plazas and city gardens. Builds large, communal nests. Unconfirmed record in Jamaica. **ID** Medium-sized, dull-plumaged, thrush-sized passerine with short deep bill strongly curved on culmen. **Ad** is olive-brown above and pale buffy below, boldly but variably streaked brown, entire throat sometimes dusky; iris red; bill dull horn-coloured to dull yellowish. **Juv** differs by having throat and foreneck almost entirely dark brown with only faintly paler edges, and rump buffish (not greenish). **Voice** Varied short, harsh notes, sometimes likened to those of Common (European) Starling. Distinctive alarm a musical whistle dropping in pitch, often given in chorus by group-members. No true song. **SS** Unique: nothing is truly similar, but compare white-eyed and longer-tailed Pearly-eyed Thrasher, which is restricted to E Dominican Republic. **TN** Birds from Gonâve I (off W Haiti) described as race *oviedo*, but appear inseparable from those elsewhere.

BOMBYCILLIDAE
Waxwings
3 extant species, 1 in region

Cedar Waxwing *Bombycilla cedrorum* LC
Fairly common to rare and irregular winter visitor (**GA, O**), vagrant (**Puerto Rico, LA**). **L** 15.5–18 cm. Two subspecies, of which nominate occurs in region. Forest edge, open woodland, farmland, gardens, parks and urban areas to mid altitudes. Irregular visitor, most frequent in Cuba, where fairly common in some years; rarer elsewhere. Recorded mostly Oct–Apr, exceptionally until late May. Typically in flocks (of up to 50) that visit trees or shrubs with ripe fruit. **ID** Medium-sized, sleek passerine with conspicuous backward-pointed crest. **Ad** has black mask sharply outlined by thin white line. Back cinnamon-brown and wings grey-brown, with small red wax-like tips to secondaries (hard to see). Rump and short tail grey, latter with black subterminal band and conspicuous bright yellow (or orange) tip. Chin and throat black. Rest of underparts tan, fading to whitish undertail-coverts. ♀ has black below restricted to chin. **Juv** is duller and greyer, with shorter crest, paler throat and diffusely streaked underparts, but similar face pattern. **Voice** Clear, short, thin trill "sreee". Also an unmusical "che-che-check". **SS** Ad unmistakable. Streaky imm less distinctive, but note crest, black mask and lores, and yellow-tipped tail.

Northern Wheatear
leucorhoa

Ruby-crowned Kinglet
calendula

Palmchat

Cedar Waxwing
cedrorum

not all figures to scale

247

PLOCEIDAE
Weavers
124 extant species, 3 introduced in region

Yellow-crowned Bishop *Euplectes afer* — LC
Introduced, uncommon and local resident (**Jamaica**, **Puerto Rico**), vagrant (**Guadeloupe**, **Barbados**).

L 10–11 cm. Three subspecies, of which nominate occurs in region. Native to sub-Saharan Africa. Initially reported in Puerto Rico in 1971 and Jamaica in 1988, both due to deliberate release or escapee cagebirds; considered established but local on both islands, although species has apparently not been seen in Puerto Rico since 2017 hurricane. Nesting reported on Martinique in 1980s (but no further details), and records on Guadeloupe since 1999 and Barbados since 1998; latter said to represent wandering birds from elsewhere in West Indies. Specimen from W Cuba, collected Oct 1959, originally identified as Northern Red Bishop. Areas with tall grass and reeds near freshwater bodies. Often in flocks. Breeding ♂ raises and puffs plumage during courtship. **ID** Strikingly bicoloured plumage (ad br ♂) and very short tail. **Ad br ♂** has stunning black-and-yellow plumage, wing feathers boldly fringed whitish; bill black. **Ad ♀/non-br ♂** have upperparts mostly brown mottled darker, underparts mainly buff; yellowish supercilium contrasts with dark brown eyestripe and ear-coverts. Breast and crown finely streaked. Bill dark brown with paler mandible. **Juv** as ad non-br, but has broad buffy-edged feathers. **Voice** Song, given from perch or in courtship, gargling notes followed by "zzeeet zzzeeet zzzeeet" trills, ending with chipping "rik-rik-rik". Similar to Northern Red Bishop. Repetitive rattles in flight. Alarm a harsh "chuk". **SS** Ad br ♂ unique. Much commoner (in Puerto Rico) ♀ and non-br ♂ Northern Red Bishop have paler brown eyestripe and cheeks, and minimal yellow in supercilium. Grasshopper Sparrow has whitish central crown-stripe. Compare Bobolink and various Cardinalidae with plumages other than br ♂.

Northern Red Bishop *Euplectes franciscanus* — LC
Introduced, local but fairly common to very rare resident (**Jamaica**, **Puerto Rico**, **Guadeloupe**, **Martinique**).

L 11 cm. Monotypic. Native to sub-Saharan Africa. Sugarcane and rice fields with grassy borders, parks and even urban areas. Forages mostly on ground. Often flocks of up to 50. Breeding ♂ raises and puffs plumage in courtship. Fairly common but currently very local in Puerto Rico (present since 1960s) and Jamaica (released 1988), far less so on Guadeloupe and Martinique (first seen 1982). Isolated records on Vieques I, Virgin Is (St Croix) and Barbados. Could hybridize with Yellow-crowned Bishop. **ID** Short-tailed bishop with highly contrasting (ad br ♂) plumage. **Ad br ♂** has very distinctive black-and-red plumage, browner on back and wings, latter with bold whitish feather edges. Tail usually concealed by red uppertail-coverts. Bill black. **Ad ♀/non-br ♂** are very similar to respective plumages of Yellow-crowned Bishop, but eyestripe and cheeks paler brown, supercilium usually buffier and shows much less obvious and more restricted streaking below. **Juv** as ad non-br but has broad buff feather edges. **Voice** Song of thin squeaky notes followed by guttural "zee-zee-zee" buzz and sizzling sounds. Rattle call in flight. Contact a high-pitched "tsip"; a harsh "chak" in alarm. **SS** Grasshopper Sparrow has yellow spot near bill and central whitish crown-stripe. Compare other, less common wintering sparrows. **AN** Orange Bishop.

Village Weaver *Ploceus cucullatus* — LC
Five subspecies divided into four subspecies groups, of which one occurs in region.

Village Weaver *Ploceus (cucullatus) cucullatus*
Introduced, common (**Hispaniola**) to locally common resident (**Martinique**).

L 17 cm. Two subspecies in group, of which nominate occurs in region. Native to sub-Saharan Africa. Dry forest, second growth, scrub, agriculture, rice fields, coastal vegetation, villages and gardens, to 600 m on Hispaniola, where widespread and common (but apparently declining), and reportedly established as early as 1780s (Lever 2005). Also breeds on Martinique, where first seen pre-1963, and recently (since 1998) reported on Guadeloupe. In flocks, sometimes large. Noisy colonies of spherical woven nests; ♂ performs spread-wing, flapping display. Feeds on seeds and grain. **ID** Chunky weaver with conical bill. **Ad ♂** is orange-yellow with black hood to breast, and chestnut-brown nape. Wings dark brown with feathers boldly yellow-fringed. Eyes red; bill blackish. **Ad ♀** is yellowish green, brightest on face and breast, with yellow supercilium. Mantle and wings darker, underparts paler. **Juv** is similar to ad ♀ but eyes brownish, and bill paler and duller. **Voice** Steady high-pitched chatter with musical whistles. Also a harsh alarm and unpleasant scratchy notes. **SS** Note red eyes in ad. On Hispaniola, ad ♂ Baltimore Oriole (rare migrant) also black and orange-yellow, with slimmer longer bill and longer tail. ♀ Antillean Siskin smaller, with paler and smaller bill, and no supercilium.

ESTRILDIDAE
Waxbills
141 extant species, 11 introduced in region

Orange-cheeked Waxbill *Estrilda melpoda* — LC
Introduced, fairly common (**Puerto Rico**) or locally common to uncommon resident (**Guadeloupe**, **Martinique**).

L 10 cm. Monotypic. Native to sub-Saharan Africa. Tall seeding grass, sugarcane, farms, thickets and urban areas in Puerto Rico, where perhaps two separate introductions (initially in first half of 1800s, then 1950s), but first reported in 1870s (Lever 2005). Less numerous on Guadeloupe and Martinique, although well established on first-named at least. Forages in pairs/small groups and, outside nesting season, larger flocks. Takes seeds while perched but mostly on ground, often with other waxbills. **ID** Handsome with conspicuous orange face. **Ad ♂** has deep orange to red bill. Crown and nape grey, back and wings brown, uppertail-coverts red, tail blackish brown. Throat whitish, rest of underparts mostly pale grey, except whitish lower belly/undertail-coverts, with small pale orange lower belly patch. **Ad ♀** is very similar but belly patch yellowish. **Juv** as ad ♀ but rump reddish brown, face paler and bill black. **Voice** Calls weak and lispy or squeaky, nasal and strident, "zee", a testy "chi-dee-chi". Song a variable medley, e.g. "de-de-sweea, sweea, sweea". **SS** Black-rumped Waxbill has red eyestripe rather than orange face, and black rather than red rump. On Martinique see Common Waxbill.

♂ breeding / ♀ / non-breeding ♂

Yellow-crowned Bishop
afer

♂ breeding / ♀ / non-breeding ♂

Northern Red Bishop

♂ / ♀

Village Weaver
cucullatus

not to scale

Orange-cheeked Waxbill

249

Black-rumped Waxbill *Estrilda troglodytes* LC
Introduced, locally common to uncommon resident (**Guadeloupe**, **Martinique**).

L 10 cm. Monotypic. Native to sub-Saharan Africa. Abandoned cultivation, tall grass near water or sugarcane fields, and edges of swamps. Feeds on seeds while perched or on ground, in pairs and small groups, sometimes large flocks, often with other waxbills. Locally common on Guadeloupe (where present since late 1960s), especially in E; uncommon and local on Martinique. Formerly present in coastal lowlands of Puerto Rico (probably since 1960s, first reported 1971), but no longer established. Also reported St Thomas (Virgin Is) but probably never established. **ID** Small waxbill with conspicuous red eyestripe. **Ad** ♂ has lores and eyestripe red, forehead to upperparts pale grey-brown, finely barred dark grey on back. Rump/tail black, latter with white edges. Cheeks whitish. Below whitish tinged pink, very faintly and finely barred (hard to see). Vent white. Bill pinkish red. **Ad** ♀ lacks pink tones below. **Juv** is similar to ad ♀, but eyestripe blackish and fainter. Upperparts paler or duller and unbarred, and bill black. **Voice** Loud, explosive "tche-tcheeer!" like bullet ricocheting off rock. Also harsh "chuur", "chew-tch-tch", nasal downslurred "jeeeu" and drawn-out, rising "chihooee". **SS** Ad Orange-cheeked Waxbill has orange cheeks and reddish rump in all plumages. On Martinique, very similar Common Waxbill has paler rump/tail and black vent. Dull juv similar to other ♀ or juv waxbills, finches or grassquits, but note neater plumage and faint blackish eyestripe contrasting with paler cheeks.

Common Waxbill *Estrilda astrild* LC
Introduced, locally common to rare resident (**Martinique**).

L 9.5–13 cm. Fifteen subspecies, of which at least one (unknown) occurs in region. Native to sub-Saharan Africa. Most introductions elsewhere in world refer either to nominate or *jagoensis*. Tall grass, marshes, abandoned cultivation and grassy areas in towns, in lowlands of Martinique, where reported since at least 2007. Occasionally mentioned for Puerto Rico (firstly by Blake 1975) but no evidence for this. Takes grass seeds while perched, but also on ground. Gregarious, foraging in groups of up to 50. **ID** Comparatively small-bodied but long-tailed waxbill, very similar to Black-rumped. **Ad** ♂ differs in brown (rather than black) rump and tail, black (not white) undertail-coverts, and bright pink central belly. **Ad** ♀ has less pink in plumage, smaller belly patch and fuscous undertail-coverts. **Juv** has bill black, underparts buffier with pinkish cast to belly, and red eyestripe paler and narrower (juv Black-rumped has blackish eyestripe and no pink below). **Voice** Call a sharp "jip"; song 2–3 sharp notes followed by rising bubbling sound, "ti-cket please!" or "di-di-di-JEEE". **SS** See also Orange-cheeked Waxbill. Juv finches and grassquits lack any pink. Compare Red Avadavat.

Red Avadavat *Amandava amandava* LC
Three subspecies divided into two subspecies groups, of which one occurs in region.

Red Avadavat *Amandava (amandava) amandava*
Introduced, common to uncommon or rare resident (**Hispaniola**, **Puerto Rico**, **Guadeloupe**, **Martinique**).

L 9.5–10 cm. Two subspecies in group, of which nominate occurs in region. Native to S & SE Asia. Grassy edges of freshwater marshes, sugarcane fields and weedy drainage canals. Typically in flocks, usually on ground under tall grass, thus harder to see than other small seed-eaters. Uncommon on Guadeloupe (first reported 1965), rare on Martinique (since 1967) and in Puerto Rico (since late 1960s? but no longer established), and recorded just twice in Dominican Republic (small flock in 1997 and single in 2008). **ID** Handsome waxbill with spotted wings. **Ad br** ♂ is mostly deep red with rows of small white spots on wings, rump, flanks and sides. Tail, lower belly and vent black. Lores black and bill red. **Ad non-br** ♂ and ♀ mostly greyish brown above, paler below with whitish throat, red rump, white spots on wings and rump, and dark eyestripe. **Juv** is similar to ad non-br but lacks red in plumage, has buff-coloured wing spots, unmarked grey face, grey to buffy-white underparts and pinkish-black bill. **Voice** Musical "sweet" and "sweet-eet" calls. Song a melodious high-pitched descending whistled twitter, beginning with a contact call and ending in a quiet trill. **SS** Ad br ♂ unmistakable; other plumages distinguished by small spots on wings. See Checkered and Chestnut Munias.

Bronze Mannikin *Spermestes cucullata* LC
Introduced, common resident (**Puerto Rico**), vagrant (**Virgin Is**).

L 9 cm. Two subspecies, of which one (probably nominate) occurs in region. Native to sub-Saharan Africa. Grassy fields, urban parks and rice fields, mostly in coastal lowlands below 300 m. Long-standing introduction in Puerto Rico (including Vieques I), where already considered abundant in 1860s. Single records (one involving small flock) on St Croix (Virgin Is) and Culebra. Situation on Martinique unclear, but does not appear to be established. Forages in pairs and small groups, larger outside nesting season. Feeds on grass seeds both while perched and on ground. **ID** Very small mannikin with bicoloured bill and dark hood. **Ad** has brownish-black hood with browner hindneck and neck-sides. White barring on rump. Tail blackish. Rest of underparts white, with prominent dark brown barring on sides, flanks and vent, meeting that on rump. Bill black with pale bluish-grey mandible. **Juv** is mostly uniform brown above, with paler underparts, sometimes showing faint hood. Bill all black. **Voice** Varied vocabulary includes a coarse "crrit", wheezy "tsek" or "chik-chik" and "chi, chi chi chi chu chuu". Song a quiet "chi, chu, chi, chu, cheeri-hit, chu" lasting 1–2 seconds. **SS** Note tiny size. Ad Tricoloured Munia has chestnut back and tail, and black belly, while Checkered Munia has browner upperparts and checkered flanks and lower breast. Juv of these species warmer cinnamon or chestnut above than juv of present species. No overlap with Chestnut Munia or Yellow-bellied Seedeater. Female grosbeaks and buntings obviously larger.

Indian Silverbill *Euodice malabarica* LC
Introduced, locally common resident (**Puerto Rico**), vagrant (**Virgin Is**).

L 11 cm. Monotypic. Native to S Asia. Arid scrub, cultivation, grassy coastal areas and gardens, mostly in lowlands. In Puerto Rico (probably introduced 1960s), especially common around San Juan and on S & SW coasts. In Virgin Is, recorded only on St Croix. Feeds mainly on ground, usually in small to mid-sized flocks. **ID** Rather drab waxbill with long, pointed tail. **Ad** has mostly dull brown upperparts, faintly scaled crown and greyish-white face. Faintly barred, light buff flanks. White lower back, rump, undertail-coverts and central uppertail-coverts contrast with black outer tail-coverts and blackish tail. Bill dark grey to blackish, with mandible pale blue-grey. ♀ has shorter tail. **Juv** as ad ♀, but rump and undertail-coverts mottled brown, even shorter and rounder tail, unbarred flanks and grey bill. **Voice** Calls include a quick, double-noted "chit-tit" or "tsiptsip", or single "tseep", sometimes repeated. Song a series that rises then falls, rapidly repeated in a trill. **SS** All plumages distinguished by prominent pale rump/lower back and undertail-coverts, and dark tail. Additionally, ad has long pointed tail.

Black-rumped Waxbill

♂ variant

Common Waxbill
astrild

Red Avadavat
amandava

Bronze Mannikin
cucullata

Indian Silverbill

251

Scaly-breasted Munia *Lonchura punctulata* `LC`
Eleven subspecies divided into two subspecies groups, of which one occurs in region.

Checkered Munia *Lonchura (punctulata) punctulata*
Introduced, locally common to locally uncommon resident (**GA**, **LA**). **L** 10–12.5 cm. One subspecies in group. Native to S & SE Asia. Grassy borders of sugarcane fields and other agriculture, parks in urban areas and sites with seeding grass, in lowlands. Typically in flocks, foraging on ground or on seed heads, and can be a pest in rice-growing areas. Locally common in Cuba (since 1990s), Jamaica (1989), Hispaniola (1978), Puerto Rico (first definitely recorded 1971) and Vieques I, Virgin Is (St Croix and Anegada, perhaps only vagrants), St Kitts, Antigua, Guadeloupe (1984), Dominica and Martinique (reportedly 1995). Spreading rapidly in some areas, with records N to Eleuthera (in Bahamas). **ID** Distinctive mannikin with boldly checkered underparts. **Ad** has deep cinnamon hood and upperparts, darker foreface, chin and throat. Rump and tail paler yellowish cinnamon or orangey. Below white, profusely scaled or scalloped brownish black (central belly often all white). Bill blackish with paler mandible. **Juv** has brownish-buff upperparts and buff to whitish underparts; bill black. **Voice** A soft, plaintive whistle "peeet" that drops in pitch and fades at end. **SS** Ad unmistakable. Dull juv very similar to other juv mannikins and munias. Juv Chestnut and Tricoloured Munias have slightly paler bill, more cinnamon upperparts and slightly darker, buffier underparts. Bronze Mannikin marginally smaller and duller, browner, with no cinnamon or buffy cast on plumage. Compare larger ♀ grosbeaks and buntings. **TN** We have found no evidence that race *topela* (part of the Scaly-breasted Munia group) has been introduced in region, despite some suggestions to contrary. **AN** Nutmeg Mannikin.

Tricoloured Munia *Lonchura malacca* `LC`
Introduced, locally common to uncommon resident (**GA**). **L** 11–12 cm. Monotypic. Native of Indian Subcontinent. Usually in tall grass, sugarcane fields, marshes, rice plantations and agricultural fields that have seeded, mostly in lowlands. Typically in flocks, sometimes large (and a potential pest) and with Checkered or other munias, but Tricoloured appears dominant. Forages on ground and seed heads. Locally common Cuba (first reported 1990) and Hispaniola (1982), uncommon Puerto Rico (probably late 1960s) and very local Jamaica (1988) and Martinique (1980s, but not established), with several 1990s records on Grand Cayman and recently (2017/18) on Grand Bahama and Great Inagua, Bahamas, and Providenciales, Caicos Is (eBird). Some populations may reflect natural dispersal within Antilles, rather than separate introductions. **ID** Well-patterned mannikin with black hood and cinnamon upperparts in ad. **Ad** has glossy black hood to upper back and breast. Rest of upperparts and tail cinnamon-chestnut. Breast and flanks white, sharply demarcated from black upper breast, central belly patch and undertail-coverts. Bill pale greyish. **Juv** has cinnamon-brown upperparts and buffy underparts. Bill dull greyish black. **Voice** Call a thin, nasal honk, less plaintive, clearer and more melodious than Checkered Munia. Very thin song virtually inaudible to human ear. **SS** Ad unmistakable. Dull juv very similar to other juv mannikins, but has warmer upperparts, buffier underparts, slightly paler bill, and usually seen with more distinctive ad. Compare larger ♀ grosbeaks and buntings, and allopatric Yellow-bellied Seedeater. **AN** Black-headed Munia.

Chestnut Munia *Lonchura atricapilla* `LC`
Introduced, locally uncommon resident (**NE Cuba**, **Jamaica**, **Martinique**). **L** 11–12 cm. Eight subspecies, of which at least one (probably nominate) occurs in region. Native to S & SE Asia. First recorded Jamaica in 1988, Martinique sometime before 2005 and at a single site in Cuba since at least 2013 (Navarro Pacheco 2018), but perhaps present on other islands too (most observers pay little attention to these birds). Tall grass bordering dense vegetation, marshes and coasts, occasionally grassy areas in parks and urban areas, in lowlands. Takes seeds while perched, but also on ground. Forages in small groups, sometimes larger flocks. **ID** Elegant, conspicuously patterned munia with pale greyish-blue bill. **Ad** has glossy black hood, sharply demarcated from cinnamon-chestnut body. Tail and rump more cinnamon-orange; central belly black. **Ad** ♀ has hood slightly less glossy and belly patch smaller. **Juv** has warm brown upperparts and buff underparts; bill blackish or dark grey. **Voice** Calls a thin, nasal honk, also "peet" or "pink! pink!". Song virtually inaudible to human ear. **SS** Ad unmistakable. Juv almost identical to Tricoloured Munia (which see). Compare other juv mannikins and munias, and larger ♀ grosbeaks and buntings. **TN** Sometimes considered conspecific with Tricoloured Munia. **AN** Chestnut Mannikin.

White-headed Munia *Lonchura maja* `LC`
Introduced, rare resident (**Martinique**). **L** 11–11.5 cm. Monotypic. Native to S & SE Asia. Introduced Martinique sometime prior to 2007, where flocks of up to 40 seen. Also once La Désirade, Guadeloupe, Oct 2011 (eBird). Grasslands, sugarcane fields, marshes and grassy coastal areas, mostly in lowlands. Behaviour very similar to other munias. **ID** Mid-sized munia with pale head, and pale blue-grey bill. **Ad** has white head often with pinkish to brownish cast, especially on nape, grading into chestnut-brown upperparts, coppery or cinnamon-brown rump and tail. Throat pale buffy, belly grey and undertail-coverts black. Black eye conspicuous in white head. **Juv** is overall brown, slightly paler and buffier below, with sandy-white chin and throat. Bill blackish grey. Could hybridize with Chestnut Munia. **Voice** Song a series of clicks and drawn-out "weeeee heeheeeheeeheeeheee". Call a soft "preet" or "prit"; that of ♂ higher-pitched and longer than ♀. **SS** Ad is only munia with white head. Juv from other juv munias by whitish throat.

Java Sparrow *Lonchura oryzivora* `EN`
Introduced, locally fairly common resident (**Puerto Rico**). **L** 15–16 cm. Monotypic. Native of Indonesia. Introduced Puerto Rico in late 1950s/early 1960s, and well established around San Juan and environs. Also reportedly introduced to Jamaica in early 20th century, but died out c. 1946. Urban areas with short grass, e.g. sports fields, large lawns and grassy parks. Behaviour similar to other munias; recorded in flocks of up to 100. **ID** Large-bodied, with very big pinkish-red bill. **Ad** has black head with obvious white cheeks. Pale bluish-grey upperparts and breast, pinkish-grey belly, white undertail-coverts, and black rump and tail. Narrow reddish eye-ring and pink legs. ♀ has slightly smaller bill, and narrower and duller red eye-ring. **Juv** has greyish-brown upperparts, buff cheeks, white throat, indistinctly streaked buff breast, whitish belly and vent, and buff eye-ring. Bill usually pinkish brown. **Voice** Calls include a hard, metallic chink and low churring "tup, t-luk" or "ch-luk". Song begins with bill-clicks, a complex jingle and rattle ending in a whistle. **SS** Ad unmistakable. Juv distinguished by relatively massive (usually pinkish-brown) bill, buff cheeks and pink legs.

Checkered Munia
(Scaly-breasted Munia)

Tricoloured Munia

Chestnut Munia
atricapilla

White-headed Munia

Java Sparrow

253

VIDUIDAE
Whydahs
20 extant species, 1 introduced in region

Pin-tailed Whydah *Vidua macroura* — LC
Introduced, uncommon and local resident (**Puerto Rico**).
L 11–12 cm (breeding ♂ 30–32 cm). Monotypic. Native to sub-Saharan Africa. First reported Puerto Rico in 1971 (probably introduced the previous decade) but numbers have crashed since 2017 hurricane; reached Vieques I in 1988. Probably parasitizes Orange-cheeked Waxbill. Two recent (May and Jul 2013) records in Dominican Republic (Latta & Rodríguez 2018). Fields with short grass, cultivation and gardens, mostly on coasts, but also into mountains. Alone or with conspecifics, or in mixed-species flocks, foraging on ground. ♂ sings and displays on favoured perch, and is aggressive towards other birds. **ID** Small finch with stubby bill. **Ad br** ♂ is piebald with very long, narrow tail plumes (c. 20 cm); bill red. **Ad non-br** ♂ has mostly reddish-brown upperparts streaked blackish, and bold facial stripes. Tail mostly brown, underparts largely buffy white, streaked brownish on flanks; bill still red. **Ad br** ♀ as non-br but face stripes dark brown and tawny-buff, and bill black. **Ad non-br** ♀ is similar but bill red. **Juv** has crown, upperparts and tail uniform mouse-brown, with paler cheeks and obscure pale superciliary. Throat whitish and rest of underparts pale buff. Bill black. **Voice** Song a jerky series of harsh single notes with sibilant quality and sparrow-like chirps. Also an emphatic "sweet". **SS** Ad br ♂ unmistakable. Other plumages more sparrow-like, but note small size and red bill in all except br ♀ (but as species often occurs in small groups latter identifiable by association). Juv similar to other small finches or sparrows, and best identified by accompanying ad in flocks.

PASSERIDAE
Old World Sparrows
43 extant species, 1 vagrant in region

House Sparrow *Passer domesticus* — LC
Twelve subspecies divided into two subspecies groups, of which one occurs in region.

House Sparrow *Passer (domesticus) domesticus*
Introduced, fairly common (**Cuba**, **Puerto Rico**) to local and uncommon resident (rest of **GA**, **LA**, **O**), vagrant (**Barbados**).
L 16–18 cm. Six subspecies in group, of which at least one (possibly nominate) occurs in region. Native to Old World. Strictly commensal, from isolated farms to urban centres, e.g. parks, squares, docks and airports. Now widespread and still expanding, but remains rare or absent on some smaller islands. Initially introduced Cuba in 1850s, then New Providence (Bahamas) in 1875, Jamaica in 1903/04 and Virgin Is in 1953, from where perhaps spread naturally to Puerto Rico (first seen 1972) and Hispaniola (1976). Guadeloupe reached only in 1999. Has benefitted from deliberate and indirect human agency (e.g. via grain ships) but also natural colonizing abilities (e.g., to St Croix, Antigua and St Vincent), with records on Barbados perhaps emanating from South America. Primarily forages on ground. Gregarious, forming flocks year-round. **ID** Familiar Old World sparrow. **Ad br** ♂ has grey forehead and crown, large black bib, chestnut band from nape to eyes, and contrasting whitish cheeks. Back and wings chestnut streaked black, with prominent white wingbar. Lower back and rump grey. Underparts pale grey. **Ad non-br** ♂ is similar but slightly duller, with less extensive bib and chestnut partly obscured. **Ad** ♀ is mainly dull brown, with paler superciliary. Upperparts lack chestnut tones. No black bib. **Juv** is like ad ♀, but young ♂ shows hint of darker bib. **Voice** Commonest vocalization a disyllabic "chirrup" by ♂, sometimes strung together in series as a rudimentary song. **SS** Adult ♂ head pattern distinctive. Grasshopper Sparrow (never in same areas) has pale central crown-stripe and buffy underparts. Ad Rufous-collared Sparrow (Hispaniola) has grey and brown stripes on slightly crested crown, and narrow black neckband. ♀ and juv could recall other less common, vagrant sparrows, but habitat and behaviour should function as important clues.

MOTACILLIDAE
Pipits and Wagtails
67 extant species, 1 in region + 1 vagrant

Buff-bellied (American) Pipit *Anthus rubescens* — LC
Four subspecies divided into two subspecies groups, of which one occurs in region.

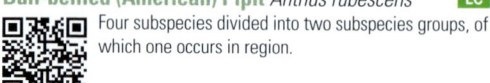

American Pipit *Anthus (rubescens) rubescens*
Very rare winter visitor (**Bahamas**), vagrant (**Cuba**, **Jamaica**, **Haiti**, **San Andrés**).
L 14–17 cm. Three subspecies in group, of which at least one (probably nominate) occurs in region. Open areas near rivers and lakes, wet fields, pastures and other level ground, in lowlands. Primarily Oct–Mar. Forages on ground, regularly bobs tail, but also wades in shallow pools. **ID** Slender, thin-billed and long-tailed terrestrial bird, which regularly bobs tail while walking. **Ad non-br** has brownish-grey upperparts faintly streaked blackish except on hindneck and rump. Buffish supercilium and malar, and two faint wingbars. Below usually uniform rich buff in autumn becoming pale grey in winter, moderately streaked blackish, especially on breast. Tail blackish brown with white outer feathers conspicuous in flight. Legs dark brown to blackish. **Ad br** (unlikely to be seen in region) has greyer upperparts and less streaked underparts. **1st-w** as ad non-br. **Voice** In flight, call a high, distinctive "sip-it" or "sip" or "tseep", sometimes rapidly repeated: "si-si-si-si-sif". **SS** Note white outer tail feathers in flight; also slender bill and body, longish legs and tail, and habit of bobbing tail when walking. Only other pipit claimed in region is Sprague's (which see).

Sprague's Pipit *Anthus spragueii* — VU
Hypothetical (**Bahamas**).
L 15–17 cm. Monotypic. Three reports: Grand Bahama, photo, Oct 1966 (Bond 1967, Kale *et al.* 1969); sight, Andros, Apr 1971 (Bond 1972), Eleuthera, Sept 1971 (Connor & Loftin 1985). We treat as hypothetical because Bond (1982: 2) subsequently considered the photo-documented record to be 'unidentifiable'. Most likely to occur in open weedy fields, preferring taller grasses than American Pipit. Secretive and hard to see, forages on ground, rarely in open. Does not pump tail while walking. **ID** Very similar to American Pipit. **Ad** differs from latter by pinkish or yellowish (not blackish) legs, warmer-coloured upperparts (including rump) more heavily streaked blackish, with scaly appearance to back, less richly coloured underparts with unstreaked flanks, more prominent dark eye (looks 'isolated' within head), more extensive white on outer rectrices and different behaviour. **1st-w** is basically identical to ad (though usually retains some unmoulted juv outer greater coverts/tertials). **Voice** Calls an explosive squeaky "sweep" or "speep", often doubled, or, when flushed, "speep-beep-beep". **SS** Compare American Pipit.

♂ breeding ♀

Pin-tailed Whydah

♂ ♀

House Sparrow
domesticus

non-breeding

breeding

American Pipit
(Buff-bellied Pipit)
rubescens

Sprague's Pipit

255

White Wagtail *Motacilla alba* `LC`
Eleven subspecies divided into nine subspecies groups, of which one occurs in region.

White-faced Wagtail *Motacilla (alba) alba*
Vagrant (**Barbados**).
L 16.5–18 cm. Three subspecies in group, of which probably nominate occurs in region. Ad non-br/1st-w, Jan 1987 (*AB* 41: 335, Ingels *et al.* 2010). Conspicuous and confident, in open areas, often around habitation. Feeds on ground, walks with jerky head movements, and pumps tail up and down near-constantly. **ID** Distinctive black, grey and white terrestrial bird with long black-and-white tail. **Ad non-br** has grey crown and upperparts, white forehead and two white wingbars. White chin to upper breast contrast with black breast-band. Rest of underparts white, washed greyish on breast-sides and flanks. Bill and legs black. **Ad br** is even more distinctive but unlikely to occur in region. **1st-w** has brownish-grey head, with dusky moustachial. Greyish-white underparts, sometimes tinged buffy, with narrow dark grey-brown gorget. **Voice** Call in flight, a high-pitched "chissik" or monosyllabic "zit" or "psit". **SS** Unmistakable in region. **TN** Other subspecies groups, namely Siberian Wagtail *M. a. ocularis* (breeds N & NE Siberia and W Alaska) and Black-backed Wagtail *M. a. lugens* (breeds SE Russia to N Korea and Japan) have wandered to E USA and potentially could occur in region.

FRINGILLIDAE
Finches
211 extant species, 6 in region + 3 vagrants + 1 introduced

Jamaican Euphonia *Euphonia jamaica* `LC`
Common resident (**Jamaica**).
L 11 cm. Monotypic. Endemic. Woodland, borders, open or shrubby areas with trees, gardens and orchards, at all elevations but commoner in hilly lowlands. Forages at upper levels. Active and can be acrobatic. May forage in small flocks, or gather with other birds at food. **ID** Small, rather dull euphonia with very stubby blackish bill (paler mandible). **Ad** ♂ is greyish blue overall, with darker blue wings and tail, and slightly paler underparts. Centre of lower breast and belly, and vent yellow. Narrow area at wingbend (usually hidden) also yellow. **Ad** ♀ has head and underparts bluish grey; back, wings, rump, flanks and tail dull olive. Throat whitish, paler than head and breast. **Imm** is similar to ad ♀, but slightly duller. **Voice** A staccato churring, somewhat similar to motor starting. Also a pleasant, squeaky whistle and other chuckling, mewing, harsh, chattery or squeaky notes. **SS** Only euphonia in range. Orangequit has longer, slightly decurved bill.

Hispaniolan Euphonia *Euphonia musica* `LC`
Locally common resident (**Hispaniola**).
L 10–12 cm. Monotypic. Endemic. Dense humid and wet forest, particularly with mistletoe, from lowlands to at least 2300 m. Less common in pine forest and rare in dry forest or arid scrub below 600 m. Also edges, clearings and shade-coffee plantations, and on Gonâve I at sea level. More local in Haiti. Usually in pairs or small groups, foraging high in canopy, where easily overlooked despite colourful appearance. Occasionally joins mixed-species flocks. **ID** Small, compact euphonia, with distinctive sky-blue crown and nape. **Ad** ♂ has blackish-violet upperparts, chin, throat and head-sides, with orangey-yellow forehead, rump and underparts. Stubby black bill, with base of mandible blue-grey. **Ad** ♀ is duller overall, with mostly greenish upperparts and yellowish-green underparts, but has blue crown and nape, and yellowish forehead and rump. **Juv** as ad ♀ but slightly duller. **Voice** Calls include a rapid, subdued, almost tinkling "ti-tit", hard, metallic "chi-chink" and plaintive "wheee" like some *Myiarchus* flycatchers but more melodious. Jumbled, tinkling song mixed with explosive notes. **SS** Unmistakable in range. Other euphonias are allopatric. **TN** Usually treated as conspecific with Puerto Rican and Lesser Antillean Euphonias, under name Antillean Euphonia *E. musica*.

Puerto Rican Euphonia *Euphonia sclateri* `LC`
Locally common resident (**Puerto Rico**).
L 10–12 cm. Monotypic. Endemic. Also introduced to Vieques I, but no longer present there (Gemmill 2015). Humid montane forest, shade-coffee plantations and lowland dry scrub-forest. Apparently performs seasonal altitudinal migrations. Feeds mainly on mistletoes. Other habits similar to Hispaniolan Euphonia. **ID** Similar to allopatric Hispaniolan Euphonia. All plumages differ mainly by slightly smaller average size and shorter tail. **Ad** ♂ further differs by bright yellow (vs. orange-yellow) underparts and rump, and bright yellow (vs. violet-black) throat. **Ad** ♀ is duller overall than ♂, with mostly greenish upperparts and yellowish-green underparts, except blue crown and nape, and yellowish forehead and rump. **Juv** presumably as ad ♀ but slightly duller. **Voice** Virtually identical to Hispaniolan Euphonia. **SS** Unmistakable in range. **TN** See Hispaniolan Euphonia.

Lesser Antillean Euphonia *Euphonia flavifrons* `LC`
Uncommon resident (**LA**).
L 10–12 cm. Monotypic. Endemic. Occurs on Anguilla, Saint Barthélemy, Barbuda, Antigua, St Kitts & Nevis, Montserrat, Guadeloupe (where dramatic decline evident), Dominica, Martinique, St Lucia, St Vincent and Grenadines, and Grenada; vagrants reported on several additional islands. Most frequent in canopy of moist montane forest. Less often in woodland at lower and middle elevations. Feeds mostly on mistletoes. Habits presumably like other euphonias. **ID** Small, rather dull, female-plumaged euphonia with blue hood. **Ad** ♂ has mainly olive upperparts with sky-blue crown and nape, some blue on mantle. Very narrow blackish band separates tiny yellow forehead from blue crown. Small yellow rump. Olive-yellow below, slightly yellower on throat and belly. **Ad** ♀/**juv** are apparently duller, with paler blue hood. **Voice** Apparently identical to Hispaniolan and Puerto Rican Euphonias. **SS** Blue hood and chunky shape distinctive. **TN** See Hispaniolan Euphonia.

White-faced Wagtail
(White Wagtail)
alba

breeding

non-breeding

Jamaican Euphonia

Hispaniolan Euphonia

Puerto Rican Euphonia

Lesser Antillean Euphonia

House Finch *Haemorhous mexicanus* LC
Vagrant (**Cuba**).

L 12.5–15 cm. Thirteen subspecies, of which one (presumably *frontalis*) occurs in region. ♂ photographed and sound-recorded, La Habana, Dec 2018–Feb 2019, at least (eBird). With records on Florida Keys, vagrancy (either natural or ship-assisted) is clearly possible, but a local escapee cannot be eliminated (especially given location). **ID** Relatively small finch with fairly long, notched tail, short, rounded wings and rounded or domed head with short bill and curved culmen. **Ad** ♂ has wine-red forehead and supercilium, crown and upperparts pale brown, spotted darker, chin to mid-breast light russet-red, below buff or pinkish-buff, streaked brown on flanks. **Yellow morph** has red areas orange. **Ad** ♀ is pale grey-brown above, finely streaked darker on face, mantle and scapulars, rump unstreaked grey-brown; wing-coverts have pale whitish-buff tips; buffish white below, extensively streaked darker. **Juv** is like ♀ (not always separable); young ♂ acquires breeding plumage as 2nd-w. **Voice** Song, by ♂ in flight or from perch, a slow, disjointed jumble of hoarse and musical notes, with downslurred conclusion, "whee-er" or "che-er"; ♀ may give simpler version. Call "cheep" or similar, often given in flight or in series, and "fillp" or "fiidllp" when perched; also sharp "chirp" recalling House Sparrow. **SS** Should be unmistakable in West Indian context, but bear in mind possibility of (other?) escaped cagebirds.

Redpoll *Acanthis flammea* LC
Five subspecies divided into three subspecies groups, of which one occurs in region.

Common Redpoll *Acanthis (flammea) flammea*
Hypothetical (**Bahamas**).

L 12.5–14 cm. Two subspecies in group, of which one (presumably nominate) may occur in region. Claimed once: Eleuthera, Jan 1986 (Bond 1986, *AB* 40: 339). Jamaica records (Raffaele *et al.* 1998) erroneous (Levy in prep.). Most likely to appear in lightly wooded areas. **ID** Small streaky finch with pale wingbars, short conical yellowish bill and slightly notched tail. **Ad non-br** has small deep red cap and black lower forehead, lores and chin. Narrow pale supercilium and thin blackish eyestripe. Above grey-brown heavily streaked black. Rump paler streaked dusky. Two white wingbars. Below whitish with well-streaked flanks. **Ad br** ♂ (unlikely in region) has breast and rump tinged pink. **Ad br** ♀ is like ad non-br. **Juv** as ad non-br, but lacks any red or pink in plumage. **Voice** Characteristic flight call a rapid, reeling series of "chi" notes, sometimes a "tsooee". **SS** Red cap of ad distinctive. Juv may recall a number of streaky-plumaged sparrows or siskins (which see). **TN** North American authorities currently recognize each of the three subspecies groups at species level.

Hispaniolan Crossbill *Loxia megaplaga* EN
Uncommon and local resident (**Hispaniola**), vagrant (**Jamaica**).

L 15–16 cm. Monotypic. Endemic. Montane pine forest, usually at 1500–2600 m, but to 540 m in non-breeding season. Often quiet and secretive while feeding on pine cones. May form flocks and wander in search of food. Declining. At least three reported Jamaica, Dec 1970–Apr 1971 (Bond 1972). **ID** Two broad white wingbars on blackish wings, and uniquely crossed bill tips. **Ad** ♂ is dusky brown with variable pale red wash, most intense on head, upper back and upper breast. **Ad** ♀ has finely streaked breast, often with yellow wash to foreparts and yellowish rump. **Juv** is like ♀ but browner and more heavily streaked, especially breast. **Voice** Noisy in flocks. High-pitched, emphatic, repeated "chu-chu-chu-chu" recalling keys of an electric typewriter being struck. During breeding season also a soft, whistling warble. **SS** Nothing similar in range and habitat.

Pine Siskin *Spinus pinus* LC
Three subspecies divided into two subspecies groups, of which one occurs in region.

Pine Siskin *Spinus (pinus) pinus*
Vagrant (**Bahamas**).

L 11–13 cm. Three subspecies, of which one (presumably nominate) occurs in region. Three, Grand Bahama, Jan 2011 (*NAB* 65: 359). Irruptive; most likely to appear in coniferous woodland, but also possible in deciduous or even semi-open areas. **ID** Small to medium-sized, slender-billed, short-tailed and drab-coloured finch. **Ad** ♂ is heavily streaked above and below, with poorly defined supercilium, thin dark eyestripe, yellow or pale yellow bases to outertail feathers, and wing-coverts tipped pale buff or warm buff-brown to yellowish. Small square yellow patch at base of outer primaries (often concealed). In flight broad yellowish wingbar across bases of flight feathers. Some are greener above, paler and less streaked below, with more yellow in wings and tail. **Ad** ♀ usually has paler or whiter tips to wing-coverts and smaller area of pale yellow at base of outer primaries. **Juv** resembles ad, but usually more narrowly streaked on slightly paler buffish-brown upperparts, tips of larger wing-coverts whitish, and underparts pale buffish and more thinly streaked. **Voice** Calls include a husky or buzzing "chee-ee", loud "clee-ip" or "chlee-it", more nasal and rising "sweeeet" and shorter chatter or twitter, "tit" or "ji-ji-ji" or "bid bid". **SS** Yellow in flight feathers, heavily streaked breast, and lack of red on crown and black on throat separate from Common Redpoll (vocalizations also useful). Compare much more boldly patterned American Goldfinch and Antillean Siskin.

American Goldfinch *Spinus tristis* LC
Vagrant (**Bahamas**, **Cuba**).

L 11.5–13 cm. Four subspecies, of which nominate occurs in region. Reported mid Oct–Jan, also Apr, on Bimini, Grand Bahama, Abaco and Eleuthera, as well as twice in Cuba. Weedy fields, thickets, second growth and orchards in lowlands. Flight undulating. **ID** Small to mid-sized finch with conical bill and forked black-and-white tail. **Ad non-br** has brownish or greyish upperparts, whitish rump and pale yellowish face. Can show black on crown. Black wings with white wingbar and fringes, and yellow shoulders. Throat to neck-sides yellow, dull grey or buffish-brown breast and flanks. Bill dark grey-brown. ♀ lacks any black on forecrown and yellow shoulders. **Ad br** ♂ is bright yellow with black cap, white rump and undertail-coverts. Wings more contrasting. Bill orangey to pinkish. **Ad br** ♀ has mostly dull olive upperparts and yellowish underparts, white wingbar, rump and vent. **Ad non-br** is generally duller but has black wings with very broad pale wingbar and yellow shoulder (more obvious in ♂) and ♂ has brighter yellow face and slightly greyish shawl (but no black on head). **Juv** is similar to ad non-br ♀, but slightly buffier. **Voice** Contact call a short series of light twittering notes "tsee-tsi-tsi-tsit" or "ti-dee-did-di", often given in flight. Alarm a high-pitched "ch-ween". **SS** Ad br ♂ distinctive. Conical bill distinguishes from all superficially similar vireos or warblers. ♀ Black-backed Goldfinch *S. psaltria* (see Appendix 2) has yellow (not whitish) vent. See also sparrows, grassquits and Saffron Finch.

House Finch
frontalis

♀
♂

Common Redpoll
(Redpoll)
flammea

♀
♂

♂ ♀
Hispaniolan Crossbill
not to scale

E

variant

♀ breeding

♂

Pine Siskin
pinus

American Goldfinch
tristis

259

Red Siskin *Spinus cucullatus* EN
Introduced (**Puerto Rico**).

L 10–11 cm. Monotypic. Native to N South America. Introduced in 1920s or 1930s, apparently always rare and local, and recent sightings in SE Puerto Rico kept confidential given species' extraordinary rarity now (consequently we do not map). Dry woodland and scrub in foothills. Pairs or small flocks. Apparently also once present in parts of N Cuba (specimens from 19th century and 1930s), but whether deliberately introduced or escaped cagebirds is impossible to divine. **ID** Small, bright-coloured finch with sharply pointed bill and notched tail. **Ad** ♂ has jet-black hood. Rest of upperparts scarlet with grey or blackish feather bases. Paler red or pinkish lower back and rump. Black wings show prominent orange patch and orange on primaries. Below scarlet with whitish belly and pale pinkish vent. **Ad** ♀ is mostly greyish, faintly streaked dusky on back, with rump and breast-band reddish orange. Wings duller. **Juv** as ♀, but paler or greyer, with reduced orange or red. **Voice** Song a long series of repeated high-pitched twitters and trills lasting up to 2–3 minutes. **SS** Other species with partly or largely red plumage are larger or possess different shape.

Antillean Siskin *Spinus dominicensis* LC
Locally common resident (**Hispaniola**).

L 11–12 cm. Monotypic. Endemic. Edges of moist broadleaf and pine forest, nearby grassy clearings and weedy patches at 500–3000 m, but most numerous at 1000–2500 m. Apparently wanders to edges of cultivation and dry scrub-forest at lower elevations, post-breeding and in cold weather. Forages on ground or in low-growing vegetation, also in trees, usually in loose groups of up to 25 birds. **ID** Small, chunky siskin with conical, pale yellow bill. **Ad** ♂ has black head and throat bordered by bright yellow collar and neck-sides. Above olive-green with blackish wings, feathers fringed whitish to olive or yellowish. Notched black tail has two prominent yellow patches. **Ad** ♀ is olive-green above with pale yellowish rump and yellowish-white underparts streaked greyish. Can show ill-defined wingbars. **Juv** is like ♀ but duller and more boldly streaked above and below. **Imm** ♂ is more yellow on underparts than imm ♀. **Voice** Song a low bubbling trill or jumble of notes, to North American ears recalling American Goldfinch. Calls include a soft low "chut-chut" and higher-pitched "swee-ee" or "swee-ee seee-ip". **SS** Note range and habitat. Ad ♂ distinctive. ♀ recalls ♀ Village Weaver (p. 248), which is larger with yellow supercilium and darker, more massive bill. ♀ Yellow-faced Grassquit has dark bill, and lacks wingbars or streaks below.

CALCARIIDAE
Longspurs
6 extant species, 2 vagrants in region

Lapland Longspur *Calcarius lapponicus* LC
Vagrant (**Cuba**).

L 15.5–17 cm. Three subspecies, of which one (probably nominate) occurs in region. Ad br ♂, photographed, Zapata Peninsula, May 2016 (Martínez et al. 2016). Could be found on ploughed fields, stubble and open grassland, along coasts, even shores of lakes and rivers, and other open ground. **ID** Fairly large stocky bunting with relatively short tail, deep-based, short and triangular bill, long primary extension, and white edges to tail. **Ad br** ♂ has solid black head and breast, long white supercilium reaching well down neck-sides, bright chestnut shawl and heavily black-streaked upperparts. **Ad non-br** ♂ has dark crown with sandy median crown-stripe, rufous-buff nape, and distinctive face pattern of warm buff outlined black at edge of ear-coverts. Above streaked rusty and brown, wings more extensively rufous, with bright rusty-edged greater coverts and tertials, and two narrow white wingbars. Throat whitish, upper breast with variable blackish (often forming breast-band), below white, streaked brownish on flanks. Bill yellowish. **Ad non-br** ♀ is generally somewhat paler, nape browner or greyer (not rufous), little or no black on breast. Bill pale brown. **Voice** Distinctive "tew" call and a dry rattle, both given in flight; resembles Snow Bunting. **SS** Compare various North American sparrows, but this species should prove difficult to confuse given reasonable views. **TN** The Cuban bird was considered closest to proposed race *subcalcaratus* (described from Greenland), which subspecies is not accepted under our classification.

Snow Bunting *Plectrophenax nivalis* LC
Vagrant (**Bahamas**).

L 14–18 cm. Four subspecies, of which nominate occurs in region. ♂ specimen, Cat I, Dec 1963 (Paulson 1966; LSUMZ 145075). Usually associated with beaches, especially pebbly ones, in winter. **ID** Unmistakable: black-and-white bunting with robust conical bill and relatively short, notched tail. **Ad non-br** ♂ is largely white, washed rusty brown, especially on nape, crown, ear-coverts and breast. Black feathers on back edged frosty brown. Primaries and central rectrices black. Bill yellowish. **Ad non-br** ♀ is smaller than ♂, duller and darker on upperparts, especially on tail and rump. Cheeks brown. **Voice** Call notes a descending "cheeew" and "tweet". **SS** None.

PASSERELLIDAE
New World Sparrows
145 extant species, 10 in region + 4 vagrants

Chipping Sparrow *Spizella passerina* LC
Very rare passage migrant and winter visitor (**Bahamas**, **Cuba**), vagrant (**Cayman Is**).

L 12–14 cm. Five subspecies, of which at least one (probably nominate) occurs in region. Mainly Abaco, Eleuthera, Grand Bahama and New Providence, Oct–Apr. Just one report Cayman Is, Nov 2009 (eBird). Pastures, open grassy fields, bushy thickets, coastal areas and open woodland (even pines) to 1150 m in Cuba. Forages principally on or near ground, often near edges of fields. **ID** Small, slim sparrow with long, notched tail. **Ad non-br** is very similar to ad Clay-colored Sparrow, but has less contrasting head pattern including buffier and duller supercilia, less contrasting ear-coverts without dark outline, fainter moustachial, dull chestnut crown, greyer underparts and grey (not buffy or brown) rump. **Ad br** (less likely to be seen in region) has bright rufous crown, grey cheeks, greyish-white supercilium and black eyestripe. **1st-y** is similar to ad non-br but slightly duller, with browner (less rufous or chestnut) crown and buffier underparts. **Juv** is similar to 1st-y but has face, breast and flanks heavily streaked blackish. **Voice** Call a thin, clear "tseep" or dry "chip". Song a thin, dry rattle on one pitch (unlikely to be heard in region). **SS** Ad br has distinctive rufous crown and eyestripe. Grasshopper Sparrow (commoner in Cuba) has shorter tail and yellow spot in front of eye. See Clay-colored, Savannah and larger Lark Sparrows.

Red Siskin

Antillean Siskin

Lapland Longspur
lapponicus

Snow Bunting
nivalis

Chipping Sparrow
passerina

261

Clay-colored Sparrow *Spizella pallida* LC
Rare passage migrant and winter visitor (**Bahamas**, **Cuba**, **Cayman Is**).

L 12–13.5 cm. Monotypic. Records late Sept–late Feb in Bahamas (Abaco, Andros, Eleuthera, Grand Bahama, Mayaguana, New Providence, Providenciales, San Salvador, Treasure Cay), W & C Cuba (where most numerous) and Grand Cayman (Nov 2011, Oct 2013). Coastal thickets, around salt pools, grassy and bushy areas, open woodland, rarely far from coasts. Forages low down or on ground. **ID** Small, slim sparrow with contrasting head pattern and long, brownish, notched tail. **Ad** has brown crown streaked black, and buffy-white to whitish median crown-stripe. Pale lores, broad whitish supercilia and brown cheeks with thin dark outline. Broad pale stripe between ear-coverts and blackish moustachial. Grey nape contrasts with buffy-brown upperparts, boldly streaked black on back, unstreaked on rump. Two inconspicuous buffy wingbars. Underparts unstreaked whitish with ochre breast. Bill dull flesh with darker culmen. **1st-y** is buffier, with duller face and finely streaked breast. **Voice** Call a sharp "tsip", or "seep". **SS** Distinctive head pattern. Non-br Chipping Sparrow has less contrasting face pattern and grey (not buffy-brown) rump contrasting with back. Compare Savanna, larger Lark and Grasshopper Sparrows (latter commoner in Cuba, and shorter-tailed).

Lark Sparrow *Chondestes grammacus* LC
Very rare passage migrant and winter visitor (**Cuba**), vagrant (**Bahamas**, **Jamaica**).

L 14.1–18.3 cm. Two subspecies, of which nominate occurs in region. Reported late Aug–late Mar, in Bahamas (Abaco, Eleuthera, Grand Bahama, Harbour I, New Providence, South Bimini), Cuba (including several offshore cays) and Jamaica (Feb 1993). Open areas with scattered bushes. Not shy, forages mainly on or near ground. **ID** Large sparrow, with bold head pattern, and much white in long, squarish (not notched) tail. **Ad** has whitish median crown-stripe, rusty lateral crown-stripes becoming blacker near bill, broad whitish supercilium, white crescent below eye, chestnut ear-coverts with white spot at rear, white submoustachial, and prominent black malar becoming broader on neck. Boldly streaked blackish brown on mantle. Two indistinct whitish wingbars. Tail with prominent white corners and edges, obvious in flight. Prominent black breast spot. Undertail white with narrow blackish base. Bill pale horn to greyish. **1st-y** has less distinct head pattern, chestnut replaced by brown, and buffish, heavily streaked breast and flanks. **Voice** Call a sharp, distinctive, parulid-like "tslip" or metallic "cheep", often given in flight. **SS** Large size, face and tail patterns, and breast spot unique. Compare smaller Clay-colored, Savannah and Chipping Sparrows.

Grasshopper Sparrow *Ammodramus savannarum* LC
Uncommon to rare winter visitor (**Cuba**, **O**), locally common resident (rest of **GA**).

L 12–12.5 cm. Twelve subspecies, of which four (*pratensis*[A] winters Bahamas, Cuba and Cayman Is, nominate[B] Jamaica, *intricatus*[C] Hispaniola, *borinquensis*[D] Puerto Rico including Vieques I) occur in region. Weedy open fields, tall-grass pastures, savannas and rice fields, to 400 m. Unobtrusive, forages on or near ground; usually in small colonies. Winter visitors uncommon in Cuba, rare in Bahamas and Cayman Is (Sept–early May). Resident races locally common but declining. Vagrant (race *pratensis*?) Swan Is, Nov 1958 (Bond 1959). **ID** Small, flat-headed, short-tailed sparrow. **Ad** has pale buffy median crown-stripe and dark brown lateral crown-stripes. Supercilia buffish with golden patch anteriorly. Thin dark brown post-ocular stripe and narrow white eye-ring. Above mostly greyish streaked blackish. Throat whitish; below washed pale buffy. Bill flesh-horn to grey with darker culmen; legs pinkish. **Juv** has indistinct median crown-stripe. Breast and flanks streaked brown. **Voice** Two distinct songs: a long, thin, insect-like buzz followed by a hiccup, "zzzzzzz-hic", and thin, high-pitched twitter or tinkling song, like fairy bells. Call an insect-like "kr-r-it" or "tillic". **SS** Compare non-br ♂ Yellow-crowned and Northern Red Bishops (mainly Puerto Rico), which lack single central crown-stripe. Savannah Sparrow more heavily streaked than imm Grasshopper. On Hispaniola, larger Rufous-collared Sparrow has slightly crested crown, rufous collar and is confined to mountains. See other rare or vagrant sparrows.

Green-tailed Towhee *Pipilo chlorurus* LC
Vagrant (**Cuba**).

L 15.7–18 cm. Monotypic. Ad ♂ specimen, Casilda, SC Cuba, Jan 1964. Recent report, Cayo Coco, lacks any details (Kirkconnell *et al.* in press). Favours shrubby areas, thickets and similar. Forages principally on ground and low in vegetation. **ID** Relatively large sparrow with green on wings and long tail, and striking head pattern. **Ad** has blackish forehead, rusty cap and white supraloral, submoustachial and throat, separated by black malar. Above olive-grey with olive-green to yellowish-green fringes on wing and tail. Whitish on belly, buff-washed rear flanks and vent. Bill dark blue-grey. **Juv** is browner and duller with no rusty crown. Crown, back, breast and flanks heavily streaked dark brown. Throat whitish with dark brown malar. Two indistinct wingbars. **Voice** Call a sharp "keek", or high-pitched "tseee"; also cat-like "meeeow". **SS** Ad distinctive. Juv less so, but note long tail, with green on tail and wings, and malar stripe.

Zapata Sparrow *Torreornis inexpectata* VU
Locally common to rare endemic (**Cuba**).

L 16.5 cm. Three subspecies, all of which (nominate[A] Zapata Swamp, W Cuba, *varonai*[B] Cayo Coco and Cayo Romano, off NC Cuba, *sigmani*[C] coastal Guantánamo, SE Cuba) occur in region. Endemic. Uncommon to rare in seasonally flooded sawgrass prairies of Zapata Swamp, but common in dry to semi-wet open forest (Cayo Coco), thorn scrub, dry forest and cactus thickets (coastal Guantánamo). Forages on ground. On Cayo Coco observed foraging on trails and open ground in forest; in Zapata in dense grass and more difficult to see. Often in twos or threes. **ID** Large-headed, moderately long-tailed, plump sparrow with conical blackish bill. **Ad** has chestnut crown with narrow grey median stripe, grey face and supercilia, and narrow brown eyestripe. White submoustachial, bold dark malar and narrow moustachial. Nape grey, rest of upperparts olive-grey with back streaked brown. Throat white, pale grey upper breast; below mostly dull yellow with yellow-olive flanks. **Juv** as ad but duller. **GV** Race *varonai* has darker rufous crown and bolder face pattern; *sigmani* smaller, with relatively small and paler bill, duller crown, less colour on flanks, yellow underparts paler and more extensive, and upperparts less noticeably streaked. **Voice** Songs usually duets, a highly synchronized series of high-pitched buzzes that become louder as song progresses, but slow and decrease in amplitude and frequency to end in a chatter, "tzi-tzi-tzi...tiiitziii...tik-tik-thk". Another common type comprises shorter chatter notes given by both sexes; in another duet ♂ sings buzz notes, ♀ chatter notes. Contact call of ♀ a single buzz, ♂ often 2–4 in series; also a high thin "chip", softer "pit" and metallic "oing". **SS** Combination of conical black bill, striped grey head, malar stripe and yellow underparts unique in range (and habitats). **AN** Cuban Sparrow.

Clay-colored Sparrow

Lark Sparrow
grammacus

Grasshopper Sparrow
savannarum

Green-tailed Towhee

not to scale

Zapata Sparrow
inexpectata

263

Dark-eyed Junco *Junco hyemalis* LC
Fifteen subspecies divided into five subspecies groups, of which one occurs in region.

Slate-colored Junco *Junco (hyemalis) hyemalis*
Vagrant (**O**, **GA**, **Virgin Is**).
L 13–17 cm. Three subspecies in group, of which at least one (probably nominate) occurs in region (*carolinensis* perhaps also occurs). Specimen, New Providence, Nov 1959 (Bond 1960), Grand Bahama, Nov 1969 (Bond 1972) and Jan 2011 (*NAB* 65: 358); Cuba, Nov 2002 (Mitchell 2009); Cayman Is, Nov 2017 (eBird); Jamaica, Jan–Apr 1971 (Smith 1971); Puerto Rico, Oct 1963 (Raffaele 1989); and St Thomas, Virgin Is, Nov 1928 (Holt 1932), the latter on a ship. Also, unconfirmed, San Salvador, Bahamas (White 1998). Coastal thickets. Forages on or near ground, where hops and occasionally runs. **ID** Medium-sized, dark, with white outer tail feathers and pale pinkish bill. **Ad** ♂ is blackish grey except white lower breast to vent. Outer tail feathers white. Wing feathers show paler edges and tips. **Ad** ♀ is paler and duller, often washed brown above. Edges of wing feathers browner. **Juv** is like ad ♀, but even more brownish, often with some streaks below. **GV** Race *carolinensis* slightly larger and paler (more blue-grey) than nominate, also bill darker. **Voice** Call a simple "tit tit tit", also smacking "tack tack tack". Song unlikely to be heard in region. **SS** Pinkish bill and white on outer rectrices distinguish all plumages.

Rufous-collared Sparrow *Zonotrichia capensis* LC
Locally common resident (**Hispaniola**).
L 15–16 cm. Twenty-five subspecies, of which one (*antillarum*) occurs in region. Scrub, streamside thickets and understorey of montane pine forest and its edges, 900–2450 m. One record from SW Dominican Republic (outside main range) and another (probably an escapee) from Cuba, ad ♀, Nov 1935 (Kirkconnell *et al.* in press). Retiring, feeds on ground, or low in trees and bushes, often in pairs. **ID** Handsome sparrow with peaked crown and conical bill. **Ad** has grey median crown-stripe and black lateral crown-stripes. Narrow rufous nape and neck-sides often extending to breast-sides. Above brown coarsely streaked blackish, tail and wing feathers edged rufous, indistinct whitish wingbars and some yellow at wingbend. Throat white and broad black breast-band. Below greyish white, variable buffy flanks and vent. Bill greyish. **Juv** is duller, without black or rufous markings, or black breast patch, has less distinct head pattern and well-streaked underparts. **Voice** Song an accelerating trill, "whis-whis-whis-whis-whiswhisu-whiswhis". Call a sharp "chip" or "chink". **SS** Note habitat and range. Boldly patterned ad distinctive. Juv could recall smaller Grasshopper Sparrow, but latter not in montane forests. Song and Lincoln's Sparrows streaked above and below. House Sparrow usually at much lower elevations.

White-throated Sparrow *Zonotrichia albicollis* LC
Vagrant (**Bahamas**, **Puerto Rico**).
L 15–17 cm. Monotypic. At least six records (several photo-documented): Grand Bahama, Nov 1997 (*FN* 52: 133) and Jan 2012 (eBird), Treasure Cay, Mar 2013, Abaco, May 2013 (*NAB* 66: 569); white-striped ad, Puerto Rico, Jan–Feb 1971 (Raffaele 1981, considered perhaps ship-assisted by Bond 1982) and Dec 1989 (*AB* 44: 1014). Usually in wooded edges and clearings. Forages mainly on ground, often in leaf litter. **ID** Fairly chunky sparrow with conspicuous head pattern and long, slightly notched tail. **Ad** has narrow white median crown-stripe bordered by black stripes, broad white or brownish-tan supercilia (white-stripe and tan-stripe 'morphs') becoming yellow in front and above eye. Intermediates occur, some with tan median crown-stripe. Upperparts rusty, streaked dark. Two faint whitish wingbars. Throat white (occasionally a faint dark malar) contrasting with greyish and sometimes faintly streaked breast (black line can separate throat and breast). Flanks pale grey to brown, sometimes faintly streaked. Belly and vent whitish. Bill dark with paler mandible. **1st-w** has median crown-stripe indistinct, duller greyish eyebrow with yellow reduced or absent, dull pale greyish throat and heavily streaked breast and sides. **Voice** Song a clear, loud whistle characteristically starting with lower note, followed by 3–4 higher, wavering notes, or less frequently first note higher than others. Call a distinctive "tseet"; also quiet "tip" notes, and louder "pink" alarm. **SS** Head-on, might be confused with imm Pin-tailed Whydah. White-crowned Sparrow has broader, bolder median crown-stripe, a less contrasting, duller throat, and more erect posture. See also House and smaller Grasshopper Sparrows.

White-crowned Sparrow *Zonotrichia leucophrys* LC
Scarce passage migrant and rarer winter visitor (**Bahamas**, **Cuba**), vagrant (**Jamaica**, **Cayman Is**).
L 14–17 cm. Five subspecies, of which two (*gambelii*, nominate) occur in region. Early Oct–early May, mostly Oct–Nov and Mar. Race *gambelii* little more than vagrant (five records in Cuba). Exceptional away from Bahamas and Cuba, e.g. Jamaica, Jan–Apr 1968 (Downer 1968); five, Little Cayman, May 1998 (Bradley 2000). Scattered trees, edges, brushy fields, wooded gardens and coastal thickets. Can appear in pairs or small groups, forages on or near ground. **ID** Fairly large sparrow with long brown tail and bold head pattern. **Ad** has broad white median crown-stripe reaching almost to bill, and black lateral crown-stripes. Broad white supercilium, narrow black eyestripe and grey face-sides. Above grey-brown, boldly streaked black. Brown-rufous edges to wing-coverts and faint whitish wingbars. Grey throat, buff-washed flanks and buff vent. Bill reddish pink to yellow. **Juv** has rusty-brown and buff crown-stripes. **GV** Race *gambelii* relatively large but small-billed (orange-yellow with dark tip) and has pale grey supraloral. **Voice** Call a hard "pink" or "tsit". Song variable but distinctive, characteristically two (1–4) clear whistles, the second slightly lower than first, followed by three descending buzzy or husky ones, "dear-dear buzz buzz buzz". **SS** Sharply contrasting black and white head-stripes distinctive in ad. Compare smaller Chipping and Clay-colored, and larger Lark Sparrows. See vagrant White-throated Sparrow.

Vesper Sparrow *Pooecetes gramineus* LC
Vagrant (**Bahamas**, **Cayman Is**).
L 14–18 cm. Three subspecies, of which presumably nominate occurs in region. Grand Bahama, Nov 1959 (Bond 1960), Dec 1962 (Bond 1964) and photo, Nov 2015 (*NAB* 70: 131); photo, Grand Cayman, Oct 2012 (*NAB* 67: 175). Dry open fields, thickets, brushy second growth and coastal areas. Runs or hops on ground. **ID** Medium-large streaky sparrow with fairly large, conical bill. **Ad** is generally similar to Savannah Sparrow, but slightly larger, with bold white eye-ring, white outer edges and corners to notched tail (conspicuous in flight), and no evident supercilia or yellow on lores. Diagnostic small chestnut shoulder not always visible. **Juv** is similar to ad but buffier overall. **Voice** Call a sharp "chirp". Song sweet and musical, usually 2–4 long clear notes, often with downward slurs, followed by shorter flute-like trills, typically rising then falling in pitch. **SS** White eye-ring and fringes to tail, and chestnut shoulder patch distinguish it from other sparrows, but see Savannah, Lincoln's and Song Sparrows.

♀

♂

Slate-colored Junco
(Dark-eyed Junco)
hyemalis

Rufous-collared Sparrow
antillarum

white-stripe 'morph'

tan-stripe 'morph'

White-throated Sparrow

leucophrys

gambelii

White-crowned Sparrow

Vesper Sparrow
gramineus

265

Savannah Sparrow *Passerculus sandwichensis* `LC`
Two subspecies placed in separate subspecies groups, of which one occurs in region.

Savannah Sparrow
Passerculus (sandwichensis) sandwichensis
Common to uncommon (**Bahamas, Cuba**) or rare passage migrant and winter visitor (**Cayman Is**), vagrant elsewhere.
L 11.4–15.4 cm. One subspecies in group. Recorded mid Oct–late Apr. Vagrants: ad ♀, Haiti, Mar 1928 (Keith *et al.* 2003); Swan Is, specimen plus two trapped, Nov 1958 (Bond 1959); San Andrés, Oct 2001 (Salaman *et al.* 2008); Jamaica, Apr 2019 (R. Hoyer). Open fields, pastures, bushy savannas, coastal thickets and at edges of ponds, occasionally parks, in lowlands. Usually in small groups of up to c. 20, feeding mostly on ground. **ID** Small to medium-sized, very variable, streaked sparrow with conical bill and relatively short, notched tail. **Ad** has dark brown to buffish-brown crown with very narrow pale median crown-stripe and usually yellow (or buff) lores and supercilium, paler at rear. Crown can look slightly peaked. Dark eyestripe, grey-brown ear-coverts and narrow dark moustachial and malar stripes. Sandy brown above heavily streaked darker. Sometimes has paler outermost tail feathers. Neatly dark-streaked upper breast and flanks, sometimes an indistinct central breast spot. Legs pink. **Juv** is usually buffier overall, with reduced (or no) yellow in supercilium. **Voice** Usually silent outside breeding season, but calls include thin "seet", "chip" or "tzip". **SS** All plumages of Grasshopper Sparrow have flatter head, more golden supercilium and no malar; additionally, juv has finer, paler streaks on underparts, and ad lacks any streaks below. Vagrant Vesper Sparrow lacks bold supercilium.

Henslow's Sparrow *Passerculus henslowii* `LC`
Hypothetical (**Bahamas**).
L 12.1–13.3 cm. Two subspecies, of which one (unknown) might occur in region. Eleuthera, Sept 1972 (Connor & Loftin 1985). Secretive and difficult to flush, often preferring to run away and, if flushed, only flies a short zigzag before abruptly dropping into cover. **ID** Small sparrow, with short tail, large bill and flat head. **Ad** has buffy-olive median crown-stripe, dark brown lateral crown-stripes, broad buffy-olive supercilia and dark brown eyestripe, moustachial and malar. Nape buffy olive, mantle and upper back dark brown, edged buff, lower back and rump buffy olive with brown centres. Tail blackish with rufous edges, many wing feathers edged rufous, yellow at bend of wing. Throat white to pale buff, breast and flanks thinly streaked brown, belly whitish, vent buffy. **Juv** is rather buffier, lacks moustachial and malar stripes, and has little or no streaking on breast and flanks. **Voice** Call "tsip". **SS** Ad Grasshopper Sparrow paler and relatively unstreaked, without distinct face markings but has bright yellow supraloral and no olive tinge to head. Juv Grasshopper streaked below but again lacks olive tinge to head and dark chestnut-rufous in tertials. **TN** North American authorities usually place in genus *Centronyx*.

Song Sparrow *Melospiza melodia* `LC`
Twenty-five subspecies divided into five subspecies groups, of which one occurs in region.

Eastern Song Sparrow *Melospiza (melodia) melodia*
Hypothetical (**Bahamas, Dominican Republic**).

L 12–17 cm. Three subspecies in group, of which at least one (presumably either or both *atlantica* and nominate) occurs in region. Records (none documented beyond field notes): Abaco, Jan 1984 (*AB* 38: 362), Eleuthera, Oct–Apr in 1970s, including Dec 1976, 1977 and 1978 (*AB* 31: 904, 32: 908, 33: 685, Connor & Loftin 1985), Grand Bahama, Dec 1962 (Bond 1964); Dominican Republic, Nov 1997 (Rimmer & McFarland 1998). Open and semi-open areas with thickets or wooded edges; once at 1900 m (Hispaniola). Less shy than Lincoln's Sparrow, often feeds on ground with tail cocked. Pumps tail up and down in flight. **ID** Medium-large sparrow, similar to Lincoln's Sparrow. **Ad** differs by larger size, with slightly longer, rounded tail, thicker, bolder dark malar, more boldly streaked plumage, with streaks on breast often coalescing into central spot, less grey on face and without buffy-washed breast and sides. **Juv** is similar to ad but buffier overall, with finer streaking and slightly barred tail. **Voice** Call a distinctive "tchenk", "tchip" or "chimp"; also a "tseep" note. **SS** Compare Lincoln's Sparrow, which is commoner in Bahamas. Equally rare Swamp Sparrow lacks distinct malar and moustachial, and has unstreaked greyish underparts.

Lincoln's Sparrow *Melospiza lincolnii* `LC`
Rare passage migrant and winter visitor (**Bahamas, Cuba**), vagrant (**Jamaica, Hispaniola, Puerto Rico, San Andrés**).
L 11.5–14.5 cm. Three subspecies, of which nominate occurs in region. Present mid Oct–early Apr in Bahamas and Cuba. Vagrants (?): eight records between 1963 and 2012 in Jamaica; two records involving at least ten individuals in Haiti, early 1980s (Keith *et al.* 2003); Dominican Republic, Nov 2010 (Ortiz *et al.* 2012); Puerto Rico, Dec 1923, late Apr 1980 (Raffaele 1989), Mar 2003, Dec 2008, Feb 2015, Feb 2018; San Andrés, Nov 2017 (eBird). Thickets and borders, and highland broadleaf forest, especially around clearings, to 2100 m. Forages alone and easily overlooked due to skulking habits. Pumps tail in flight. **ID** Small to mid-sized, unobtrusive sparrow. **Ad** has narrow grey central crown-stripe and brown to rufous-brown lateral stripes. Broad supercilium, ear-coverts and neck-sides grey. Malar buffy, surrounded by narrow black stripes. Above greyish brown, finely streaked black. Breast and sides buffish, finely streaked black. Central belly unstreaked whitish. **1st-w** is similar but crown is brown or greyish brown and streaked, and supercilium buffish. **Voice** Calls a sharp, low-pitched "chip" and soft, high-pitched "zeet". Song (unlikely to heard in region) a rich warble similar to that of Northern House Wren. **SS** Song Sparrow is slightly larger, overall more boldly streaked, with streaks on breast often forming central spot, has less grey on face and lacks buffy wash on breast.

Swamp Sparrow *Melospiza georgiana* `LC`
Three subspecies divided into two subspecies groups, of which one occurs in region.

Swamp Sparrow *Melospiza (georgiana) georgiana*
Very rare winter visitor (**Bahamas**).
L 12–15 cm. Two subspecies in group, of which nominate occurs in region. Ten records: Abaco, two, Dec 2006 (*NAB* 61: 348), Dec 2011 (eBird), Nov 2012 (*NAB* 67: 176) and Nov 2016 (eBird); Bell I, May 1983 (Buden & Sprunt 1993); Grand Bahama: two, Oct 2016 (eBird); Mayaguana, Apr (Brudenell-Bruce 1975) and Nov 1962 (Buden & Sprunt 1993); specimen plus sight, New Providence, Nov 1960 (Bond 1962); photo, San Salvador, Jan 2013 (eBird). Marshes, thickets and brushy areas. Forages mostly on or near ground, especially in reedbeds. Often pumps tail in flight. **ID** Medium-sized, rather stocky

Savannah Sparrow

Henslow's Sparrow

Eastern Song Sparrow
(Song Sparrow)
melodia

Lincoln's Sparrow
lincolnii

first-winter

breeding

Swamp Sparrow
georgiana

sparrow with much rufous on wings. **Ad non-br** has face, supercilium and neck-sides grey. Post-ocular stripe dark brown and ear-coverts greyer, sometimes faintly outlined ochre. Thin dark moustachial, pale submoustachial, and sometimes narrow dark malar. Upperparts and tail mostly rufous, mantle and scapulars boldly streaked darker. Chin and throat whitish, sometimes flecked brown. Breast greyish, can be faintly streaked. Flanks beige and thinly streaked; belly and vent mostly white. **Ad br** has rufous to dark brown crown, black forehead and slightly brighter rufous upperparts. **1st-w** is similar to ad non-br but crown blackish or darkly streaked, breast and throat unstreaked grey, and cheeks buffish. **Juv** is similar but is more streaked below. **Voice** Call a metallic "chink". **SS** Compare Song, Lincoln's, Chipping and Savannah Sparrows.

TERETISTRIDAE
Cuban Warblers
2 extant species, 2 in region

Yellow-headed Warbler *Teretistris fernandinae* LC
Common to fairly common resident (**Cuba**).
L 13 cm. Monotypic. Endemic. Principally found in semi-deciduous and evergreen woodland, but also pine forest, second growth and coastal thickets, to at least 700 m. Reaches E as far as Sierra de Trinidad (Cienfuegos province) in S Cuba, and to around Itabo and Hicacos Peninsula, Matanzas province, in N. Sight records from Jardines de la Reina archipelago (off S Cuba) extremely doubtful. Small flocks often associate with Cuban Pewee and Cuban Vireo. **ID** The (for a warbler) somewhat long and slightly decurved bill is characteristic of the newly erected family Teretistridae (for this and next species). **Ad** has yellow hood (tinged olive on crown and nape), contrasting grey upperparts and greyish-white underparts; yellow eye-ring. **Juv** is apparently undescribed. **Voice** Song a series of buzzy grating notes interspersed by sweeter, more musical ones. Characteristic call a rapid, high-pitched, staccato chattering, or rasping trill, usually repeated several times; also other buzzy and grating notes. **SS** Only likely to be confused with Oriente Warbler, in areas where their ranges come close; apparent hybrids between them show scattered yellow feathering on midline of belly plus grey central crown and nape. Prothonotary Warbler (a scarce migrant in Cuba) has primarily yellow underparts, the colour of head is deeper and more vibrant, with greener upperparts, bluer wings and white in tail.

Oriente Warbler *Teretistris fornsi* LC
Common to fairly common resident (**Cuba**).
L 13 cm. Monotypic. Endemic. Most numerous in coastal thickets (including on several large cays off NC Cuba) and semi-deciduous and evergreen woodland, but also montane forest to 1970 m. Reaches W as far as Salinas de Bidos, Itabo, Matanzas province, where it appears to occur within 10 km of Yellow-headed Warbler, while on S coast they seemingly overlap between Trinidad and Cienfuegos, with some records of apparent hybrids. Often in small, very active flocks. **ID** Note characteristic bill shape. **Ad** has crown, nape and upperparts grey, face, throat and most of underparts yellow, becoming white at rear, with dull brownish wash on flanks; yellow eye-ring. **Juv** is apparently undescribed. **GV** Birds from above 1400 m on Pico Turquino, in SW of range, recently proposed as race *turquinensis* (Garrido 2000a), being allegedly slightly larger and longer-tailed, with darker, more sooty-grey crown and upperparts, and grey (rather than brownish) wash on rear flanks. **Voice** Less harsh than Yellow-headed Warbler. Song a series of buzzy notes interspersed by sweeter ones, very similar to its congeneric but slightly more monotonous. Calls include a sharp "tchip".

SS Yellow-headed Warbler chiefly differs in its all-yellow head. Nashville, Hooded and Wilson's Warblers all have olive-coloured upperparts and wings; ♀/1st-y Canada Warbler (very rare in Cuba) has grey, not yellow, cheeks and has some dark markings on breast. See Prothonotary Warbler (p. 288).

ICTERIDAE
New World Blackbirds
114 extant species, 25 in region + 5 vagrants + 1 introduced

Yellow-breasted Chat *Icteria virens* LC
Very rare passage migrant and winter visitor (**Bahamas**, **Cuba**), vagrant elsewhere.
L 19 cm. Two subspecies, of which nominate occurs in region. Most records N Bahamas and Cuba, in Aug–May (especially Oct/Nov and Feb–May), but also vagrant to Grand Cayman, Dominican Republic and Jamaica. Occurs in low dense vegetation, being difficult to spot. **ID** Stocky, warbler-like bird (long treated as a parulid) with relatively stout bill and long tail. **Ad br** has dark olive-green crown and upperparts, grey head-sides with black lores, short white supercilium, white eye-crescents and moustachial, bright yellow throat and breast, and white rear underparts; black bill. ♀ is slightly duller, with greyish lores, and browner bill with flesh-coloured mandible base. **Ad non-br** is marginally duller and bill is dark brown. **1st-y** resembles ad, but is even duller. **Voice** Calls include harsh grating "chack", a mew reminiscent of Grey Catbird and sharp "kuk-luk-kuk". **SS** Unmistakable, if seen well.

Yellow-headed Blackbird LC
Xanthocephalus xanthocephalus
Vagrant (**GA**, **O**).
L ♂ 26.5 cm; ♀ 21.5 cm. Monotypic. Records in Bahamas (four in Oct, at least one each Nov, Dec–Feb, Mar and Sept: Kale *et al.* 1969, Miller 1978, *FN* 52: 396, *NAB* 52: 396, 53: 111, 62: 173, 65: 183, eBird), Cuba (five, Oct–Dec), Cayman Is (three, Aug–Oct: Bradley 2000, eBird) and Puerto Rico (1985, 1997 or 1998, Jan 1999, 2006, Nov 2013: eBird). Associated with marshy wetlands, but in West Indies mostly recorded in farmland and other open areas. **ID** Distinctive blackbird. **Ad br** ♂ has black mask, bright yellow head, throat and breast, otherwise black with white wing patch. **Ad br** ♀ has pale yellow supercilium and moustachial stripe, whitish throat, yellow breast; rest of plumage brown to brownish black. **Ad non-br** has crown and nape feathers with yellowish-brown tips. **Juv** has cinnamon-buff head and underparts, dark mask, brownish upperparts with cinnamon tips and edges, two pale wingbars; ♂ paler, more tawny above with whiter wingbars than ♀, which has darker back and buff wingbars. **Voice** Call a hoarse "chuk" or "chek", deep and hollow in quality. **SS** Compare Yellow-hooded Blackbird (vagrant to Barbados).

Bobolink *Dolichonyx oryzivorus* LC
Uncommon to rare passage migrant (**throughout**).
L 16–18 cm. Monotypic. Regular but not common in Bahamas, Cuba, Jamaica and Cayman Is, less frequent in Puerto Rico, Virgin Is, Guadeloupe and Barbados, rare on Hispaniola, St Barthélemy, Antigua and Dominica, and very rare on rest of Lesser Antilles. More regular Aug–Dec, less so Feb–Jun, exceptionally reported late Jun. Gathers in large flocks at rice fields, pastures and grassy areas (once 3000–5000 on Grand Bahama). **ID** Emberizid-like icterid with long pointed wings and sharply pointed tail feathers. **Ad br** ♂ is mostly black with large but variable pale buffish crown and nape, long buff streaks on upper-

Yellow-headed Warbler × Oriente Warbler

Oriente Warbler

Yellow-headed Warbler

Yellow-breasted Chat
virens

Yellow-headed Blackbird

Bobolink

269

parts, pale grey rump and white in wings. **Ad br** ♀ has dark crown with pale buff central stripe, long broad supercilium bordered by dark post-ocular stripe, buff upperparts broadly streaked dusky brown, whitish throat, creamy-buff underparts partly streaked dusky; bill pinkish. **Ad non-br** ♂ is like br ♀, sometimes with blackish spots on throat; in Feb–Mar generally appears intermediate between br and non-br. **Ad non-br** ♀ is brighter, upperparts having warmer golden-buff or orangey fringes, face and underparts richer orange-buff, bill brighter pink. **Juv** resembles non-br ♀, but upperparts have thin buff fringes (look scaly), and underparts entirely buff and unstreaked except slightly on breast-sides. **Voice** Common call a distinct "pink" note. **SS** Unlikely to be confused; ♀ Red-winged Blackbird is heavily streaked with longer, pointed bill.

Eastern Meadowlark *Sturnella magna* NT
Sixteen subspecies divided into three subspecies groups, of which two occur in region.

Eastern Meadowlark *Sturnella (magna) magna*
Vagrant (**Bahamas**, **Caymans Is**).
L 20–24 cm. Fourteen subspecies in group, of which nominate occurs in region. Andros, photo, early Dec 2014 (*NAB* 69: 308); Cayman Brac, photo, Sept 1987 (Bradley 2000). Could occur in grassland and pastures with scattered trees. Perches on wires and fences, but usually forages on ground. **ID** Stocky with long bill. **Ad** has blackish crown with pale buff to white centre, long white supercilium, yellow patch in front of eye, blackish eyestripe behind it, whitish malar; yellow underparts with V-shaped black breast-band, dark-streaked buffy flanks and lower belly. ♀ is duller than ♂. **Juv** (unlikely in region) is similar to ad but paler, with more prominent pale fringes above, throat and underparts paler, black breast-band replaced by dark streaking. **Voice** Common call a sharp "dzert" or "jerzík"; also a loud rattle in alarm and "weet" in flight. **SS** Cuban Meadowlark is marginally smaller with better-streaked underparts and on average darker upperparts.

Cuban Meadowlark *Sturnella (magna) hippocrepis*
Fairly common resident (**Cuba**).
L 23 cm. One subspecies in group. Endemic. Throughout mainland Cuba, I of Pines and several N coast cays. Open grassland, savannas, dry marshes and pastures with scattered trees, from sea level to mid elevations. Alone or in pairs, foraging on ground but perching on fences and wires. **ID** Structure as Eastern Meadowlark. **Ad** differs from Eastern Meadowlark only in darker upperparts with narrower pale fringes to feathers, more vivid yellow breast and more heavily streaked underparts. **Juv** is similar to ad, but overall paler and has a grey throat. **Voice** Song comprises sweet slurred whistles, on average with shorter and less slurred notes than Eastern Meadowlark and typically ending in a buzz or rattle. ♀ commonly produces a bubbling chatter. Calls not known to differ from those of Eastern Meadowlark. **SS** Plumage unique in restricted range. Even in flight is distinctive, low on whirring wings with short glides, and much white showing in outertail.

Jamaican Oriole *Icterus leucopteryx* LC
Fairly common resident (**Jamaica**, **San Andrés**).
L 21 cm. Three subspecies, of which all three (nominate[A] Jamaica, *bairdi* extinct Grand Cayman, *lawrencii*[B] San Andrés) occur or did occur in region.

Endemic. Common throughout Jamaica (especially in lowlands) and on San Andrés, in most of types of woodland, but rare in mangroves. On Grand Cayman last recorded in 1967, and now believed extinct. Forages in pairs or family groups. **ID** Olive-yellow oriole with black mask and throat. **Ad** is mostly olivaceous yellow, brighter and yellower on belly, black wings with striking white patch; black tail. ♀ is perhaps slightly duller. **Imm** is dull olivaceous green, with reduced black on head but more on wings (less white), and tail dusky green. **Juv** lacks all-black throat, bib and lores, and white in wings is reduced to two wingbars. **GV** Race *lawrencii* brighter than nominate, more yellow-green, becoming lemon-yellow on underparts; very different *bairdi* had olive-yellow replaced by vivid orange-yellow. **Voice** Song, by both sexes, 2–3 musical, modulated whistles, e.g. "tie-tiewu", "tie-tiewu-tiewu" and similar (often transliterated as "Auntie Katie"), in rapid succession. Call a single or two-note whistle, or a short chatter. **SS** ♂ is distinctive; ♀ can resemble ♀/imm Baltimore Oriole which have orange-yellow belly and two white wingbars (rather than large wing patch).

Yellow Oriole *Icterus nigrogularis* LC
Hypothetical (**Grenada**).
L 20–21 cm. Four subspecies, of which potentially any one could occur in region. Only record, Jul 1955 (Devas 1970). The brief description is more suggestive of nominate race than the arguably more expected *trinitatis*. In native range inhabits arid and semi-arid woodland and scrub, also mangroves. **ID** Chunky oriole with stout bill. **Ad** ♂ is mostly orange-yellow with black lores and large black bib reaching breast, as well as black wings and tail, relieved only by white wingbar and fringes to flight feathers (most extensive in nominate race of mainland South America, least in *trinitatis* of Trinidad and NE Venezuela). **Ad** ♀ is tinged olivaceous over yellow areas. **Juv** is duller and greenish, without black bib, and has greenish tail. **Voice** Song relatively soft, involving repeated flute-like notes, buzzes and harsher notes. Alarm call "chet-chet-chet". **SS** Baltimore Oriole and Venezuelan Troupial are both more orangey, with black head and back, and larger white patch on wing.

Baltimore Oriole *Icterus galbula* LC
Fairly common passage migrant and rarer winter visitor (**GA**, **O**), vagrant (**LA**).
L 18–22 cm. Monotypic. Most frequent in Cuba, Bahamas and Jamaica, rarer Hispaniola, Puerto Rico and some Virgin Is, and only vagrant in Lesser Antilles. Observed Sept–May in wide variety of wooded habitats, from canopy of humid forest to second growth, coffee and cacao plantations, and gardens. **ID** Distinctive black-and-orange oriole. **Ad** ♂ has black head, upper back, throat and upper breast, otherwise bright orange; white wingbars and flight feathers broadly edged white. **Ad** ♀ is variable, some very like ♂, but usually olive-brown to orange-brown on head and upperparts, brownish lesser coverts, sometimes tinged yellow or orange, median coverts broadly tipped white (not orange), underparts more orange-yellow. **Juv** is similar to dull ♀, but head and upperparts usually olive tinged orange, throat and underparts pale golden-orange, flanks olive. **Voice** Common calls include a chatter and ascending whistle. **SS** See Jamaican Oriole. Venezuelan Troupial is larger with more extensive black on breast, orange hindneck, more white in wings and yellow iris. ♀ Scarlet Tanager (p. 320) is more greenish, with stubbier bill and no white in wings; Village Weaver (p. 248) is chunkier with heavy bill and shorter tail.

Eastern Meadowlark
magna

Cuban Meadowlark
(Eastern Meadowlark)

Jamaican Oriole
leucopteryx

nigrogularis

trinitatis

Yellow Oriole

♂

♀ variant

♀

Baltimore Oriole

271

Bullock's Oriole *Icterus bullockiorum* LC
Vagrant (**Bahamas**).

L 19–23 cm. Two subspecies, of which one (presumably nominate) occurs in region. Videotaped, Grand Bahama, Feb 2001 (*NAB* 55: 238, Mlodinow 2002). Several principally W USA species have reached the Bahamas and/or Cuba, including this one. **ID** Fairly typical, sexually dimorphic oriole. **Ad** ♂ has crown to back black, black lores and thin black stripe from chin to upper breast. Rest of head and body orange; wings and tail mainly black, but median and greater coverts mostly white, and outer rectrices orange with black tips. In fresh (non-br) plumage has narrow greyish feather tips on head and back, and broader white wing edgings. **Ad** ♀ is considerably duller, with head and facial markings olivaceous, head and upper breast duller yellow-orange, belly greyish white, back and wings paler and greyer, without white wing panel, but does have wingbar on median coverts, and sometimes a faint second white wingbar on greater coverts. Some birds show black on throat. **Juv** is similar to ♀; young ♂ acquires ad plumage via intermediate stages. **Voice** Song, by both sexes but presumably unlikely to be heard in region, a short series of whistles, less musical than Baltimore Oriole, sometimes with notes repeated 2–5 times. Calls include clicks, harsh "chuck" sounds and a short rattle. **SS** Black-capped and -backed ♂ should be distinctive; ♀/young might be confused with similar plumages of Baltimore Oriole, but latter shows two, more obvious, white wingbars on generally blacker wings, more dark centres above and yellow feathering is typically more intensely coloured. Hybridizes extensively with Baltimore Oriole across USA, and reports of vagrants should clearly establish purity of individuals concerned.

Venezuelan Troupial *Icterus icterus* LC
Three subspecies divided into two subspecies groups, of which one occurs in region.

Venezuelan Troupial *Icterus (icterus) icterus*
Introduced resident (**Puerto Rico**), vagrant (**LA**).
L 23–27 cm. Two subspecies in group, of which both (*ridgwayi*, nominate) occur in region. Introduced Puerto Rico, where common in SW and uncommon on Mona I, Culebra, and on Water I and St Thomas (Virgin Is); has wandered to (or escaped on) Jamaica, St John (Virgin Is), Antigua, Barbuda and Dominica. Apparently just one record of nominate race; a ♀ specimen, Grenada, 1929 (Bond 1956). Arid scrub. Usually alone or in pairs. **ID** Large attractive oriole with long stout bill and piercing yellow eyes. **Ad** has black head, throat and breast, tail, wings and upperparts, otherwise orange-yellow, including hindneck; large white patch in wings. **Juv** is patterned like ad, but duller, more yellowish and brownish black, with bare ocular patch smaller and greyer. **Voice** Loud rich melodious phrases with 2–4 elements, e.g. "troo-pear-pééá". Oft-repeated. Mimicry of other birds recorded. Calls include nasal notes and pleasant whistles. **SS** See Baltimore Oriole. **AN** Formerly known simply as Troupial.

Hooded Oriole *Icterus cucullatus* LC
Five subspecies divided into two subspecies groups, of which one occurs in region.

Eastern Hooded Oriole *Icterus (cucullatus) cucullatus*
Vagrant (**Cuba**).
L 18.5–20 cm. Three subspecies in group, of which one (presumably nominate) occurs in region. Four records, three of them specimens (two undated), in Nov and Mar (Kirkconnell *et al.* in press). Scattered trees and coastal scrub. **ID** Small orange-yellow oriole with somewhat curved bill. **Ad** ♂ is mostly deep orange with black face to breast, upper back and tail, white shoulder patch and narrower white wingbar. **Ad** ♀ has crown and nape greyish olive-yellow, upperparts olive, greyer on mantle, rump olive-green, yellowish face and underparts, belly paler, flanks washed grey; two white wingbars. **Juv** is like ♀ but much duller, olive-brown above and light olive-yellow below, wingbars buffish. **Imm** ♂ resembles ♀, but has blackish lores and bib. **Voice** Calls include whistled "wheet", hard "chit" and fast chatter. **SS** ♀ confusable with ♀ Orchard Oriole, which has shorter, thicker and less decurved bill, and shorter tail.

Orchard Oriole *Icterus spurius* LC
Rare passage migrant (**Cuba**), vagrant elsewhere.

L 15–17 cm. Monotypic. In Cuba, mainly Oct/Nov and Mar–May, but may occasionally overwinter (Kirkconnell *et al.* in press); basically vagrant, with records also in Jan and Jul, in Bahamas, rest of Greater Antilles, Cayman Is, Montserrat, Guadeloupe and Swan Is (Levesque *et al.* 2012, Oppel & Boatswain 2013, eBird). Open woodland and gardens. **ID** Small, slender, short-tailed oriole. **Ad br** ♂ is mostly black with deep chestnut underparts, rump and shoulders, flight feathers edged white. **Ad non-br** ♂ has upperparts tipped olive or pale chestnut, underparts tipped yellowish. **Ad** ♀ has olive-green upperparts and bright greenish-yellow underparts, breast tinged ochraceous; narrow white wingbars, edges of flight feathers white. **Juv** is like ♀ but crown washed yellowish, nape and back browner, wingbars buffish. **Imm** ♂ is like ♀, but has black face and bib. **Voice** Call a sharp "chuck"; chattering alarm call. **SS** Compare Eastern Hooded Oriole. Slightly larger Baltimore Oriole has longer bill and orange-yellow underparts; imm Cuban Oriole has uniform wings, yellowish-green patches on shoulders, rump and undertail-coverts, and olive-coloured back.

Bahama Oriole *Icterus northropi* CR
Very rare resident (**Bahamas**).

L c. 20–22 cm. Monotypic. Endemic. Confined to Andros (including Mangrove Cay), where probably fewer than 250 individuals. Until early 1990s, also on Abaco. Pine and broadleaf woodland, gardens; favours mature coppice with thatch palms (*Leucothrinax morrisii*) and introduced coconut palms (*Cocos nucifera*). **ID** Slender oriole with no white wingbars. **Ad** ♂ is mostly black with yellow underparts, rump and shoulders. **Ad** ♀ has black areas duskier. **Juv** has olive upperparts, yellower rump/uppertail-coverts, yellow throat, dull olive breast, rest of underparts dull yellow; two wingbars. **Imm** has olive-grey upperparts, head washed yellowish, lower back to uppertail-coverts greenish yellow (brightest on rump), greenish-yellow underparts, ♂ with variable black on throat. **Voice** Song comprises 6–11 emphatic whistles. **SS** Baltimore Oriole has orange (not yellow) plumage and white wing patches. **TN** In past often treated as conspecific with Cuban, Hispaniolan and Puerto Rican Orioles (the single species being named Greater Antillean Oriole), but morphological and behavioural studies suggest that all merit species rank.

♀ variant

♀

Venezuelan Troupial
icterus

♂

Bullock's Oriole
bullockiorum

♀

♂

Eastern Hooded Oriole
(Hooded Oriole)
cucullatus

♀

♂

Orchard Oriole

Bahama Oriole

E

273

Cuban Oriole *Icterus melanopsis* LC
Common resident (**Cuba**).
L c. 20 cm. Monotypic. Endemic. Throughout Cuba (including I of Pines and larger N cays) in forest edge, woodland, parks and gardens, to 1300 m. Forages mostly in pairs or family groups. **ID** Slender-bodied oriole. **Ad** is almost all black, with yellow rump, shoulders, thighs and small spots on undertail-coverts. ♀ is slightly duller than ♂. **Juv** has olive upperparts, olive-green underparts becoming yellower on vent, wing feathers with greenish-olive edges. **Imm** is similar, but has black lores, chin and throat. **Voice** Song of descending whistles over narrow pitch range; some phrases may suggest species' Cuban vernacular name, "solibio". Common calls a sharp "kit" and nasal "whep". **SS** Other orioles recorded in Cuba generally show white in wings, among other differences. **TN** See Bahama Oriole (p. 272).

Martinique Oriole *Icterus bonana* VU
Uncommon resident (**Martinique**).
L 18–21 cm. Monotypic. Endemic. Occurs in mangroves, dry forest on limestone soils, humid forest and plantations; to 700 m. Forages in pairs or small family groups. **ID** Slim-bodied oriole. **Ad** has chestnut head and most of breast; black upper back, wings and tail; orange-tawny rump, shoulders and rear underparts. ♀ is slightly duller than ♂. **Juv/Imm** are undescribed. **Voice** Song consists of soft warbles. Call a harsh, scolding "cheeu". **SS** Only oriole in range, but compare vagrant Baltimore Oriole: ♂ has more orange overall, including on tail, and black head.

Puerto Rican Oriole *Icterus portoricensis* LC
Fairly common resident (**Puerto Rico**).
L 22 cm. Monotypic. Endemic. Forest and edges of many types, also parks and gardens, especially with palms. Usually in pairs or family groups. **ID** Slender oriole with mainly black plumage. **Ad** has yellow rump, shoulders, thighs and vent; uppertail-coverts and rear undertail-coverts black with yellow tips. **Juv** has olive-green upperparts with reddish tinge on head, and mostly dull yellow underparts, but washed rufescent on breast. **Voice** Rather high-pitched series of ascending and descending whistles, mixed with buzzes and warbles. Common call a sneezing "chk". **SS** Yellow-shouldered Blackbird (very small range within Puerto Rico) is all black with yellow shoulders. **TN** See Bahama Oriole (p. 272).

Montserrat Oriole *Icterus oberi* VU
Rare resident (**Montserrat**).
L 21 cm. Monotypic. Endemic. Now restricted to Centre Hills and South Soufriere Hills, with total population fewer than c. 700 individuals. Mesic to humid tropical forest, especially in ravines ('ghauts') with stands of *Heliconia caribaea*, mainly at 400–900 m. Generally forages in pairs, mostly in understorey, sometimes higher. **ID** Only endemic Caribbean oriole with marked sexual dichromatism. **Ad** ♂ is mostly black with yellow rump, shoulders and underparts below breast. **Ad** ♀ has olive crown, upperparts and tail, brownish-olive wings with two pale bars, and olive-yellow underparts. **Juv** resembles ♀. **Voice** Loud series of well-spaced whistles. Common call a harsh "chrr". **SS** Only oriole in its range.

Hispaniolan Oriole *Icterus dominicensis* LC
Fairly common resident (**Hispaniola**).
L 20–22 cm. Monotypic. Endemic. Widespread (including all of the large satellite islands) in broadleaf forest and shade-coffee plantations with palms, to 1100 m; rarer in xeric woodland, scrub and gardens, and almost never seen in pine forest. Sometimes in large groups. **ID** The common oriole on Hispaniola. **Ad** is mostly black with yellow rump, shoulders, lower belly and thighs. ♀ is duller than ♂. **Imm** has dark olive-green head to breast and upperparts including tail, chestnut-tinged forehead and breast, greenish-olive edges to wing feathers, and underparts below breast greenish yellow. **Voice** Series of high-pitched whistles. Common calls a sharp "kt" and harsh "chrr". **SS** Other *Icterus* are very rare in Hispaniola, but compare Baltimore and Orchard Orioles. Tawny-shouldered Blackbird (in Haiti) is all black with reddish-yellow shoulders. **TN** See Bahama Oriole (p. 272).

St Lucia Oriole *Icterus laudabilis* NT
Uncommon to scarce resident (**St Lucia**).
L 20–22 cm. Monotypic. Endemic. All types of woodland, including dry and coastal scrub, banana plantations, mangroves and humid forest, to 700 m. Forages in pairs and groups of up to ten. **ID** Distinctively patterned oriole. **Ad** is mostly black with orange-yellow rump, shoulders and underparts below breast. ♀ has rump and underparts yellow, rather than orange. **Juv** has cinnamon head, black face and throat, and blackish wings with two white wingbars. **Imm** ♂ resembles ♀, but black and orange-yellow areas much duller, tinged brownish or olive, wing feathers with olive edges. **Voice** Song comprises up to seven notes, most of them loud ascending whistles. Call a harsh "chwee". **SS** Only oriole in its range.

Cuban Oriole

Martinique Oriole

Puerto Rican Oriole

Montserrat Oriole

Hispaniolan Oriole

St Lucia Oriole

275

Jamaican Blackbird *Nesopsar nigerrimus* EN
Uncommon resident (**Jamaica**).

L 18 cm. Monotypic. Endemic. Wet montane forest with abundant epiphytes, especially bromeliads (in which it regularly feeds), *Phyllogonium* mosses and tree-ferns. Also wet limestone forest, particularly in Cockpit Country. Absent or rare in drier sclerophyll forest on windy slopes and ridges. Mostly 500–2200 m; down to 210 m outside nesting season. Arboreal and typically unobtrusive. Often alone, or in pairs. **ID** All-dark blackbird with slender sharp-pointed bill and relatively short tail and legs. **Ad** is black with slight blue gloss. **Juv** resembles ad, but heavily tinged brown and lacks gloss. **Voice** Song a quick warble followed by 3–5 nasal buzzes covering wide frequency range. Common call a sharp "chek". **SS** Greater Antillean Grackle is obviously larger with strong purple gloss, longer V-shaped tail, and yellow iris, while Shiny Cowbird has shorter, more conical bill and is not strictly arboreal; neither is forest-dependent. ♂ Jamaican Becard is stockier with stubbier bill.

Tawny-shouldered Blackbird *Agelaius humeralis* LC
Common (**Cuba**) to rare resident (**Haiti**), vagrant (**Cayman Is**).

L 19–22 cm. Two subspecies, of which both (nominate[A] mainland Cuba and larger cays, plus W Haiti, *scopulus*[B] Cayo Cantiles, E of I of Pines) occur in region. Endemic. Open woodland and edges, farmland, gardens and rice fields; in Haiti (where confined to NW and sometimes said to have been introduced in 1920s/30s) along sloughs and channels in dry open woodland. Undocumented report from NW Dominican Republic (Latta *et al.* 2006). Forages on ground, but also in bushes and trees. Forms large flocks, mainly in non-breeding season. Vagrant Grand Cayman, Jun 2014 and Mar–May 2015 (*NAB* 68: 563, 69: 510). **ID** Mid-sized blackbird. **Ad** ♂ is black glossed bluish, with tawny shoulders. **Ad** ♀ is duskier and lacks gloss; tawny shoulders smaller. **Juv** is dusky black, with little or no tawny in wing. **GV** Race *scopulus* smaller and thinner-billed than nominate, with reduced shoulder patch. **Voice** Emits 1–2 protracted buzzing notes, sometimes introduced by shorter and higher-pitched buzz; pairs occasionally duet. Calls include "chuk" or "cheek", plus nasal and metallic notes. **SS** Shoulder patch sometimes difficult to see, thus resembles ♀ of larger and restricted-range Red-shouldered Blackbird which lacks epaulets and has longer bill and more pointed tail feathers. Compare Cuban Blackbird.

Yellow-shouldered Blackbird *Agelaius xanthomus* EN
Rare and local resident (**Puerto Rico**).

L 20–23 cm. Two subspecies, of which both (nominate[A] mainland Puerto Rico, *monensis*[B] Mona I and Monito I, off W Puerto Rico) occur in region. Endemic. Primarily nests on SW coast on small mangrove islands, and visits nearby pastures and palm plantations, including Puerto Rico royal and coconut palms; on Mona I frequents coastal cliffs. Forages in trees, probing and gaping among bark and epiphytes, but also on ground. Usually seen in small groups but larger flocks in non-breeding season. Formerly throughout Puerto Rico, but numbers much reduced, at least in part due to parasitization by Shiny Cowbirds. **ID** Easily identified icterid within its small range. **Ad** is black glossed bluish, with yellow shoulder patch. ♀ is slightly smaller than ♂. **Juv** is duller, more brownish grey than blackish, with much smaller and paler shoulders. **GV** Race *monensis* has paler shoulders, sometimes yellowish white. **Voice** Varied vocalizations. A rasping nasal "nyaaaaaa", sometimes by several individuals in chorus. Common call "chuk"; alarm "ct-zeee". **SS** Puerto Rican Oriole has yellow rump, lower belly and thighs.

Red-winged Blackbird *Agelaius phoeniceus* LC
Twenty-two subspecies divided into two subspecies groups, of which one occurs in region.

Red-winged Blackbird *Agelaius (phoeniceus) phoeniceus*
Locally common resident (**NW Bahamas**), hypothetical (**Puerto Rico**).

L ♂ 22.7 cm; ♀ 18.5 cm. Nineteen subspecies in group, of which one (*bryanti*) occurs in region. Marshes, swamps and around ponds on New Providence, Andros, Grand Bahama, Eleuthera and Abaco; also Bimini, Berry Is and Cay Sal Bank. Highly gregarious. Undocumented reports, by experienced observers, SW Puerto Rico, Jan 1990 (*AB* 44: 336), and on Mona I, Sept 2018 (eBird). **ID** Mid-sized blackbird. **Ad** ♂ is dull black with red shoulder patch edged yellow. **Ad** ♀ has dark brownish upperparts and rather whitish underparts, both heavily streaked dark brown; conspicuous pale supercilium. **Juv** resembles ♀, but pale areas more yellowish, with broad buffish fringes to upperparts, including wing feathers. **Imm** ♂ is streaked like ♀, but blackish overall and has small shoulder patch. **Voice** ♂ song variable, a few short notes followed by long, loud nasal trill. ♀ has two songs: a chatter or 'rattle' that can form one part of duet; and series of harsher notes ('growl'). Calls include low "chek" or "kek"; alarm a buzzing "zeer" or whistled "cheet", also descending "teeew". **SS** Shiny Cowbird has heavier bill and ♂ lacks red shoulder patch, ♀ and imm have finer and less distinct streaking below.

Red-shouldered Blackbird *Agelaius assimilis* LC
Fairly common but very local resident (**Cuba**).

L ♂ 22 cm; ♀ 20 cm. Monotypic. Endemic. Occurs only in W Cuba and at one site on I of Pines. Freshwater wetlands with sawgrass, *Phragmites*, *Sagittaria lancifolia*, *Scirpus* and *Typha*, although it also forages in rice fields and other agriculture. Mostly feeds on ground, sometimes with Cuban and Tawny-shouldered Blackbirds outside breeding season. Typically sings from bush or tree. **ID** The Cuban equivalent of the familiar Red-winged Blackbird. **Ad** ♂ is black with slight bluish gloss and fairly small deep red shoulder patch edged yellow. **Ad** ♀ is all black but less iridescent than ♂, and lacks coloured shoulder. **Imm** ♂ has reduced and paler shoulders; ♀ is more brownish. **Voice** Rather shrill creaking "o-wi-hiiii" repeated quite frequently, sometimes in duet. Call "chuk" or "chek", sometimes in short series. **SS** Both sexes of Tawny-shouldered Blackbird have shoulder patch, but not always visible. ♂ Shiny Cowbird is glossy purplish with heavier bill. Cuban Blackbird larger and lacks shoulder patch. **TN** Until mid 1990s was considered a race of Red-winged Blackbird.

not to scale

Jamaican Blackbird

Tawny-shouldered Blackbird
humeralis

Yellow-shouldered Blackbird
xanthomus

Red-winged Blackbird
bryanti

Red-shouldered Blackbird

277

Giant Cowbird *Molothrus oryzivorus* `LC`
Vagrant (**Barbados**).
L ♂ 34 cm; ♀ 29 cm. Two subspecies, of which one (probably nominate) occurs in region. ♀, photographed, Mar 2000–Mar 2001 (*NAB* 55: 384, Buckley *et al.* 2009). Usually in open habitats, including livestock pastures. **ID** Very large, long-winged and long-tailed cowbird with neck ruff, making head appear small. **Ad** ♂ is all black with bronze to bluish iridescence, inconspicuous black frontal casque covered by feathers; iris yellow to orange. **Ad** ♀ is smaller and duller, with less expansive ruff. **Juv** is blackish and has pale yellow bill that eventually becomes black; brown iris. **Voice** Usually silent, but can emit sharp "chuk". **SS** Shiny Cowbird much smaller with quite different shape; Barbados Grackle has V-shaped tail.

Shiny Cowbird *Molothrus bonariensis* `LC`
Common to very common resident (**throughout**).
L 17–21.5 cm. Seven subspecies, of which one (*minimus*) occurs in region. First recorded in Caribbean in 1860, on Vieques I, off E Puerto Rico, and spread slowly W through Greater Antilles (reaching Cuba only in early 1970s), as well as N through Lesser Antilles from Grenada, where initially recorded in 1901, in some cases aided by deliberate introductions (Post & Wiley 1977). Almost all open habitats, including agricultural land. In small to large groups. **ID** Smaller black icterid with conical bill. **Ad** ♂ is glossy purple-black, with wings and tail glossed bluish and greenish. **Ad** ♀ has dusky brownish-grey upperparts, sometimes with slight bluish gloss, and indistinct pale supercilium. **Juv** is like ♀. **Voice** ♂ has 'gurgling' song also incorporating thin whistles or warbles; in flight a fast, high-pitched warbling and twittering, introduced by thin descending whistle. ♀ song a rattle. Common call "chuck", by both sexes. **SS** Cuban Blackbird, Greater Antillean Grackle and Carib Grackle are larger with heavier bills. Yellow-shouldered, Tawny-shouldered and Red-winged Blackbirds have coloured epaulets. Jamaican Blackbird has longer and slender bill. ♀ Brown-headed Cowbird is greyer than ♀ Shiny Cowbird, also has whitish throat and lacks supercilium, but often hard to distinguish. ♀ Red-shouldered Blackbird has slender bill and lacks glossy purple plumage.

Brown-headed Cowbird *Molothrus ater* `LC`
Rare non-breeding visitor (**Bahamas**), vagrant (**Cuba**).
L ♂ 18–20 cm; ♀ 16–18 cm. Three subspecies, of which nominate occurs in region. Records in Bahamas (Abaco, Crooked I, Eleuthera, Grand Bahama, Great Inagua, New Providence, San Salvador), Aug–Apr; twice in Cuba, Jan and Feb (Bond 1961). Unconfirmed record Turks & Caicos Is. In almost all open habitats, e.g. grassland, wooded edges, agricultural fields, orchards, suburban areas and towns. Forages on ground, cocking tail as it does so. **ID** Smaller, terrestrial icterid. **Ad** ♂ is mostly black with greenish iridescence and dark brown head. **Ad** ♀ is brownish overall, darker on upperparts, with indistinctly paler superciliary; slightly paler below, especially on throat. **Juv** is brownish like ♀, but feathers of upperparts edged buff, throat greyish white to buff white, and underparts streaked dusky brown. **Voice** ♂ song is two low-pitched warbles followed by thin descending whistle; ♀ call a dry chattery warble or rattle. **SS** See Shiny Cowbird.

Cuban Blackbird *Ptiloxena atroviolacea* `LC`
Common resident (**Cuba**).
L 25–28 cm. Monotypic. Endemic. Widespread, in all manner of open woodland, parks, suburban gardens and villages, below 1000 m. Strangely absent from I of Pines, but occurs on a number of N coast cays. In pairs and small flocks, but sometimes joins other icterids, such as Greater Antillean Grackle or Tawny-shouldered Blackbird, especially at roosts. **ID** Square-shaped tail and relatively short bill. **Ad** is black overall, with strong blue and purple iridescences on body, and greenish-blue gloss on wings. ♀ is slightly duller than ♂. **Juv** is much duller, being brownish black and unglossed. **Voice** Song varied, probably by both sexes, a mix of musical notes and some nasal sounds. Musical "twee-o" and "twee-te-to" calls. **SS** Shiny Cowbird and ♀ Red-shouldered Blackbird are smaller, and both are much less abundant (with the blackbird confined to small areas of W Cuba); equally common Greater Antillean Grackle has V-shaped tail and yellow eyes. Compare also widespread Tawny-shouldered Blackbird (epaulets not always easy to see).

♂ variant

♀

♂

Giant Cowbird
oryzivorus

♀

♂

Shiny Cowbird
minimus

♀

♂

Brown-headed Cowbird
ater

Cuban Blackbird

279

Rusty Blackbird *Euphagus carolinus* VU
Hypothetical (**Bahamas**).

L ♂ 23 cm; ♀ 21 cm. Two subspecies, of which one (perhaps nominate on geographical grounds) occurs in region. Grand Bahama, Oct 1967 (Brudenell-Bruce 1975), Andros, May–Jun 1970 (*FN* 24: 595). Cuban reports withdrawn or unacceptable (Kirkconnell *et al.* in press). Usually found near water, in wooded swamps and riparian vegetation along creeks, but also in open fields near wetlands. **ID** Largely black icterid with piercing yellow eyes. **Ad br** ♂ is uniformly black with rather slight greenish or bluish gloss. **Ad br** ♀ is slate-grey, usually with darker upperparts and bluish-green gloss. **Ad non-br** is black with broad paler feather fringes, rusty brown on crown, nape and ear-coverts, with marked buff-brown supercilium, buff-brown on underparts. ♀ is similar to ♂, but fringes more cinnamon, and rump and underparts barring slate-grey. **Juv** is dull greyish, back and throat washed brown, wing feathers narrowly edged pale brown; dark iris soon becomes paler. **Voice** Common call "chuk"; alarm "chip". **SS** Shiny Cowbird is smaller with dark iris; vagrant Brewer's Blackbird has purplish gloss (not green) and heavier-looking bill.

Brewer's Blackbird *Euphagus cyanocephalus* LC
Hypothetical (**Bahamas**).

L ♂ 23 cm; ♀ 21 cm. Monotypic. Sight record: Grand Bahama, Nov 1959 (Bond 1960, Hundley 1962). Likely in open areas including farms, pastures or suburban areas. Usually forages on ground. **ID** Not easily separated from Rusty Blackbird, although present species appears more grackle-like on ground. **Ad** ♂ is glossy black, with purple iridescence on head and neck, greenish to bluish iridescence over rest of plumage; iris yellow. **Ad** ♀ is dark greyish brown, with slightly paler supercilium, marginally paler below, darker on belly, has slight violaceous gloss on head, greenish gloss on back; iris dark brown (yellow in ♀ Rusty). **Juv** is like ♀, but more uniform, browner and without gloss. **Voice** Call "chuk"; scolding "tschup". **SS** Rusty Blackbird has slightly thinner-based bill (which consequently looks longer) and shorter tail with a less elegant gait (crouching with legs slightly bent). ♂ Rusty is much less boldly glossed than ♂ Brewer's, and ♀ Rusty is obviously greyer, especially on rump. In non-br plumages, Rusty typically has much more boldly and extensively rufous-fringed feathers (tertials nearly always so), and contrasting black lores; tertials never rufous-fringed in Brewer's, which only ever shows many pale fringes in early autumn.

Common Grackle *Quiscalus quiscula* NT

Three subspecies divided into two subspecies groups, of which one occurs in region.

Purple Grackle *Quiscalus (quiscula) quiscula*
Hypothetical (**Bahamas**).

L ♂ 27.2 cm; ♀ 26 cm. Two subspecies in group, of which nominate occurs in region. Sight records from North Andros, date unknown, and Grand Bahama, Mar 2003 (eBird). In USA found in open areas with scattered trees, swamps and urban areas. Forages on ground, but readily perches in trees and bushes. **ID** Medium-large, with long wedge-shaped tail; plumage all black and glossy. **Ad** ♂ has strong purple gloss on head, dark green gloss on back, purple-blue iridescence on belly and blue-green gloss on tail (keel-like shape); iris pale yellow. **Ad** ♀ is smaller, less glossy, with shorter tail ('normal-looking'). **Juv** is like ♀, but duller and distinctly browner, with dark eyes. **Voice** Call a harsh, dry "kerrr", also thin "zweeej" and low "kek" in flight. **SS** Other blackbirds within its range have rather shorter and square (not V-shaped) tails; vagrant Boat-tailed Grackle larger with brown eyes.

Carib Grackle *Quiscalus lugubris* LC

Eight subspecies divided into two subspecies groups, both of which occur in region.

Carib Grackle *Quiscalus (lugubris) lugubris*
Common to very common resident (**LA**).

L ♂ 26 cm; ♀ 21.7 cm. Six subspecies in group, of which three (*guadeloupensis*[A] Montserrat, Guadeloupe, Marie Galante, Dominica and Martinique, *inflexirostris*[B] St Lucia, *luminosus*[C] Grenadines and Grenada) occur in region. Open woodland and scrub, pastures, agricultural fields, and urban and suburban areas. Forages mostly in flocks on ground, sometimes associating with Shiny Cowbird. **ID** Conical bill, piercing yellow eyes and V-shaped tail. **Ad** ♂ (*guadeloupensis*) is black with strong violet gloss, reduced on underparts; gloss on wing-coverts more violet-blue, that on flight feathers greener. **Ad** ♀ is pale brown above, with pale buff lores and conspicuous buffy supercilium, light buffish below, with whitish chin and throat, streaked breast. **Juv** ♂ is dark brown, streaked below, with dark eyes, but soon acquires black feathering and paler iris. **Juv** ♀ is similar to ad, but warmer brown, and retains dark iris longer than juv ♂. **GV** Race *luminosus* has relatively longer, narrow bill, ♂ has gloss more extensive than *guadeloupensis*, ♀ is similar to ♂ but unglossed, with paler throat and browner underparts, juv more obviously streaked below than those of other races; *inflexirostris* is slightly larger and ♀ brown above with darker rump, indistinct lighter supercilium, wings and tail blackish with paler fringes, buff-brown below, slightly paler on throat and greyer on flanks. **Voice** Noisy, with frequent singing. Gives a series of squeaky notes; on Martinique a descending, *Molothrus*-like whistle followed by a complex warble, and other populations ascribed to *guadeloupensis* and on St Lucia (*inflexirostris*) are similar, but different from those on Grenada and Grenadines (*luminosus*) (Bond 1965). Calls include "chk", loud whistles and rattles. **SS** Shiny Cowbird is rather smaller with thinner bill and square-ended (not V-shaped) tail. **TN** A molecular study by Humphries *et al.* (2019) found little or no concordance between traditional subspecies limits and genetic relationships.

Barbados Grackle *Quiscalus (lugubris) fortirostris*
Abundant resident (**S LA**), introduced (**N LA**).

L ♂ c. 24 cm; ♀ c. 21 cm. Two subspecies in group, of which both (*contrusus*[A] St Vincent, *fortirostris*[B] Barbados) occur in region. Endemic. Open woodland and scrub, pastures, agricultural fields, urban and suburban areas. Race *fortirostris* reportedly introduced to Antigua, Barbuda, St Kitts and St Croix, has spread to other islands N to Anguilla, but most of these unconfirmed as to race (Buckley *et al.* 2009). Same race apparently an occasional vagrant to St Vincent, but no recent reports. Behaviour not known to differ from Carib Grackle. **ID** Shape and structure as Carib Grackle, but bill short and thick with gradually decurving culmen, and virtually no sexual dimorphism. **Ad** ♂ is black with virtually no gloss. **Ad** ♀ is very similar to ♂ but somewhat paler. **Juv** ♂ is not known to differ from juv Carib Grackle. **GV** Race *contrusus* ♀ is even darker, being blackish brown above and below. **Voice** Not well described, but generally similar to Carib Grackle; however Bond (1965) reported that race *fortirostris* differs from that of *contrusus*, which is more similar to race *luminosus* of Carib Grackle. **SS** See Carib Grackle. **TN** See Carib Grackle.

breeding

Rusty Blackbird
carolinus

Brewer's Blackbird

Purple Grackle
(Common Grackle)
quiscula

Carib Grackle
guadeloupensis

Barbados Grackle
(Carib Grackle)
fortirostris

281

Boat-tailed Grackle *Quiscalus major* LC
Local resident (**N Bahamas**).

L ♂ 40 cm; ♀ 27.5 cm. Four subspecies, of which at least one (probably *westoni*) occurs in region; one of yellow-eyed races (*torreyi* or *alabamensis*) perhaps also present. Records from Abaco, Grand Bahama, Eleuthera and, especially, New Providence, most since 2011, but reports date from 2003 (eBird) and first documented record in Dec 2005 (*NAB* 60: 302); appears to be fairly well established. Has colonized probably due to frequency of cruiseliners visiting islands from Florida. One possible record (race *torreyi*) from Puerto Rico, also in 2003. Typically occupies brackish to freshwater marshes along coast, and estuaries, but also readily accepts man-modified habitats. **ID** Large grackle with long V-shaped tail. **Ad** ♂ is all black with violet-glossed head, blue-green gloss on body, and duller green gloss on wings and tail. **Ad** ♀ is much smaller due to shorter tail, with dark brown upperparts, slight greenish gloss on wings and tail, ill-defined supercilium and pale buff to warm cinnamon underparts (throat palest). **Imm** ♂ has shorter tail than ad ♂ and more restricted and greener gloss. **Imm** ♀ is somewhat warmer brown, more frequently tinged cinnamon than ad ♀, wing-coverts sometimes fringed rufous, breast can be indistinctly streaked. **GV** Some observers have reported yellow-eyed birds, presumably either the migratory race *torreyi* (comparatively short-tailed) or *alabamensis* (deep-billed and long-tailed). **Voice** Call a harsh "jeeb-jeeb-jeeb", but also various whistles and harsher rattles. Song, by both sexes (♀ more rarely) during display, a series of sharp "kip" notes followed by grating rattles, ending with further "kip" notes. **SS** Purple Grackle is smaller with a quite different tail, and always has yellow eyes. See very similar Great-tailed Grackle (not yet reported in Bahamas).

Greater Antillean Grackle *Quiscalus niger* LC
Common to very common resident (**GA**, **Cayman Is**), hypothetical (**Virgin Is**).

L 25–30 cm. Seven subspecies, of which all (*caribaeus*[A] extreme W Cuba and I of Pines, *gundlachii*[B] elsewhere in Cuba including many cays, *caymanensis*[C] Grand Cayman, *bangsi*[D] Little Cayman, *crassirostris*[E] Jamaica, nominate[F] Hispaniola including its major satellites, *brachypterus*[G] Puerto Rico and Vieques I) occur in region. Endemic. Observed in all manner of habitats, from mangroves, marshes, savannas, coconut plantations, open fields, agricultural land and light woodland, to parks and gardens; often roosts in towns and cities. To 1000 m in Cuba and even as high as 2180 m on Hispaniola. Several reports from Virgin Is, but apparently none documented. Commonly seen in flocks, foraging on ground. **ID** Large, with long V-shaped tail and piercing yellow eyes. **Ad** ♂ is entirely black with strong purple gloss on head and upperparts, and greenish gloss on flight feathers and tail. **Ad** ♀ is smaller than ♂ due to shorter tail, less glossy, with shorter tail that is not V-shaped. **Juv** is brownish black, lacks iridescence, and has dark brown eyes. **GV** Race *gundlachii* (both sexes) glossed violaceous; *caribaeus* smaller with more bluish gloss (the ranges of these two require much better elucidation); *caymanensis* much smaller, with blue gloss tinged purplish, uppertail-coverts glossed greenish blue, wings glossed bronze-green; *crassirostris* similar to *gundlachii* but has shorter and thicker bill, ♂ with violet sheen, becoming blue on tail-coverts (above and below) and belly, and wing-coverts glossed greenish bronze, ♀ somewhat duller; *bangsi* smaller than previous with longer bill, ♂ bluer, ♀ more brownish; *brachypterus* resembles *crassirostris*, but smaller, with uppertail-coverts glossed purplish (not blue). **Voice** Song varies considerably among islands, but (for a grackle) is relatively musical, a metallic "cling-cling-cling" and flute-like whistles. Call "chuk" or "chuk-chuk", or "chin-chin-chilin"; also high "wee-si-si" and harsh notes. **SS** Other grackles in range lack V-shaped tail. Shiny Cowbird much smaller with dark eyes and shorter bill. Yellow-shouldered and Tawny-shouldered Blackbirds are smaller with coloured epaulets. Jamaican Blackbird smaller with dark eye, and is exclusively forest-based.

Great-tailed Grackle *Quiscalus mexicanus* LC
Eight subspecies divided into two subspecies groups, of which one occurs in region.

Great-tailed Grackle *Quiscalus (mexicanus) mexicanus*
Local and uncommon resident (**GA**, **O**).

L ♂ 43 cm; ♀ 33 cm. Six subspecies in group, of which nominate occurs in region. Recent colonist, probably ship-bourne. First recorded Dominican Republic, around Santo Domingo, in 2007 (Mejía *et al.* 2009) with breeding recorded 2011 (Paulino *et al.* 2013); in Jamaica, also recently established (first report 2005) and apparently breeding, mainly in environs of Kingston but also reported at Montego Bay and other coastal areas (eBird); in Puerto Rico, records from several different areas, including San Juan and Arecibo, on N coast, and Aguirre State Forest in S (eBird). Also present on San Andrés (where large numbers counted) and Providencia since pre-2006. In USA found in marshes, coastal mudflats and lagoons, mangroves and riparian habitat, but also in savannas, open fields, agricultural land, light woodland, and parks and gardens in towns. **ID** Large, with big bill and long V-shaped tail. **Ad** ♂ has violet iridescence on head, back, throat and breast, more bluish over rest of body and greenish on flight feathers; iris yellowish white to pale yellow. **Ad** ♀ is much smaller due to shorter tail, dark brown above, somewhat paler on head, and throat and breast warmer brown, contrasting with darker belly. **Juv** is similar to ♀, dark brown above and buffish brown with dark streaks below, eyes dark. **Voice** Song quite different to Boat-tailed Grackle. First part of ♂ song consists of hard notes, the sound resembling cracking of twigs, followed by "chewechewe" and shorter sequence of twig-cracking notes, ending with loud "chawee" notes; song varies in pitch. ♀ produces a chatter. Calls include rising whistles (diagnostic), rattles and guttural sounds; some ♂ calls bugle-like, somewhat musical; all rather similar to Boat-tailed Grackle, but latter generally flatter in tone. **SS** Greater Antillean Grackle is much smaller with shorter tail and bright yellow (not pale) eyes. Smaller (but still large) Boat-tailed Grackle (established in N Bahamas) has rounder head shape, obvious forehead angle, shorter and narrower tail, shorter wings and relatively longer legs; violet iridescence is also less extensive in ♂ and can appear thicker-necked in display, while ♂ Boat-tailed also shows less gloss above and paler underparts. ♀ Boat-tailed Grackle is generally warmer and paler than same sex of present species, usually has dark eyes (except race *torreyi*) and a much-reduced malar stripe, and shares some of the structural features that separate ♂♂.

♂ torreyi

♀ westoni

Boat-tailed Grackle

♀

Greater Antillean Grackle
niger

♂

♀

Great-tailed Grackle
mexicanus

283

Yellow-hooded Blackbird *Chrysomus icterocephalus* `LC`
Two subspecies placed in separate subspecies groups, of which one occurs in region.

Yellow-hooded Blackbird
Chrysomus (icterocephalus) icterocephalus
Vagrant (**Barbados**).
L c. 17–19 cm. One subspecies in group. Ad ♂, collected, Sept 1887 (Buckley *et al.* 2009). In N South America inhabits marshes, but also humid savannas, mangroves and agricultural land. **ID** Unmistakable icterid with conical bill. **Ad** ♂ is mostly glossy black with bright yellow head and breast, and black lores. **Ad** ♀ has yellow supercilium and throat to upper breast, rest of face and crown citrine, dull greyish-olive upperparts with dusky streaking, and olive underparts. **Juv** is like ♀ but duller, with less yellow on head, paler upperparts, and obscurely streaked buffish underparts. **Imm** ♂ is like ♀, but brighter and more extensively yellow on throat, and may show black feathers both above and below. **Voice** Calls include a "chek" and a descending whistle. **SS** Very distinctive.

PARULIDAE
New World Warblers
122 extant species, 50 in region + 4 vagrants

Ovenbird *Seiurus aurocapilla* `LC`
Common passage migrant and winter visitor (**GA, O**), uncommon to vagrant (**LA**).
L 14–16.5 cm. Three subspecies, all of which (nominate, *cinereus, furvior*) occur in region. Present mid Aug–late May (exceptionally Jun), being common in W half of Caribbean basin, including Bahamas, Cayman Is, Greater Antilles and San Andrés, but much less numerous (or only vagrant to some islands) in Lesser Antilles. Very few confirmed records of races *furvior* and *cinereus* (all specimens). Recorded to 1250 m in Cuba and 2100 m on Hispaniola, in all manner of wooded areas, including pines, even semi-desert scrub, coastal thickets and areas of scattered trees (at least on migration), foraging exclusively on floor. Often first detected by rustle of leaves as it works. **ID** Distinctive, largely terrestrial warbler with characteristic head-bobbing, tail-flicking gait. **Ad** has dark olive-green head and upperparts, broad black-bordered orange central crown-stripe, white eye-ring, broad black malar, and white underparts boldly streaked black on breast and flanks. **1st-y** is similar to ad, but has narrow rusty-edged greater coverts and tips to tertials when fresh. **GV** Race *cinereus* slightly paler and greyer above than nominate; *furvior* browner above than nominate, with heavier streaking below, orange crown-stripe duller and black lateral crown-stripes broader. **Voice** Usual call a sharp, dry "chip" or "tsuk", often repeated rapidly when agitated; flight call a thin, high-pitched "seee". Rarely heard song (in late winter/spring) 8–10 loud, emphatic double notes rising in volume and usually in pitch, "tee-cher tee-cher tee-cher...". **SS** Far rarer Wood Thrush is larger and lacks black crown-stripes. Compare Northern and Louisiana Waterthrushes (both have obvious pale supercilia and slimmer-bodied appearance, among other differences).

Worm-eating Warbler *Helmitheros vermivorum* `LC`
Fairly common to uncommon passage migrant and winter visitor (**GA, O**), principally vagrant (**LA**).
L 11–13 cm. Monotypic. Generally fairly common across N & W of region, in mid Aug to early May, but mainly vagrant to Lesser Antilles (even on Guadeloupe, where regular mist-netting efforts). Favours woodland with dense understorey, secretively foraging in tangles of dead leaves at all levels, usually alone or in pairs, often loosely associated with an Ovenbird or Swainson's Warbler, but sometimes with mixed-species flocks. Territorial and uses same sites across seasons. Recorded to 2100 m on Hispaniola, but mainly below 1000 m. **ID** Distinctive by rich buff head with broad bold black stripes; appears somewhat flat-headed and short-tailed. **Ad** has narrower blackish eyestripe (than lateral crown-stripe), uniform olive-brown upperparts, rich buff throat and breast, becoming paler below. **1st-y** has rusty-fringed tertials when fresh. **Voice** Rather quiet. Usual call "zeep-zeep", often given several times in succession; a sharp, sweet "chip", similar to Swainson's Warbler but softer, is rarely heard in winter, as is the buzzing trill used as song. **SS** Somewhat uniform plumage relieved by bold head-stripes should prove unmistakable; also relatively unmarked Swainson's Warbler is much less arboreal, with stronger bill, creamy-buff supercilium and chestnut-brown crown.

Louisiana Waterthrush *Parkesia motacilla* `LC`
Fairly common (**GA, O**) to uncommon passage migrant and winter visitor (**LA**).
L 14.5–16 cm. Monotypic. Present Jul–Apr, mainly in W half of basin, in Bahamas, Greater Antilles and Cayman Is, but progressively less common further E & S, including across Lesser Antilles (where merely vagrant on many islands). Usually by streams, lagoons and even pools of rain water in woodland, preferring running water at higher elevations, to 1150 m (in Cuba) and 1800 m (on Hispaniola). **ID** Constantly bobs tail up and down, slowly and emphatically. **Ad** has dark grey-brown head and upperparts, with long bicoloured supercilium, buff in front of eye and pure white behind it, white underparts, washed buff on flanks and vent, and diffusely streaked brown on breast and flanks. **1st-y** is similar to ad but has narrow rusty-tipped tertials and greater coverts when fresh. **Voice** Usual call a sharp resonant "chink", similar to Northern Waterthrush but louder and more emphatic, also slightly lower-pitched and less metallic. **SS** Northern Waterthrush slightly smaller with less heavy bill, uniformly buff and even-width supercilium (not white and flaring behind eye), more uniformly buffish ground colour below, more distinct streaking extending to throat, and much duller pink legs. Compare Ovenbird.

Northern Waterthrush *Parkesia noveboracensis* `LC`
Fairly common to uncommon passage migrant and winter visitor (**throughout**).
L 12.5–14 cm. Three subspecies, of which two (nominate, *notabilis*) occur in region. Generally fairly common throughout West Indies in Jul–May (exceptionally from late Jun), in mangroves, evergreen and semi-deciduous woodland, coastal scrub and gallery forest, near pools and streams, where forages on or near ground, often on fallen logs. To 1640 m on Hispaniola. Race *notabilis* apparently much rarer than nominate. **ID** Like Louisiana Waterthrush, constantly bobs tail up and down like a pipit or Spotted Sandpiper. **Ad** has dark olive-brown head and upperparts, long pale buff supercilium, off-whitish to buffy-white underparts, distinctly dark-streaked breast and flanks, with usually finer streaks on throat. **1st-y** is similar to ad,

284

Yellow-hooded Blackbird

Ovenbird
aurocapilla

Worm-eating Warbler

Louisiana Waterthrush

Northern Waterthrush
noveboracensis

but has narrow rusty-tipped tertials and greater coverts in fresh plumage. **GV** Race *notabilis* generally has greyer upperparts, whiter underparts and supercilium, with slightly larger bill. **Voice** Usual call a sharp, metallic "chink". Rarely heard song (in late winter/spring) 3–4 short but loud emphatic notes ending in downslurred flourish, "swee swee swee chit chit weedleoo". **SS** See Louisiana Waterthrush and Ovenbird.

Bachman's Warbler *Vermivora bachmanii* **CR(PE)**
Formerly not uncommon winter visitor (**Cuba**).

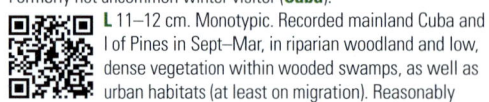

L 11–12 cm. Monotypic. Recorded mainland Cuba and I of Pines in Sept–Mar, in riparian woodland and low, dense vegetation within wooded swamps, as well as urban habitats (at least on migration). Reasonably common until 1940s, with last well-documented records in W Cuba in 1962, 1964 and perhaps 1966, with at least eight subsequent claims not accepted (last in 1988). Dedicated searches of former wintering areas in 1987–89 drew a blank (Hamel 1989). Expansion of sugarcane industry in Cuba, which accelerated post-Revolution, perhaps a causal factor in species' decline to apparent extinction. Single record in Bahamas, on Cay Sal (Riley 1905). **ID** Striking *Vermivora* with slender, slightly decurved bill. **Ad** ♂ has black crown, throat and breast, olive-green upperparts, yellow forehead, chin, malar and underparts. Yellow wingbend very distinctive. White patches in tail like those of Northern Parula. **Ad** ♀ has pale eye-ring, yellowish forehead, grey crown and nape, olive back, a grey or dusky throat, and whitish vent; lacks black patches and yellow wingbend. **1st-y** is like ad but ♂ has forehead washed olive, no black on forecrown and much duller breast marking; ♀ is very drab (can almost entirely lack yellow pigments), without white in tail and has whitish eye-ring (Dunn & Garrett 1997). **Voice** Poorly known, but a low hissing "zeep e eep" reported. **SS** ♂ Hooded Warbler has yellow face surrounded by black hood and ♀ less black on head; Oriente Warbler lacks black patches and has grey upperparts. ♀ (especially 1st-y) of present species might require careful separation from resident race of Yellow Warbler, as well as Orange-crowned, Tennessee and Nashville Warblers.

Golden-winged Warbler *Vermivora chrysoptera* **NT**
Generally rare passage migrant or winter visitor (**GA**, **O**), vagrant (**LA**).

L 12–12.5 cm. Monotypic. Very uncommon passage migrant (Sept–Nov, Mar–May) and even rarer in winter (Dec–Feb), mainly in Greater Antilles (but virtually no recent records in Cuba). Vagrant Virgin Is, St Kitts, Antigua, Barbados, St Vincent and San Andrés. From gardens and light woodland (Cuba) to montane forest (Puerto Rico); easily overlooked as usually forages in canopy. **ID** Distinctive by virtue of bright golden-yellow crown and wing patches. **Ad** ♂ has broad white supercilium (wider behind eye), black ear-coverts and throat separated by broad white stripe, grey upperparts and tail, and pale greyish-white underparts. **Ad** ♀ is similar but duller, with grey (not black) ear-coverts and throat, upperparts tinged olive. **1st-y** is similar to respective ad, but has broad olive tertial fringes, grey upperparts washed brownish, chin usually whitish and white in outertail feathers often reduced or less pure. **Voice** Calls probably indistinguishable from those of Blue-winged Warbler. **SS** Very distinctive in all plumages. Hybridizes regularly with Blue-winged Warbler, producing two phenotypes: commoner 'Brewster's Warbler' (records Cuba and Hispaniola) with white underparts and narrow black eyestripe (♀ is duller), and much rarer 'Lawrence's Warbler' (no records), which has yellow underparts, a dark throat and blackish ear-coverts. A backcross or second-generation (or higher) hybrid recorded on St Martin (Brown & Collier 2005b).

Blue-winged Warbler *Vermivora cyanoptera* **LC**
Uncommon passage migrant and winter visitor (**GA**, **O**).

L 11–12 cm. Monotypic. Present late Aug–early May in Bahamas, Greater Antilles, Cayman and Virgin Is, but records exceptional elsewhere (e.g. Guadeloupe, San Andrés). Humid forest and edges, well-treed areas including *Casuarina* windbreaks, foraging mostly in midstorey. **ID** Small, delicate but beautifully plumaged wood-warbler. **Ad** ♂ has bright yellow forehead, throat and underparts contrasting with short narrow black eyestripe, olive-green upperparts, blue-grey wings with two white wingbars, white vent. **Ad** ♀ is similar to ♂ but slightly duller, with less yellow on head. **1st-y** is similar to ad ♀ but even duller with less yellow on head in ♂, whereas 1st-y ♀ has forehead virtually concolorous with rest of upperparts. **Voice** Usual call a sharp, rather musical "tchip". Flight call a thin, buzzy "zwee". **SS** Prothonotary Warbler lacks white wingbars and black eyestripe. Compare Golden-winged Warbler and note potential for hybrids between them (described under latter).

Black-and-white Warbler *Mniotilta varia* **LC**
Common (**GA**, **O**) to fairly common passage migrant and winter visitor (most of **LA**).

L 11–13 cm. Monotypic. Present mid Jul to late May, but mainly Aug–Apr throughout region, being one of the most regularly encountered parulids; status in Lesser Antilles varies by island (only vagrant on some of southernmost). Forest and woodland, mangroves, parks and gardens; to 2600 m on Hispaniola. Forages at all levels, by creeping up, down and around trunks and heavier branches. Regularly joins mixed-species flocks. **ID** Unmistakable, stocky, relatively short-tailed but quite long-billed warbler. **Ad br** ♂ is brightest and purest black and white, with black throat and ear-coverts. **Ad non-br** ♂ is marginally duller, with throat mottled whitish. **Ad** ♀ is duller overall, with white of underparts less pure, grey ear-coverts, whitish throat, and flanks streaking greyer. **1st-y** ♂ in autumn is like ad ♀ but averages brighter, with flanks streaking blacker, by spring resembles ad ♂ but duller and remiges, tail, primary-coverts and alula worn and brown. ♀ in fresh plumage is relatively dull, washed pale buff on lores, ear-coverts and underparts, and flanks streaking greyer and less distinct than ad. **Voice** Call a sharp, hard "tick", becoming a chatter in alarm; flight call a soft, thin "seet" or "seet-seet", often also given when foraging. **SS** Uniquely black-and-white tree-creeping warbler, but compare ad ♂ Blackpoll Warbler (very different face pattern), Arrowhead Warbler (Jamaica only) and Elfin Woods Warbler (Puerto Rico).

Bachman's Warbler

Golden-winged Warbler

'Brewster's Warbler'

Blue-winged Warbler

Black-and-white Warbler

287

Prothonotary Warbler *Protonotaria citrea* LC
Uncommon passage migrant to vagrant (**throughout**) and winter visitor (**Guadeloupe**).
L 13–15 cm. Monotypic. Generally uncommon passage migrant, especially to coasts, throughout region, Aug–Nov and Feb–Apr, in mangroves, riparian and open woodland, swampy areas and scattered trees near flowing water, even semi-xerophytic habitats; regularly overwinters in mangroves on Guadeloupe, and occasionally does so elsewhere, e.g. in Puerto Rico. Forages at low to mid levels, sometimes with mixed-species flocks. **ID** Strikingly coloured parulid. **Ad** ♂ is mostly brilliant golden-yellow with prominent black eye, green upperparts, blue-grey wings and tail (latter with white patches). Belly slightly paler than breast, vent white. **Ad** ♀ is considerably duller, especially on head, crown and nape heavily washed olive (not contrasting with mantle), duller yellow throat and breast, less white in tail. **1st-y** ♂ is similar to ad ♀ but averages brighter with more white in tail; ♀ is rather duller than ad, with heavier olive wash to head and breast, and again less white in tail. **Voice** Call a loud ringing "tsip" or "chip", and softer, more sibilant "psit"; flight call a long, thin, clear "seeep", sometimes given when perched. **SS** Blue-winged Warbler has two wingbars and black eyestripe; the Cuban endemics, Yellow-headed and Oriente Warblers have pale grey underparts or grey crown, respectively.

Swainson's Warbler *Limnothlypis swainsonii* LC
Uncommon to rare winter visitor (**GA**, **O**).
L 13–14 cm. Monotypic. Uncommon in Cuba and Jamaica, rarer in Bahamas, Cayman Is, rest of Greater Antilles and San Andrés; early Sept–mid Apr. Presence best revealed via mist-net surveys. Forests with dense understorey and extensive leaf litter; also canebrakes, thickets and swampy borders; recorded to 2025 m on Hispaniola, but usually much lower. Retiring and typically observed on or very close to ground, often in company with an Ovenbird or Worm-eating Warbler. Highly philopatric to wintering grounds (McNicholl 1992). **ID** Distinctive but relatively featureless parulid with long, deep-based and pointed bill. **Ad** has warm brown crown, long whitish supercilium, dark eyestripe, brownish-olive upperparts and whitish underparts. **1st-y** has pale yellowish-washed underparts. **Voice** Usual call a loud, sweet "sship", similar to Worm-eating or Prothonotary Warblers, but louder and more forceful; flight call a high-pitched, thin, slightly buzzy "swees", sometimes doubled. **SS** Worm-eating Warbler has black crown- and eyestripes.

Tennessee Warbler *Leiothlypis peregrina* LC
Fairly common to rare passage migrant and winter visitor (**GA**, **O**), vagrant (**LA**).
L 10–13 cm. Monotypic. Principally moves through W of region, in Bahamas, Cayman Is, Cuba and San Andrés (locally abundant on last two), much less commonly across rest of Greater Antilles, and only vagrant in Lesser Antilles; mid Sept to early May, with incidence of midwinter records apparently increasing in recent decades, at least in Cuba. Forest edge, woodland, shade-coffee plantations, gardens and taller scrub, generally arboreal (often in midstorey to canopy); to mid elevations. Sometimes with mixed-species flocks. **ID** Rather short-tailed and point-billed parulid. **Ad br** has olive-green upperparts, darker wings and tail, contrasting grey head with slight dark eyestripe and whitish supercilium; white underparts. ♀ is duller, with head washed olive, supercilium and underparts washed yellowish, especially breast. **Ad non-br** is similar to br ♀ but even duller, some with mostly pale yellowish underparts, contrasting white vent, head more or less concolorous with upperparts, and yellowish supercilium. **1st-y** has slightly more obvious wingbars and more yellowish underparts (sometimes to vent). **Voice** Call a soft, sharp "tsit"; flight call a thin, clear "see". **SS** Longer-tailed and more patterned Orange-crowned Warbler has faint supercilium, thin eyestripe and yellow (not white) vent. All vireos have heavier bills and some possess much more obvious wingbars. **TN** North American authorities usually place in genus *Oreothlypis*.

Orange-crowned Warbler *Leiothlypis celata* LC
Rare passage migrant and winter visitor (**Bahamas**), vagrant elsewhere.
L 11.5–13 cm. Four subspecies, of which nominate occurs in region. Records Oct–early May. Rare winter visitor to Bahamas, more or less only vagrant in Cuba (where one record perhaps attributable to race *lutescens*) and elsewhere in N of region. Open woodland and coastal scrub, even parks and gardens, mostly foraging at low to middle levels. **ID** Rather dull parulid, with variable orange crown patch (seldom visible in field). **Ad** has greyish-olive upperparts, often washed slightly greyer on head. Faint pale supercilium, thin dark eyestripe and narrow pale eye-ring. Paler greyish-olive underparts, brighter yellowish on vent, faint olive streaks on breast-sides. ♀ is slightly duller. **1st-y** is mostly brownish olive-grey, paler below, with darker mottling on throat and breast, narrow yellowish wingbars. **GV** Race *lutescens* smaller and much brighter, quite green above and yellow below. **Voice** Call a hard, sharp "tek" or "chet"; flight call a thin high "seee". **SS** Compare Tennessee Warbler, and Philadelphia and Eastern Warbling Vireos, both of which have white supercilia and grey crown. **TN** North American authorities usually place in genus *Oreothlypis*.

Nashville Warbler *Leiothlypis ruficapilla* LC
Rare passage migrant and winter visitor (**Bahamas**, **Cuba**, **Cayman Is**), vagrant elsewhere.
L 11–12 cm. Two subspecies, of which nominate occurs in region. Records Sept–Apr; vagrant San Andrés, Jamaica, Hispaniola, Puerto Rico and reported Guadeloupe. Coastal thickets and other wooded areas, foraging at all levels, but mainly fairly low, in outer branches. **ID** Small, rather dainty wood-warbler. **Ad br** ♂ has olive-green upperparts, contrasting grey head and neck, bold white eye-ring, dull whitish lores, and semi-concealed rufous crown patch. Yellow underparts, small whitish area on belly. **Ad br** ♀ shows less contrast between head and upperparts, and paler yellow underparts. **Ad non-br** is very slightly duller. **1st-y** is duller than ad non-br; ♀ dullest of all, with upperparts washed brownish and underparts pale buffy yellow, often with whitish throat. **Voice** Usually quiet in region. **SS** Oriente Warbler (Cuba only) has grey upperparts. ♀/1st-y Mourning and Connecticut Warblers larger with all-grey hood to throat, and solid yellow belly. Northern Parula has wingbars. **TN** North American authorities usually place in genus *Oreothlypis*.

Prothonotary Warbler

Swainson's Warbler

Tennessee Warbler

Orange-crowned Warbler
celata

Nashville Warbler
ruficapilla

289

Virginia's Warbler *Leiothlypis virginiae* `LC`
Vagrant (**Bahamas**).

L 11–12 cm. Monotypic. Grand Bahama, Mar 1993 (Smith *et al.* 1994) in mangroves, and Abaco, Aug 2018 (eBird); report from W Cuba, also in mangroves, Mar 1989, unacceptable (Kirkconnell *et al.* in press). Winters in S & W Mexico, where usually occurs low in dense semi-arid scrub in highlands. **ID** A rather shy, mostly grey parulid. **Ad** ♂ has mid-grey upperparts and paler grey underparts, with contrasting yellow patches on breast and vent, and pale olive-yellow rump. Bold white eye-ring, pale grey lores and semi-concealed rufous crown patch. **Ad** ♀ is slightly duller with less yellow on breast. **1st-y** ♂ is browner than ad ♂ with less chestnut in crown. **1st-y** ♀ is even duller than ad ♀, with little or no yellow on breast, and no rufous on crown. **Voice** Call a dry sharp "tink" or "chink"; flight call a high, clear "seet". **SS** Shorter-tailed Nashville Warbler is generally less grey (always with some olive-green in upperparts), but extent of yellow in plumage can be very similar due to variation in present species. **TN** North American authorities usually place in genus *Oreothlypis*.

Semper's Warbler *Leucopeza semperi* `CR`
Exceptionally rare, perhaps extinct, resident (**St Lucia**).

L 14.5 cm. Monotypic. Endemic. Perhaps locally common in 19th century but very rare by 20th century, with only five certain records since 1920s (and none since 1961), all from ridge between Piton Flore and Piton Canaries. Unconfirmed, but probably reliable, sightings in May 1989 at Gros Piton, Sept 1995 at Piton Flore, and in 2003. Lower montane and montane rainforests and elfin woodland with undisturbed understorey. **ID** Dull sexually monomorphic parulid with long, deep-based, rather pointed bill. **Ad** has dark grey head and upperparts, paler supercilium, browner wings and tail, whitish underparts with extensive grey-brown wash on breast-sides and flanks; bill dark greyish horn with mainly flesh-coloured mandible; legs flesh-coloured. **Juv** is undescribed, but 1st-y is browner above, washed pale buff below. **Voice** Song undescribed. Only documented calls a soft "tuck-tick-tick-tuck" and a chattering in alarm. **SS** Should be unmistakable.

Connecticut Warbler *Oporornis agilis* `LC`
Very rare passage migrant to vagrant (**GA**, **LA**, **O**).

L 13.5–15 cm. Monotypic. Vagrant or very rare Bahamas, Cayman and Virgin Is (St Croix), throughout Greater Antilles (mostly on Hispaniola, almost certainly due to incidence of mist-netting work in appropriate habitats), St Barthélemy, Guadeloupe and Barbados, with records in Sept–early Nov and Apr–May, once Jan (Puerto Rico). All types of woodland (including mangroves) with dense cover, favouring wet thickets, foraging on or near ground, mostly by walking rather than hopping. **ID** Complete pale eye-ring, slightly thrush-like shape and terrestrial habits. **Ad** has grey hood to upper breast, olive-green upperparts, yellow underparts. ♀ is duller with brownish-grey hood, paler throat, and paler yellow below. **1st-y** ♂ in autumn resembles ad ♀, but eye-ring creamier and less whitish, by spring resembles ad ♂ but averages slightly duller. **1st-y** ♀ is relatively dull in autumn, with brownish-olive hood and pale buff eye-ring, much like ad by spring. **Voice** Typically quiet, but gives a loud sharp, nasal "plink" or high-pitched, buzzy "zee". **SS** ♀/1st-y Mourning Warbler has shorter undertail-coverts (thus longer-looking tail), strongly yellow-tinged throat (whitish or pale tawny in present species, and paler and duller below this), with at most a narrower and usually incomplete white eye-ring (many none at all, but note eye-ring in Connecticut can also be broken, albeit typically remains more obvious). Gait on ground also differs (Mourning hops, Connecticut walks).

Mourning Warbler *Geothlypis philadelphia* `LC`
Rare passage migrant or vagrant (**GA**, **O**).

L 13–15 cm. Monotypic. Very rare passage migrant in W & N of region, where apparently regular only on Turks & Caicos and San Andrés, Sept–early Nov and Apr–May. Has apparently overwintered once, at high altitude in SE Dominican Republic (Keith *et al.* 2003). Woodland and scrub with dense undergrowth, thickets and swamp edges. Forages on or near ground. **ID** Largely terrestrial parulid. **Ad** ♂ has dark grey hood to upper breast, heavy black mottling on breast, olive-green upperparts, yellow underparts, black maxilla and pink mandible. **Ad** ♀ has paler grey hood, lighter throat, no black on breast, and narrow broken whitish eye-ring. **1st-y** ♂ in autumn resembles ad ♀, but has yellowish-washed throat and often faint black mottling on breast-sides; by spring like ad ♂ but slightly duller and can have remnants of eye-ring. **1st-y** ♀ is duller than 1st-y ♂ in autumn, with brownish-olive hood, but like ad ♀ by spring. **Voice** Call a sharp, unmusical "jik" or "chit"; flight call a sharp "zeee", less buzzy than Connecticut Warbler. **SS** See Connecticut Warbler.

Kentucky Warbler *Geothlypis formosa* `LC`
Very rare passage migrant and even rarer winter visitor (**GA**, **LA**, **O**).

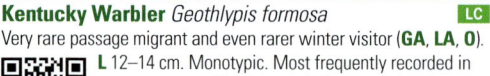

L 12–14 cm. Monotypic. Most frequently recorded in westernmost archipelagos—Bahamas, Cuba, Cayman Is and San Andrés—in Aug to Apr; basically only vagrant in Lesser Antilles (except Guadeloupe, where mist-netting efforts most intense). Wetter forest, mangroves and second growth with dense understorey, to mid elevations, foraging on ground and near it. **ID** Striking parulid with distinctive head pattern. **Ad** ♂ has bold yellow 'spectacles' surrounded by black crown, face and neck-sides, olive-green upperparts, uniform yellow throat and underparts. **Ad** ♀ is duller on head, with mostly olive crown. **1st-y** ♂ is duller than ad with black on head further restricted or lacking altogether; ♀ has black replaced by greyish olive. **Voice** Call a low, sharp "chup", given persistently when agitated; flight call a loud buzzy "zeep". **SS** Face pattern very distinctive; Hooded Warbler has variable black hood, all-yellow face and white tail spots; Canada Warbler has white vent in all plumage; and 1st-y ♂ Common Yellowthroat lacks yellow 'spectacles' and uniform yellow underparts.

Virginia's Warbler

Semper's Warbler

Connecticut Warbler

Mourning Warbler

Kentucky Warbler

291

Bahama Yellowthroat *Geothlypis rostrata* LC
Fairly common to uncommon resident (**Bahamas**).
L 15 cm. Four subspecies, of which all (nominate[A] New Providence, N Bahamas, *tanneri*[B] Grand Bahama and Abaco, N Bahamas, *exigua*[C] Andros, WC Bahamas, *coryi*[D] Eleuthera and Cat I, EC Bahamas) occur in region. Endemic (three reports of vagrants in Florida, USA). Dense scrubby understorey of open pine forest, especially where bracken and thatch palm (*Leucothrinax morrisii*) dominate; also, less commonly, in other woodland types and low coppice. Usually not in damp or marshy habitats. Rare or close to extinct on New Providence and Andros, and uncommon on Eleuthera. Sight record on Little Inagua not generally accepted. **ID** Large, somewhat sluggish warbler. **Ad** ♂ has broad black mask, mid-grey crown, yellowish olive-green upperparts, yellow throat and underparts, slightly paler belly, washed olive on flanks. **Ad** ♀ has brownish-washed crown with greyish ear-coverts and narrow greyish-white supercilium. **Juv** is undescribed. **1st-y** in autumn is slightly duller than ad, with marginally browner upperparts and more buffish underparts, ♂ probably has less distinct head pattern. **GV** Race *exigua* slightly smaller and smaller-billed than nominate, with darker upperparts, ♂ has darker grey crown and narrower, usually broken forecrown band; *coryi* brightest, with crown mostly olive, narrow forecrown band mainly or entirely yellow, less olive on flanks; *tanneri* intermediate between *coryi* and nominate in head pattern, and has heavier olive wash on flanks than former. **Voice** Song a loud "witchity witchity witchit", very similar to Eastern Yellowthroat but richer and slower-paced. Usual call a rather sharp "tuck", less harsh than Eastern Yellowthroat; also gives dry rattle. **SS** Eastern Yellowthroat smaller with less heavy and shorter bill, less yellow on underparts (extends to belly in Bahama), pale frontal band usually more obvious, and appears less lethargic; young Bahama acquires ad plumage more swiftly.

Common Yellowthroat *Geothlypis trichas* LC
Thirteen subspecies divided into three subspecies groups, of which one occurs in region.

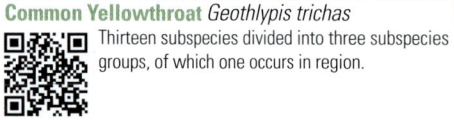

Eastern Yellowthroat *Geothlypis (trichas) trichas*
Fairly common passage migrant and winter visitor (**GA**, **O**), vagrant (**LA**).
L 11.5–13 cm. Four subspecies in group, of which three (nominate, *typhicola*, *ignota*) occur in region. Generally common and widespread across N & W of region including Bahamas, Greater Antilles, Cayman Is, Virgin Is and San Andrés; mainly vagrant in Lesser Antilles. Present Sept–May (rarely Jun). Most abundant in wet areas with grass and bushes, including swamps and marshes, usually foraging on or near ground, but also found in all manner of dry scrubby growth, including forested areas, to 2300 m (Hispaniola). Race *ignota* apparently very rare in region; *typhicola* more numerous, but still outnumbered by nominate. **ID** Similar to but more numerous and widespread version of Bahama Yellowthroat. **Ad** ♂ has broad black mask bordered above by greyish white, olive-green crown to upperparts, yellow throat, breast and vent, and whitish belly. **Ad** ♀ has rufous-tinged olive crown, mottled olive ear-coverts, and indistinct pale eye-ring and short supercilium. Yellow on breast slightly paler than ♂. **1st-y** ♂ in autumn has few black feathers on mask, otherwise like ad ♀, by spring resembles ad ♂ but retains some buff in eye-ring. **1st-y** ♀ is relatively dull, with pale buffy-yellow throat/breast and vent. **GV** Race *typhicola* has relatively smaller bill and browner-washed flanks than nominate; *ignota* has browner upperparts than nominate, warm brown flanks and brighter yellow breast. **Voice** Usually emits dry, husky "tjip" call, also a dry rattle; flight call a low-pitched, buzzy, unmusical "zeet". In late winter/spring, may sing loud and rollicking "witchity witchity witchity witch", with distinctive rhythm. **SS** See Bahama Yellowthroat. Both latter and present species in plumages other than ad ♂ may require separation from Connecticut, Mourning and Kentucky Warblers: the first-named is larger, chunkier and longer-winged, with a more obvious eye-ring, and is typically terrestrial; Mourning is more uniformly yellow below (including throat and belly), shorter-tailed and has thin, complete eye-ring; and Kentucky is even more consistently bright yellow on underparts, brighter olive-green above and has a distinctive yellow supraloral (latter sometimes shown, to lesser extent, by some ad ♂ Eastern Yellowthroats).

Whistling Warbler *Catharopeza bishopi* EN
Rare resident (**St Vincent**).
L 14.5 cm. Monotypic. Endemic. Primary rainforest and palm brakes at 300–600 m, less commonly to 1100 m in elfin forest, and humid secondary forest and edges. Forages alone or in pairs, hopping slowly through understorey or acrobatically searching undersides of leaves. **ID** Distinctive warbler with long tail frequently held cocked. **Ad** has blackish-grey head, broad breast-band and upperparts, bold white eye-ring, whitish loral spot and underparts, and dark grey flanks. **Juv** has head and body mostly dusky brown, with little contrast between upper- and underparts, lacks eye-ring and breast-band. **1st-y** has dark olive-brown head and upperparts, narrower and buffier eye-ring, cinnamon-buff below with indistinct olive-brown breast-band. **Voice** Series of short, rich whistles starts softly and relatively low-pitched, rises rapidly in pitch with crescendo effect. Usual call a soft, low-pitched "tuk" or "tchuk"; also harsher "tuk" when agitated. **SS** Nothing similar within its range.

Plumbeous Warbler *Setophaga plumbea* LC
Common resident (**Guadeloupe**, **Dominica**).
L 12–14 cm. Monotypic. Endemic. Dry lowland scrub-forest, montane forest, elfin forest with dense understorey, occasionally mangroves, and especially numerous in swamp forest. Generally common, but uncommon on Marie-Galante and perhaps extirpated on Terre de Haut (Les Saintes, Guadeloupe). Forages in lower growth. **ID** Grey parulid that regularly flicks tail. **Ad** is grey above with white supercilium broken over eye, white spot below eye, two wingbars, white-tipped tail, and whitish underparts with greyer flanks. **Juv** is much greener above with a yellowish supercilium and heavily saturated underparts. **1st-y** is like ad, but has greyish-olive upperparts, with buffy-white face markings and underparts, and buffy wingbars. **Voice** Short and simple but quite melodic "pa-pi-a" or "de-de-diu". Usual calls include short "chek" and loud rattle. **SS** Nothing similar within its range.

rostrata ♀

coryi ♂

Bahama Yellowthroat

trichas ♀

rostrata ♂

♂

ignota ♂

Eastern Yellowthroat
(Common Yellowthroat)

Whistling Warbler

Plumbeous Warbler

293

Elfin Woods Warbler *Setophaga angelae* **EN**
Uncommon and local resident (**Puerto Rico**).
L 12.5–13.5 cm. Monotypic. Endemic. Mainly undisturbed, closed-canopy humid lower montane, montane and elfin forests with many vines, high subcanopy and sparse understorey, at 370–1030 m. Forages largely in canopy; may accompany mixed-species flocks when not nesting. Remarkably, went undiscovered until 1971. **ID** Distinctive black-and-white parulid. **Ad** has black head with white lores, narrow white supercilium, broken white eye-ring, white band around rear ear-coverts and nuchal line. Black upperparts with white wingbars and tertial spots. White below, boldly streaked black on throat, breast and flanks. ♂ has more streaked underparts. **Juv** is unstreaked below. **1st-y** has ad pattern, but greyish-olive head and upperparts, with wingbars and tertial spots tinged yellowish, throat and underparts pale olive-yellow with indistinct darker streaks. **Voice** Series of short, rapidly delivered unmusical notes on single pitch, increasing in volume and ending in short series of double notes; resembles song of Bananaquit. Usual call a short metallic "chip", also similar to Bananaquit. **SS** Black-and-white Warbler creeps up and down tree trunks and has black and white stripes on crown. Compare ad ♂ Blackpoll Warbler.

Arrowhead Warbler *Setophaga pharetra* **LC**
Locally common resident (**Jamaica**).
L 12.5–13 cm. Monotypic. Endemic. In breeding season mainly in humid montane forest, also using lowland humid forest at other times. Forages by gleaning at all levels on branches, leaves and vines. **ID** Small, heavily streaked parulid. **Ad** has head and upperparts streaked black and white, two white wingbars, dark tail, black arrowhead streaks on throat to flanks, and whitish vent. ♀ is slightly duller. **Juv** has unstreaked underparts and brownish-olive upperparts, yellowish eye-ring and relatively indistinct wingbars. **1st-y** has all white parts greyer. **Voice** Series of high-pitched squeaky notes, e.g "sww-sw-swee-sww-sw-swee-sww-sw-swee-swee-swee". Usual call a high-pitched, metallic "tic", repeated regularly. **SS** Black-and-white Warbler has different head pattern (bolder, broader stripes, rather than narrow streaks) and creeps around tree trunks. Young birds from non-ad ♂ Blackpoll Warbler by fine yellowish-buff streaks on head and upperparts, less distinct wingbars, uniform underparts, and lack of distinct eyestripe and supercilium. Jamaican Vireo similar to 1st-y but lacks eye-ring, dark eyestripe and white in tail.

Hooded Warbler *Setophaga citrina* **LC**
Uncommon and local passage migrant, even rarer winter visitor (**GA**, **O**), mainly vagrant (**LA**).
L 12–14 cm. Monotypic. Present early Aug to mid May (mainly Sept–Apr). Most frequent in N & W of region, in Bahamas, some of Greater Antilles, Cayman and Virgin Is, and San Andrés, but only casual on most of Lesser Antilles. Humid forest with abundant understorey, and mangroves. Forages at lower levels by flycatching or gleaning. **ID** Restless parulid that flicks and fans its tail constantly, revealing much white in outer rectrices. **Ad** ♂ has yellow face and forehead highlighting dark eye and lores, virtually surrounded by black hood; olive-green upperparts and yellow underparts. **Ad** ♀ is variable, but always considerably duller on head with less white in tail. **1st-y** ♂ is very similar to ad ♂ but shows pale fringes to black of hood. **1st-y** ♀ is similar to ad ♀, but almost always lacks black on head. **Voice** Call a loud, sharp, metallic "chink" or "tchip". **SS** ♀ Eastern Wilson's Warbler is smaller and squatter, has yellow (not dark) lores and lacks white in tail. Compare rarer Kentucky Warbler.

American Redstart *Setophaga ruticilla* **LC**
Common winter visitor and passage migrant (**throughout**).
L 11–13 cm. Monotypic. Present late Jul to late May (though few arrive before mid Sept and most have left by Apr), being common virtually throughout region, albeit less so in Lesser Antilles. Has bred, apparently several times, in W Cuba (Kirkconnell & Garrido 1996) and there are occasional summer records from elsewhere (*NAB* 62: 630). Woodland and scrub, including parks and almost any areas with trees, including towns and mangroves, to at least 1500 m. Some degree of habitat segregation by age and sex. Forages at all levels, but mainly in subcanopy, by gleaning or acrobatic flycatching. **ID** Restless parulid with long frequently fanned tail highlighted by large orange (♂) or yellow (♀) patches. **Ad** ♂ is mostly glossy black, with white rear underparts, and bright orange on wings and breast-sides. **Ad** ♀ has grey head, olive upperparts and whitish underparts, orange being replaced by yellow. **1st-y** ♂ in autumn resembles ad ♀ but breast-sides more orange-yellow, by spring sports scattered black feathers. **1st-y** ♀ is duller than ad, with very little (sometimes no) yellow in wing, paler yellow breast-sides and more uniform greyish-olive head and upperparts. **Voice** Call a sharp, sweet "chip" or "tsip", similar to but thinner than Northern Yellow Warbler; also a clear, penetrating and rising "sweet". Song (occasionally heard in region) a short variable series of high-pitched buzzy notes ending in a lower (occasionally higher) note, but this sometimes lacking entirely. **SS** Very distinctive.

Kirtland's Warbler *Setophaga kirtlandii* **NT**
Rare winter visitor (**Bahamas**), vagrant (**Cuba**, **Jamaica**).
L 14–15 cm. Monotypic. Entire population winters in Bahamas (including Turks & Caicos), mainly Oct–Apr. Vagrant: Cuba (offshore cays) Nov 2004 (Parada Isada 2006), Feb 2017 (eBird); Jamaica, Feb 2019. Possible hybrid between Kirtland's and Blackburnian Warblers, Dominican Republic, Oct 1997 (Latta & Parkes 2001), and two possible records of present species, Mar 1977 and Mar 1985 (Keith *et al.* 2003). Primarily C Bahamas in low, dense broadleaf scrub with scattered taller trees, also open Caribbean pine (*Pinus caribaea*) woodland, where forages in lower strata. **ID** Somewhat robust parulid that vigorously pumps tail. **Ad** ♂ has grey head, black lores and white eye-crescents, grey upperparts streaked black, two indistinct white wingbars, and yellow underparts fading to white on vent, black streaks on breast-sides and flanks. **Ad** ♀ is similar but duller with grey (not black) lores, brownish-washed upperparts and paler yellow below. **1st-y** is slightly duller than ad ♀, with brownish head and upperparts, pale buffy-yellow throat and underparts, and often some indistinct dark streaking on throat, breast and flanks. **Voice** Call a loud, smacking, relatively low-pitched "tchip"; flight call a thin, high-pitched "zeet". **SS** Magnolia Warbler smaller, non-br ♂ or 1st-y has yellow rump, greener back, bolder wingbars, complete eye-ring and species does not tail-pump. Yellow-throated and Bahama Warblers have black mask, white supercilia and forage in midstorey to canopy.

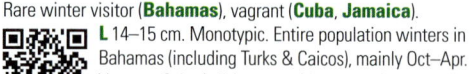

Elfin Woods Warbler

Arrowhead Warbler

Hooded Warbler

American Redstart

Kirtland's Warbler

Cape May Warbler *Setophaga tigrina* `LC`
Fairly common to uncommon winter visitor and passage migrant (**GA, LA, O**).
L 12.5–14 cm. Monotypic. Most of population winters in Greater Antilles, where present mid Sept–mid May (at least once in Jun); rarer in Lesser Antilles, where only vagrant in far S. Open second-growth forest, plantations, edges, parks, gardens and mangroves, favouring areas with numerous flowering trees, as mainly feeds on nectar. **ID** Rather slender-billed parulid. **Ad** ♂ has orange-chestnut face with rich yellow surround, dark olive crown, olive upperparts heavily black-streaked, olive-yellow rump, prominent white wing patch, white in outer tail, yellow throat and underparts heavily streaked black. **Ad** ♀ lacks orange-chestnut on face and has white wingbars (no patch). **1st-y** is duller than ad, ♀ often very dull, mostly grey-brown above and off-white below, with indistinct streaking on underparts and relatively faint wingbars. **Voice** Call a very high-pitched, thin "sip"; flight call a soft buzzy "zeet". Song may be heard in spring, a disyllabic "seetee seetee seetee seetee seetee". **SS** Magnolia Warbler has grey (not yellow) neck patch and a white band on most rectrices. Myrtle Warbler browner above with better-defined yellow rump and small yellow patches on sides. Compare Bay-breasted Warbler.

Cerulean Warbler *Setophaga cerulea* `VU`
Rare passage migrant to vagrant (mainly **GA, O**).
L 10–13 cm. Monotypic. Little more than a vagrant over most of region, with almost no records in Lesser Antilles; probably most frequently recorded on San Andrés and, to lesser extent, Cuba; reported early Aug–late Nov and Apr. Wooded areas, favouring tall trees, where forages in canopy. **ID** Rather short-tailed, plump-looking parulid. **Ad** ♂ has head and upperparts deep cerulean-blue streaked black, two white wingbars, and white underparts with narrow breast-band and streaks on flanks black admixed blue. **Ad** ♀ has turquoise-blue head and upperparts, pale yellowish-white supercilium and white wingbars, whitish underparts with yellow-tinged throat and breast, faint greyish flank streaks. **1st-y** ♂ resembles ad ♀, but has bluer upperparts, especially rump, more heavily streaked mantle, and whiter underparts. ♀ has much greener upperparts and yellower underparts than ad. **Voice** Call a sharp, emphatic and quite musical "chip"; flight call a loud, buzzy "zzee". **SS** Generally unmistakable if seen well. ♀ and young Cerulean could be confused with 1st-w ♀ Blackburnian Warbler, which is longer-tailed, has different head pattern (e.g. supercilium joining pale area on neck-sides), pale lines on back and buffier (less yellow) underparts. Tennessee Warbler lacks bold wingbars and streaks on underparts and has different bill shape. Compare even rarer Black-throated Grey Warbler.

Northern Parula *Setophaga americana* `LC`
Fairly common passage migrant and winter visitor (**throughout**).
L 10.5–12 cm. Monotypic. Widespread but generally less numerous in Lesser Antilles (especially in S); late Jul to mid May, with bulk of passage Sept–Oct and mid Mar–early Apr. Varied wooded habitats, including deciduous forest, plantations, parks and gardens, and scrub, to at least 1700 m. Mostly forages in canopy, but can descend lower. **ID** Small neatly proportioned warbler. **Ad** ♂ has blue-grey head and upperparts, with large yellowish-green patch on mantle, blackish lores, prominent white eye-crescent, two white wingbars, yellow throat and breast, narrow blue-grey band on upper breast with diffuse rufous band below, some red on sides; rest of underparts white. **Ad** ♀ is slightly duller and lacks blue-grey and rufous breast-bands. **1st-y** is even duller than ad with remiges edged greenish (not blue-grey), ♀ can have less distinct head pattern, relatively little yellow below and faint mantle patch. **Voice** Call a high, sharp "tsip" or "chip"; flight call a high, weak, descending "tsif". **SS** ♀ similar to Adelaide's Warbler, which has yellow supercilium and lacks yellowish-green patch on mantle. Much rarer Nashville Warbler lacks wingbars, has complete eye-ring and more extensively yellow underparts. Compare Yellow-throated and Bahama Warblers.

Magnolia Warbler *Setophaga magnolia* `LC`
Fairly common to uncommon passage migrant and winter visitor (**throughout**).
L 11–13 cm. Monotypic. Fairly common passage migrant and winter visitor across N & W of region, including Bahamas, Cayman and Virgin Is, Greater Antilles and islands in SW Caribbean, but generally uncommon in Lesser Antilles. Present mid Sept–late May. Mostly in lowlands, including open woodland, shade-coffee plantations, mangroves and occasionally gardens, to at least 850 m. Forages mainly at low to mid levels, gleaning and flycatching. **ID** Stocky parulid with white spots on tail forming band when spread. **Ad br** ♂ has blue-grey crown, broad white supercilium and black mask, narrow white crescent below eye, blackish upperparts, yellow rump and prominent white wing patch. Yellow throat, narrow black breast-band, broad black streaks on flanks and white vent. **Ad br** ♀ has grey crown and nape, black-streaked olive upperparts, less bold streaks on underparts. **Ad non-br** ♂ is duller, with grey mask, nape and breast-band, no supercilium, greenish upperparts, two white wingbars, slightly less bold streaks on flanks. **Ad non-br** ♀ has grey ear-coverts, underparts less streaked, narrow grey breast-band. **1st-y** is similar to non-br ♀ but even duller with unstreaked underparts. **Voice** Call a rather dry, nasal "clenk" or "tzek", which differs from calls of congeners. **SS** White median tail-band diagnostic. Myrtle Warbler also has some white on tail but a white or pale grey (not yellow) throat and yellow below limited to sides; larger Kirtland's Warbler (note range in region) has dark rump and incomplete eye-ring.

Cape May Warbler

Cerulean Warbler

Northern Parula

Magnolia Warbler

297

Bay-breasted Warbler *Setophaga castanea* **LC**
Generally rare passage migrant (**GA**, **O**), vagrant (**LA**).
L 13–15 cm. Monotypic. Rare passage migrant through westernmost islands in region, e.g. Bahamas, Greater Antilles, Cayman Is and San Andrés (where apparently fairly common), Sept–Nov and, fewer, Mar–May; mainly vagrant elsewhere. Wide variety of wooded habitats including open areas with scattered trees and gardens. Forages in canopy by gleaning or flycatching, and frequently joins mixed-species flocks. **ID** Parulid with strikingly different plumages. **Ad br** ♂ has black forecrown and face, chestnut crown, creamy-yellow neck-sides, grey upperparts streaked black, two white wingbars, large white tips to outer tail, chestnut on throat and upper breast extends to flanks. **Ad br** ♀ has only suggestion of ♂ head pattern and chestnut below confined to breast-sides. **Ad non-br** ♂ has olive-green head and upperparts, obscure supercilium, heavy black streaking on upperparts, pale buffy-white throat and underparts, much chestnut on flanks. **Ad non-br** ♀ is like non-br ♂ but crown and upperparts only faintly streaked, flanks warm buff, sometimes with chestnut feathers. **1st-y** is duller than ad, ♀ dullest, with only faint streaking above and off-white underparts without chestnut. **Voice** Calls a loud sweet "chip", very similar to Blackpoll Warbler, and thin, high-pitched "see" or "tseet". **SS** Non-br Blackpoll Warbler has paler legs, finely streaked underparts, white vent, and slightly less distinct eyestripe and supercilium.

Blackburnian Warbler *Setophaga fusca* **LC**
Uncommon to rare passage migrant (**GA**, **O**), vagrant (**LA**).
L 11–12 cm. Monotypic. Uncommon to rare passage migrant through N & W of region, mainly Bahamas, Cuba and San Andrés; only vagrant in E Greater Antilles and Lesser Antilles. Recorded Aug–Dec (mainly Sept–Oct) and Feb–May (especially Apr). Wooded areas, including conifers at mid elevations, usually in canopy, foraging by gleaning or flycatching; occasionally joins mixed-species flocks. **ID** Strikingly beautiful parulid. **Ad br** ♂ has black crown and ear-coverts, bright orange forecrown, supercilium and throat, black upperparts with white 'tramlines', broad white wing patch, much white in tail, off-white underparts tinged orange on breast, streaked black on flanks. **Ad br** ♀ has less intense orange tracts, paler head and upperparts, olive-brown streaked black, and two white wingbars. **Ad non-br** ♂ is like br ♀. **Ad non-br** ♀ is even duller. **1st-y** is duller than ad, ♀ has orange areas of head and underparts pale peachy buff, and flank streaks olive and rather indistinct. **Voice** Call a sharp, very high-pitched "tsip" or "chip". **SS** 1st-y Black-throated Green Warbler has buff-coloured throat and breast, and lacks whitish stripes on back. Compare Yellow-throated and Bahama Warblers.

(American) Yellow Warbler *Setophaga petechia* **LC**
Forty-three subspecies divided into four subspecies groups, of which two occur in region.

Northern Yellow Warbler *Setophaga (petechia) aestiva*
Fairly common to uncommon passage migrant (**throughout**).
L 11.5–12.5 cm. Nine subspecies in group, of which two or three (*rubiginosa*, *aestiva*, perhaps *amnicola*) occur in region. Passage migrant and perhaps winter visitor throughout, to all manner of open wooded habitats (including mangroves), but status inadequately known on many islands (e.g. just two records on Guadeloupe) due to confusion with resident races of Golden Warbler. Present at least mid Aug–mid Nov and late Feb–early May. Forages at all levels, by gleaning or flycatching. **ID** Almost uniform yellow warbler. **Ad br** ♂ (*aestiva*) has bright yellow head and underparts, with forecrown and ear-coverts tinged golden-orange, yellow-tinged olive-green upperparts, breast and flanks with broad rufous streaks. **Ad br** ♀ has faintly streaked underparts. **Ad non-br** ♂ is duller, ♀ is much duller. **1st-y** ♂ is similar to ad ♀, and ♀ is duller pale olive above with greyish wash on crown, no streaks below. **GV** Race *rubiginosa* (vagrant to Cuba) greyer above and paler below, ♂ has crown and nape concolorous with upperparts; *amnicola* (not yet confirmed to occur?) only slightly duller and darker than *aestiva*, ♂ generally duller yellow below with streaks somewhat darker and thinner, ♀ greyer above. **Voice** Call a loud, emphatic "tship". **SS** Golden Warbler has chestnut crown and dark olive upperparts. Eastern Wilson's Warbler has black crown and plain underparts. Saffron Finch (p. 332) is larger with heavier bill and occurs in grassland.

♀ breeding
♂ non-breeding
♂ breeding

Bay-breasted Warbler

first-winter

♀/♂ non-breeding

Blackburnian Warbler

♂ breeding

♂ rubiginosa

Northern Yellow Warbler
(American Yellow Warbler)

♀

aestiva

♂

299

Golden Warbler *Setophaga (petechia) petechia*
Common to fairly common resident (**throughout**).
L 13–13.5 cm. Eighteen subspecies in group, of which 14 (nominate[A] Barbados, *flaviceps*[B] Bahamas, *gundlachi*[C] Cuba, I of Pines and many offshore cays; *eoa*[D] Cayman Is and Jamaica, *albicollis*[E] Hispaniola, including Tortue and Î-à-Vache, *solaris*[F] Gonâve and Petite Gonâve, off Haiti, *chlora*[G] Cayos Siete Hermanos, off N Hispaniola, *bartholemica*[H] Puerto Rico, Virgin Is and S to Montserrat and Antigua, *melanoptera*[I] Guadeloupe S to Dominica, *ruficapilla*[J] Martinique, *babad*[K] St Lucia, *alsiosa*[L] on Grenadines, *armouri*[M] Providencia, *flavida*[N] San Andrés) occur in region. Near-endemic; also on Florida Keys (USA), Cozumel (Mexico), Leeward Antilles and coastal NE Venezuela (plus associated islands). Throughout region, but less common in N Bahamas. Mainly mangroves during breeding season, but nests in all habitats on Guadeloupe, and elsewhere will use all types of coastal scrub, freshwater marshes and riparian growth at other times; *ruficapilla* on Martinique in montane forest above 300 m. Forages at all levels by gleaning alone, and rarely follows mixed-species flocks. **ID** Bright, yellow-coloured warbler. **Ad** ♂ has dark rufous-chestnut crown, olive-green upperparts tinged yellowish. Throat and underparts bright golden-yellow, breast and flanks with well-defined dark rufous streaks. **Ad** ♀ lacks rufous cap, nape and ear-coverts yellowish green, unstreaked underparts or only faintly streaked rufous. **Juv** is probably pale olive-grey or olive-brown, slightly paler below and lacks yellow tones. **GV** Race *gundlachi* ♂ has darker olive-green upperparts, including crown (latter sometimes yellow-tinged, occasionally slightly rufous), ♀ has dull grey-olive upperparts, pale whitish to yellowish below; *flaviceps* like *gundlachi*, but yellower on head; *eoa* also like *gundlachi*, but ♂ slightly more rufous on crown (often to ear-coverts) and more obscurely streaked below; *albicollis* duller, paler below, ♂ crown slightly darker; *chlora* has chestnut crown darker, upperparts darker green; *solaris* paler than last, brighter and yellower above than *albicollis*; *bartholemica* relatively long-billed, ♂ has pale orange-rufous crown (more mottled in Virgin Is), moderately streaked underparts; *melanoptera* like last, but ♂ has slightly darker crown, narrowly streaked underparts, ♀ orange-tinged crown; *ruficapilla* ♂ has head and throat dark rufous-chestnut; *babad* ♂ like *gundlachi*, with paler crown; *alsiosa* ♂ has forehead golden-yellow; *armouri* ♂ has yellow crown (faint rufous), streaks form rufous breast patch; *flavida* ♂ has pale orange crown with heavier streaks below. **Voice** Song very variable, but typically 3–5 high-pitched "swee" notes on one pitch followed by short staccato warble, e.g. "sweet sweet sweet I'm so sweet". Call a dry strong "chip". **SS** See Northern Yellow Warbler. **TN** Some authors recognize race *cruciana* (Puerto Rico and Virgin Is) from *bartholemica*.

Chestnut-sided Warbler *Setophaga pensylvanica*   LC
Uncommon to rare and local passage migrant and winter visitor (**O**, **GA**), mainly vagrant (**LA**).
L 12–13 cm. Monotypic. Most reports from westernmost islands including Bahamas, Cuba and San Andrés (Pacheco 2012), in Sept–Nov and Feb–May, becoming rarer further E and only vagrant to most of Lesser Antilles. Few midwinter records, e.g. just one in Cuba (Kirwan *et al.* 2001), but regular at this season on Puerto Rico and Guadeloupe. Open woodland, edges and gardens with trees. **ID** Distinctive in all plumages. **Ad br** ♂ has yellow crown, black eyestripe, white forehead and face-sides, narrow black moustachial joining broad chestnut stripe on flanks, otherwise white underparts. Upperparts olive-green, heavily streaked black on mantle, two yellowish-white wingbars. **Ad br** ♀ has olive-green crown, greyer eyestripe and moustachial, less chestnut on flanks. **Ad non-br** ♂ has lime-green crown and upperparts, pale greyish-white face and underparts, prominent white eye-ring, two pale wingbars, no moustachial but extensive chestnut on flanks. **Ad non-br** ♀ is similar but with less chestnut (sometimes none) on flanks.
1st-y averages slightly duller with little or no chestnut on flanks.
Voice Call a rather husky "tchip"; flight call a distinctive, rather rough or burry "breet". **SS** Even non-br plumage distinctive by virtue of lime-green upperparts, conspicuous eye-ring and chestnut on flanks.

Blackpoll Warbler *Setophaga striata*   NT
Fairly common to uncommon passage migrant (**throughout**).
L 13–15 cm. Monotypic. Commonest in N & W of region, but regular migrant even across Lesser Antilles; Aug–Dec (mainly Sept–Nov) and late Mar–early Jun (mostly late Apr/early May). Occasional midwinter records in Cuba. Open woodland and coastal vegetation, foraging at middle to high levels, occasionally flycatching. Small groups may join mixed-species flocks. **ID** Most plumages relatively undistinctive. **Ad br** ♂ has black cap, streaky black chin and malar, contrasting white face, grey upperparts tinged olive and heavily streaked black, two white wingbars, and white underparts boldly streaked black on flanks. **Ad br** ♀ has short whitish supercilium, whitish face faintly mottled darker, streaks on sides duskier than ♂. **Ad non-br** has yellowish supercilium, dusky eyestripe, yellowish throat/breast, whitish belly and white vent, with bold blackish/dusky streaks on flanks. ♀ has less prominent streaking above and on flanks. **1st-y** is duller with relatively indistinct dark streaking above and blurred olive streaks on breast-sides. **Voice** Call a loud "chip" and flight call a buzzy "zeet"; very similar to Bay-breasted Warbler (probably indistinguishable). **SS** Ad non-br/1st-y Bay-breasted Warblers have unstreaked underparts, buffy (not white) vent and black (not pale) legs. Pine Warbler (breeds Bahamas and Hispaniola, vagrant elsewhere) lacks streaking above and has black (not yellowish) legs and feet.

Golden Warbler
(American Yellow Warbler)
petechia

Chestnut-sided Warbler

Blackpoll Warbler

301

Black-throated Blue Warbler
Setophaga caerulescens **LC**

Fairly common passage migrant and winter visitor (**GA**, **O**), mainly vagrant (**LA**).

L 12–14 cm. Two subspecies, of which both (nominate, *cairnsi*) occur in region. Present Aug–May (mostly late Sept–late Apr), with at least one Jul record. Varied wooded and scrub habitats, including primary forest, open deciduous woodland, plantations and gardens, mainly foraging at low to mid levels; to 2450 m (on Hispaniola) but usually lower. ♂♂ prefer more heavily forested localities. Often rather tame. **ID** White wing flash very distinctive. **Ad** ♂ has dark blue crown and upperparts, black throat and face, white underparts with broad black streak on flanks, and extensively white outer tail. **Ad** ♀ has dark brownish-olive head and upperparts, narrow whitish supercilium and lower eye-crescent, pale buffy underparts, less white on wing and tail than ♂, and wing patch can be pale buff. **1st-y** ♂ has greenish (not blue) edges to remiges; ♀ slightly browner above and buffier below, wing patch usually indistinct or lacking. **GV** Race *cairnsi* (much less numerous) ♂ has brighter and darker blue upperparts, often blackish mantle and larger white wing patch; ♀ usually darker and browner above with paler upperparts. **Voice** Call a soft flat "stip" or "tik"; flight call a prolonged "tseet". **SS** ♂ unmistakable; ♀ distinguished by narrow supercilium and white spot on wings, but compare Orange-crowned and Tennessee Warblers. ♀ might also be confused with Bananaquit, but note latter's shorter tail and short decurved bill.

Palm Warbler *Setophaga palmarum* **LC**

Two subspecies placed in separate subspecies groups, both occur in region.

Western Palm Warbler *Setophaga (palmarum) palmarum*

Common passage migrant and winter visitor (**GA**, **O**), vagrant (**LA**). **L** 12–14 cm. One subspecies in group. Ubiquitous winter visitor and passage migrant across Bahamas, Greater Antilles, Swan, Cayman and Virgin Is, but only vagrant to Lesser Antilles. Records in every month (even early Jun and late Jul). Open, usually dry, areas with short grass and scattered bushes; also mangroves, open pine woodland and scrub; common in urban areas, even being observed on building sites in downtown Havana. Generally below 500 m, but to 2400 m on Hispaniola. Forages mainly on ground or in low shrubs. **ID** Terrestrial parulid with tail-wagging habit; ageing and sexing very difficult in field. **Ad br** has rufous crown, pale yellowish supercilium, dark eyestripe, grey-brown head-sides and upperparts (latter streaked dark), olive-yellow rump, yellow throat and vent, otherwise whitish underparts streaked darker on breast and flanks. **Ad non-br** lacks rufous cap and yellow throat, and is less heavily streaked. **Voice** Call a sharp "tsik" or "tsup"; flight call a high-pitched "seet" or "see-seet". **SS** Prairie Warbler (also common in N of region and habitually tail-pumps) has blackish face markings and all-yellow underparts heavily streaked on sides; ♀/1st-y Cape May Warbler have white vent and do not pump tail; waterthrushes are also ground-dwellers but are more strikingly marked and bob their rear bodies (not just their tails).

Eastern Palm Warbler *Setophaga (palmarum) hypochrysea*

Very rare winter visitor (**GA**, **O**).

L 12–14 cm. One subspecies in group. Status very poorly known (potentially much overlooked and consequently not mapped here), with a few records Bahamas (e.g. USNM 274855, eBird), Cuba (Kirkconnell *et al.* in press) and Puerto Rico (Barnés 1947). However, mist-net surveys in SW Dominican Republic found that 2% of all Palm Warblers were this taxon (Latta *et al.* 2006) and Bond (1972) reported four ringed on Grand Bahama in just ten days in autumn 1971; given species' overall abundance, comparatively large numbers conceivably winter in region. Monroe (1968) opined that USNM 111240, from Swan Is, is perhaps this taxon and not Western Palm (as labelled). Records span late Oct–Mar. In same habitats as Western Palm Warbler. **ID** Very similar to Western Palm Warbler in plumage and behaviour. **Ad br** (vs. latter) has all-yellow ground colour to underparts with broad, bold chestnut streaking on sides and flanks, and warmer brown-tinged upperparts. **Ad non-br** differs from Western Palm in having supercilium and entire underparts yellowish (i.e. vent contrasts but little), and wingbars and tertial fringes are richer, more reddish brown. **Voice** Calls (to the experienced ear) sound a little sharper and higher-pitched than Western Palm Warbler. **SS** See Western Palm Warbler.

Olive-capped Warbler *Setophaga pityophila* **LC**

Fairly common resident (**N Bahamas**, **Cuba**).

L 12.5–13 cm. Monotypic. Endemic. Confined to N Bahamas (Grand Bahama, Abaco) and mainland W & E Cuba, in open pine (especially *Pinus caribaea*) forest and pine barrens, to c. 850 m in Cuba. Forages in midstorey to canopy, but occasionally descends to ground when nesting. **ID** Distinctive pine-inhabiting wood-warbler. **Ad** has most of crown yellowish olive, rest of upperparts slate-grey to plumbeous. White wingbars and outer tail, yellow throat and breast irregularly bordered by blotchy black streaks, whitish underparts with brownish-olive flanks. ♂ is brighter than ♀. **Juv** has brownish head and upperparts, indistinct whitish wingbars, paler underparts. **1st-y**, especially ♀, is often duller than ad. **Voice** Variable series of rather shrill whistles (usually 7–9), descending in pitch and delivered fairly slowly, "wisi-wisi-wisi-wiseu-wiseu". Call "tsip-tsip-tsip", repeated frequently. **SS** Yellow-throated and Bahama Warblers have bold white supercilia, black mask, white neck-sides and longer bills. **TN** Based on mensural characters (shorter tail), Garrido (2000b) argued for continued recognition of separate Cuban and Bahaman (*bahamensis*) races. Lovette *et al.* (1998) found little genetic structure to support this.

Black-throated Blue Warbler
caerulescens

Western Palm Warbler
(Palm Warbler)

Eastern Palm Warbler
(Palm Warbler)

Olive-capped Warbler

Pine Warbler *Setophaga pinus* `LC`
Fairly common resident (**Bahamas**, **Hispaniola**), vagrant elsewhere.
L 12.5–14.5 cm. Four subspecies, of which three (nominate vagrant, *achrustera*[A] NW Bahamas, *chrysoleuca*[B] Hispaniola) occur in region. Common in Caribbean pine forest in N Bahamas (Grand Bahama, Abaco, Andros, New Providence); on Hispaniola (common in Dominican Republic, rare in Haiti) in high-elevation pine forest at 700–2600 m. Vagrant elsewhere: Cuba (ascribed to nominate race) and rest of Greater Antilles, Cayman Is, San Andrés (Pacheco 2012) and several of Lesser Antilles; Oct–Apr. Forages at all levels, including ground, and may join mixed-species flocks. **ID** Fairly large-billed warbler. **Ad** ♂ (*chrysoleuca*) has olive-green head and upperparts, broken yellow eye-ring, white wingbars, extensively white outer tail, yellow underparts, white belly and vent, and dusky streaks on breast-sides. **Ad** ♀ has grey-green head and upperparts, pale greenish-yellow throat and breast, indistinctly streaked breast-sides, less white in tail. **Juv** is very drab, with grey-brown head and upperparts, buffy wingbars, and pale buffish-white underparts faintly mottled olive on breast. **1st-y** is duller, ♀ typically resembles juv but lacks mottling on breast. **GV** Both nominate and race *achrustera* are duller than *chrysoleuca*, especially those in Bahamas and particularly ♀♀. **Voice** Song (given year-round) a rapid, musical trill on single pitch. Call a sweet, sharp "chip"; flight call a slightly buzzy "zeet". **SS** White-eyed Vireo has stronger bill, distinct yellow eye-ring, white iris and unstreaked underparts; Yellow-throated Vireo has yellow eye-ring and no streaks on flanks. Orange-crowned Warbler lacks wingbars; Bay-breasted and Blackpoll Warblers have streaked upperparts and shorter-looking tails.

Audubon's Warbler *Setophaga auduboni* `LC`
Three subspecies divided into three subspecies groups, of which one occurs in region.

Audubon's Warbler *Setophaga (auduboni) auduboni*
Vagrant (**O**, **GA**).
L 14–15 cm. One subspecies in group. Records: Bahamas (Providenciales, Mar 1985, Grand Bahama, Oct 2017), W Cuba (Nov 2017), Jamaica (Dec 2005) and undocumented Grand Cayman (Dec 2017) (Aldridge 1987, Graves 2006, eBird). Could occur in any wooded habitat including scrub, thickets and gardens. **ID** Relatively large warbler with conspicuous yellow throat and rump. **Ad br** ♂ has grey crown with bright yellow central patch, blackish lores, conspicuous broken white eye-ring, grey upperparts streaked black, white on wing and outer tail, black breast-band, bright yellow patch on breast-sides, streaked black on flanks. **Ad br** ♀ is duller with paler yellow on crown and throat. **Ad non-br** is duller with brown-washed upperparts and fewer streaks below. **1st-y** is similar to non-br, but ♀ can be duller. **Voice** Call short, sharp "chep" or "chup" notes, softer, slightly higher-pitched and less emphatic than Myrtle Warbler. **SS** Myrtle Warbler has white (not yellow) throat, less white on wings forming two distinct bars, narrow white supercilium and black cheeks. Smaller Magnolia Warbler has yellow underparts. **TN** Usually treated as conspecific with Myrtle Warbler, under name Yellow-rumped Warbler.

Myrtle Warbler *Setophaga coronata* `LC`
Fairly common passage migrant and winter visitor (**GA**, **O**), vagrant (**LA**).
L 14–15 cm. Two subspecies, of which nominate occurs in region. Basically common but somewhat erratic visitor to N & W of region, mid Sept to early May; rare in Lesser Antilles (mostly vagrant). Wide range of habitats, including forest and edges, woodland, scrub, thickets and gardens, but also mangroves; to 2450 m. Forages at all levels, including ground; often in small groups of up to 50. **ID** Similar to, but much more numerous than, Audubon's Warbler. **Ad br** ♂ has grey crown with bright yellow central patch, white supercilium, black cheeks with white crescent below eye, grey upperparts streaked black, white wingbars and much white on outer tail. White throat, black breast-band, bright yellow breast-side patches, streaked black on flanks. **Ad br** ♀ has greyer ear-coverts, faint yellow crown and breast-band reduced to heavy streaking at sides. **Ad non-br** (both sexes) washed brownish on head and upperparts. **1st-y** ♀ is often very dull, with grey-brown head and upperparts streaked darker, narrow whitish supercilium, pale throat extending to neck-sides, off-white underparts faintly streaked darker on flanks, yellow on breast-sides virtually lacking but conspicuous yellow rump. **Voice** Virtually identical to Audubon's Warbler. **SS** See Audubon's Warbler; Magnolia Warbler has yellow (not white) throat and bold white median tail-band. Other parulids with yellow rumps have this area much paler/duller. **AN** Yellow-rumped Warbler (when treated as conspecific with Audubon's Warbler).

pinus

chrysoleuca

Pine Warbler

breeding

non-breeding

Audubon's Warbler

breeding

non-breeding

Myrtle Warbler
coronata

305

Yellow-throated Warbler *Setophaga dominica* `LC`
Fairly common passage migrant and winter visitor (**GA**, **O**), vagrant (**LA**).

L 13–14 cm. Monotypic. Frequent visitor to most islands in N & W of region, including Bahamas, Greater Antilles, Cayman Is and San Andrés; vagrant in Lesser Antilles. Present mid Jul–late Apr. Uses wide variety of habitats, including open woodland, second growth and gardens with tall trees, also mangroves. Forages mainly in canopy and regularly visits flowering trees. **ID** Boldly patterned warbler with notably long bill. **Ad** has black face and forecrown, white supercilium, lower eye-crescent and neck-sides, yellow supraloral and throat to breast. Rear crown and upperparts grey, streaked black on former. Two white wingbars, extensive white on outer tail; underparts streaked black on flanks. **Ad** ♀ is slightly duller, forecrown usually grey, streaked black, with less white in tail. **1st-y** is slightly duller than ad, ♀ can have brownish upperparts and pale yellow throat. **Voice** Call a loud, sweet "chip"; flight call a loud, clear, high-pitched "see". **SS** Bahama Warbler has longer bill, narrower white supercilium behind eye and smaller white area behind black cheeks, far more extensive yellow underparts blending into whitish vent and streaks on breast-sides. Olive-capped Warbler lacks white supercilium and white neck-sides; bulkier Kirtland's Warbler bobs its tail and lacks white neck-sides.

Bahama Warbler *Setophaga flavescens* `NT`
Fairly common resident (**Bahamas**).

L 13–14 cm. Monotypic. Endemic. Confined to Grand Bahama and Abaco, in Caribbean pine (*Pinus caribaea*) woodland. Yellow-throated Warblers wintering (Jul–Apr) on same islands also use pines but routinely occur in other habitats. Often forages by climbing up and down trunks like nuthatch (Yellow-throated, Olive-capped and Pine Warblers also feed on trunks, but never as consistently). **ID** The longest-billed parulid. **Ad** has white supercilium, yellow supraloral and grey cheeks. Bright yellow throat and underparts with black-streaked flanks and white vent. Dull brown upperparts. **Juv** has greyish olive-brown head and upperparts, darker-streaked mantle, buffy-white wingbars and underparts, olive-brown mottling on throat/upper breast. **1st-y** is slightly duller than ad, ♀ can have brownish upperparts and pale yellow throat. **Voice** Song a series of loud, clear whistles that descend in pitch and end with flourish. Call a loud, sweet "chip"; flight call a loud, clear, high-pitched "see". **SS** See Yellow-throated Warbler, including separation from Olive-capped and Kirtland's Warblers.

Vitelline Warbler *Setophaga vitellina* `NT`
Common resident (**Cayman Is**, **Swan Is**).

L 13 cm. Three subspecies, all of which (*nelsoni*[A] Swan Is, nominate[B] Grand Cayman, *crawfordi*[C] Little Cayman and Cayman Brac) occur in region. Endemic. Arid scrub-woodland, coastal scrub, clearings and logged areas of dry forest, sometimes in urban areas on Cayman Is. Almost nothing known, even current status, for race *nelsoni*. Relatively tame, forages at low to mid levels. **ID** Mostly yellow warbler with distinctive face pattern. **Ad** has olive-green crown, yellow supercilium and large yellow patch below eye, olive-green eyestripe and lower border to ear-coverts. Upperparts bright olive-green, white-tipped tail, bright yellow below, olive on flanks forming indistinct, blurred streaks. ♀ is slightly duller, with more uniformly yellow ear-coverts. **Juv** has pale greyish or greyish-brown head and upperparts, slightly darker mantle, obscure pale buff wingbars, whitish throat, yellowish-tinged belly, greyish-brown wash on flanks. **GV** Race *crawfordi* slightly paler and yellower above with less obvious head pattern, lacks faint olive flank streaking; *nelsoni* intermediate between other races in head pattern and upperparts colour. **Voice** Series of 4–5 wheezy, slightly grating notes rising in pitch, like Prairie Warbler but delivery slower. Call also similar to Prairie Warbler. **SS** Prairie Warbler slightly smaller with more conspicuously streaked flanks, streaked mantle, and bolder face pattern.

Prairie Warbler *Setophaga discolor* `LC`
Common winter visitor and passage migrant (**GA**, **O**), generally rare (**LA**).

L 11–12 cm. Two subspecies, of which both (nominate, *paludicola*) occur in region. Almost entire population winters in Caribbean. Widespread across Bahamas, Greater Antilles, Cayman and Virgin Is, but mainly rare or vagrant in Lesser Antilles (uncommon Guadeloupe); late Jul–mid May. All manner of dry scrub, edges and clearings, open second-growth forest with well-developed understorey, gardens and mangroves; to 1140 m, but mainly at low elevations. Forages mainly at low to middle levels in scrub, by gleaning or flycatching. **ID** Small parulid with tail-bobbing habit. **Ad br** ♂ has olive-green crown and nape, yellow supercilium and large yellow patch below eye, black eyestripe and lower border to ear-coverts, olive-green upperparts, mantle streaked chestnut, two narrow yellowish wingbars, white outer tail, and black streaks on flanks. **Ad br** ♀ is duller with less striking head pattern and less distinct streaks on mantle and flanks, less white in tail. **Ad non-br** is similar to ♀. **1st-y** is duller than ad; ♀ often with greyish-olive head and upperparts, lacking chestnut streaks, more whitish face, and indistinct olive flank streaks. **GV** Race *paludicola* (apparently rare winter visitor to Cuba) slightly duller than nominate, ad ♂ olive-grey above with less distinct chestnut streaking, paler yellow below with less distinct flanks streaking. **Voice** Call a low, smacking "tcheck"; flight call a thin "seep". **SS** Facial pattern distinctive in all plumages, although more subtle in 1st-y. Brighter Palm Warblers (share tail-bobbing habit) have yellow vent, yellow-green rump, long supercilium and less white in tail. Compare Vitelline Warbler within its restricted range.

Adelaide's Warbler *Setophaga adelaidae* `LC`
Common (**Puerto Rico**) to rare resident (**Virgin Is**).

L 12–13.5 cm. Monotypic. Endemic. Lowland dry scrub-forest and forest edge in Puerto Rico including Culebra and Vieques I, and has recently (since Mar 2012) colonized St Thomas and St John (Virgin Is). Replaced, albeit with some overlap, in humid montane forest by Elfin Woods Warbler. Forages in canopy. **ID** Striking parulid. **Ad** has short yellow supercilium edged black above, becoming white behind eye, black lores, broken yellow eye-ring, grey crown and upperparts, two white wingbars, extensive white in outer tail, yellow throat, and white lower belly and vent. ♂ averages brighter than ♀. **Juv** has brownish-grey head, browner mantle, buffy-white wingbars, yellow areas pale yellowish white, whiter rear underparts, with dusky spotting on breast-sides. **1st-y** ♀ is dullest, often lacking black on head, and upperparts washed olive. **Voice** Two song types: a variable trill, often ascending or descending in pitch, and a slightly more complex and lower-pitched version. Usual call "chick". **SS** ♀ Northern Parula has yellowish green on upperparts and lacks yellow supercilium. Compare Canada Warbler. **TN** Formerly considered conspecific with Barbuda and St Lucia Warblers.

Yellow-throated Warbler

Bahama Warbler

E

Vitelline Warbler
vitellina

♂ breeding

♀ /ad non-breeding

Prairie Warbler
discolor

A B C
E

Adelaide's Warbler
E

307

Barbuda Warbler *Setophaga subita* NT
Common resident (**Barbuda**).

L 12–13.5 cm. Monotypic. Endemic. Inhabits lowland dry shrubby forest, edges and scrub. Forages by gleaning in higher strata. **ID** Brightly patterned arboreal warbler. **Ad** has grey crown, neck-sides and upperparts tinged brown, short yellow supercilium, dusky lores, broad yellow crescent below eye, two pale wingbars and extensive white on outer tail. Yellow underparts, becoming white on lower belly and vent. **Juv** is undescribed, but presumably similar to juv Adelaide's Warbler. **Voice** Poorly known but recalls Adelaide's Warbler, albeit considered more musical. Call similar to latter. **SS** Should prove unmistakable within very restricted range; ♀ Northern Parula has yellowish green on upperparts and lacks yellow supercilium. **TN** See Adelaide's Warbler.

St Lucia Warbler *Setophaga delicata* LC
Fairly common resident (**St Lucia**), vagrant (**Martinique**).

L 13.5 cm. Monotypic. Endemic. From relatively dry scrub to montane rainforest, but generally most numerous at middle and high elevations. Once on Martinique; a singing ♂, Aug 1985 (Belfan & Conde 2016). Forages by gleaning, mainly at lower levels, often in undergrowth. **ID** Small, boldly patterned warbler. **Ad** has narrow black lateral crown-stripe and lores, short but broad supercilium, bright yellow eye crescent. Rest of head and upperparts clean bluish grey with two white wingbars and extensive white on outer tail. Bright yellow below, becoming white on lower belly and vent. **Juv** is presumably similar to juv Adelaide's and Barbuda Warblers. **Voice** Voice not known to differ from Adelaide's and Barbuda Warblers, but robust studies are lacking. **SS** Compare ♀ Northern Parula and Canada Warbler (both only vagrants on St Lucia). **TN** See Adelaide's Warbler.

Black-throated Grey Warbler *Setophaga nigrescens* LC
Vagrant (**Cuba**).

L 12–13 cm. Two subspecies, of which nominate occurs in region. Ad ♀/1st-w ♂, Cayo Coco, Oct 1997 (Wallace *et al.* 1999). Usually in dry open woodland with dense understorey. **ID** Small, generally dull-coloured wood-warbler with conspicuous yellow supraloral spot. **Ad** ♂ has black head and throat with broad white supercilium behind eye and broad white submoustachial. Grey upperparts streaked black, two white wingbars, extensive white on outer tail, white underparts streaked black on flanks. **Ad** ♀ is similar, but has greyer head, throat mottled whitish. **1st-y** is slightly duller than ad; ♀ often has throat mostly white with faint black mottling. **Voice** Call a dull, flat "tup"; flight call a clear, high-pitched "see". **SS** Both Black-and-white and Blackpoll Warblers lack yellow supraloral spot and have different head patterns, former with white crown-stripes and latter white on head-sides.

Townsend's Warbler *Setophaga townsendi* LC
Vagrant (**Bahamas, Cuba, Cayman Is**).

L 12–13 cm. Monotypic. Records in Bahamas (Grand Bahama, Apr 1984 and Oct 2009, Grand Turk, Jan 1986), SC Cuba (Feb 2015) and Cayman Is (Feb 2014). Also mentioned for Cayo Coco (Cuba) by AOU (1998) but presumably in error. Usually in forest, favouring conifers. Forages in canopy, occasionally lower. **ID** Small, strikingly patterned warbler. **Ad** ♂ has black crown, throat and ear-coverts, latter broadly surrounded by yellow. Yellow lower eye crescent, green upperparts streaked black, two white wingbars, extensive white on outer tail. Yellow breast, white rear underparts, boldly streaked black on flanks. **Ad** ♀ is similar but black replaced by greenish to dark olive, chin and most of throat yellow, slightly less white in tail. **1st-y** is duller than ad, ♀ with only indistinct streaking on breast-sides. **Voice** Call a sharp, high "tchip", very similar to Black-throated Green Warbler but perhaps slightly sharper and higher-pitched; flight call a thin, high "see". **SS** ♀/1st-y Black-throated Green Warbler lack olive ear-coverts and yellow throat and breast; ♀/1st-y Blackburnian Warbler have black (not olive-green) and streaked upperparts.

Golden-cheeked Warbler *Setophaga chrysoparia* EN
Hypothetical (**Virgin Is**).

L 12–14 cm. Monotypic. One report: ♂, St Croix, Nov 1939–Jan 1940 (Beatty 1943); seen only twice during this period and supported only by briefest of field notes, thus we treat as hypothetical. Usually in forests, favouring conifers. Forages high in canopy. **ID** Boldly patterned warbler. **Ad** ♂ has solid yellow cheeks and supercilia separated by narrow black eyestripe and surrounded by black crown, neck and throat (and upperparts), grey wings with two white bars, tail with extensive white on outer feathers, underparts white, heavily streaked black on flanks. **Ad** ♀ has olive crown and upperparts heavily streaked black, throat mottled whitish and less white in tail. **1st-y** is duller than ad, ♀ is quite dull, with crown and upperparts only faintly streaked and black mottling on underparts restricted to sides of throat and breast. **Voice** Call a sharp, high "tsip" or "tchip"; flight call a thin, high-pitched "see". **SS** 1st-y Black-throated Green Warbler always has much broader and duller, dark olive-green eyestripe (not blacker and more distinct), diffuse but broad olive lower border to cheeks, lacks streaks on upperparts, and has yellowish vent.

Black-throated Green Warbler *Setophaga virens* LC
Fairly common passage migrant and winter visitor (**GA, O**), mainly vagrant (**LA**).

L 12.7–13 cm. Monotypic. Most numerous in Cuba, but regular across most of Greater Antilles, Bahamas, Cayman Is, but very rare elsewhere including Virgin Is and Lesser Antilles (regular only in Guadeloupe); mid Sept–mid May. Favours forest and woodland, includes pines, but also shade-coffee plantations and taller scrub, to 2060 m (Hispaniola). Forages in all strata, by gleaning and flycatching. Usually alone and often territorial in winter. **ID** Small delicate-billed warbler. **Ad** ♂ has olive ear-coverts surrounded broadly by yellow, unstreaked olive-green crown, nape and upperparts, grey wing with two white bars, grey tail with extensive white on outer feathers, black throat and upper breast, rest of underparts white, broad black streaks on flanks, vent washed yellowish. **Ad** ♀ has throat mottled yellowish white, less white on tail. **1st-y** is duller than ad, ♀ with throat mostly whitish, whereas ♂ has chin to upper breast heavily mottled dark. **Voice** Call a sharp, high-pitched "tchip"; flight call a thin and high-pitched "see". **SS** See Townsend's and Golden-cheeked Warblers (both vagrants to region) compared to which yellow vent of present species is always diagnostic. ♀ Blackburnian Warbler has yellow throat and breast, plus whitish stripes on upperparts.

Barbuda Warbler

St Lucia Warbler

Townsend's Warbler

Black-throated Grey Warbler
nigrescens

Black-throated Green Warbler

Golden-cheeked Warbler

309

Canada Warbler *Cardellina canadensis* `LC`
Very rare passage migrant (**Cuba**, **Cayman Is**), vagrant elsewhere. **L** 12–15 cm. Monotypic. Very rare Cuba and Cayman Is, mere vagrant elsewhere, e.g. Bahamas (Abaco, Andros, Eleuthera, Exumas, Grand Bahama, New Providence) and San Andrés; mainly mid Sept–early Nov (occasionally Aug), and Feb–May, with two midwinter records. Open vegetation with scattered trees, often near swamps or other standing water. Forages at low to mid levels. **ID** Small, slim-looking warbler with necklace of bold black streaks on breast. **Ad** ♂ has black forehead and face with yellow lores, variably yellow-and-white eye-ring, bluish-grey crown, ear-coverts and upperparts, streaked black on crown. Yellow below, fading to white on vent. **Ad** ♀ has head and upperparts dull grey, black on head replaced by olive-grey, and streaks greyish and quite indistinct. **1st-y** ♂ is like ad ♀ but generally brighter, often with fairly bold blackish streaks on breast. ♀ is relatively dull, with no black on head, yellowish-olive forehead, distinct olive wash on head and upperparts, greyish breast streaks very indistinct. **Voice** Call a sharp "tik" or "chik", also softer, slightly lisping "tsip" and flight call a high-pitched "zzee". **SS** ♀ can recall Kentucky Warbler, but it lacks any streaking below and has more greenish upperparts; Nashville Warbler also has olive-green on upperparts, bright yellow vent, and again lacks breast streaking. Oriente Warbler (endemic to Cuba) has yellow cheeks and no streaks on breast.

Wilson's Warbler *Cardellina pusilla* `LC`
Three subspecies divided into two subspecies groups, both of which occur in region.

Western Wilson's Warbler *Cardellina (pusilla) pileolata*
Vagrant (**Cuba**). **L** 11.5–12 cm. Two subspecies in group, of which one (*pileolata*) occurs in region. Specimen, Guanahacabibes Peninsula, Oct 1999 (Llanes Sosa *et al.* 2016). Could occur in all manner of woodland and tall scrub, usually in areas with dense undergrowth. **ID** Very similar to Eastern Wilson's Warbler. Ageing and sexing difficult. **Ad** ♂ has solid black cap, yellow forehead and face highlighting dark eye, olive-green nape and upperparts, darker wings and tail; uniform yellow below. **Ad** ♀ lacks solid black cap, varying from a few feathers to (occasionally) a full cap, but flecked green. **1st-y** is similar to ad, ♀ nearly always lacks black on crown. **Voice** Call a fairly low-pitched "tchep" or "timp", often sounding quite nasal; flight call a sharp, downslurred "tsip". **SS** Eastern Wilson's Warbler is slightly smaller (in biometrics) and less bright, ♀ typically has even less black on crown. ♀/1st-y Yellow Warbler lack prominent supercilium, have eye-ring, bold black eye and more greenish upperparts. ♀/1st-y Hooded Warbler larger with darker lores, white tail patches and lack supercilium. Compare duller Orange-crowned Warbler, which never has such a bland-looking facial expression.

Eastern Wilson's Warbler *Cardellina (pusilla) pusilla*
Rare passage migrant (**Cuba**, **Cayman Is**), vagrant elsewhere. **L** 11.5–12 cm. One subspecies in group. Rare passage migrant Cuba and Cayman Is, mainly early Sept–mid Nov and mid Feb–mid Apr, but increasing evidence for at least sporadic wintering in region. Vagrant elsewhere, E to Puerto Rico (Raffaele 1989) and S to Swan Is (Aceituno & Medina 2009). Woodland and tall scrub, usually with dense undergrowth. **ID** Small, plump-looking parulid, very similar to Western Wilson's Warbler. **Ad** ♂ has solid black cap, yellow forehead and face highlighting dark eye, olive-green upperparts, darker wings and tail; uniform yellow underparts. **Ad** ♀ has cap smaller and less shiny, either mainly limited to forecrown or admixed olive at rear, but occasionally no cap at all. **1st-y** is similar to respective ad, but ♂ always has scattered olive feathers in crown and ♀ olive-washed underparts, forehead (which hardly contrasts with crown) and ear-coverts, and appearance of yellow eye-ring. **Voice** Calls do not differ significantly from Western Wilson's Warbler. **SS** See Western Wilson's Warbler.

PHAENICOPHILIDAE
Hispaniolan Tanagers
4 extant species, 4 in region

Black-crowned Palm-tanager `LC`
Phaenicophilus palmarum
Common resident (**Hispaniola**).
L 17–18 cm. Monotypic. Endemic. Found in dry to humid zones across most of Hispaniola (including Saona I), except SW Haiti, especially in thickets, gardens, trees and forest of all types, including mangroves and pines, but mainly in lowlands, to c. 2500 m. In pairs or small family groups, sometimes with mixed-species flocks, foraging from ground to treetops. Moves rather deliberately, often flicking tail. **ID** A robust-bodied 'tanager' that is one of commonest Hispaniolan endemics. **Ad** has striking black-and-white head pattern, grey nape, bright yellowish-olive upperparts, diffuse white throat and central underparts (sometimes suffused greyish) with deeper grey breast-sides, flanks, belly and vent. Iris dark red-brown. Bill rather long, strong and sharply pointed. **Juv** usually has black areas replaced by dusky grey, and whitish foreparts tinged buff. **Voice** Song varies but typically comprises jumbled squeaky notes that grow louder, then diminish and slow to short "chit" notes; sometimes contains more raspy and buzzy phrases. Frequently gives nasal, buzzy "pe-u", and higher, more penetrating "tseep", often doubled; also a low "chep". **SS** Highly distinctive over most of range. Only likely to be confused with Grey-crowned Palm-tanager, which see.

Grey-crowned Palm-tanager `NT`
Phaenicophilus poliocephalus
Common but local resident (**Hispaniola**).
L 17–18 cm. Three subspecies, of which all three (*coryi*[A] Gonâve I, nominate[B] Tiburón Peninsula, Haiti, occasionally SW Dominican Republic, plus Grande Cayemite I, *tetraopes*[C] Î-à-Vache, off SW Hispaniola) occur in region. Endemic. Forested areas, including thickets, semi-open areas, mangroves and gardens, from sea level to 2400 m. **ID** Similar to Black-crowned Palm-tanager, especially young birds. **Ad** has yellow-green upperparts and grey crown, nape and underparts, but most striking is black mask within which there are three white spots, and highly contrasting but very narrow white chin and throat. Iris dark brown. Bill as Black-crowned Palm-tanager. **Juv** is like ad, but duller. **GV** Race *tetraopes* slightly paler than nominate, especially grey of hindcrown, nape and underparts, with paler green back, whiter belly, and bill slightly longer; *coryi* slightly larger and paler than nominate, with whitish central underparts and vent, and some (but not all) ♂ have small white spot on central crown. **Voice** Call a brief "peee-u", often doubled. Breeding ♂ gives long musical song and apparently also a canary-like whisper song. **SS** Only overlap with Black-crowned Palm-tanager is apparently in extreme SW Dominican Republic, in westernmost Sierra de Bahoruco (where present species persistently but irregularly reported) and near Jacmel Depression, Haiti, where they hybridize (Latta *et al.* 2006). Vs. ad Black-crowned Palm-tanager, note differences in white areas on head, grey (not black) crown and much more sharply defined (but

Canada Warbler

Western Wilson's Warbler
(Wilson's Warbler)
pileolata

Eastern Wilson's Warbler
(Wilson's Warbler)

Black-crowned Palm-tanager

Grey-crowned Palm-tanager
poliocephalus

311

smaller) white throat. Not all juv Black-crowned have black crown, but always much more diffuse white throat.

White-winged Warbler *Xenoligea montana* VU
Fairly common to uncommon and local resident (**Hispaniola**).
L 13–14.5 cm. Monotypic. Endemic. Undisturbed humid, montane broadleaf forest with dense understorey, sometimes in pine forests provided there is significant lower growth, at 875–2000 m, but mostly above 1300 m. Typically alone or in pairs, often within mixed-species flocks, usually also containing Green-tailed Warbler. Forages at all levels. Undoubtedly has declined significantly and now rare in many areas. **ID** Jizz almost like a cross between a tanager and warbler. **Ad** has head and upper mantle grey, with blackish lores, narrow white eye-crescents and whitish supraloral stripe. Rest of upperparts mainly bright green, with long white stripe on closed wing; tail grey, white spots at tips of outer feathers. Below whitish, with pale grey wash on sides. Bill quite stout, bluish grey with blackish culmen. **Juv** has brownish-grey head, greyish above tinged olive-brown, below off-white with some brown tones, and paler bare parts. **Voice** Song a short series of high-pitched squeaky notes, sometimes accelerating at end. Calls include low chatters and thin "tseep". **SS** Most likely to be confused with Green-tailed Warbler, which lacks any white in wing, tail and on foreface, and has red eye, while present species is less likely to feed close to or on ground. **AN** White-winged Warbler-tanager, Hispaniolan Highland Tanager.

Green-tailed Warbler *Microligea palustris* LC
Fairly common but local resident (**Hispaniola**).
L 12–14.5 cm. Two subspecies, of which both (nominate[A] highlands of Hispaniola, *vasta*[B] xeric lowlands of SW Dominican Republic) occur in region. Endemic. Montane broadleaf and pine forests with dense, undisturbed understorey, to 2925 m; race *vasta* in semi-arid xeric scrub in lowlands. Also on I Beata and I Saona. Notably rare and local in Haiti, but generally more numerous than White-winged Warbler in rest of range. Forages mainly alone or in pairs, but routinely joins mixed-species flocks, in dense undergrowth and thickets. **ID** A long-tailed warbler-like bird with slender body and fairly robust-looking bill. **Ad** has grey head and upper mantle, contrasting olive-green upperparts, white crescents above and below eye, and underparts pale greyish, whiter on belly. Iris ruby-red; bill and legs blackish to grey. **Juv** is undescribed. **1st-y** is similar to ad, but slightly duller with brown iris. **GV** Race *vasta* paler overall and more extensively whitish below. **Voice** Short rasping and squeaking notes frequently accelerated to form what may be song. Latter described also as high-pitched "sip sip sip". Readily responds to 'pishing' like true warblers. **SS** Only really likely to be confused with White-winged Warbler (which see). **TN** Population recently found in xeric lowlands of NW Haiti may represent an additional, undescribed, race; birds in lowlands of SE Dominican Republic also unascribed racially. **AN** Green-tailed Ground-tanager.

SPINDALIDAE
Spindalises
4 extant species, 4 in region

Western Spindalis *Spindalis zena* LC
Five subspecies divided into five subspecies groups, of which four occur in region.

Bahamas Green-backed Spindalis
Spindalis (zena) townsendi
Fairly common resident (**Bahamas**).
L 15 cm. One subspecies in group. Endemic (a few records in SE Florida). Occurs in a variety of wooded habitats, especially pine forests (both native and introduced) on Grand Bahama, Little Abaco, Abaco and offshore cays. **ID** Smallest spindalis, bill short and conical. **Ad** ♂ differs from Bahamas Black-backed Spindalis in having nuchal collar tawny-rufous (not tawny), larger yellow throat, back dark olive-green variably mixed dusky brown (darkest on Abaco), lower back darker tawny-orange, and scapulars more dusky. **Ad** ♀ is very similar to Bahamas Black-backed but always lacks blackish spots on crown, and blackish shaft-streaks on mantle are rarely visible. **Voice** Vocalizations poorly known, and probably similar to those of Bahamas Black-backed Spindalis. **SS** Wholly distinctive within limited range. **AN** Western Stripe-headed Tanager (also applicable to all Western Spindalis races).

Bahamas Black-backed Spindalis *Spindalis (zena) zena*
Fairly common to common resident (**Bahamas**).
L 15 cm. One subspecies in group. Endemic (wanders to S Florida). Occurs on Berry Is, Andros, Green Cay, New Providence, Eleuthera, Cat I, Exuma, Long I, Acklin, Mayaguana, and Turks & Caicos Is. In similar habitats to Bahamas Green-backed Spindalis. **ID** Typical spindalis. **Ad** ♂ has long white supercilium, broad white submoustachial and white chin, and bright yellow central throat. Collar and highly variable breast-band chestnut and rich yellow. Mantle black, lower back tawny-yellow, rump darker golden-rufous. Wing-coverts look mostly white and tertials broadly edged white. Mid-breast yellow, becoming white below. **Ad** ♀ is greyish olive above, with obscure whitish submoustachial. Rump paler than back, greater coverts dusky, broadly edged whitish, tertials dusky black, broadly edged whitish olive, and underparts unstreaked dirty greyish to buffy whitish. **Juv** is similar to ♀ but much duller, with minimal pattern on foreparts. **Voice** Song high, thin, sibilant and rather variable, e.g. "see-tee" doublets mixed with high "seet" and reedy "deet", sometimes with twittering notes and usually delivered from atop tall tree. Flight song a high, sibilant "seeet sit-t-t-t-t". High, thin quality of vocalizations often makes bird difficult to locate. Calls include high thin "seeip" and soft "tsit-tsit-tsit". **SS** Nothing similar occurs in C & S Bahamas. **TN** With Bahamas Green-backed Spindalis, sometimes considered to represent a species apart from other members of Western Spindalis complex. All of latter previously considered conspecific with Hispaniolan, Puerto Rican and Jamaican Spindalis, under name Stripe-headed Tanager (Garrido *et al.* 1997b).

Cuban Spindalis *Spindalis (zena) pretrei*
Common to fairly common resident (**Cuba**).
L 15 cm. One subspecies in group. Endemic. Semi-deciduous and evergreen woodland, coastal thickets, mangroves, riparian and mixed woodland, to 1974 m on mainland Cuba, I of Pines and many cays (where often very numerous). Regularly observed in small, single-species flocks. **ID** Basic structure like other spindalis. **Ad** ♂ differs from other members of Western Spindalis complex in its bright olive-green back, brighter rump, paler tawny-yellow nuchal collar, paler tawny breast, chestnut lesser coverts and median coverts broadly tipped olive-green. **Ad** ♀ differs in having long dull whitish supercilium, some yellow on underparts and often a weak greyish malar stripe. **Juv** is much like ♀, but even duller. **Voice** Songs comprise very thin high-pitched notes (similar to Cuban Bullfinch) with variable phrases and patterns of delivery. In other contexts gives weak "tsee" (sometimes doubled), an intensive "chip" while foraging and prolonged "seeee" prior to song bouts. **SS** Like other members of Western Spindalis complex, even ♀/juv

White-winged Warbler

Green-tailed Warbler
palustris

Bahamas Green-backed Spindalis
(Western Spindalis)

Bahamas Black-backed Spindalis
(Western Spindalis)

Cuban Spindalis
(Western Spindalis)

313

(much less ♂) very difficult to mistake within range. **TN** Race *pinus*, described from I of Pines, inadmissible.

Grand Cayman Spindalis *Spindalis (zena) salvini*
Fairly common resident (**Cayman Is**).
L 15 cm. One subspecies in group. Endemic. Basically confined to Grand Cayman; one record on Little Cayman (Apr–Jun 1985). In all habitats except dry shrublands and coastal *Rhizophora* mangrove; behaviour similar to other spindalis. **ID** Like other Western Spindalis races. **Ad** ♂ is much like Cuban Spindalis, but (among other slight differences) is slightly larger, with marginally darker rump, strongly bicoloured nuchal collar, more tawny-orange and extensive breast-band, and more extensively white flanks and lower underparts. **Ad** ♀ differs little from ♀ Cuban Spindalis, only in marginally greyer (less green) upperparts and more conspicuous whitish malar. **Juv** is much like ♀. **Voice** Song short-lived, of simple notes but complex syllables, with emphasis on final note; similar to other races of Western Spindalis. **SS** None.

Hispaniolan Spindalis *Spindalis dominicensis* LC
Locally common resident (**Hispaniola**).
L 16–16.5 cm. Monotypic. Endemic. Occurs over most of Hispaniola and Gonâve I, mainly in montane broadleaf and pine forests to 2450 m, and is uncommon below 700 m and in more xeric habitats. In pairs and small groups, but congregates locally in larger numbers at ripe fruit. Forages at all levels. **ID** Differs from similar Jamaican Spindalis in smaller size, yellow hindcollar and small rufous shoulder patch. **Ad** ♂ has black head with bold white markings, brilliant yellow nape tinged orange, and primarily yellow back with darker mantle. Tail and wings black with white fringes. Below yellow, washed reddish brown on breast. **Ad** ♀ has greyish-olive head and mantle, yellower on rear upperparts, and smaller wing-coverts mostly dull olive. Malar and throat whitish with ill-defined dusky spots and hint of dusky malar; below whitish with ill-defined dusky streaks, cleanest on belly and vent. **Juv** is similar to ♀ but much duller, with black on head obscure. **Voice** Dawn song a thin, high-pitched whistle or prolonged weak, sibilant "tsee see see see". ♂ sings from exposed perch or lower, inside dense thickets. ♀ gives whisper song of jumbled notes like ♂. Commonest call a high "thseep"; also a more drawn-out "seeee" sometimes followed by ticking notes. **SS** None in range. **TN** See Bahamas Black-backed Spindalis.

Puerto Rican Spindalis *Spindalis portoricensis* LC
Fairly common resident (**Puerto Rico**).
L 16.5–17 cm. Monotypic. Endemic. Occurs at all elevations in Puerto Rico, but not resident on any of its satellites (just one record Vieques I), in forests, suburban gardens and plantations. Behaviour very similar to other spindalis. **ID** Distinctive from all spindalis, except Hispaniolan, in that ♀ has streaked underparts. **Ad** ♂ is most like previously conspecific Jamaican Spindalis, but is smaller, with orangey hindcollar, rufous patch on wingbend and less black on lower throat. **Ad** ♀ is very like also formerly conspecific ♀ Hispaniolan Spindalis, but streaking below less smudgy, has white head markings and often some yellowish on central underparts. **Juv** is like ♀ but much duller, with black pattern on head obscure. **Voice** Song a thin, high, squeaky or wiry "zeé-tit-zeé-tittit-´zeé", the "zeé" notes sound like air being inhaled; or a thin high "tswee, tswee, tsweey". Two variant calls: a thin trill like a tiny hammer, and a short twittering; also a soft "tweep". **SS** Unmistakable within range. **TN** See comments under Bahamas Black-backed Spindalis.

Jamaican Spindalis *Spindalis nigricephala* LC
Common resident (**Jamaica**).
L 18 cm. Monotypic. Endemic. From sea level to highlands, in moist to wet forests, but much more numerous in hills and mountains, and local on N & SW coasts. Typically observed in pairs or family groups, sometimes congregating at fruiting trees. **ID** Highly distinctive spindalis by virtue of more ♂-like ♀ plumage. **Ad** ♂ easily identified in range by black head with bold white malar and superciliary, and black wings broadly fringed white; underparts largely orange-yellow, with narrow orange breast. **Ad** ♀ has greyish head with much less distinct markings than ♂, greenish-yellow breast and belly, small orange patch in central breast, greenish-grey upperparts, yellowish rump, and dark grey wings edged white. **Juv** is much duller than ♀. **Voice** Notably quiet. Long-sustained whisper song is a phrase c. 4 seconds long repeated over and over, e.g. "chu wheet, chee see whee see, chu wheet". Soft "seep" notes, often heard in flight, and a high, fast "chi-chi-chi-chi-chi" while foraging in groups. ♂ also gives churrs or rattles. **SS** Wholly distinctive on Jamaica. **TN** See Bahamas Black-backed Spindalis.

Grand Cayman Spindalis
(Western Spindalis)

Hispaniolan Spindalis

Puerto Rican Spindalis

Jamaican Spindalis

315

NESOSPINGIDAE
Puerto Rican Tanager
1 extant species, 1 in region

Puerto Rican Tanager *Nesospingus speculiferus* LC
Locally common resident (**Puerto Rico**).

L 18–20 cm. Monotypic. Endemic. Prefers humid pre-montane and montane forest, but also second growth, palm forests, thickets and shade-coffee plantations at lower elevations, c. 200–1330 m. An important, central constituent of mixed-species flocks, which travels in pairs or small groups. Has apparently expanded its range in recent decades, having previously been confined to a handful of upland forests. **ID** Large, rather drab tanager-like bird, with contrasting white throat and fairly stout bill. **Ad** has dusky-brown hood, narrow blackish-brown moustachial and dull olive-brown upperparts, with small white spot at base of primaries. Throat white, contrasting with slightly duller breast and lower underparts are faintly streaked and smudged brown, heaviest on lower breast. **Juv** is more brownish than ad, especially below, and lacks white wing spot. **Voice** Noisy. Often gives loud, sharp "chewp" or "chuck", sometimes extended into a chatter of varying length, "chi-chi-chit"; also "tsweep, tsweep", which may be a song. A soft short twitter, and thin sigh like heavy exhalation. **SS** Size and plumage should render species impossible to confuse.

CALYPTOPHILIDAE
Chat-tanagers
2 extant species, 2 in region

Western Chat-tanager *Calyptophilus tertius* VU
Uncommon and local resident (**Hispaniola**).

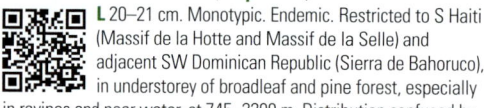

L 20–21 cm. Monotypic. Endemic. Restricted to S Haiti (Massif de la Hotte and Massif de la Selle) and adjacent SW Dominican Republic (Sierra de Bahoruco), in understorey of broadleaf and pine forest, especially in ravines and near water, at 745–2200 m. Distribution confused by fact that birds in E Sierra de Bahoruco are more similar to Eastern Chat-tanager and sing similarly, but populations of latter in Sierra de Neiba have Western-like song. Shy and often difficult to see, spending much time close to ground. **ID** Dull with rather long and pointed bill, large, strong legs and feet, and long tail. **Ad** has small, inconspicuous rusty loral spot, rest of head mainly rich dark olive-brown, becoming blackish on crown. Above deep rufescent-brown. Below white, sides and lower underparts heavily washed greyish brown; yellowish underwing-coverts often visible at wingbend. **Juv** is undescribed. **Voice** Song, by both sexes year-round, mostly at dawn and repeated many times, "wee-chee-chee-chee", quite similar to Eastern Chat-tanager, but weaker and buzzier. Makes "chip-chip" while foraging. **SS** Eastern Chat-tanager smaller, with brighter yellow pre-ocular spot and yellow eye-ring, less rufescent above, more extensive yellow-orange patches on wingbend and underwing-coverts. **TN** Often treated as conspecific with Eastern Chat-tanager, and despite molecular findings and morphological evidence this may still be appropriate, but differences appear to be strong. Proposed race *selleanus* (described from Morne Malanga, Haiti) long considered a synonym of nominate, but has been suggested it differs vocally, as well as being larger and darker (Latta *et al.* 2006).

Eastern Chat-tanager *Calyptophilus frugivorus* NT
Uncommon to rare and local resident (**Hispaniola**).

L c. 17–19 cm. Three subspecies, of which all (nominate[A] Dominican Republic, in Cordillera Central and Samaná Peninsula on NE coast, *abbotti*[B] Gonâve I, *neibae*[C] Sierra de Neiba, in W Dominican Republic and possibly adjacent Haiti) occur in region. Also, race unknown, in E Sierra de Bahoruco and Sierra de Martín García. Endemic. Usually encountered on ground or close to it. Dense undergrowth of montane forest, often along streams and ravines, and in areas dominated by invasive fern, *Dicranoteris pectinada*, from sea level on Gonâve I (where inhabits dense semi-arid scrub) to at least 2000 m in Cordillera Central, but primarily above 1000 m. Uncommon and increasingly local, especially in Sierra de Neiba, and no recent reports of race *abbotti* or from Samaná Peninsula. **ID** Overall jizz and structure much like Western Chat-tanager. **Ad** has crown dusky brown, with narrow and rather inconspicuous broken yellow eye-ring and narrow but more prominent yellowish loral mark. Otherwise mainly dark olive-brown above; below white, sides and lower underparts heavily washed greyish brown, yellowish underwing-coverts often visible at wingbend. **Juv** is undescribed. **GV** Race *neibae* smaller and darker than nominate, with rufescent tail; *abbotti* slightly smaller and more greyish brown than nominate. **Voice** Like Western Chat-tanager (from which song differs only subtly) heard mostly at dawn. Songs differ geographically; nominate race gives a low slurred or whistled "swerp, swerp, chip, chip, chip…", sometimes accelerating to a chatter; that of *neibae* similar but more varied, "weet-weet-werp chip-cheep-sweet…", sometimes ending in short trill. Call a sharp "chin chin chin", also most frequently heard in early morning. **SS** Only really likely to be confused with Western Chat-tanager (which see) and their respective ranges might still require some clarification. **TN** See Western Chat-tanager. All populations in highlands of W & C Dominican Republic should perhaps be ascribed to race *neibae*.

CARDINALIDAE
Cardinals
52 extant species, 7 in region + 3 vagrants + 1 introduced

Rose-breasted Grosbeak *Pheucticus ludovicianus* LC
Uncommon passage migrant and winter visitor (**GA**, **O**), mainly vagrant (**LA**).

L 18–20 cm. Monotypic. Present late Sept–early May. Locally common on autumn passage in W Cuba. Scrub, edges and humid broadleaf forest, at all elevations. Small groups or lone birds forage in upper strata, sometimes with mixed-species flocks. Agile, sometimes even acrobatic. **ID** Robust, large- and triangular-billed passerine. **Ad** ♂ has black hood, white lower back and rump, black-and-white wings and tail, rose-red breast, with some black streaks at sides. Underwing-coverts rose-pink. Bill pinkish white to slate-grey. **Ad** ♀ has brown crown with broad buff-brown central stripe, broad whitish supercilium and pale off-white crescent below eye. Above mid-brown streaked paler and darker brown; white in wing reduced and buffier. Throat whitish, buffier below, with narrow but profuse blackish-brown streaks reaching flanks. **Juv** is like ♀ but has cinnamon wingbars and tawny fringes above, underwing-coverts pink (♂) or orange-yellow (♀); 2nd-y ♂ is like ad but has variable brown fringes, especially on head. **Voice** Call a sharp, metallic "chink" or harsh "squack". **SS** ♂ unmistakable; ♀/imm ♂ of vagrant Black-headed Grosbeak very similar, but generally less streaked below (confined to sides and flanks); bill looks bicoloured. Dickcissel smaller, non-br ♂/♀ have weaker head pattern, paler and buffier overall, with longer bill; habitat and habits differ. ♀ Hispaniolan Crossbill superficially similar to ♀, but duller with yellowish rump, blacker wings and obviously different bill.

Puerto Rican Tanager

Western Chat-tanager

neibae

abbotti

♀

Rose-breasted Grosbeak

♂

frugivorus

Eastern Chat-tanager

317

Black-headed Grosbeak *Pheucticus melanocephalus* LC
Vagrant (**Cuba**).

L 18–20.5 cm. Two subspecies, of which one (presumably nominate) occurs in region. Subad ♂, La Habana, May 2007 (Garrido & Kirkconnell 2008). Could occur in any wooded area. Habitat and habits as Rose-breasted Grosbeak. **ID** Structurally recalls Rose-breasted Grosbeak. **Ad** ♂ has black hood, orange-brown hindneck, black mantle with orange-brown and whitish fringes, orange-brown rump. Wings blackish with much white, outertail feathers also extensively white. Most underparts orange-brown, underwing-coverts bright yellow. Maxilla slate-brown, mandible bluish white. **Ad** ♀ has broad buffy central crown-stripe, pale supercilium, streaked upperparts, less white in wings; below brownish yellow, sides and flanks variably but faintly blackish streaked; underwing-coverts bright yellow. **Juv** resembles ♀, but has cinnamon-fringed wing-coverts, upperparts edged tawny. **Voice** Calls include sharp "pik", less sharp than Rose-breasted Grosbeak. **SS** ♀ Rose-breasted Grosbeak is more heavily streaked below, ground colour paler, with pale horn bill and more subdued head pattern. Imm ♂ can have streakier breast, but underwing-coverts pinkish.

Dickcissel *Spiza americana* LC
Rare passage migrant (**Cuba**, **O**), vagrant elsewhere.

L 15–16 cm. Monotypic. Recorded mostly Sept–Dec and Feb–May (twice Jan in Puerto Rico), mainly in Bahamas, Cuba and Cayman Is; only vagrant elsewhere in Greater and Lesser Antilles, and S to San Andrés. Open grassland with scattered trees. Can form large flocks in rice, sugarcane and other agriculture, foraging largely on seeds. **ID** Boldly marked passerine with pale conical bill. **Ad br** ♂ has long bright yellowish supercilium becoming near-whitish at rear, broad white submoustachial yellowish in centre, black 'V' on throat and central foreneck, yellow chest with variable black spotting that can coalesce into larger patch. Rest of underparts grey, suffused yellow, but whiter on flanks to vent. **Ad non-br** ♂ has back suffused brownish and yellow less bright. **Ad** ♀ is smaller, black on throat reduced to two narrow streaks, throat whitish, becoming yellowish on central breast, supercilium and submoustachial much less prominent than ♂, back broadly streaked blackish. **1st-w** ♂ has little black on chest, narrow submalar, less yellow on underparts, none on crown, reduced supercilium. **Voice** Call a sharp, dry "click"; flight call a buzzy "bzzrt" on migration. **SS** House Sparrow lacks yellowish wash on breast and supercilium, has whitish wingbar, smaller bill, no pale submalar or narrow dusky malar.

Indigo Bunting *Passerina cyanea* LC
Fairly common passage migrant and winter visitor (**GA**, **O**), mainly vagrant (**LA**).

L 14–15 cm. Monotypic. Present late Sept–mid May, mainly in N & W of region, across Bahamas, Greater Antilles, Cayman and Virgin Is, and islands in SW of region. One anomalous record suggestive of breeding in Cuba (Rodríguez Castaneda & Wiley 2015). Wooded edges, clearings, thickets, scrub, rice fields, to at least 1300 m; singles or small groups forage low or on ground. Can be rather shy. **ID** Stunning blue ♂, warm brown ♀. **Ad br** ♂ is mostly deep blue; below paler; bill blackish above, blue-grey below. **Ad br** ♀ has warm brown upperparts, some blue on shoulders and indistinct pale buff wingbars. Face buffy brown, chin and throat dull whitish, chest warmer with faint dusky streaking, belly brownish white. **Ad non-br** ♂ is similar to ♀, but has some blue on head and body. **Juv** is generally dark brown, paler on lower breast and belly, ♂ can have blue edges to tail. **Voice** Calls include an emphatic "twit" and similar notes. Song a cheerful series of "swee" notes (rarely heard in region). **SS** Larger Blue Grosbeak has bigger head and bill, bold wingbars, ♀ and non-br ♂ warmer brown below. Lazuli Bunting (vagrant) has paler but more distinct wingbars, both sexes duller and buffier overall, and in ♀ plain breast contrasts with greyish throat.

Blue Grosbeak *Passerina caerulea* LC
Scarce passage migrant and even rarer winter visitor (**GA**, **O**), vagrant (**LA**).

L 15–19 cm. Seven subspecies, of which one (*caerulea*) occurs in region. Mainly Sept–Apr (exceptionally from Aug and until early May), principally in N & W of region; vagrant in Lesser Antilles. Wooded edges, pines, gardens, parks, coastal thickets and areas with scattered trees, to at least 1050 m. Often forages near ground, but assumes prominent perches. Flicks tail regularly. **ID** Heavy bill, large head, tail long and rounded. **Ad br** ♂ is blue, with small black mask, two prominent rufous-chestnut wingbars, bill blackish above, silvery below. **Ad non-br** ♂ is mostly patchy blue and reddish brown, darker on forecrown and above, wings somewhat duller; bill horn. **Ad** ♀ has mid-brown upperparts, dusky-streaked mantle and scapulars, with bluish-tinged lower back and rump, paler and buffier underparts, deeper-toned breast and two dull wingbars, upper more chestnut, lower tawny. **Juv** is all brown, but ♂ has traces of blue at bases of rectrices, and acquires blue wing feathers with age. **Voice** Calls include a metallic "tink", buzzy "bzzzt", and high and rolling "preet". **SS** Indigo Bunting is uniformly blue in br plumage, with no wingbars; non-br and ♀ duller than corresponding plumages in Blue Grosbeak, with much fainter and narrower wingbars, some fine streaks on breast to flanks, whitish throat, smaller bill. See also Lazuli Bunting, vagrant to Cuba, with distinct but paler wingbars, duller and buffier overall, buffy breast, smaller bill.

Lazuli Bunting *Passerina amoena* LC
Vagrant (**Cuba**).

L 13–14 cm. Monotypic. All records in W & C Cuba: specimens, Mar 1960 and Oct 2003 (Kirkconnell *et al.* in press); and individuals caught by bird-trappers, Nov 2012 and Dec 2013 (Rodriguez Castaneda *et al.* 2017). Overgrown fields and thickets. Mainly feeds on ground. **ID** Striking ♂ is unmistakable, other plumages similar to Indigo Bunting. **Ad br** ♂ has blue hood and rump, darker mantle, back and scapulars, two white wingbars, upper one broad, orange-chestnut breast, belly white. **Ad non-br** ♂ is similar to br ♂, but duller. **Ad** ♀ has greyish-brown upperparts, greyer or bluish rump, greyish throat, richer buff breast, pale buffy-grey belly. Lesser coverts tinged bluish, narrow but distinct whitish-buff wingbars, upper broader. **Juv** is all brown, duller than ♀. **Voice** Calls include short "chip", a short trill in alarm, and loud, strident "eeee". **SS** ♀ and non-br ♂ Indigo Bunting are warmer brown, with faintly streaked breast to flanks, and narrower and less distinct wingbars.

Black-headed Grosbeak
melanocephalus

Dickcissel

Indigo Bunting

Blue Grosbeak
caerulea

Lazuli Bunting

319

Painted Bunting *Passerina ciris* `LC`
Two subspecies placed in separate subspecies groups, of which one occurs in region.

Eastern Painted Bunting *Passerina (ciris) ciris*
Fairly common (**Cuba**, **O**) to rare passage migrant and winter visitor (rest of **GA**), vagrant (**San Andrés**).
L 13–14 cm. One subspecies in group. Mainly in Greater Antilles (but not recorded Haiti). Present late Sept–late Apr, in woodland, coastal thickets, open areas with scattered trees and bushes, grassy meadows and cultivation, to 1100 m. Small, loose flocks forage low down or on ground, sometimes with Indigo Buntings. **ID** More frequently seen ♀/juv are plain greenish and yellowish, with narrow pale eye-ring. **Ad** ♂ has deep blue head, rose-pink eye-ring and rump, bright green upperparts, rose-red underparts, slightly paler lower down; bill dark. Slightly duller in winter. **Ad** ♀ has olive-green upperparts, brighter on rump, bluish-grey suffusion to wing-coverts. Below dull yellowish green, greyer on breast, brighter yellow on belly; narrow yellowish eye-ring. **Juv** has greenish-grey upperparts, buffy-green underparts, narrow whitish eye-ring. **Voice** Calls a loud "chip" and short series of high "pik" notes. **SS** Stunningly coloured ♂ unmistakable. Other buntings in region are less green and possess wingbars.

Scarlet Tanager *Piranga olivacea* `LC`
Uncommon to scarce passage migrant (**GA**, **O**), mainly vagrant (**LA**).
L 17–18 cm. Monotypic. Passage late Sept–early Nov (exceptionally Dec) and mid Feb–early May; records throughout region, but mainly in N & W (e.g. Bahamas, Greater Antilles, San Andrés), being only vagrant to most of Lesser Antilles. Woodland, parks and gardens with trees. Sometimes in small groups, occasionally joins mixed-species flocks. Forages in upper strata, where often remains rather concealed. **ID** Gaudy br ♂ unmistakable; other plumages olive-yellow and less obvious. **Ad br** ♂ is bright red with jet-black scapulars, wings and tail. **Ad non-br** ♂ is dark olive above, olive-yellow below, brighter on throat and lower belly. Retains black wings. **Ad br** ♀ has dark greyish-olive upperparts, paler olive-yellow underparts; wings mostly dusky with yellow-olive edges. **1st-y** ♂ is much like non-br ♂, but has browner flight feathers and sometimes shows vaguest hint of wingbars. **Voice** Harsh, double-noted "chik-burrr" is most commonly heard call in winter. **SS** Superficially similar Summer Tanager has wings and tail nearly concolorous with body in all plumages, and is larger-billed. Much rarer Western Tanager sports wingbars in all plumages, and much darker back in ♂ or paler rump in ♀.

Summer Tanager *Piranga rubra* `LC`
Scarce passage migrant and even rarer winter visitor (**GA**, **O**), vagrant (**LA**).
L 17–18 cm. Two subspecies, of which nominate occurs in region. Present early Sept to mid May, and only regularly recorded in Bahamas, Cuba, Cayman Is, Jamaica and San Andrés, but merely a vagrant from Hispaniola E throughout Lesser Antilles. Scattered trees, heavier woodland, shade-coffee plantations, also parks and gardens; to mid elevations. Often in pairs, sometimes with mixed-species flocks. Forages in upper strata. **ID** Heavy pale bill; occasionally looks slightly crested. **Ad br** ♂ is mostly red, head, throat and underparts brightest; bill pale to dusky horn. **Ad** ♀ has yellowish-olive upperparts, yellower below, sometimes tinged buff, cinnamon or orange. **Ad non-br** ♂ is similar to ♀, but is blotched irregularly with red throughout, especially head and foreparts. **1st-y** ♂ is similar to ad non-br but by spring usually has virtually all-red head; ♀ often has dull orangish wash on belly. **Voice** Loud, staccato "pik-tup", "pik-ti-tup" or "pik-a-tuptup". Seldom sings on wintering grounds; a slow, musical carolling. **SS** Scarlet Tanager has dark wings even in ♀ and non-br ♂ plumages, and consistently shows olive in plumage (not shown by Summer Tanager). Vagrant Western Tanager has wingbars in all plumages, blacker wings and back, yellow rump. Both species are smaller-billed.

Western Tanager *Piranga ludoviciana* `LC`
Vagrant (**Bahamas**, **Cuba**, **Dominican Republic**).
L 17–18 cm. Monotypic. Bahamas, on New Providence, Sept (year unknown), Sept 1961 (Bond 1963), Cuba (♂ trapped, Jan 1978, held in captivity for one month; another, date unknown), and Dominican Republic (♂, early spring 1996, ♂ and ♀, Dec 1999). In addition, Cory (1892) reported ♀ specimen (no longer extant or reidentified?), Watling's I, S Bahamas, Oct 1891. Could occur in any area with trees.
ID Wing-barred *Piranga* with pale edges to tertials and flight feathers. **Ad br** ♂ has crown, face and throat red, fading to bright yellow on neck and breast. Mantle and scapulars black, lower back to uppertail-coverts bright yellow. Wings mainly black except bright yellow shoulders and two wingbars, upper yellowish white, lower white. Below bright yellow. **Ad non-br** ♂ has much duller red on head, more confined to face, back somewhat mottled black. **Ad** ♀ occurs in two morphs: yellow morph similar to ♂, but duller overall, no red on head, back and wings grey, weaker wingbars; grey morph is greyer, duller and more uniform. **1st-w** is similar to ♀. **Voice** Call a loud "pit-tick" or "pit-er-ick"; also a burry "tu-weep". **SS** Summer Tanager lacks wingbars and is larger-billed. Wingbars in Hispaniolan Crossbill are bolder, etc. See Scarlet Tanager.

Eastern Painted Bunting
(Painted Bunting)

Scarlet Tanager

breeding

Summer Tanager
rubra

breeding

first-winter

Western Tanager

yellow morph

grey morph

321

Northern Cardinal *Cardinalis cardinalis* `LC`
Nineteen subspecies divided into two subspecies groups, of which one occurs in region.

Northern Cardinal *Cardinalis (cardinalis) cardinalis*
Introduced resident (**Bahamas**, **Swan Is**).
L 22–23.5 cm. Nineteen subspecies, of which at least one (perhaps *floridanus*) occurs in region. Present on Swan Is since Jun 1996, probably originally escapees, with small numbers evidently breeding (Anderson *et al.* 1998, Aceituno & Medina 2009); nearest mainland population in Belize. Small numbers also on Cat I, in Bahamas, where bred in 2008 (*NAB* 68: 493) and present several years; reported on Grand Bahama and Green Turtle Cay. Formerly also introduced on Barbados, but not seen since 1930s. Sightings on both Grand Cayman and Cayman Brac. Lightly wooded areas. **ID** Obviously crested, robust-billed passerine. **Ad** ♂ has face and underparts bright red, darker on nape, back and rump. Lores, forehead and most of throat black; bill orange-red. **Ad** ♀ is mid-brown above, underparts buffier, long crest strongly tinted pinkish red, with some pinkish feathers in face, and mask dull greyish black; bare parts much as ♂. **Juv** is similar to ♀, but little or no red in crest, bill blackish or dusky grey, and ♂ can have some red on chest or flanks. **Voice** Song typically a series of loud, cheerful monosyllabic whistles, "wheet-wheet-wit-wit-wit-wit" and so on, a dozen or so notes at a time, or disyllabic whistles, "wheata-wheata-wheata" etc.; or more complex whistles. Contact call a liquid "chip" and, among other vocalizations, an agonistic "chuck, pee-too". **SS** Unmistakable.

THRAUPIDAE
Tanagers
408 extant species, 20 in region + 2 vagrants + 2 introduced

Swallow Tanager *Tersina viridis* `LC`
Vagrant (**Cayman Is**).
L 14–15 cm. Three subspecies, of which one (presumably *occidentalis*) occurs in region. One record, attributed either to vagrant or (less likely?) an escaped cagebird: Grand Cayman, Apr 1982 (Bradley 2000); the bird was apparently photographed, but sex involved is not stated. Elsewhere, mostly in canopy, forest edges and adjacent clearings, using exposed perches. Engages in poorly known seasonal or long-distance movements. Just outside our region, imm ♂, on Bonaire, Feb 2008 (Prins *et al.* 2009). **ID** Turquoise ♂, with black mask and white central belly; green-and-yellow ♀. **Ad** ♂ is mostly bright turquoise-blue, mask and bib black, central belly to vent white, flanks with coarse black barring. **Ad** ♀ is mainly bright green; forehead, face and throat dingy yellowish barred and scaled brown, central belly to vent pale yellow, flanks coarsely barred dusky green and yellow. **Juv** ♂ is like ♀, but progressively shows blue in plumage, even when belly still yellowish and barred as ♀.
Voice High-pitched, buzzy and metallic "tz'sink", sometimes repeated at short intervals. ♂ song more variable, but unlikely to be heard in region. **SS** Confusion unlikely.

Red-legged Honeycreeper *Cyanerpes cyaneus* `LC`
Uncommon and local resident (**Cuba**), vagrant? (**Jamaica**).
L 11–13 cm. Eleven subspecies, of which one (*carneipes*) occurs in region. Principally in woodland and pine forest, but also areas with viny tangles, especially nectar-rich Majagua (*Hibiscus* sp.) and bottlebrush (*Callistemon citrinus*) trees, to 1200 m but mainly at mid elevations. Occasionally enters suburban areas. Suggested as having been introduced (Garrido 2001), but no real evidence for this supposition exists. Ad ♂ specimen from Kingston, Jamaica, May 1890, speculated to have been an escapee (Scott 1893). Pairs or small groups, very active at flowering trees, inserting bill into flowers or taking small invertebrates from underside of foliage, also joins mixed-species flocks. **ID** Deep purple ♂ has yellow underwings; ♀ mostly green. Legs red; bill long, slender and decurved. **Ad br** ♂ mostly bright purplish blue, with crown contrastingly azure-blue, black mantle, wing-coverts and tail; underwing-coverts and underside of remiges bright yellow. **Eclipse** ♂ is like ♀, but wings and tail black, underwing yellow. **Ad** ♀ has olive-green upperparts, faint whitish supercilium, dusky-green eyestripe, dull greyish-white to yellowish-white underparts faintly and narrowly streaked dusky greenish; legs brighter when breeding. **Juv** ♂ is similar to ♀ but gradually acquires black wings and tail, later showing patchy blue. **Voice** Quite noisy, but calls weak and buzzy, including high, thin "tseet"; other nasal notes and a more inflected "tsirp". **SS** Unmistakable in Cuba.
TN Claim (Hilty 2011) that proposed race *ramdseni* is synonym of nominate race (from N South America) wholly improbable; rather it should be subsumed within *carneipes*, but its validity might yet be re-evaluated.

Lesser Antillean Saltator *Saltator albicollis* `LC`
Fairly common resident (**LA**).
L 21–22 cm. Two subspecies, of which both (*albicollis*[A] Martinique and St Lucia, *guadelupensis*[B] Guadeloupe and Dominica) occur in region. Endemic. Dry to humid forest and edges, adjacent clearings, gardens and mangroves; to c. 500 m. Pairs forage together, mostly atop trees and bushes, and perch high to sing. Single record from Nevis either a vagrant or escaped cagebird. **ID** Dull olive-green, with narrow white supercilium, streaked underparts and dark bill with orange tip. **Ad** is olive-green above, except greyer rump and tail; carpal yellow-green, throat whitish, submoustachial streak blackish and underparts dull olive with faint dark streaking mostly on breast-sides; flanks greyer, vent buffier. **Juv** is similar, but has duller face pattern and weaker streaking. **GV** Race *guadelupensis* separated on basis of darker upperparts, and tawnier or yellower underparts, but perhaps better treated as synonym of nominate. **Voice** A complex, melodic series of rather loud, harsh, rising and falling notes. **SS** ♀ Rose-breasted Grosbeak (very rare in Lesser Antilles) has heavier and paler bill, strongly streaked underparts with hint of pink, two whitish wingbars, whiter and broader supercilium, and no submoustachial streak. Juv might resemble ♀ bunting or *Passerina* grosbeak, but all lack whitish supercilium, bicoloured bill, green-olive upperparts, grey tail, etc. **TN** Formerly considered conspecific with extralimital Streaked Saltator *S. striatipectus* (Central and South America).

Northern Cardinal

Swallow Tanager
occidentalis

Red-legged Honeycreeper
carneipes

breeding

Lesser Antillean Saltator
albicollis

323

Bananaquit *Coereba flaveola* LC
Forty-one subspecies divided into three subspecies groups, all three occur in region.

Bahama Bananaquit *Coereba (flaveola) bahamensis*
Fairly common resident (**Bahamas**), rare visitor (**Cuba**).
L 10–11 cm. Two subspecies in group, of which one (*bahamensis* Bahamas SE to Great Inagua and Grand Turk) occurs in region. Near-endemic (also occurs in SE Mexico). Occupies wide array of habitats, like nominate subspecies group; regularly in parks and gardens. Very active and nimble, feeds mostly on nectar, also berries and insects. Increasing records (all presumed to be this race) in Cuba since mid 1990s, largely from N coast and offshore cays, but no evidence of breeding. Cuba is only significant land area in West Indies from which Bananaquit is basically absent. **ID** Very similar to Greater Antillean Bananaquit. **Ad** has crown and upperparts uniform dark grey, with long broad white supercilium, smaller white wing spot, black bill with red gape, whitish throat and breast extending to head-sides, yellow mid-breast to upper belly, below white. **Juv** is duller and dingier with weaker supercilium. **Voice** Several ascending ticks followed by rapid, unmusical, buzzy clicking; often sounds more twittering. **SS** Unmistakable in range.

Greater Antillean Bananaquit *Coereba (flaveola) flaveola*
Common to abundant resident (**GA**).
L 10–11 cm. Four subspecies in group, all of which (*sharpei*[A] Cayman Is, nominate[B] Jamaica, *bananivora*[C] Hispaniola including Gonâve I, Petite Cayemite I and Î-à-Vache, *nectarea*[D] Tortue I, off Haiti) occur in region. Virtually ubiquitous from humid and pine forests to dry scrub, gardens, shrubbery, mangroves, etc., to 2500 m on Hispaniola. Behaviour as Bahama Bananaquit. **ID** Short, sharply pointed, decurved bill. **Ad** is blackish over most of upperparts, with bright yellow rump, prominent white supercilium, white patch at base of remiges. Throat dark greyish, underparts yellow, lower belly whiter. **Juv** is similar to ad, but paler and dingier, with duller supercilium. **GV** Race *bananivora* differs from nominate in being slightly paler overall, above blackish slate (not deep black), throat less slaty, below paler yellow, smaller white wing spot; *nectarea* very similar, but throat and foreneck slightly darker; *sharpei* has somewhat longer bill, smaller and duller yellow rump, and white tail tips. **Voice** Typically short, weak, high-pitched, often ascending series of unmusical buzzes, "chip" notes and insect-like hisses, often ending in short trill; weak "tsit" foraging notes. Notable variation regionally and even individually. **SS** Confusion unlikely on any island.

Common Bananaquit *Coereba (flaveola) bartholemica*
Common and widespread resident (**Puerto Rico**, **LA**, **O**).
L 10–11 cm. Thirty-five subspecies in group, of which ten (*tricolor*[A] Providencia, *oblita*[B] San Andrés, *portoricensis*[C] Puerto Rico, *sanctithomae*[D] Vieques and Culebra, off Puerto Rico, and Virgin Is, *newtoni*[E] St Croix, *bartholemica*[F] Anguilla S to Dominica, *martinicana*[G] Martinique and St Lucia, *barbadensis*[H] Barbados, *atrata*[I] St Vincent, *aterrima*[J] Grenada and Grenadines) occur in region. Habitat and behaviour as Greater Antillean Bananaquit. Vagrant *martinicana* seen on Guadeloupe and several records of *bartholemica* on Martinique. **ID** Generally resembles nominate subspecies group. **Ad** displays strong geographic variation (see below); generally has blackish tinged olive upperparts, bright yellow rump, prominent white supercilium, white wing spot, dark greyish throat, yellow underparts, whiter lower belly and tail tips. **Juv** is paler and dingier than ad. **GV** Race *portoricensis* has darker grey throat and large white tail tips; *martinicana* a longer bill with bold red gape, broad white supercilium, blacker head-sides, central throat whitish, small or no white wing spot and no white tail tips; *barbadensis* much like *martinicana* but smaller white throat; *bartholemica* has throat slate-coloured and again lacks wing spot; *newtoni* like *portoricensis* but has much darker throat and more olive-yellow rump; *sanctithomae* also similar but brighter below, cleaner yellow flanks; *aterrima* has two colour morphs, one mainly sooty black with slight greenish-yellow wash on uppertail-coverts, breast and lower underparts, other black above, with slate-grey throat to breast, dull yellow belly, white vent, large white tail tips (dark morph predominates on Grenada, pale morph on Grenadines); *atrata* very like *aterrima* but larger with longer, heavier bill; *tricolor* and *oblata* similar to Greater Antillean Bananaquit, but have pale grey throat to breast (*oblita* also has yellow of underparts paler and duller, bill shorter). **Voice** Rather variable as in Greater Antillean Bananaquit, but less musical, and has more ascending, faster and notably more insect-like "buzzz" ending in short trill. **SS** Confusion unlikely anywhere in range.

Yellow-faced Grassquit *Tiaris olivaceus* LC
Common and widespread resident (**GA**).
L 9.3–11.5 cm. Five subspecies, of which two (*olivaceus*[A] Cuba, I of Pines and many offshore cays, Cayman Is, Jamaica and Hispaniola including all major satellites, *bryanti*[B] Puerto Rico and E satellites) occur in region. Open woodland, pine forest with abundant undergrowth, plantations, second growth, coastal thickets, savannas and cultivation, to 1300 m in Cuba and at least 2500 m on Hispaniola. Occurs alone, in pairs or small groups, sometimes with other grassquits and other seed-eating birds. Perches atop grass or on ground to feed; sings from trees. **ID** Small and short-tailed, with small conical bill. **Ad** ♂ has dull olive upperparts, bold deep orange-yellow supercilium, crescent below eyes and chin, face and breast blackish, sides and rest of underparts olive-grey. **Ad** ♀ is similar but duller greenish, with no black on face and breast; supercilium, face markings and chin paler yellow. **Juv** resembles ♀ but duller. **GV** Race *bryanti* rather smaller than nominate, brighter olive-green above and more yellowish below. **Voice** Song a long, weak, buzzy and high-pitched trill; calls include quiet "tick" or "tsit" notes. **SS** Rich yellow face markings unique, but ♀ Cuban Grassquit has dull yellow crescent around ear-coverts and rustier face, ♂ has rich yellow crescent outlining black mask. ♀ Black-faced (Hispaniola, Jamaica) and Yellow-shouldered Grassquits (Jamaica) lack yellowish face markings, among other distinguishing features.

Bahama Bananaquit
(Bananaquit)
bahamensis

Greater Antillean Bananaquit
(Bananaquit)
flaveola

aterrima

pale morph

dark morph

Common Bananaquit
(Bananaquit)
bartholemica

Yellow-faced Grassquit
olivaceus

325

Orangequit *Euneornis campestris* `LC`
Locally common resident (**Jamaica**).

L 14 cm. Monotypic. Endemic. Humid forest and edges, shade-coffee plantations, to 1500 m. Singles or pairs forage for nectar, also fruit and insects, sap from sapsucker holes, etc.; regularly in canopy but lower at edges and often with other species. **ID** Small, finch-like bird, with longish, slightly decurved bill. **Ad** ♂ is mainly grey-blue with orange-red throat. **Ad** ♀ has crown olive-grey, upperparts and tail warmer and browner, below dull greyish white and faintly streaked, flanks to vent washed pale buff. **Juv** is similar to ♀. **Imm** ♂ acquires blue patches with age. **Voice** Thin, high-pitched "swee" or "fi-swee". **SS** ♂ unmistakable and well-named. Markedly stubbier-billed Jamaican Euphonia has superficially similar pattern to ♀, but has greyer-green upperparts, yellower vent and sides, and bluish-grey wash to head. ♀ grassquits have much shorter, conical bills, are drabber overall, and occupy different foraging strata. ♀ might resemble some warblers and vireos, with Blue Mountain Vireo superficially similar to ♀, but underparts yellower, upperparts browner; behaviour and structure differ.

Puerto Rican Bullfinch *Melopyrrha portoricensis* `LC`
Two subspecies placed in separate subspecies groups, of which both occur in region, but apparently only one is extant.

Puerto Rican Bullfinch
Melopyrrha (portoricensis) portoricensis
Common resident (**Puerto Rico**).

L 16.5–19 cm. One subspecies in group. Endemic. Montane humid forest to lowland forest, also tropical deciduous forest, dry coastal thickets, occasionally mangroves; to 1000 m. Arboreal, but often feeds on ground. **ID** Large and stocky, with short heavy bill. **Ad** ♂ is mostly black, with orange-rufous forecrown and sides of crown, chin, throat, breast and vent. **Ad** ♀ is similar, but duller black. **Juv** is deep olive-brown, with orange-rufous undertail-coverts. **Voice** Gives 2–10 rising whistles, followed by a buzz; louder "check" calls; also loud "coochi, coochi, choochi". **SS** The only West Indian bullfinch in Puerto Rico, but Lesser Antillean Bullfinch is spreading W (as far as St John, in Virgin Is); it is smaller, with weaker bill, rufous confined to throat, ♀ paler overall, with richer orange-rufous in wings and tail-coverts. **TN** Believed extinct race *grandis* (St Kitts Bullfinch), not definitely seen since 1920s (two reports in last three decades), larger and thicker-billed, with rufous parts darker and less extensive throat patch; perhaps a separate species (Garrido & Wiley 2003).

Greater Antillean Bullfinch *Melopyrrha violacea* `LC`
Fairly common resident (**Bahamas**, **GA**).

L 13.2–17.5 cm. Five subspecies, all of which (*violacea*[A] most larger islands of Bahamas, *ofella*[B] Caicos Is, *affinis*[C] Hispaniola and Gonâve I, Î-à-Vache, Beata, Catalina and Saona I, *maurella*[D] Tortue I, off NW Hispaniola, *ruficollis*[E] Jamaica) occur in region. Borders and dense bushy thickets, dry coastal scrub to humid forest, woodland and gardens; to almost 2100 m. In singles or pairs, mostly arboreal, but sometimes feeds low. **ID** Sturdy, with heavy bill, and some rufous in face and vent. **Ad** ♂ is mostly black with short rufous-orange supercilium, throat and undertail-coverts. **Ad** ♀ is duller black, especially on upperparts. **Juv** is like ♀, but browner and duller, rufous-orange parts smaller, especially on throat. **GV** Race *ofella* smaller than nominate; *affinis* also smaller, and glossier; *maurella* glossier but larger than *affinis*, *ruficollis* larger than nominate, duller and greyer, with paler rufous throat and supercilium. **Voice** Song a repeated trill of insect-like "t'zeet" notes. Call a thin "spit". **SS** Confusion unlikely; grassquits are smaller, with notably less robust bills; Yellow-faced is superficially similar in pattern, but yellow and green vs. rufous-orange and black. **TN** Proposed race *parishi* (Î-à-Vache) supposedly differs in being smaller, but measurements belie this; birds on Beata (and previously Catalina) sometimes allotted to this race. Those on Gonâve and Saona sometimes placed in *maurella*.

Cuban Bullfinch *Melopyrrha nigra* `NT`
Fairly common resident (**Cuba**).

L 14–15 cm. Monotypic. Endemic. Mangroves and semi-deciduous forest to evergreen and open woodland, second growth, coastal scrub, pine forest with bushy undergrowth, and montane serpentine shrubwoods; to 1300 m. Often joins mixed-species flocks of warblers, in all strata. ♂ sings from atop tree or bush. Declining locally due to cagebird trade. **ID** Small dark finch, deep-billed, with white on edge of wing. **Ad** ♂ is black with blue gloss, white alula, edges of primary-coverts and outer webs of outer primaries, plus white underwing-coverts and axillaries. **Ad** ♀ is duller black or more slate, and paler on belly. **Juv** resembles ♀, but has little, if any, white in wings, and bill is paler. **Voice** Song a thin melodious trill or warble, ascending and descending in pitch. Calls buzzing "chip" or doubled, more staccato "chi-dip", also thin "tsee". **SS** Confusion unlikely; vagrant Black-faced Grassquit ♂ is considerably smaller, with thin and more conical bill, dark olive upperparts, tail and lower flanks; behaviour and habitat also differ. **TN** See Grand Cayman Bullfinch.

Grand Cayman Bullfinch *Melopyrrha taylori* `NT`
Very uncommon to locally common resident (**Cayman Is**).

L 14–15 cm. Monotypic. Endemic. Only on Grand Cayman; single record from Little Cayman, Sept–Oct 1998, either this species or Cuban Bullfinch. In wide range of wooded habitats, including parks, gardens and mangroves. All strata, including ground, and usually in flocks. **ID** Small dark finch, bill deep and silvery grey, with white on edge of wing. **Ad** ♂ is dull blackish slate, with white fringes to outer primaries, alula and edges to some wing-coverts, axillaries and underwing-coverts. **Ad** ♀ is similar, but more bicoloured and paler overall. **Juv** resembles ♀, but more olive-tinged head and paler rear upperparts, brownish fringes to flight feathers, belly tinged cinnamon and little, if any, white in wings. **Voice** Song a long series with buzzy quality; begins as a trill "zee-zee-zee", falls briefly then rises again, reaching very high pitch and barely audible at end. Also insect-like "chi-p" and "zee zee" calls, the first note high-pitched. Less complex in structure and shorter than Cuban Bullfinch, with fewer elements, and lower-pitched. **SS** Nothing similar in range. **TN** Formerly considered conspecific with Cuban Bullfinch (Garrido *et al.* 2014) and retained as such by some authors.

Orangequit

Puerto Rican Bullfinch

Greater Antillean Bullfinch
violacea

Cuban Bullfinch

Grand Cayman Bullfinch

327

Yellow-shouldered Grassquit   LC
Loxipasser anoxanthus
Fairly common resident (**Jamaica**).

L 10.2–11.5 cm. Monotypic. Endemic. Dry to humid forest borders, adjacent woodland and gardens, to c. 1800 m; apparently moves lower in winter. Pairs to small family groups often low to ground, and tend to avoid mixed flocks. **ID** Small, tail rather short, but bill thick and head appears large for size of bird. **Ad** ♂ has black head, nape and most underparts, greenish lower belly and flanks, rusty vent, upperparts and tail dull green, with bright yellow shoulders. **Ad** ♀ has head and breast grey, above greenish with yellow wash, richer on shoulders; belly and flanks greyish green, pale rusty vent. **Juv** resembles ♀, but little yellow on shoulders. **Voice** Short, descending series of "chi" notes, with an echo-like quality. **SS** ♂ Black-faced Grassquit lacks yellow shoulders and rusty undertail-coverts; ♀ has whitish undertail-coverts and no yellow in wings; bill more conical and pointed. See Yellow-faced Grassquit, but ♀ has at least some yellow in foreface and supercilium, and no yellow on wing-coverts or pale rusty vent. Juv of larger and bulkier Greater Antillean Bullfinch has rusty undertail-coverts, but is richer brown, with no yellow on wing-coverts, has rufous-orange supercilium and heavier bill.

Cuban Grassquit *Phonipara canora*   LC
Fairly common to rare resident (**Cuba**), introduced (**Bahamas**).

L 11.5 cm. Monotypic. Endemic. Coastal scrub, semi-deciduous woodland, thickets, pine forests, bushes near cultivation and other areas with scattered trees, below 900 m. Perhaps formerly on I of Pines. Island-wide across mainland, but cagebird trade has decimated numbers locally. Established on New Providence (Bahamas) after cagebirds released there, Mar 1963 (but not mapped here). Feeds on ground, often in small groups with Yellow-faced Grassquit; groups split into pairs during breeding season. **ID** Small, with short tail, bold head pattern. **Ad** ♂ has black face and throat, outlined by broad yellow band that is broken on central throat. Crown and upperparts, including tail, olive-green. Black breast, below olive-grey. **Ad** ♀ is duller and lacks black; collar paler yellow. **Juv** resembles ♀, but even duller. **Voice** Raspy trill with rich, harmonic syllables and other more buzzy notes; also longer song of many pure-toned syllables. Call a soft "chip" or high "tsit", often repeated. **SS** Smaller and more petite ♀ Yellow-faced Grassquit has at least some yellow in foreface and supercilium, but none around ear-coverts and throat; ♂ has yellow in face and throat, but pattern is very different. **TN** North American authorities usually place in genus *Tiaris*.

Barbados Bullfinch *Loxigilla barbadensis*   LC
Very common resident (**Barbados**).

L 15.5 cm. Monotypic. Endemic. Ubiquitous throughout Barbados, even heavily urbanized areas. Often in small flocks, regularly with Black-faced Grassquits; from ground level to bush tops, mostly on ground. Sings from exposed perch. **ID** Small and dull, with relatively short rounded bill. **Ad** has greyish-brown head, slightly browner crown, brownish-grey upperparts, with rustier wings and tail. Chin whitish, throat and most of underparts grey, but vent pale cinnamon (sometimes cream to whitish, perhaps ♀ or juv). **Juv** resembles ad. **Voice** Song a repeated set of short sibilant whistles, "sip sip sip sip sip" or "tse tse tse tse tse". **SS** Distinctive as nothing truly similar in range. ♀ Black-faced Grassquit is duller olive-grey overall, with longer more pointed bill, mandible more flesh-coloured, reddish gape and pink (not black) legs. ♀ and juv ♀ Lesser Antillean Bullfinch (a vagrant to Barbados) very similar. **TN** Formerly considered conspecific with Lesser Antillean Bullfinch (Buckley & Buckley 2004).

Lesser Antillean Bullfinch *Loxigilla noctis*   LC
Common and widespread resident (**LA**).

L 14–15.5 cm. Eight subspecies, all of which (*noctis*[A] Martinique, *ridgwayi*[B] St Thomas and St Croix E to Antigua, *coryi*[C] Saba, St Eustatius, St Kittis, Nevis and Montserrat, *desiradensis*[D] La Désirade, *dominicana*[E] Guadeloupe S to Dominica, *sclateri*[F] St Lucia, *crissalis*[G] St Vincent, *grenadensis*[H] Grenada) occur in region. Endemic. Varied habitats, from tropical humid forest to deciduous woodland, second growth, gardens, sometimes dry scrub and mangroves; to 900 m. Pairs or small groups, outwith mixed flocks; mostly arboreal but occasionally descends low and can be tame. Apparently engages in some seasonal movements or tends to wander, arrived c. 1960 in Virgin Is, where found primarily in dry scrub and may continue to expand (e.g. recently reported on uninhabited Great Tobago), and three records on Barbados. **ID** Quite robust, mostly black ♂, ♀ brownish with rustier wings. **Ad** ♂ is black, with small rufous-chestnut patch in front of eye, and rufous-chestnut throat; bill black. **Ad** ♀ has greyish-brown head and upperparts, tinged slightly rusty on wing- and tail-coverts; below greyer; bill brown, with paler mandible. **Juv** resembles ♀. **GV** Race *sclateri* smaller than nominate; *crissalis* has more extensive rufous throat, vent more chestnut-rufous; *grenadensis* has smaller throat patch than *crissalis*, vent partly black; *dominicana* has rufous vent admixed black; *desiradensis* smaller than *dominicana* (but perhaps requires reassessment); *ridgwayi* like *dominicana* but smaller, with larger bill and feet, plumage more greyish black, below more slate-grey; and *coryi* slightly darker. **Voice** Quite variable; a crisp trill of 5–10 "tseep" notes, usually ending in sharp "chuck"; also a harsher "chuk", thin "tseep" that can run into longer twitter; and other buzzy notes. **SS** Confusion unlikely, but see St Lucia Black Finch, of which ♂ lacks rufous on face or throat, and has pink legs; ♀ has browner upperparts, head more contrasting grey, browner buff underparts, bill longer and more conical, legs pink. Blue-black and Black-faced Grassquits (Grenada) are smaller, differ in habitat choice and behaviour; ♂ Blue-black is much glossier blue, ♀ is streaked above and below; ♂ Black-faced has olive upperparts, ♀ is duller greyish olive, legs pink. **TN** See Barbados Bullfinch. Some races, e.g. *desiradensis*, might require reconsideration.

Yellow-shouldered Grassquit

Cuban Grassquit

Barbados Bullfinch

noctis

ridgwayi

Lesser Antillean Bullfinch

329

St Lucia Black Finch *Melanospiza richardsoni* `EN`
Uncommon to rare and local resident (**St Lucia**).

L 13–14 cm. Monotypic. Endemic. Tropical evergreen and semi-deciduous forest, and edges, to 950 m (most frequent in montane areas). Occurs in La Sorcière and Edmond Forest Reserves. Regularly in pairs, mostly terrestrial and fond of dense thickets. Poorly known. **ID** Thickset, with pink legs, fairly large conical black bill, large flesh-coloured legs. **Ad** ♂ is deep black. **Ad** ♀ has greyish head, rich brown upperparts, wings and tail; dull light brown underparts, rustier on vent; bill greyish horn. **Juv** is like ♀. **Voice** Song a high-pitched, burry, loud and sibilant "tseéééééwww- swisiwis-tew", the initial note an explosive whistle with near-electric quality. Call a sharp "tiiip!". **SS** Lesser Antillean Bullfinch rather bulkier, with smaller bill and darker legs; ♂ has dark rufous-chestnut throat, ♀ has browner head, duller upperparts but rustier wing-coverts, greyer and duller underparts. Black-faced Grassquit is found in more open country, and is smaller; ♂ has dark olive upperparts, wings and tail, ♀ mostly dull olive.

Black-faced Grassquit *Melanospiza bicolor* `LC`
Fairly common resident (**GA**, **LA**, **O**).

L 10.2–11.5 cm. Eight subspecies, of which four (*bicolor*[A] Bahamas and islands off N Cuba, *marchii*[B] Jamaica and Hispaniola, *omissa*[C] Puerto Rico SE to Grenada, *grandior*[D] San Andrés, Providencia and St Catalina) occur in region. Dry woodland and cactus scrub, grasses and shrubs, roadsides, parks, gardens and plantations, to at least 1800 m (Hispaniola). Status in Cuba confused and poorly known. Mostly in small groups feeding on ground or among low grasses and weeds. Often outnumbers Yellow-faced Grassquit in urban and cultivated areas, except on Hispaniola. **ID** Small and dark, relatively short-tailed, with small, conical bill. **Ad** ♂ has black head and most of underparts, lower flanks dark olive; above dark olive, suffused black on nape; bill black, gape vivid pink-red when breeding, legs flesh. **Ad** ♀ is drab olive-green, browner or greyer below, especially lower belly; dark maxilla, paler mandible, pinkish-red gape. **Juv** is like ♀ but has paler bill for first c. 3 months. **GV** Race *omissus* smaller than nominate; *marchii* similar, but ♂ has less extensive black underparts; *grandior* like *omissus*, but larger with brighter olive-green upperparts. **Voice** Buzzy series of short, emphatic "tse" or "chit-tse" notes; call a soft "tsip". **SS** Yellow-faced Grassquit often consorts with Black-faced; ♀ and juv ♂ show at least some yellow in foreface and supercilium, and lack pinkish-red gape (difficult to see). Both Lesser Antillean and Barbados Bullfinches more arboreal, larger, with shorter bills, some orange-rusty in wings and tail-coverts (except mostly black ♂ Lesser Antillean). ♀ St Lucia Black Finch also larger, with rusty wash below, browner upperparts, greyer head. See ♀ and juv ♂ Yellow-bellied Seedeater (Grenada and Grenadines) with all-dark thicker bill, yellowish-white belly, and no reddish gape. **TN** North American authorities usually place in genus *Tiaris*.

Blue-black Grassquit *Volatinia jacarina* `LC`
Fairly common migrant breeder (**Grenada**, **Carriacou**).

L 8.7–10.9 cm. Three subspecies, of which one (*splendens* Grenada) occurs in region. Agricultural areas, hedges and roadsides throughout Grenada, and recently photographed on Carriacou (Coffey & Ollivierre 2019). Forages low down, often in pairs or small groups with other seed-eating birds and grassquits. ♂ has an attractive display, singing incessantly from an exposed perch and jumping up several cm on rapidly flapping wings. Apparently departs Grenada post-breeding; recorded mostly May–Oct, but at least once in late Mar. Two undocumented reports in Grenadines. **ID** Small, with longish conical bill. **Ad** ♂ mostly black glossed dark blue, less iridescent on flight feathers and tail, some white on axillaries, underwing-coverts and bases of remiges (visible in display). **Ad** ♀ is more or less warm brown above, with two indistinct wingbars. Underparts pale whitish buff and more boldly streaked than upperparts, especially on flanks and breast; outertail feathers tipped buff when fresh. **Juv** resembles ♀, but wings and tail darker, more extensively streaked below. Young ♂ acquires blackish feathers with age. **Voice** Song a rapid, rather buzzy trill, with emphatic "tseep" first and a buzz at end; call sharp "check". **SS** Black-faced Grassquit has less conical bill and pinkish legs, dull olive upperparts and is always unstreaked. Only other small, black passerine in range is Lesser Antillean Bullfinch, which is larger, with heavier bill, chestnut chin and vent, ♀ unstreaked warmer brown; habits and habitat also differ.

Lined Seedeater *Sporophila lineola* `LC`
Vagrant (**Guadeloupe**).

L 10–11 cm. Monotypic. ♂ photographed, Sept 2017 (eBird). Given species' relative lack of popularity among cagebird enthusiasts, its well-known migration pattern that occasionally reaches N as far as Costa Rica, and that date of only West Indian record matches this, we have no hesitation in treating the Guadeloupe sighting as a wild bird. Could appear in any scrubby or grassy habitat. **ID** Tiny seedeater with thick bill and very distinctive ♂ plumage. **Ad** ♂ has most of head and throat black, broad white central crown-stripe and large white triangular patch on face-sides bordered by black line. Upperparts black, with narrow white rump and white on primary bases. Underparts whitish, except black thighs; bill black. **Ad** ♀ is dull brownish olive above, paler below, with buffy breast, paler creamy throat, and pale buff lower breast to vent; maxilla dark, mandible horn-coloured. **Juv** is undescribed. **Voice** Calls "check", a dull trilling "ki-rik" and strident and repeated "peeu". **SS** Ad ♂ unmistakable. ♀ potentially impossible to separate from other seedeaters, but much browner above and more buffy below than ♀ grassquits.

Yellow-bellied Seedeater *Sporophila nigricollis* `LC`
Fairly common migrant breeder (**Grenada**, **Grenadines**).

L 8.5–10.3 cm. Three subspecies, of which nominate occurs in region. Agricultural fields, pastures, roadsides and shrubby clearings. Often in small groups, but congregates in larger parties and with other seed-eaters; feeds in grass. Present Mar–Nov; apparently migrates to South America post-breeding, but also wanders locally elsewhere in extensive range. Vagrant to St Vincent and St Kitts. **ID** Small seedeater, with thick bill and distinctly rounded culmen. **Ad** ♂ has black hood, dull olive upperparts and pale yellowish underparts; bill greyish. Some have greyer upperparts and whiter underparts. **Ad** ♀ is nondescript, mostly warm buffy-brown above, and on throat and breast, with paler belly and vent; bill and legs blackish. **Juv** is similar to ♀, sometimes warmer below. **Imm** ♂ acquires blackish feathers on face and throat with age. **Voice** Sweet, musical series of warbles and twitters, ending in more buzzy notes; calls dry "chip" and buzzier "jit" notes. **SS** ♂ Black-faced Grassquit has all-black underparts, darker and more pointed bill; ♀ duller olive overall, with pinkish legs, paler horn bill. ♀ Blue-black Grassquit is boldly streaked, with longer conical bill.

bicolor

Black-faced Grassquit

St Lucia Black Finch

marchii

Blue-black Grassquit
splendens

Lined Seedeater

Yellow-bellied Seedeater
nigricollis

331

Chestnut-bellied Seed-finch *Sporophila angolensis* LC
Introduced (**Martinique**).
L 10.6–12.4 cm. Two subspecies, of which one (presumably *torrida*) occurs in region. Shrubby clearings, agricultural land, young second growth and even forest edges. Introduced c. 1992. Singles or pairs, occasionally with grassquits. Feeds from ground to mid-strata, mainly on seeds. **ID** Mid-sized, heavy-billed finch; bicoloured ♂. **Ad** ♂ has head, breast, upperparts and wings glossy black, small white patch at base of primaries, and deep chestnut underparts. Wing-linings white, extending to marginal-coverts. Black bill. **Ad** ♀ is mostly warm brown, paler and more cinnamon below, and chin and central belly whitish buff; bill dark. **Juv** resembles ♀.
Voice Variable and musical series of whistles interspersed with chatters and trills. **SS** Black-faced Grassquit has small bill, black underparts, no white wing spot, ♀ much duller overall, with no white in wing. ♂ Lesser Antillean Bullfinch is larger, more arboreal, blacker overall, with chestnut only on vent, face and chin; ♀ duller brown, with pale chestnut undertail-coverts, face and chin. **TN** Previously considered conspecific with extralimital Thick-billed Seed-finch *Sporophila funerea* (SE Mexico to W Ecuador).

Saffron Finch *Sicalis flaveola* LC
Five subspecies divided into two subspecies groups, of which one occurs in region.

Saffron Finch *Sicalis (flaveola) flaveola*
Introduced, locally common resident (**Jamaica**, **Puerto Rico**).
L 13.5–15 cm. Three subspecies in group, of which one (presumably nominate) occurs in region. Well-established introduction in Jamaica (since 1820s) and Puerto Rico (since c. 1960). Single records: Dominican Republic, Nov 1993 (suspected escapee) and E Cuba, Oct 1996 (Garrido 1997), although there are other vaguer mentions for the latter country, both historical and more modern (Navarro Pacheco 2018). Cultivation, gardens, roadsides, parks. Pairs to small groups, mainly terrestrial, but perches atop bushes or trees to sing. **ID** Stocky, well-proportioned finch. **Ad** ♂ is mostly rich yellow, with brighter orange forecrown. Upperparts faintly streaked. Belly slightly paler than rest of underparts. Tail dark with broad yellowish edges. Bill dusky above, mandible mostly yellowish. **Ad** ♀ is duller yellow overall, little orange on crown, and paler yellow belly. **Juv** has greyish-brown head and upperparts, faintly dark-streaked, olive-yellow rump, faint whitish supercilium, slightly darker eyestripe, pale yellow breast, sides and vent, becoming brighter with age.
Voice Lively, leisurely, semi-musical song comprising well-spaced whistles, other sweet, clear notes and a few harsher ones; calls include an upslurred "whip" and short "tuup". **SS** Dickcissel is bulkier, heavier-billed, notably streaked and browner on upperparts, more streaked on sides to flanks, with distinct face pattern. In Puerto Rico (and perhaps elsewhere), escaped Yellow-fronted Canary *Crithagra mozambica* (see Appendix 2) is more olive above than juv Saffron Finch, with obvious yellow supercilium and blackish malar. Introduced Yellow-crowned Bishop (p. 248) differs in habitat and behavior; ♀ and non-breeding ♂ darker and streakier above than young Saffron, with bold buffy supercilium and buffier underparts.

Grassland Yellow-finch *Sicalis luteola* LC
Eight subspecies divided into four subspecies groups, of which one occurs in region.

Grassland Yellow-finch *Sicalis (luteola) luteola*
Uncommon to locally common resident (**LA**).
L 9.8–12.5 cm. Three subspecies in group, of which nominate occurs in region. First noted in region 1890s, on Barbados, from where believed to have colonized other islands, reaching St Lucia in 1942, Martinique in 1951, St Vincent in 1971 and Guadeloupe in 1983; it now occurs as far S as Grenada and Mustique (1924) and N to Antigua (since 1965). Often said to be introduced, Buckley *et al.* (2009) argued that it is equally or more likely a natural colonist. Open grassy fields, edges of cultivation. Forms flocks of up to 50, feeding on ground, in low vegetation or on grass stems. **ID** Small and compact yellow finch; bill short, legs pale, longish wings. **Ad** ♂ has olive crown and upperparts, with fine dark streaks, narrow yellowish eye-ring and supercilium, yellow neck-sides. Throat and malar brighter yellow, below duller, washed olive on breast-sides to flanks, belly slightly paler. **Ad** ♀ has paler brown upperparts, dark malar, less yellow on face and neck-sides, duller yellow underparts. **Juv** is similar to ♀, but upperparts buffier, with stronger malar, finely streaked breast, and belly even duller. **Voice** Fast, buzzy series of trilled notes, canary-like, each trill at slightly different frequency from previous one and with different cadence; flight song longer than perched song. Call rather explosive, a disyllabic "pit-tchew!". **SS** On Martinique, the larger, longer-billed and longer-legged Village Weaver (introduced, p. 248) is superficially similar. Introduced Northern Red Bishop (Martinique and Dominica) is only superficially similar; ♀ and non-br ♂ heavily streaked above, face pattern distinctive, tail much shorter, etc.

St Vincent Tanager *Tangara versicolor* LC
Fairly common resident (**St Vincent and Grenadines**).
L 14–15 cm. Monotypic. Endemic. Most wooded areas throughout St Vincent, including montane rainforest, and commoner at higher elevations. Habits not known to differ from Grenada Tanager. **ID** Unique tanager similar to Grenada Tanager. **Ad** ♂ differs in its pale chestnut crown, more russet washed bluish underparts, buffier, less greenish tone to hindcollar, mantle, back and rump, and bluer tinge to pale turquoise-green wing and tail fringes. **Ad** ♀ is duller, with greener upperparts and greyer underparts. **Juv** is much duller, with hint of dark mask and little chestnut on crown. **Voice** Like Grenada Tanager. **SS** Unmistakable. **TN** Considered by most authors as conspecific with Grenada Tanager, under name Lesser Antillean Tanager. More information is required from the Grenadines to establish whether it is the present species alone that occurs there.

Grenada Tanager *Tangara cucullata* LC
Fairly common resident (**Grenada**).
L 14–15 cm. Monotypic. Endemic. Arid scrub to moist forest, forest edges, second growth, parks and gardens, at all elevations. Pairs or small groups gather at fruiting trees; rather noisy and very active. **ID** Only buffy-green tanager in range with bluish wings and chestnut cap. **Ad** ♂ has dark maroon-chestnut crown, blackish mask, blue-green wash on ear-coverts, pale yellowish-buff upperparts tinged greenish. Wings and tail blackish, with bold bluish-green edges. Underparts lavender admixed buff, spotted dark on chin and throat. **Ad** ♀ is duller, underparts greenish grey. **Juv** is even duller than ♀, with only hint of mask and little or no chestnut on crown. **Voice** Song a weak, high-pitched series of "weet" notes followed by a clear twitter; simpler "chirp" calls. **SS** Unmistakable. **TN** See St Vincent Tanager.

Chestnut-bellied Seed-finch
torrida

Saffron Finch
flaveola
adult
juvenile

Grassland Yellow-finch
luteola

St Vincent Tanager

Grenada Tanager

333

APPENDIX 1. REGIONAL CHECKLIST

We include all species that have been convincingly shown to have occurred in the wild within the West Indies. Following previous authors, the Virgin Is are considered a single unit, while records for St Vincent and the Grenadines and Grenada are assigned with respect to political boundaries, so, for example, Carriacou is included as a dependency of Grenada rather than within the Grenadines.

APPENDIX KEY

Bah	Bahamas
T&C	Turks and Caicos Islands (UK)
Cub	Cuba
Cay	Cayman Islands (UK)
Jam	Jamaica
Hai	Haiti
DR	Dominican Republic
PR	Puerto Rico (USA)
VI	Virgin Islands (USA / UK)
Ang	Anguilla (UK)
StM	Saint Martin (France / Netherlands)
StB	Saint Barthélemy (France)
Sab	Saba (Netherlands)
StE	Sint Eustatius (Netherlands)
K&N	Saint Kitts and Nevis
A&B	Antigua and Barbuda
Mon	Montserrat (UK)
Gua	Guadeloupe (France)
Dom	Dominica
Mar	Martinique (France)
StL	Saint Lucia
V&G	Saint Vincent and the Grenadines
Gre	Grenada
Bar	Barbados
Ave	Isla de Aves (Venezuela)
Swa	Swan Islands (Honduras)
Pro	Providencia and Santa Catalina (Colombia)
San	San Andrés (Colombia)

Endemic (E). Used for species which are restricted to a single geographical unit, whether a country, territory, island or the region as a whole.

Near-endemic (NE). Species with a breeding distribution largely restricted to the West Indies, only a minor proportion of which extends to adjacent areas.

Regular (R). Denotes species that occur regularly, whether as residents, breeding or non-breeding visitors, or passage migrants.

Vagrant (V). A species outside its usual range. Lack of comparable data between islands does not permit us to use a statistical definition. Instead, we have tried to use this category to convey the sense that such species are not usual members of the avifauna, but rather birds out of place, and should therefore not be expected on any given visit. Nevertheless, in order to assign a status of V rather than R, we have examined a large number of published and unpublished sources to understand frequencies of encounter as well as trends over time. Typically a vagrant is represented by fewer than five recent records, perhaps as many as double this number over a period of 100 years. The borderline between V and R is imprecise and apt to change over time, so a species like Lesser Black-backed Gull was until recently a vagrant, but is already a fairly frequent non-breeding visitor and will soon be a regular component of the avifauna almost throughout the region. It merits underlining that overall status in the region is defined by local patterns of occurrence. Consequently, a species listed as vagrant in all relevant territories will be considered similarly at the regional level, despite the possibility that it is represented by more records than some more regularly occurring species with restricted ranges in the West Indies.

Hypothetical (H). Denotes species for which no verifiable proof (specimen, photograph, sound recording) of occurrence in the region exists. Our criteria at the sub-regional level are slightly different in that, unless new data exist to change our view, we typically regard records accepted by Bond or a national checklist as sufficient evidence for occurrence in a country, territory or island, on the premise that they have already undergone rigorous review. Those published in regional compilations in *NAB* and *Cotinga* are treated similarly unless there are grounds to suppose otherwise, such as an expression of a degree of uncertainty by the observers or editors. Other published or electronic (principally eBird) records are accepted when supported by incontrovertible evidence of the type specified above.

Introduced (I). Species which have been deliberately or unintentionally introduced and maintain a *self-sustaining breeding population*. We do not cover exotic species that have unequivocally failed to become established, although some of these are covered under Appendix 2. Note that species introduced to one island may well be regarded vagrants on neighbouring islands, while not necessarily establishing a population that would qualify them as I. Exotics that have reached the region no more than a few times (Mute Swan, Egyptian Goose, Common Myna) are listed in Appendix 2.

Extinct (Ex) is applied to species which are no longer extant in a particular territory. On a regional level, we follow BirdLife International and IUCN categories.

NATIONAL REVIEWERS

Bahamas: Bruce Purdy
Turks and Caicos Islands: Bruce Purdy
Cayman Islands: Peter Davey (National Trust for the Cayman Islands)
Jamaica: Catherine Levy
Haiti: Sean Christensen
Puerto Rico: Sergio A. Colón López
Virgin Islands: Robert L. Norton, Lisa Yntema (USVI)
Saint Martin: Adam C. Brown (Environmental Protection in the Caribbean [EPIC])
Saint Kitts and Nevis: Michael H. Ryan, Percival Hanley
Dominica: Niels Larsen, Bertrand Jno Baptiste
Saint Lucia: Lenn Isidore
Saint Vincent and the Grenadines: Juliana Coffey
Grenada: Anthony Jeremiah, Juliana Coffey
Swan Islands: John van Dort
San Andrés, Providencia and Santa Catalina: Thomas Donegan, Steven L. Hilty

Common name	Regional status	Bah	T&C	Cub	Cay	Jam	Hai	DR	PR	VI	Ang	StM	StB	Sab
Cracidae														
Rufous-vented Chachalaca	I													
Numididae														
West African Guineafowl	I			I			I	I	I	I		I		
Odontophoridae														
Eastern Bobwhite	I	I					I							
Black-breasted Bobwhite	I			I				I	I	I				
Crested Bobwhite	I	?							Ex(I)					
Phasianidae														
Indian Peafowl	I	I												
Red Junglefowl	I	I			I			I	I					I
Grey-rumped Pheasant	I	I		I		I		I						
Anatidae														
White-faced Whistling-duck	V			V				V	V					
Northern Black-bellied Whistling-duck	R+I	V+I		R										
Southern Black-bellied Whistling-duck	R	V			V	V	V		V	V			V	
West Indian Whistling-duck	E	R	R	R	R	R	R	R	R	V	V			
Fulvous Whistling-duck	R+I	R	R	R	V	V	R	R	R	V			R	
Masked Duck	R	V	V	R	V	V	R	R	R	R				
Ruddy Duck	R+I	R	R	R	V	R	R	R	R	R	R	R	V	
Tundra Swan	V			V				V	V					
Pale-bellied Brent Goose	V													
Black-bellied Brent Goose	V	?V							V					
Canada Goose	V	V	V	V	V			V	V					
Snow Goose	V	V		V	V			V	V	V				
Ross's Goose	V	V												
Greater White-fronted Goose	V	V		V										
Long-tailed Duck	V									V				
Surf Scoter	V			V										
White-winged Scoter	H	H		H										
Bufflehead	V	V		V	V			V			V			
Hooded Merganser	V	V	V	V	V	V	V	V	V	V				
Goosander / Common Merganser	H			H										
Red-breasted Merganser	R	R	V	R	R			V	V	V	V			
Orinoco Goose	V				V									
Common Shelduck	V													
Muscovy Duck	I			I							I			
Wood Duck	R+I	R		R	I+V	V		V	V		V			V
Common Pochard	V													
Redhead	R	R	V	R	V		V	V						
Canvasback	V	V		V				V	V					
Ring-necked Duck	R	R	R	R	R	R	R	R	R	R	R	R	R	

Common Name	StE	K&N	A&B	Mon	Gua	Dom	Mar	StL	V&G	Gre	Bar	Ave	Swa	Pro	San
Cracidae															
Rufous-vented Chachalaca									I						
Numididae															
West African Guineafowl		I	I												
Odontophoridae															
Eastern Bobwhite															
Black-breasted Bobwhite		Ex(I)													
Crested Bobwhite									Ex(I)						
Phasianidae															
Indian Peafowl															
Red Junglefowl	I				I		I		I						I?
Grey-rumped Pheasant															
Anatidae															
White-faced Whistling-duck					V						V				
Northern Black-bellied Whistling-duck															
Southern Black-bellied Whistling-duck		V	V		R	V	R	V	V	V	R				
West Indian Whistling-duck		Ex	R	V	R	V	V			V	V				
Fulvous Whistling-duck		H	R		R	V	V	V	V	V	R+I	V			
Masked Duck		V	V		R	V	R	R		V	R		V		
Ruddy Duck		R	V		R		V	V	V	R	V+I			V	
Tundra Swan		H	V												
Pale-bellied Brent Goose											V				
Black-bellied Brent Goose															
Canada Goose															
Snow Goose					V										
Ross's Goose															
Greater White-fronted Goose															
Long-tailed Duck															
Surf Scoter															
White-winged Scoter															
Bufflehead															
Hooded Merganser			V		V		V				V				
Goosander / Common Merganser															
Red-breasted Merganser														V	V
Orinoco Goose											V				
Common Shelduck							V				V				
Muscovy Duck							I								
Wood Duck		V			H						I				
Common Pochard											V				
Redhead															
Canvasback			V												
Ring-necked Duck		V	V		R	V	V	V	V	V	R			V	V

337

Common name	Regional status	Bah	T&C	Cub	Cay	Jam	Hai	DR	PR	VI	Ang	StM	StB	Sab
Tufted Duck	V								V					
Greater Scaup	V	V	V	H		V		H		V				
Lesser Scaup	R	R	R	R	R	R	R	R	R	R	R	R	R	
Garganey	V								V					
Northern Shoveler	R	R	R	R	R	R	R	R	R	R	V	V	V	
Cinnamon Teal	V	V		V		V			V	V				
Blue-winged Teal	R	R	R	R	R	R	R	R	R	R	R	R		V
Gadwall	R	R	R	R	V	V	V		V	H				
Eurasian Wigeon	V	V		H			V	V	V	V	V			
American Wigeon	R	R	R	R	R	R	R	R	R	R	R	R		V
Mallard	R+I	V	V	R	V+I	V	H	V	V	V		V	V	
American Black Duck	V	V	V						V	V		V		
Mottled Duck	V	V		V										
White-cheeked Pintail	R	R	R	R	V	H	R	R	R	R	R	R		
Northern Pintail	R	R	R	R	R	V	R	R	R	R	V	V	R	
Eurasian Teal	V	V							V					
Green-winged Teal	R	R	R	R	R	V	V	V	V	R	R	R	R	
Podicipedidae														
Least Grebe	R	R	R	R	V	R	R	R	R	R				V
Pied-billed Grebe	R	R	R	R	R	R	R	R	R	R	R	R		
Red-necked Grebe	H	H												
Black-necked Grebe / Eared Grebe	V	V												
Phoenicopteridae														
American Flamingo	R	R	R	R	V	R	R	R	R	V	V			
Phaethontidae														
Red-billed Tropicbird	R	V		V			V	R	R	R	R	R	R	
White-tailed Tropicbird	R	R	R	R	R	R	R	R	R	R	R	R	R	
Columbidae														
Rock Dove / Rock Pigeon	I	I	I	I	I	I	I	I	I	I	I	I	I	
Eurasian Collared-dove	R+I	I	I	R+I	I	I	I	I	I	I	I	I	R	
African Collared-dove	I			I			I							
White-crowned Pigeon	NE	R	R	R	R	R	R	R	R	R	R	R	V	
Scaly-naped Pigeon	NE	V	V	R		V	R	R	R	V	R	R	R	
Ring-tailed Pigeon	E				E									
Plain Pigeon	E			R		R	R	R	R					
Blue-headed Quail-dove	E			E										
Crested Quail-dove	E				E									
Grey-headed Quail-dove	E			E										
White-fronted Quail-dove	E						E							
Ruddy Quail-dove	R			R		R	R	R	R	V				
Key West Quail-dove	E	R	R	R			R	R	R					
Bridled Quail-dove	E								R	R	V	R	R	
Caribbean Dove	NE	I		R	R									
Grenada Dove	E													

Common Name	StE	K&N	A&B	Mon	Gua	Dom	Mar	StL	V&G	Gre	Bar	Ave	Swa	Pro	San
Tufted Duck					V						V				
Greater Scaup											V				
Lesser Scaup		V	V		R	R	V	V	V	V	R		V		V
Garganey					V		H?				V				
Northern Shoveler		V	R		R	V	V	V	V	V	R		H		V
Cinnamon Teal			V							?					V
Blue-winged Teal	V	R	R	R	R	R	R	R	R	R	R		R	R	R
Gadwall					H			V							V
Eurasian Wigeon			V				V			V	V				
American Wigeon		V	V		R	V	R	V	V	H	R		R		V
Mallard			V		V		V		V		I				
American Black Duck					V										
Mottled Duck															
White-cheeked Pintail		R	R		R		V		V		V				
Northern Pintail		V	V		R		R	V	V		R		R		
Eurasian Teal					V						V				
Green-winged Teal		V	V	V	R	V	V	V	V	V	R		H		V
Podicipedidae															
Least Grebe					V					H					
Pied-billed Grebe		R	R	R	R	R	R	R	V	R	R		V		V
Red-necked Grebe															
Black-necked Grebe / Eared Grebe															
Phoenicopteridae															
American Flamingo		V			V		V	V	V	V	V				
Phaethontidae															
Red-billed Tropicbird	R	R	R	R	R	R	R	R	R	R	R	R			
White-tailed Tropicbird	R	R	R	R	R	R	R	R	R	R	R				
Columbidae															
Rock Dove / Rock Pigeon	I	I	I	I	I	I	I	I	I	I	I				I
Eurasian Collared-dove	R	R	I	V	I	R	I	I	.	I	I				
African Collared-dove															
White-crowned Pigeon	R	V	R	R	R	V	R	V	V		V	V	R	R	R
Scaly-naped Pigeon	R	R	R	R	R	R	R	R	R		I				
Ring-tailed Pigeon															
Plain Pigeon															
Blue-headed Quail-dove															
Crested Quail-dove															
Grey-headed Quail-dove															
White-fronted Quail-dove															
Ruddy Quail-dove		V	R		R	R	R	R	R	R					
Key West Quail-dove															
Bridled Quail-dove	R	R	R	R	R	R	R	R							
Caribbean Dove															R
Grenada Dove										E					

Common name	Regional status	Bah	T&C	Cub	Cay	Jam	Hai	DR	PR	VI	Ang	StM	StB	Sab
White-winged Dove	R	R	R	R	R	R	R	R	R	R	R	R	R	R
Zenaida Dove	NE	R	R	R	R	R	R	R	R	R	R	R	R	R
Eared Dove	R													
Mourning Dove	R	R	R	R	R	R	R	R	R	H				H
Common Ground-dove	R	R	R	R	R	R	R	R	R	R	R	R	R	R
Pied Imperial-pigeon	I	I												
Nyctibiidae														
Northern Potoo	R			R		R	R	R	V					
Caprimulgidae														
Common Nighthawk	R	R	R	R	R	R	H	R	V	R	R		R	
Antillean Nighthawk	NE	R	R	R	R	R	R	R	R	R	R		R	
Lesser Nighthawk	H													
White-tailed Nightjar	R								V					
Jamaican Poorwill	E					E								
Least Poorwill	E						R	R						
Eastern Whip-poor-will	V	H		V		V								
Puerto Rican Nightjar	E								E					
Chuck-will's-widow	R	R	R	R	R	R	R	R	R		V	V		R
St Lucia Nightjar	E													
Cuban Nightjar	E			E										
Hispaniolan Nightjar	E						R	R						
Apodidae														
Black Swift	R			R	V	R	R	R	R	V		V	V	V
White-collared Swift	R			R		R	R	R	V					V
Lesser Antillean Swift	E													
Ashy-rumped Swift	R													
Chimney Swift	R	R	R	R	R	V	R	R	V	V	V			V
Short-tailed Swift	R								V	V				
Antillean Palm-swift	E	V	V	R	V	R	R	R	V					
Alpine Swift	V								V					
Common Swift	V								V					
Trochilidae														
White-necked Jacobin	V													
Rufous-breasted Hermit	R													
Ruby-topaz Hummingbird	V													
Green-breasted Mango	R													
Hispaniolan Mango	E						R	R						
Puerto Rican Mango	E								R	R				
Green Mango	E								E					
Jamaican Mango	E					E								
Green-throated Carib	E								R	R	R	R	R	R
Purple-throated Carib	E								V	V	V	V	R	R
Cuban Emerald	E	R		R										
Hispaniolan Emerald	E						R	R						
Puerto Rican Emerald	E								E					

Common Name	StE	K&N	A&B	Mon	Gua	Dom	Mar	StL	V&G	Gre	Bar	Ave	Swa	Pro	San
White-winged Dove	R	R	R	R	R	V	V	V		V		V	V	R	R
Zenaida Dove	R	R	R	R	R	R	R	R	R	R	R				
Eared Dove					R		V	R	R	R	R	R			
Mourning Dove					H										
Common Ground-dove	R	R	R	R	R	R	R	R	R	R	R				R
Pied Imperial-pigeon															
Nyctibiidae															
Northern Potoo															
Caprimulgidae															
Common Nighthawk		V	R	R	R		V	V			R		R	R	R
Antillean Nighthawk		V	R	R	R	R	R				V		V	?	
Lesser Nighthawk															H
White-tailed Nightjar						R									
Jamaican Poorwill															
Least Poorwill															
Eastern Whip-poor-will															
Puerto Rican Nightjar															
Chuck-will's-widow		V	V		H									H	V
St Lucia Nightjar								E							
Cuban Nightjar															
Hispaniolan Nightjar															
Apodidae															
Black Swift		V	V	R	R	R	R	R	R	R	R				
White-collared Swift		V			V		V			R	V				
Lesser Antillean Swift		V			R	R	R	R	R						
Ashy-rumped Swift										R					
Chimney Swift					V	V					V		R	V	V
Short-tailed Swift					H		V	R	R	V	V				
Antillean Palm-swift															
Alpine Swift					V			V			V				
Common Swift										H					
Trochilidae															
White-necked Jacobin										V					
Rufous-breasted Hermit										R					
Ruby-topaz Hummingbird										V					
Green-breasted Mango														R	R
Hispaniolan Mango															
Puerto Rican Mango															
Green Mango															
Jamaican Mango															
Green-throated Carib	R	R	R	R	R	R	R	R	R	R					
Purple-throated Carib	R	R	R	R	R	R	R	R	R	V					
Cuban Emerald															
Hispaniolan Emerald															
Puerto Rican Emerald															

Common name	Regional status	Bah	T&C	Cub	Cay	Jam	Hai	DR	PR	VI	Ang	StM	StB	Sab
Blue-headed Hummingbird	E													
Antillean Crested Hummingbird	E								R	R	R	R	R	R
Red-billed Streamertail	E				E									
Black-billed Streamertail	E				E									
Lyre-tailed Hummingbird	E	E												
Bahama Hummingbird	E	R	R	H										
Vervain Hummingbird	E					R	R	R	V					
Bee Hummingbird	E		V	E										
Ruby-throated Hummingbird	R	R		R	R	H	V	V	V					
Rufous Hummingbird	H	H												
Cuculidae														
Greater Ani	V									V				
Smooth-billed Ani	R	R	R	R	R	R	R	R	R	R				V
Yellow-billed Cuckoo	R	R	R	R	R	R	R	R	R	R	R	R	R	V
Pearly-breasted Cuckoo	V									V				
Mangrove Cuckoo	R	R	R	R	R	R	R	R	R	R	R	V	V	
Dark-billed Cuckoo	V													
Black-billed Cuckoo	R	R		R	R	V		V	R	V				
Chestnut-bellied Cuckoo	E					E								
Bay-breasted Cuckoo	E						R	R						
Cuban Lizard-cuckoo	E			E										
Bahama Lizard-cuckoo	E	E												
Jamaican Lizard-cuckoo	E					E								
Hispaniolan Lizard-cuckoo	E						R	R						
Gonave Lizard-cuckoo	E						E							
Puerto Rican Lizard-cuckoo	E								E	V				
Common Cuckoo	V													
Rallidae														
Yellow Rail	V	V												
Black Rail	R	V		R		R		R	R					
King Rail	R			R		V								
Clapper Rail	R	R	R	R		R	R	R	R					
Virginia Rail	R	R		V					V					
Corncrake	V													
Zapata Rail	E			E										
Southern Spotted Rail	R			R		R	H	R						
Yellow-breasted Crake	R			R		R	R	R	R					
Sora	R	R	R	R	R	R	R	R	R	R	R	R	R	
Spotted Crake	V										V			
Purple Gallinule	R	R	R	R	R	R	R	R	R	V	V	R	V	
Common Gallinule	R	R	R	R	R	R	R	R	R	R	R	R	V	
American Coot	R	R	R	R	R	R	R	R	R	R	R	R		
Aramidae														
Limpkin	R	R	V	R	V	R	R	R	R					

Common Name	StE	K&N	A&B	Mon	Gua	Dom	Mar	StL	V&G	Gre	Bar	Ave	Swa	Pro	San
Blue-headed Hummingbird						R	R								
Antillean Crested Hummingbird	R	R	R	R	R	R	R	R	R	R	R				
Red-billed Streamertail															
Black-billed Streamertail															
Lyre-tailed Hummingbird															
Bahama Hummingbird															
Vervain Hummingbird															
Bee Hummingbird															
Ruby-throated Hummingbird															
Rufous Hummingbird															
Cuculidae															
Greater Ani											H				
Smooth-billed Ani	R	V	V	R	R	R	R	R	R	V		R	R	R	
Yellow-billed Cuckoo	R	R	V	R	R	R	R	R	V	V	R	V	R	R	R
Pearly-breasted Cuckoo															
Mangrove Cuckoo		R	R	R	R	R	R	R	R			R	R	R	
Dark-billed Cuckoo									V						
Black-billed Cuckoo		V		V	V		V			V		V			
Chestnut-bellied Cuckoo															
Bay-breasted Cuckoo															
Cuban Lizard-cuckoo															
Bahama Lizard-cuckoo															
Jamaican Lizard-cuckoo															
Hispaniolan Lizard-cuckoo															
Gonave Lizard-cuckoo															
Puerto Rican Lizard-cuckoo															
Common Cuckoo											V				
Rallidae															
Yellow Rail															
Black Rail		V													
King Rail															
Clapper Rail		R	R	R	R		V								
Virginia Rail															
Corncrake					V										
Zapata Rail															
Southern Spotted Rail															
Yellow-breasted Crake															
Sora		V	R	R	R	R	R	R	R	R	R	V	V		R
Spotted Crake					V										
Purple Gallinule		V	V	R	R	R	R	V	V		R				R
Common Gallinule	V	R	R	R	R	R	R	R	R	R					R
American Coot	V	V	V	V	R	V	R	R	R	V	R		R		R
Aramidae															
Limpkin					H										

343

Common name	Regional status	Bah	T&C	Cub	Cay	Jam	Hai	DR	PR	VI	Ang	StM	StB	Sab
Gruidae														
Sandhill Crane	R	V		R										
Gaviidae														
Common Loon	V			V										
Oceanitidae														
Wilson's Storm-petrel	R	R		V			R	R	R		R	V	V	
Hydrobatidae														
Band-rumped Storm-petrel	V	V		V		V								
Leach's Storm-petrel	R	R		V		V		V	V	R			V	V
Diomedeidae														
Atlantic Yellow-nosed Albatross	V													
Black-browed Albatross	V	V												
Procellariidae														
Northern Fulmar	H	H								H				
Trindade Petrel	V	V	V						V					
Black-capped Petrel	R	R	V	R	V	R	R	R	R	V				
Jamaican Petrel	E			E										
Sooty Shearwater	R	R		V		V			R			H		
Great Shearwater	R	R		V	V		H	V	R	V				V
Scopoli's Shearwater	R	R							V					
Cory's Shearwater	R	R	H	V			V		V	V				
Cape Verde Shearwater	V													
Manx Shearwater	R	V						V	V					
Audubon's Shearwater	R	R	R	R	V	V	R	R	R	R	R	R	R	R
Bulwer's Petrel	V													
Ciconiidae														
Wood Stork	R	V		R		V		V	V	H				
White Stork	V													
Jabiru	H													
Threskiornithidae														
Roseate Spoonbill	R	R	R	R	V	V	R	R	V	V		V	V	
Eurasian Spoonbill	V													
White Ibis	R	R	V/R?	R	R	R	R	R	R					
Scarlet Ibis	V			V		V				V				
Glossy Ibis	R	R	R	R	R	R	R	R	R	V	V	V		
White-faced Ibis	V			V										
Ardeidae														
American Bittern	R	R	V	R	V	V		R	R	V		V		
Least Bittern	R	R	R	R	R	R	R	R	R	V				
Common Little Bittern	V													
Black-crowned Night-heron	R	R	R	R	R	R	R	R	R		R	R		
Yellow-crowned Night-heron	R	R	R	R	R	R	R	R	R	R	R	R	R	
Green Heron	R	R	R	R	R	R	R	R	R	R	R	R	R	
Striated Heron	V							V	V					

Common Name	StE	K&N	A&B	Mon	Gua	Dom	Mar	StL	V&G	Gre	Bar	Ave	Swa	Pro	San
Gruidae															
Sandhill Crane															
Gaviidae															
Common Loon															
Oceanitidae															
Wilson's Storm-petrel	V	V		R	V	R	H	R	R	R					
Hydrobatidae															
Band-rumped Storm-petrel			V				V				H				
Leach's Storm-petrel		V	V		R	V	R	V			R	V			
Diomedeidae															
Atlantic Yellow-nosed Albatross					V										
Black-browed Albatross							V								
Procellariidae															
Northern Fulmar															
Trindade Petrel					H										
Black-capped Petrel					R	R	R				R				
Jamaican Petrel															
Sooty Shearwater					R		V	V			R				
Great Shearwater					R	V	R	V			R				
Scopoli's Shearwater															
Cory's Shearwater			R		R	V	R	H			R				
Cape Verde Shearwater					V										
Manx Shearwater					R	V	V		V	V	R				H
Audubon's Shearwater	R	V	V	V	R	V	R	R	R	R	R			R	R
Bulwer's Petrel					V	H					H				
Ciconiidae															
Wood Stork						V									
White Stork		V					V								
Jabiru											H				
Threskiornithidae															
Roseate Spoonbill					V		V	V							V
Eurasian Spoonbill		V					H	V			V				
White Ibis						V					V			V	V
Scarlet Ibis						V	V			V					
Glossy Ibis		V	V	R	R	V	V	V	V	V	R				R
White-faced Ibis															
Ardeidae															
American Bittern				V	V		V				V	V			
Least Bittern		V			R	V	R				V				
Common Little Bittern											V				
Black-crowned Night-heron		R	R	V	R	V	R	R	R	R	R				R
Yellow-crowned Night-heron	R	R	R	R	R	R	R	R	R	R	R		R	R	R
Green Heron	R	R	R	R	R	R	R	R	R	R	R	R	R	R	R
Striated Heron					V			V	V	V					?

Common name	Regional status	Bah	T&C	Cub	Cay	Jam	Hai	DR	PR	VI	Ang	StM	StB	Sab
Squacco Heron	V													
Western Cattle Egret	R	R	R	R	R	R	R	R	R	R	R	R	R	R
Grey Heron	R													
Great Blue Heron	R	R	R	R	R	R	R	R	R	R	R	R	R	V
Cocoi Heron	V													
Purple Heron	V													
American Great Egret	R	R	R	R	R	R	R	R	R	R	R	R	R	R
Whistling Heron	V													
Reddish Egret	R	R	R	R	R	R	R	R	R			V		
Tricolored Heron	R	R	R	R	R	R	R	R	R	R		R	R	
Little Blue Heron	R	R	R	R	R	R	R	R	R	R	R	R	R	V
Snowy Egret	R	R	R	R	R	R	R	R	R	R	R	R	R	V
Little Egret	R	V				V		V	V	V		V		
Western Reef-egret	V			H				V						
Pelecanidae														
Brown Pelican	R	R	R	R	R	R	R	R	R	R	R	R	R	R
American White Pelican	R	V		R	V	V		V	V			V		
Fregatidae														
Magnificent Frigatebird	R	R	R	R	R	R	R	R	R	R	R	R	R	R
Sulidae														
Northern Gannet	R	R	V	V					V					
Red-footed Booby	R	R	H	V	R	R	R	R	R	R	R	V	R	V
Brown Booby	R	R	R	R	R	R	R	R	R	R	R	R	R	R
Masked Booby	R	V	V	V	V	R	V	R	R	R	R	R	V	V
Phalacrocoracidae														
Double-crested Cormorant	R	R	V	R	R	V	V	R	R	V		V		
Neotropical Cormorant	R	R	R	R		V	V	V	V	V		V	V	
Anhingidae														
Anhinga	R	V		R	R	V	V							
Burhinidae														
Double-striped Thick-knee	R	R				R	R							
Haematopodidae														
American Oystercatcher	R	R	R	R		V	H	R	R	R	R	R	R	V
Recurvirostridae														
American Avocet	R	V	V	R	R	V		V	R	V				
Black-winged Stilt	V													
Black-necked Stilt	R	R	R	R	R	R	R	R	R	R	R	R	R	V
Charadriidae														
Grey Plover / Black-bellied Plover	R	R	R	R	R	R	R	R	R	R	R	R	R	R
Pacific Golden Plover	V	V			V			V						
American Golden Plover	R	R	R	R	R	R	R	R	R	R	R	V	V	
Common Ringed Plover	V													
Semipalmated Plover	R	R	R	R	R	R	R	R	R	R	R	R	R	R
Little Ringed Plover	V													

Common Name	StE	K&N	A&B	Mon	Gua	Dom	Mar	StL	V&G	Gre	Bar	Ave	Swa	Pro	San
Squacco Heron					V										
Western Cattle Egret	R	R	R	R	R	R	R	R	R	R	R	R	R	R	R
Grey Heron		V		V	R		V		V		R				
Great Blue Heron	R	R	R	V	R	R	R	R	R	R	R		R	R	R
Cocoi Heron									V						
Purple Heron											V				
American Great Egret	R	R	R	R	R	R	R	R	R	R	R	R	R	R	R
Whistling Heron															V
Reddish Egret		V	V		V										V
Tricolored Heron		R	R	V	R	R	R	V	V	V	R			R	R
Little Blue Heron		R	R	R	R	R	R	R	R	R	R	R	R	R	R
Snowy Egret	V	R	R	R	R	R	R	R	R	R	R	R	R	R	R
Little Egret		V	R		R	V	V	V	V	V	R				
Western Reef-egret								V	V		V				
Pelecanidae															
Brown Pelican	R	R	R	R	R	R	R	R	R	R	R	R	R	R	R
American White Pelican			V												V
Fregatidae															
Magnificent Frigatebird	R	R	R	R	R	R	R	R	R	R	R	R	R	R	R
Sulidae															
Northern Gannet					V		V								
Red-footed Booby	V	V	V	V	R	V	V	R	R	R			R	R	R
Brown Booby	R	R	R	R	R	R	R	R	R	R	R		R	R	R
Masked Booby		V	R		R	V	V	R	V	R	R			R	R
Phalacrocoracidae															
Double-crested Cormorant		V			V								V		V
Neotropical Cormorant		V				V									V
Anhingidae															
Anhinga		V						V		V	V				
Burhinidae															
Double-striped Thick-knee											V				
Haematopodidae															
American Oystercatcher	R	V	V		R	V	R	R	R	H	V				
Recurvirostridae															
American Avocet		V	V							V			V		
Black-winged Stilt					V										
Black-necked Stilt	R	R	R	R	R	V	R	V	V	V	R		R	R	R
Charadriidae															
Grey Plover / Black-bellied Plover	R	R	R	R	R	R	R	R	R	R	R			R	R
Pacific Golden Plover		V			V					V					
American Golden Plover	V	V	R	R	R	R	R	R	R	R	V			H	
Common Ringed Plover					V					V					
Semipalmated Plover	R	R	R	R	R	R	R	R	R	R	R	R	R	R	R
Little Ringed Plover							V								

Common name	Regional status	Bah	T&C	Cub	Cay	Jam	Hai	DR	PR	VI	Ang	StM	StB	Sab
Wilson's Plover	R	R	R	R	R	R	R	R	R	R	R	R	R	
Killdeer	R	R	R	R	R	R	R	R	R	R	R	R	R	
Piping Plover	R	R	R	R		V		R	R	R	V	H		
Snowy Plover	R	R	R	R	V	V	R	R	R	R	R	V	R	
Collared Plover	R											H		
Northern Lapwing	V	V							V					
Cayenne Lapwing	R													
Jacanidae														
Northern Jacana	R			R		R	R	R	V					
Chestnut-backed Jacana	V													
Scolopacidae														
Upland Sandpiper	R	R	V	R	R	V			V	R	H	V	R	
Eurasian Whimbrel	V			V						V		V		
American Whimbrel	R	R	R	R	R	R	R	R	R	R	R	R	R	
Eskimo Curlew	R								V					
Long-billed Curlew	V	H		V		V	V		V	V				
Eurasian Curlew	H	H												
Bar-tailed Godwit	H							H						
Western Black-tailed Godwit	H													
Marbled Godwit	V	V		V	V	V	V	V	V	V				
Hudsonian Godwit	R	V	V	V	V		V	V	R	R	V	V	V	
Ruddy Turnstone	R	R	R	R	R	R	R	R	R	R	R	R	R	V
Red Knot	R	R	R	R	R	V	R	R	R	R	R	V	R	
Ruff	R	V		V	V	V			R	V	V			
Stilt Sandpiper	R	R	R	R	R	R	R	R	R	R	R	R		
Curlew Sandpiper	R			H					V	V	V			
Sanderling	R	R	R	R	R	R	R	R	R	R	R	R	R	
Dunlin	R	R	R	R	V	V		V	R	V				
Baird's Sandpiper	R	H		H	R			V	R	V				
Little Stint	V													
Least Sandpiper	R	R	R	R	R	R	R	R	R	R	R	R	R	R
White-rumped Sandpiper	R	R	R	R	R	V	R	R	R	R	R	R		
Buff-breasted Sandpiper	R	V	V	V	V	V	V	V	V	V	V			
Pectoral Sandpiper	R	R	R	R	R	R	R	R	R	R	R	R		
Semipalmated Sandpiper	R	R	R	R	R	R	R	R	R	R	R	R		
Western Sandpiper	R	R	R	R	R	R	R	R	R	R	R	R		
Short-billed Dowitcher	R	R	R	R	R	R	R	R	R	R	R	R	R	
Long-billed Dowitcher	R	R	H	R	R	V		R	R	V	V			
Wilson's Snipe	R	R	R	R	R	R	R	R	R	R	R	R	R	
Jack Snipe	V													
Wilson's Phalarope	R	V		V	V	V		V	V	V	R	V		
Red-necked Phalarope	V	V		V		V			V	V	V			
Red Phalarope	V	V		V					V	V				
Terek Sandpiper	V													
Spotted Sandpiper	R	R	R	R	R	R	R	R	R	R	R	R	R	R

Common Name	StE	K&N	A&B	Mon	Gua	Dom	Mar	StL	V&G	Gre	Bar	Ave	Swa	Pro	San
Wilson's Plover		R	R	R	R		R		R	R	R		R		
Killdeer	R	R	R	R	R	V	V	V	V	V	R		R	R	R
Piping Plover		H	V		V		V				V				
Snowy Plover		R	V		R		R				V	R	V		
Collared Plover		H			V		V	V	V	R	R				
Northern Lapwing							V				V				
Cayenne Lapwing									V	R	R				V
Jacanidae															
Northern Jacana															
Chestnut-backed Jacana			V												
Scolopacidae															
Upland Sandpiper			R	R	V	R			V	V	R				
Eurasian Whimbrel				V					H		V				
American Whimbrel	V	R	R	R	R	R	R	R	R	R	R			R	R
Eskimo Curlew					V		V		V	V	R				
Long-billed Curlew			V		V		V								
Eurasian Curlew															
Bar-tailed Godwit															
Marbled Godwit		V			V		V		V	V	V				H
Hudsonian Godwit		V	V		R	V	V	V	V		R				
Western Black-tailed Godwit		H													
Ruddy Turnstone	R	R	R	R	R	R	R	R	R	R	R	R	R	R	R
Red Knot		R	R	R	R	V	R	V		V	R				H
Ruff		V	V	V	R		V	V	V		R				
Stilt Sandpiper		R	R		R	R	R	R	R		R				R
Curlew Sandpiper		V	V		V	V	V		V	V	R				
Sanderling	R	R	R	R	R	R	R	R	R	R	R	R		R	R
Dunlin		V			V	V					V				
Baird's Sandpiper					R	V	V	V	V		V				
Little Stint			H	V							V				
Least Sandpiper	R	R	R	R	R	R	R	R	R	R	R			R	R
White-rumped Sandpiper		R	R	R	R	V	R	R	V	R	R				
Buff-breasted Sandpiper		H	V		R		R	V	V	V	R				
Pectoral Sandpiper	R	R	R	R	R	R	R	R	R	R	R		R		R
Semipalmated Sandpiper	R	R	R	R	R	R	R	R	R	R	R	R	R	R	R
Western Sandpiper			R	R		R	V	R	R	R	R			R	R
Short-billed Dowitcher			R	R	R	R	R	R	R	R	R	R			R
Long-billed Dowitcher		H	V		V		V				V				V
Wilson's Snipe	R	R	R		R	R	R	V	V	R	R		R		R
Jack Snipe											V				
Wilson's Phalarope			V		R		V			V	R		V		
Red-necked Phalarope			V		V										
Red Phalarope			V		H										
Terek Sandpiper											V				
Spotted Sandpiper	R	R	R	R	R	R	R	R	R	R	R	R	R	R	R

Common name	Regional status	Bah	T&C	Cub	Cay	Jam	Hai	DR	PR	VI	Ang	StM	StB	Sab
Green Sandpiper	H													
Solitary Sandpiper	R	R	R	R	R	R	R	R	R	R	R	R	R	R
Willet	R	R	R	R	R	R	R	R	R	R	R	R	R	
Lesser Yellowlegs	R	R	R	R	R	R	R	R	R	R	R	R	R	R
Spotted Redshank	V								H					
Common Greenshank	V								V					
Greater Yellowlegs	R	R	R	R	R	R	R	R	R	R	R	R		
Wood Sandpiper	V													
Glareolidae														
Collared Pratincole	V													
Laridae														
Brown Noddy	R	R	R	R	V	R	R	R	R	R	R	R	R	R
Black Noddy	R		H			V			R	V	V			
Common White Tern	V	V												
Black Skimmer	R	R	V	R	V	R	V	V	R	V				
Little Gull	V								V					
Sabine's Gull	V			V					V					
Atlantic Kittiwake	V	V		V		V		V	V	V				
Bonaparte's Gull	R	R	R	R	V	V	V	V	V					
Slender-billed Gull	H													
Black-headed Gull	R	V	V	V		V			V	V	V	V		
Grey-headed Gull	V													
Franklin's Gull	V	V		V	V			V	V			V	V	
Laughing Gull	R	R	R	R	R	R	R	R	R	R	R	R	R	R
Ring-billed Gull	R	R	R	R	R	R	R	R	R	V	R	R	V	
Kelp Gull	V													
Lesser Black-backed Gull	R	R	V	R		V	V	V	V	V	R	V		
American Herring Gull	R	R	R	R	R	R	R	R	R		R	V	V	
Iceland Gull	V	V												
Glaucous Gull	V													
Great Black-backed Gull	R	R	V	R		V	R	V	R	V	V	V		
Sooty Tern	R	R	R	R	V	R	R	R	R	R	R	R	R	R
Bridled Tern	R	R	R	R	R	R	R	R	R	R	R	R	R	R
Least Tern	R	R	R	R	R	R	R	R	R	R	R	R	R	V
Large-billed Tern	V			V										
Common Gull-billed Tern	R	R	R	R	R	R	R	R	R	R	H	R		
Caspian Tern	R	R	R	R	R	R	R	R	R	V	V			
Whiskered Tern	V	V												
White-winged Tern	V	V						V	V					
Eurasian Black Tern	V													
American Black Tern	R	R	R	R	R	R	R	R	R	V	V			
Roseate Tern	R	R	R	R	V	V	R	R	R	R	R	R	R	R
Common Tern	R	R	R	R	R	R	R	R	R	R	R	R	R	R
Arctic Tern	R	V		V					V	V				
Forster's Tern	R	R	R	R	R	V	H	V	R	V		V		

Common Name	StE	K&N	A&B	Mon	Gua	Dom	Mar	StL	V&G	Gre	Bar	Ave	Swa	Pro	San
Green Sandpiper					H										
Solitary Sandpiper		R	R	R	R	R	R	R	R	R	R			H	R
Willet		R	R	R	R	R	R	R	R	R	R	R		R	R
Lesser Yellowlegs		R	R	R	R	R	R	R	R	R	R	R	R	R	R
Spotted Redshank					H					V					
Common Greenshank										V					
Greater Yellowlegs		R	R	R	R	R	R	R	R	R				R	R
Wood Sandpiper					V					V					
Glareolidae															
Collared Pratincole					V					V					
Laridae															
Brown Noddy		R	R	R	R	R	R	R	R	R	R				R
Black Noddy						V					R	R			
Common White Tern															
Black Skimmer					V					V	V				
Little Gull											V				
Sabine's Gull					V										
Atlantic Kittiwake					V	V		V			V				
Bonaparte's Gull		V	V		V						V				
Slender-billed Gull			H												
Black-headed Gull		V	V		R		V	V		V	R				
Grey-headed Gull											V				
Franklin's Gull					V						V				V
Laughing Gull	R	R	R	R	R	R	R	R	R	R	R	R		R	R
Ring-billed Gull		V	V		R	V	V	V	V		R				
Kelp Gull											V				
Lesser Black-backed Gull		R	V		R	V	V	V	V	V	V				
American Herring Gull		V	V		R	V	V	V			V	V			V
Iceland Gull					V										
Glaucous Gull											V				
Great Black-backed Gull		V			V	?	V				V				
Sooty Tern	R	R	R		R	R	R	R	R	R	R	R	V/R		R
Bridled Tern	R	R	R	R	R	R?	R	R	R	R	R	R			
Least Tern	R	R	R		R	R	R	V	V	V	R				V
Large-billed Tern									V						
Common Gull-billed Tern			R		R	V	R	V	V		R				V
Caspian Tern		V	V		V	V	V	V			V			H	V
Whiskered Tern											V				
White-winged Tern					V						V				
Eurasian Black Tern											V				
American Black Tern		V	R		V	V	V		V	V	R			R	R
Roseate Tern		R	R		R	R	R	R	R	R	R	R			
Common Tern		R	R	R	R	R	R	R	R	R	R				R
Arctic Tern					R					V					
Forster's Tern			V	V	V			V							

351

Common name	Regional status	Bah	T&C	Cub	Cay	Jam	Hai	DR	PR	VI	Ang	StM	StB	Sab
Cabot's Tern	R	R	R	R	R	R	R	R	R	R	R	R	R	V
Royal Tern	R	R	R	R	R	R	R	R	R	R	R	R	R	R
Stercorariidae														
Long-tailed Jaeger	R	V		V	V	V	V	V	H					
Arctic Jaeger / Parasitic Jaeger	R	R	H	R		V	R	R	R	V				
Pomarine Jaeger	R	R	H	R		V	R	R	R	V		R	R	R
Great Skua	V													
South Polar Skua	R	V?		V				V						
Alcidae														
Little Auk / Dovekie	V	V	V	V										
Tytonidae														
American Barn-owl	R	R	R	R	R	R	R	R						
Ashy-faced Owl	E						R	R	Ex					
Lesser Antilles Barn-owl	E											V?		
Strigidae														
Cuban Pygmy-owl	E			E										
Burrowing Owl	R	R		R			R	R						
Northern Saw-whet Owl	V	V												
Stygian Owl	R			R			R	R						
Northern Long-eared Owl	V			V										
Common Short-eared Owl	R	V	V	R	R		R	R	R	V			V	
Jamaican Owl	E					E								
Puerto Rican Screech-owl	E								E	Ex?				
Bare-legged Screech-owl	E			E										
Cathartidae														
Turkey Vulture	R+I	R		R	R	I?	R	R	R	V				
American Black Vulture	V	V		V		V			V					
Pandionidae														
Western Osprey	R	R	R	R	R	R	R	R	R	R	R	R	R	R
Accipitridae														
Hook-billed Kite	R													
Cuban Kite	E			E										
Swallow-tailed Kite	R	R	V	R	R	V	V	V	V	V				V
Western Marsh-harrier / Eurasian Marsh-harrier	V							V						
Northern Harrier	R	R	R	R	R	V	R	R	R	R			?	
Sharp-shinned Hawk	R	R	R	R		V	R	R	R	V				
Cooper's Hawk	H			H										
Gundlach's Hawk	E			E										
Bald Eagle	V	V		V					V	V				
Black Kite	V	V								V				
Mississippi Kite	R			R	V	V		V	V					
Snail Kite	R			R										
Common Black Hawk	R								V					
Cuban Black Hawk	E			E	V									

Common Name	StE	K&N	A&B	Mon	Gua	Dom	Mar	StL	V&G	Gre	Bar	Ave	Swa	Pro	San
Cabot's Tern	V	V	R		R	V	R	R	R	R	V			R	R
Royal Tern	R	R	R	R	R	R	R	R	R	R	R			R	R
Stercorariidae															
Long-tailed Jaeger		V?			R	V	V								
Arctic Jaeger / Parasitic Jaeger					R	R	R	H	V	V	R			V	
Pomarine Jaeger		V	V		R	R	R	V	V		R				
Great Skua					V		H								
South Polar Skua					R	?	V			R					
Alcidae															
Little Auk / Dovekie															
Tytonidae															
American Barn-owl															I
Ashy-faced Owl															
Lesser Antilles Barn-owl					H	R			R	R					
Strigidae															
Cuban Pygmy-owl															
Burrowing Owl		Ex	Ex		Ex										
Northern Saw-whet Owl															
Stygian Owl															
Northern Long-eared Owl															
Common Short-eared Owl								H		V					
Jamaican Owl															
Puerto Rican Screech-owl															
Bare-legged Screech-owl															
Cathartidae															
Turkey Vulture														H	H
American Black Vulture										V					
Pandionidae															
Western Osprey	R	R	R		R	R	R	R	R	R	R	R		R	R
Accipitridae															
Hook-billed Kite										R					
Cuban Kite															
Swallow-tailed Kite					H		V								V
Western Marsh-harrier / Eurasian Marsh-harrier					V					V					
Northern Harrier		H	V		V	V	V	V		?	V				V
Sharp-shinned Hawk		H													H
Cooper's Hawk															
Gundlach's Hawk															
Bald Eagle															
Black Kite					V	V				V					
Mississippi Kite															V
Snail Kite															
Common Black Hawk								V	R	R					
Cuban Black Hawk															

Common name	Regional status	Bah	T&C	Cub	Cay	Jam	Hai	DR	PR	VI	Ang	StM	StB	Sab
White-tailed Hawk	H													
Ridgway's Hawk	E						Ex	E	H					
Red-shouldered Hawk	H	H												
Broad-winged Hawk	R	V		R		V		V	R		V			
Short-tailed Hawk	H			H										
Swainson's Hawk	V	V		V				V						
Red-tailed Hawk	R	R	R	R	V	R	R	R	R	R		Ex	R	R
Trogonidae														
Cuban Trogon	E			E										
Hispaniolan Trogon	E						R	R						
Meropidae														
European Bee-eater	V													
Todidae														
Cuban Tody	E			E										
Narrow-billed Tody	E						R	R						
Broad-billed Tody	E						R	R						
Jamaican Tody	E					E								
Puerto Rican Tody	E								E					
Alcedinidae														
Ringed Kingfisher	R								H					
Belted Kingfisher	R	R	R	R	R	R	R	R	R	R	R	R	R	R
Green Kingfisher	H													
Ramphastidae														
Channel-billed Toucan	I													
Picidae														
Antillean Piculet	E						R	R						
Cuban Ivory-billed Woodpecker	E			E										
Fernandina's Flicker	E			E										
Yellow-shafted Flicker	V	V												
Cuban Flicker	E			R	R									
Yellow-bellied Sapsucker	R	R	R	R	R	R	R	R	R	R		V	V	
Cuban Green Woodpecker	E			E										
Guadeloupe Woodpecker	E													
Puerto Rican Woodpecker	E								E					
Hispaniolan Woodpecker	E						R	R						
Jamaican Woodpecker	E					E								
West Indian Woodpecker	E	R		R										
Cayman Woodpecker	E				E									
Red-bellied Woodpecker	V	V												
Bahamas Hairy Woodpecker	E	E	V						V					
Falconidae														
Crested Caracara	R			R		V								
Common Kestrel	V													
American Kestrel	R	R	R	R	R	R	R	R	R	R	R	R	R	R
Merlin	R	R	R	R	R	R	R	R	R	R	R	R	R	R

Common Name	StE	K&N	A&B	Mon	Gua	Dom	Mar	StL	V&G	Gre	Bar	Ave	Swa	Pro	San
White-tailed Hawk									H						
Ridgway's Hawk															
Red-shouldered Hawk															
Broad-winged Hawk		R	R		V	R	R	R	R	R	V				H
Short-tailed Hawk															
Swainson's Hawk															
Red-tailed Hawk	R	R			R	V			V						
Trogonidae															
Cuban Trogon															
Hispaniolan Trogon															
Meropidae															
European Bee-eater									V						
Todidae															
Cuban Tody															
Narrow-billed Tody															
Broad-billed Tody															
Jamaican Tody															
Puerto Rican Tody															
Alcedinidae															
Ringed Kingfisher				H	R	R		Ex?		V	V				
Belted Kingfisher	R	R	R	R	R	R	R	R	R	R	R		R	R	R
Green Kingfisher		H													
Ramphastidae															
Channel-billed Toucan										I					
Picidae															
Antillean Piculet															
Cuban Ivory-billed Woodpecker															
Fernandina's Flicker															
Yellow-shafted Flicker															
Cuban Flicker															
Yellow-bellied Sapsucker		V			V	V							R	R	R
Cuban Green Woodpecker															
Guadeloupe Woodpecker			V		E										
Puerto Rican Woodpecker															
Hispaniolan Woodpecker															
Jamaican Woodpecker															
West Indian Woodpecker															
Cayman Woodpecker															
Red-bellied Woodpecker															
Bahamas Hairy Woodpecker															
Falconidae															
Crested Caracara															
Common Kestrel					V			V							
American Kestrel	R	R	R	R	R	R	R	R	R	R	V	R	R	V	R
Merlin		V	V		R	R	R	R	R	R	R	R	R	R	R

Common name	Regional status	Bah	T&C	Cub	Cay	Jam	Hai	DR	PR	VI	Ang	StM	StB	Sab
Bat Falcon	V													
Aplomado Falcon	V								V					
Peregrine Falcon	R	R	R	R	R	R	R	R	R	R	R	R	R	R
Cacatuidae														
White Cockatoo	I								I					
Psittacidae														
Monk Parakeet	I	I			I				I	I				
White-winged Parakeet	I								I					
Black-billed Amazon	E					E								
White-fronted Amazon	I								I					
Yellow-billed Amazon	E					E								
Cuban Amazon	E	R		R	R									
Hispaniolan Amazon	E						R	R		I?				
Puerto Rican Amazon	E								E					
Red-crowned Amazon	I								I					
Red-necked Amazon	E													
St Lucia Amazon	E													
Imperial Amazon	E													
Orange-winged Amazon	I								I	I?				
St Vincent Amazon	E													
Green-rumped Parrotlet	I					I								
Jamaican Parakeet	E					E		I?	I					
Orange-fronted Parakeet	I								I					
Brown-throated Parakeet	I								Ex(I)	I				I
Blue-and-yellow Macaw	I			I?					I?					
Red-masked Parakeet	I								I					
Cuban Parakeet	E			E										
Hispaniolan Parakeet	E						R	R	Ex					
Rose-ringed Parakeet	I					I			I		I			
Tityridae														
Jamaican Becard	E					E								
Tyrannidae														
Yellow-bellied Elaenia	R													
Chinchorro Elaenia	NE			R										
Caribbean Elaenia	NE								R	R	R	R	R	R
Hispaniolan Elaenia	E						R	R						
Large Jamaican Elaenia	E					E								
Small Jamaican Elaenia	E					E								
Great Kiskadee	V													
Tropical Kingbird	R			V	V									
Cassin's Kingbird	V			V										
Western Kingbird	R	R		V					V					
Eastern Kingbird	R	R	V	R	R	V			V					
Grey Kingbird	R	R	R	R	R	R	R	R	R	R	R	R	R	R
Loggerhead Kingbird	E	R		R	R	R								

Common Name	StE	K&N	A&B	Mon	Gua	Dom	Mar	StL	V&G	Gre	Bar	Ave	Swa	Pro	San
Bat Falcon										V	V				
Aplomado Falcon															
Peregrine Falcon	R	V	R	R	R	R	R	R	R	R	R	R	R	V	R
Cacatuidae															
White Cockatoo															
Psittacidae															
Monk Parakeet															
White-winged Parakeet															
Black-billed Amazon															
White-fronted Amazon															
Yellow-billed Amazon															
Cuban Amazon															
Hispaniolan Amazon															
Puerto Rican Amazon															
Red-crowned Amazon															
Red-necked Amazon						E									
St Lucia Amazon								E							
Imperial Amazon						E									
Orange-winged Amazon							I			I	I				
St Vincent Amazon									E						
Green-rumped Parrotlet															
Jamaican Parakeet															
Orange-fronted Parakeet															
Brown-throated Parakeet															I
Blue-and-yellow Macaw															
Red-masked Parakeet															
Cuban Parakeet															
Hispaniolan Parakeet															
Rose-ringed Parakeet							I				I				
Tityridae															
Jamaican Becard															
Tyrannidae															
Yellow-bellied Elaenia						V		R	R						
Chinchorro Elaenia														R	R
Caribbean Elaenia	R	R	R	R	R	R	R	R	R	R	R				
Hispaniolan Elaenia															
Large Jamaican Elaenia															
Small Jamaican Elaenia															
Great Kiskadee											V				
Tropical Kingbird										R					
Cassin's Kingbird															
Western Kingbird													V		
Eastern Kingbird		?			V								R	V	R
Grey Kingbird	R	R	R	R	R	R	R	R	R	R	R	R	R	V	V
Loggerhead Kingbird															

Common name	Regional status	Bah	T&C	Cub	Cay	Jam	Hai	DR	PR	VI	Ang	StM	StB	Sab
Hispaniolan Kingbird	E						R	R						
Puerto Rican Kingbird	E								E					
Giant Kingbird	E	Ex	Ex	E										
Scissor-tailed Flycatcher	R	V		R	V				V	V				
Fork-tailed Flycatcher	R	V		V	V	V			V	V		V	V	
Sad Flycatcher	E					E								
Rufous-tailed Flycatcher	E					E								
La Sagra's Flycatcher	E	R	H	R	R									
Stolid Flycatcher	E						R	R	R					
Puerto Rican Flycatcher	E								R	R				
Lesser Antillean Flycatcher	E													
Great Crested Flycatcher	R	V		R	V				V	H				
Brown-crested Flycatcher	V	V												
Grenada Flycatcher	E													
Common Vermilion Flycatcher	V			V										
Lawrence's Flycatcher	R													
Eastern Phoebe	R	R		R	V									
Acadian Flycatcher	R	R		R	V	H			V					
Yellow-bellied Flycatcher	R	H		R										
Willow Flycatcher	R			R		V			V					
Alder Flycatcher	V			V										
Least Flycatcher	V			H	V									
Western Wood-pewee	R			R		V								
Eastern Wood-pewee	R	R	R	R	R	V		V	V	V				
Cuban Pewee	E	R		R										
Jamaican Pewee	E					E								
Hispaniolan Pewee	E		V				R	R	V					
Puerto Rican Pewee	E								E					
Lesser Antillean Pewee	E													
St Lucia Pewee	E													
Vireonidae														
Philadelphia Vireo	R	R	R	R	R	V		V	V					
Eastern Warbling Vireo	R	R		R		V		V	V					
Yellow-green Vireo	R													
Yucatan Vireo	R			R										
Red-eyed Vireo	R	R	R	R	R	R	R	V	R	V		V		V
Black-whiskered Vireo	R	R	R	R	R	R	R	R	R	R	R	R		R
Yellow-throated Vireo	R	R	R	R	V	V	R	V	R		V			V
Blue-headed Vireo	R	R		R	V	V		V						
Blue Mountain Vireo	E					E								
Jamaican Vireo	E					E								
Flat-billed Vireo	E						R	R						
Bell's Vireo	V	V												
Puerto Rican Vireo	E								E					
Providence Vireo	E													

Common Name	StE	K&N	A&B	Mon	Gua	Dom	Mar	StL	V&G	Gre	Bar	Ave	Swa	Pro	San
Hispaniolan Kingbird															
Puerto Rican Kingbird															
Giant Kingbird															
Scissor-tailed Flycatcher															
Fork-tailed Flycatcher		V			V			V	R	R	R				
Sad Flycatcher															
Rufous-tailed Flycatcher															
La Sagra's Flycatcher															
Stolid Flycatcher															
Puerto Rican Flycatcher															
Lesser Antillean Flycatcher		R	R		R	R	R	R							
Great Crested Flycatcher															R
Brown-crested Flycatcher															
Grenada Flycatcher									R	R					
Common Vermilion Flycatcher															
Lawrence's Flycatcher										Ex?					
Eastern Phoebe															
Acadian Flycatcher											V			V	R
Yellow-bellied Flycatcher															V
Willow Flycatcher															R
Alder Flycatcher															V
Least Flycatcher															V
Western Wood-pewee														H	V
Eastern Wood-pewee					V						V		R	R	R
Cuban Pewee															
Jamaican Pewee															
Hispaniolan Pewee															
Puerto Rican Pewee															
Lesser Antillean Pewee		V			R	R	R								
St Lucia Pewee								E							
Vireonidae															
Philadelphia Vireo			V		H									V	R
Eastern Warbling Vireo															
Yellow-green Vireo														R	R
Yucatan Vireo															
Red-eyed Vireo					R		R	V	V		R			H	R
Black-whiskered Vireo	R	R	R	R	R	R	R	R	R	R			R	R
Yellow-throated Vireo		V	V	V	R	V		V	V	V	V		V		R
Blue-headed Vireo															
Blue Mountain Vireo															
Jamaican Vireo															
Flat-billed Vireo															
Bell's Vireo															
Puerto Rican Vireo															
Providence Vireo														E	

Common name	Regional status	Bah	T&C	Cub	Cay	Jam	Hai	DR	PR	VI	Ang	StM	StB	Sab
San Andres Vireo	E													
White-eyed Vireo	R	R	R	R	R	V	V	V	V	V		V		
Thick-billed Vireo	E	R	R	R	R		R							
Cuban Vireo	E			E										
Laniidae														
Loggerhead Shrike	V	V												
Corvidae														
Cuban Crow	E	V	R	R										
Jamaican Crow	E					E								
White-necked Crow	E						R	R	Ex					
Fish Crow	V	V												
Cuban Palm Crow	E			E										
Hispaniolan Palm Crow	E						R	R						
House Crow	V			V										
Alaudidae														
American Horned Lark	V	V												
Hirundinidae														
Northern House Martin	V													
Cliff Swallow	R	R	R	R	R	V	V	V	V	R		V	V	V
Caribbean Cave Swallow	NE	R	V	R	R	R	R	R	R	R				
American Barn Swallow	R	R	R	R	R	R	R	R	R	R	R	R	R	R
Collared Sand Martin / Bank Swallow	R	R	R	R	R	R	R	R	R	R	R			
Tree Swallow	R	R	V	R	R	R	R	R	R	R	V			
Violet-green Swallow	V								V					
Golden Swallow	E					Ex	R	R						
Bahama Swallow	E	R	V	V										
White-winged Swallow	V													
Purple Martin	R	R	R	R	R	R	V	V	R	V				
Caribbean Martin	NE	V	V	V	R	R	R	R	R	R	R	R	R	R
Cuban Martin	E	V		E		V			V					
Northern Rough-winged Swallow	R	R	R	R	R	R	R	R	R	V				
Sittidae														
Bahama Nuthatch	E	E												
Polioptilidae														
Blue-grey Gnatcatcher	R	R	R	R	R		V	V						
Cuban Gnatcatcher	E			E										
Troglodytidae														
Zapata Wren	E			E										
Northern House Wren	V	V		V										
Antillean House Wren	E													
Southern House Wren	V													
Sedge Wren	V	V												
Eastern Marsh Wren	H			H										

Common Name	StE	K&N	A&B	Mon	Gua	Dom	Mar	StL	V&G	Gre	Bar	Ave	Swa	Pro	San
San Andres Vireo															E
White-eyed Vireo					V								R		R
Thick-billed Vireo															
Cuban Vireo															
Laniidae															
Loggerhead Shrike															
Corvidae															
Cuban Crow															
Jamaican Crow															
White-necked Crow															
Fish Crow															
Cuban Palm Crow															
Hispaniolan Palm Crow															
House Crow											V				
Alaudidae															
American Horned Lark															
Hirundinidae															
Northern House Martin					V					V					
Cliff Swallow		V	V		R	V	V	R	V	H	R		R	V	R
Caribbean Cave Swallow					V		V	V	V						V
American Barn Swallow	R	R	R	R	R	R	R	R	R	R	R	R	R	R	R
Collared Sand Martin / Bank Swallow		R	V		R	R	R	V	V	V	R		R	R	R
Tree Swallow					R						V		V		H
Violet-green Swallow															
Golden Swallow															
Bahama Swallow															
White-winged Swallow							H			V					
Purple Martin		V			R							V		V	V
Caribbean Martin	R	R	R	R	R	R	R	R	R	R	R				
Cuban Martin					V										
Northern Rough-winged Swallow					V		V			V			V		V
Sittidae															
Bahama Nuthatch															
Polioptilidae															
Blue-grey Gnatcatcher															
Cuban Gnatcatcher															
Troglodytidae															
Zapata Wren															
Northern House Wren															
Antillean House Wren					Ex	R	Ex	R	R	R					
Southern House Wren											V				
Sedge Wren															
Eastern Marsh Wren															

361

Common name	Regional status	Bah	T&C	Cub	Cay	Jam	Hai	DR	PR	VI	Ang	StM	StB	Sab
Sturnidae														
Common Starling / European Starling	R+I	I	V	V	V	I			V	V				
Mimidae														
Grey Catbird	R	R	R	R	R	R	R	R	V	V				
White-breasted Thrasher	E													
St Lucia Thrasher	E													
Scaly-breasted Thrasher	E											R	R	R
Pearly-eyed Thrasher	NE	R	R			V	H	R	R	R	R	R	R	R
Brown Trembler	E									V				R
Grey Trembler	E													
Northern Mockingbird	R	R	R	R	R	R	R	R	R	R	R			
Tropical Mockingbird	R													
Bahama Mockingbird	E	R	R	R		R			V					
Brown Thrasher	V	V		V										
Turdidae														
Eastern Bluebird	V	V		V						?				
Cuban Solitaire	E			E										
Rufous-throated Solitaire	E					R	R	R						
Wood Thrush	R	V		R	V	V			V	V				
Swainson's Thrush	R	R	R	R	R	R			V	R				
Veery	R	R		R	R	R			V	V				
Grey-cheeked Thrush	R	R	H	R	R	R			V	V				
Bicknell's Thrush	R	V		R		R	R	R	R					
Eastern Hermit Thrush	R	R		V										
Lesser Antillean Thrush	E			.										
Spectacled Thrush	R											V		
Forest Thrush	E													
White-eyed Thrush	E					E								
La Selle Thrush	E						R	R						
American Robin	R	R	V	R		V			V	V				
White-chinned Thrush	E					E								
Northern Red-legged Thrush	E	E												
Western Red-legged Thrush	E			R	R									
Eastern Red-legged Thrush	E						R	R	R	H				
Muscicapidae														
Northern Wheatear	V	V		V					V					
Regulidae														
Ruby-crowned Kinglet	R	R		R		V		V						
Dulidae														
Palmchat	E						R	R						
Bombycillidae														
Cedar Waxwing	R	R	V	R	R	R	R	R	V	V				
Ploceidae														
Yellow-crowned Bishop	V+I			I		I			I					
Northern Red Bishop	V+I					I			I	I				

Common Name	StE	K&N	A&B	Mon	Gua	Dom	Mar	StL	V&G	Gre	Bar	Ave	Swa	Pro	San
Sturnidae															
Common Starling / European Starling											V				
Mimidae															
Grey Catbird					V								R	R	R
White-breasted Thrasher							E								
St Lucia Thrasher								E							
Scaly-breasted Thrasher	R	R	R	R	R	R	R	R	R	R	Ex_				
Pearly-eyed Thrasher	R	R	R	R	R	R	R	R	H		V				
Brown Trembler	V	R	V	R	R	R	R	?	R	R					
Grey Trembler							R	R							
Northern Mockingbird															
Tropical Mockingbird		V	R		R	R	R	R	R	R	V				R
Bahama Mockingbird															
Brown Thrasher															
Turdidae															
Eastern Bluebird															
Cuban Solitaire															
Rufous-throated Solitaire						R	R	R	R						
Wood Thrush															R
Swainson's Thrush		V			V						V		R	R	R
Veery		V												V	R
Grey-cheeked Thrush					V		V				V		R	H	R
Bicknell's Thrush															
Eastern Hermit Thrush															
Lesser Antillean Thrush								R	R						
Spectacled Thrush		V			R	R	R	R	R	R					
Forest Thrush				R	R	R		R							
White-eyed Thrush															
La Selle Thrush															
American Robin															
White-chinned Thrush															
Northern Red-legged Thrush															
Western Red-legged Thrush													Ex		
Eastern Red-legged Thrush						R									
Muscicapidae															
Northern Wheatear					V						V				
Regulidae															
Ruby-crowned Kinglet															
Dulidae															
Palmchat															
Bombycillidae															
Cedar Waxwing					V	V									
Ploceidae															
Yellow-crowned Bishop					I						V				
Northern Red Bishop					I		I				V				

Common name	Regional status	Bah	T&C	Cub	Cay	Jam	Hai	DR	PR	VI	Ang	StM	StB	Sab
Village Weaver	I						I	I						
Estrildidae														
Orange-cheeked Waxbill	I								I	I?				
Black-rumped Waxbill	V+I									V				
Common Waxbill	I													
Red Avadavat	I						I							
Bronze Mannikin	V+I								I	V				
Indian Silverbill	V+I								I	V				
Checkered Munia	I	I	I	I	I	I	I	I	I	V				
Tricoloured Munia	I	V	V	I	I	I	I	I	I					
Chestnut Munia	I		I	I		I								
White-headed Munia	I													
Java Sparrow	I								I					
Viduidae														
Pin-tailed Whydah	I						I	I						
Passeridae														
House Sparrow	V+I	I	I	I	I	I	I	I	I	I	I	I		I
Motacillidae														
American Pipit	R	R	V	V		V	V							
Sprague's Pipit	H	H												
White-faced Wagtail	V													
Fringillidae														
Jamaican Euphonia	E					E								
Hispaniolan Euphonia	E						R	R						
Puerto Rican Euphonia	E								E					
Lesser Antillean Euphonia	E											V	V	Ex
House Finch	V			V										
Common Redpoll	H	H												
Hispaniolan Crossbill	E						V	R	R					
Pine Siskin	V	V												
American Goldfinch	V	V		V										
Red Siskin	I			Ex(I)					I					
Antillean Siskin	E						R	R						
Calcariidae														
Lapland Longspur	V			V										
Snow Bunting	V	V												
Passerellidae														
Chipping Sparrow	R	R		R	V									
Clay-colored Sparrow	R	R	H	R	V									
Lark Sparrow	R	V		R		V								
Grasshopper Sparrow	R	R		R	R	R	R	R	R					
Green-tailed Towhee	V			V										
Zapata Sparrow	E			E										
Slate-colored Junco	V	V		H	V	V			V	V				
Rufous-collared Sparrow	R							R						

Common Name	StE	K&N	A&B	Mon	Gua	Dom	Mar	StL	V&G	Gre	Bar	Ave	Swa	Pro	San
Village Weaver					H		I								
Estrildidae															
Orange-cheeked Waxbill					I		I								
Black-rumped Waxbill					I		I								
Common Waxbill							I								
Red Avadavat					I		I								
Bronze Mannikin															
Indian Silverbill															
Checkered Munia		I	I		I	I	I								
Tricoloured Munia															
Chestnut Munia							I								
White-headed Munia					V		I								
Java Sparrow															
Viduidae															
Pin-tailed Whydah															
Passeridae															
House Sparrow	I	I	I		I	V		V	I	I	V				
Motacillidae															
American Pipit													V	V	V
Sprague's Pipit															
White-faced Wagtail											V				
Fringillidae															
Jamaican Euphonia															
Hispaniolan Euphonia															
Puerto Rican Euphonia															
Lesser Antillean Euphonia	V	V	R	V	R	R	R	R	R	R					
House Finch															
Common Redpoll															
Hispaniolan Crossbill															
Pine Siskin															
American Goldfinch															
Red Siskin															
Antillean Siskin															
Calcariidae															
Lapland Longspur															
Snow Bunting															
Passerellidae															
Chipping Sparrow															
Clay-colored Sparrow															
Lark Sparrow															
Grasshopper Sparrow													V		
Green-tailed Towhee															
Zapata Sparrow															
Slate-colored Junco															
Rufous-collared Sparrow															

Common name	Regional status	Bah	T&C	Cub	Cay	Jam	Hai	DR	PR	VI	Ang	StM	StB	Sab
White-throated Sparrow	V	V							V					
White-crowned Sparrow	R	R	R	R	V	V								
Vesper Sparrow	V	V			V									
Savannah Sparrow	R	R	V	R	R	V	V							
Henslow's Sparrow	H	H												
Eastern Song Sparrow	H	H						H						
Lincoln's Sparrow	R	R	V	R		V	V	V	V					
Swamp Sparrow	V	V												
Teretistridae														
Yellow-headed Warbler	E			E										
Oriente Warbler	E			E										
Icteridae														
Yellow-breasted Chat	R	V		R	V	V		V						
Yellow-headed Blackbird	V	V		V	V				V					
Bobolink	R	R	R	R	R	R	R	R	R	R		V	R	V
Eastern Meadowlark	V	V			V									
Cuban Meadowlark	E			E										
Jamaican Oriole	E					Ex	R							
Yellow Oriole	H													
Baltimore Oriole	R	R	R	R	R	R	R	R	R	R		V	V	
Bullock's Oriole	V	V												
Venezuelan Troupial	V+I					Ex(I)			I	I				
Eastern Hooded Oriole	V			V										
Orchard Oriole	R	V		R	V	V		V	V	V				
Bahama Oriole	E	E												
Cuban Oriole	E			E										
Martinique Oriole	E													
Puerto Rican Oriole	E								E					
Montserrat Oriole	E													
Hispaniolan Oriole	E						R	R						
St Lucia Oriole	E													
Jamaican Blackbird	E					E								
Tawny-shouldered Blackbird	E			R	V		R	H						
Yellow-shouldered Blackbird	E								E					
Red-winged Blackbird	R	R						H						
Red-shouldered Blackbird	E			E										
Giant Cowbird	V													
Shiny Cowbird	R	R		R	R	R	R	R	R	V			V	
Brown-headed Cowbird	R	R	V	V										
Cuban Blackbird	E			E										
Rusty Blackbird	H	H												
Brewer's Blackbird	H	H												
Purple Grackle	H	H												
Carib Grackle	R									I		I	I	H
Barbados Grackle	E													

Common Name	StE	K&N	A&B	Mon	Gua	Dom	Mar	StL	V&G	Gre	Bar	Ave	Swa	Pro	San
White-throated Sparrow															
White-crowned Sparrow															
Vesper Sparrow															
Savannah Sparrow													V		V
Henslow's Sparrow															
Eastern Song Sparrow															
Lincoln's Sparrow															V
Swamp Sparrow															
Teretistridae															
Yellow-headed Warbler															
Oriente Warbler															
Icteridae															
Yellow-breasted Chat															
Yellow-headed Blackbird															
Bobolink		V	R		R	R	V		V	V	R		V	V	V
Eastern Meadowlark															
Cuban Meadowlark															
Jamaican Oriole															R
Yellow Oriole										H					
Baltimore Oriole		V			R		V	V	V	V	V	V	R		V
Bullock's Oriole															
Venezuelan Troupial			V			V			I						
Eastern Hooded Oriole															
Orchard Oriole				V	V								V		H
Bahama Oriole															
Cuban Oriole															
Martinique Oriole							E								
Puerto Rican Oriole															
Montserrat Oriole				E											
Hispaniolan Oriole															
St Lucia Oriole								E							
Jamaican Blackbird															
Tawny-shouldered Blackbird															
Yellow-shouldered Blackbird															
Red-winged Blackbird															
Red-shouldered Blackbird															
Giant Cowbird											V				
Shiny Cowbird		R			R	R		R	R		R				
Brown-headed Cowbird															
Cuban Blackbird															
Rusty Blackbird															
Brewer's Blackbird															
Purple Grackle															
Carib Grackle					R	R	R	R	R		R				
Barbados Grackle	H		I	I						R		R			

367

Common name	Regional status	Bah	T&C	Cub	Cay	Jam	Hai	DR	PR	VI	Ang	StM	StB	Sab
Boat-tailed Grackle	R	R												
Greater Antillean Grackle	E		R	R	R	R	R	R	H					
Great-tailed Grackle	R			R			R	R						
Yellow-hooded Blackbird	V													
Parulidae														
Ovenbird	R	R	R	R	R	R	R	R	R	R	V	V	R	V
Worm-eating Warbler	R	R	R	R	R	R	R	R	R	R		V	V	
Louisiana Waterthrush	R	R	R	R	R	R	R	R	R	R		V	V	V
Northern Waterthrush	R	R	R	R	R	R	R	R	R	R	R	V	R	R
Bachman's Warbler	R	V		R										
Golden-winged Warbler	R	V		R	R	V		R	R	R		V		
Blue-winged Warbler	R	R		R	R	R	R	R	R	R		V	V	
Black-and-white Warbler	R	R	R	R	R	R	R	R	R	R	R	R	R	R
Prothonotary Warbler	R	R	R	R	R	V	R	R	R	R	V	V	R	
Swainson's Warbler	R	R	H	R	R	R	V	R	V	V	V			
Tennessee Warbler	R	R	R	R	R	R	H	R	V	V				
Orange-crowned Warbler	R	R		V	V	V				H				
Nashville Warbler	R	R	V	R	R	V	V	V	V					
Virginia's Warbler	V	V												
Semper's Warbler	E													
Connecticut Warbler	R	V	V	V	V	V	H	R	V	V			V	
Mourning Warbler	R	V	R	V	V	V	V	V	V	V				
Kentucky Warbler	R	R	R	R	R	V		V	R	R		V		
Bahama Yellowthroat	E	E												
Eastern Yellowthroat	R	R	R	R	R	R	R	R	R	R	V	V		
Whistling Warbler	E													
Plumbeous Warbler	E													
Elfin Woods Warbler	E								E					
Arrowhead Warbler	E					E								
Hooded Warbler	R	R	R	R	R	R		R	V	R	V	V	V	V
American Redstart	R	R	R	R	R	R	R	R	R	R	R	R	R	R
Kirtland's Warbler	R	R	R	V		V		H						
Cape May Warbler	R	R	R	R	R	R	R	R	R		R	V	V	
Cerulean Warbler	R	V		R	V	V			V					
Northern Parula	R	R	R	R	R	R	R	R	R	R	R	R	R	R
Magnolia Warbler	R	R	R	R	R	R	R	R	R	R	V	V		
Bay-breasted Warbler	R	R	R	R	R	R	V	R	R	V				
Blackburnian Warbler	R	R	V	R	R	R	H	V	V	V				
Northern Yellow Warbler	R	R	R	R	R	R	R	R	R	R	R	R		V
Golden Warbler	NE	R	R	R	R	R	R	R	R	R	R	R		
Chestnut-sided Warbler	R	R	R	R	R	R	V	R	R	R		V		
Blackpoll Warbler	R	R	R	R	R	R	R	R	R	R	R	R		R
Black-throated Blue Warbler	R	R	R	R	R	R	R	R	R	R	V	V		
Western Palm Warbler	R	R	R	R	R	R	R	R	R					V
Eastern Palm Warbler	R	V	V	V			?	R	V					

Common Name	StE	K&N	A&B	Mon	Gua	Dom	Mar	StL	V&G	Gre	Bar	Ave	Swa	Pro	San
Boat-tailed Grackle															
Greater Antillean Grackle															
Great-tailed Grackle														R	R
Yellow-hooded Blackbird									V						
Parulidae															
Ovenbird		V	V	R	R	V	R	V	V		V		R	R	R
Worm-eating Warbler			V		R						V		R	R	R
Louisiana Waterthrush		R	V	R	R	V	R	V	V		V		V	V	V
Northern Waterthrush		R	R	R	R	R	R	R	R	R	R	R	R	R	R
Bachman's Warbler															
Golden-winged Warbler		V	V					V		V					V
Blue-winged Warbler					V										V
Black-and-white Warbler	R	R	R	R	R	R	R	V	V		R		R	R	R
Prothonotary Warbler		V		V	R	V	R	V	V	R	R	V		R	R
Swainson's Warbler													R		R
Tennessee Warbler					V						V			R	R
Orange-crowned Warbler															
Nashville Warbler					H										V
Virginia's Warbler															
Semper's Warbler								E							
Connecticut Warbler					V						V				
Mourning Warbler															R
Kentucky Warbler		V	V		R	V	V	V			V			R	R
Bahama Yellowthroat															
Eastern Yellowthroat			V		R	V							R	R	R
Whistling Warbler								E							
Plumbeous Warbler					R	R									
Elfin Woods Warbler															
Arrowhead Warbler															
Hooded Warbler		V	V	V	R	V	V		V		V		R	H	R
American Redstart	R	R	R	R	R	R	R	R	R	R	R	R	R	R	R
Kirtland's Warbler															
Cape May Warbler	R	R	R	R	R	V		V	V	V	V		R	R	R
Cerulean Warbler					H					V					R
Northern Parula	R	R	R	R	R	R	R	V	V	R	V		R	R	R
Magnolia Warbler			V	R		R	V			V	V		R	R	R
Bay-breasted Warbler			V			V			V		V			R	R
Blackburnian Warbler					V	V				V	V		R	H	R
Northern Yellow Warbler	R	R	R	R	V	R	R	R	R	R	R		R	R	R
Golden Warbler	R	R	R	R	R	R	R	R	R		R			R	R
Chestnut-sided Warbler		V	V	V	R	V			V		R			V	R
Blackpoll Warbler		R	R		R	R	R	R	R	R	R	V	V	R	R
Black-throated Blue Warbler		V	V		R	V			V		V		R	R	R
Western Palm Warbler					V	V		V	V		V		R	V	R
Eastern Palm Warbler													?		

369

Common name	Regional status	Bah	T&C	Cub	Cay	Jam	Hai	DR	PR	VI	Ang	StM	StB	Sab
Olive-capped Warbler	E	R		R										
Pine Warbler	R	R		V	V	V	R	R	V					
Audubon's Warbler	V	V	V	V		H								
Myrtle Warbler	R	R	R	R	R	R	R	R	R	R	V	V		
Yellow-throated Warbler	R	R	R	R	R	R	R	R	R	V	V			
Bahama Warbler	E	E												
Vitelline Warbler	E				R									
Prairie Warbler	R	R	R	R	R	R	R	R	R	R	R	R	R	R
Adelaide's Warbler	E								R	R				
Barbuda Warbler	E													
St Lucia Warbler	E													
Black-throated Grey Warbler	V			V										
Townsend's Warbler	V	V	V	V	V									
Golden-cheeked Warbler	H								H					
Black-throated Green Warbler	R	R	R	R	R	R	R	R	R	V	V	V	V	V
Canada Warbler	R	V		R	R	V		V	V	V		V		H
Western Wilson's Warbler	V			V										
Eastern Wilson's Warbler	R	V		R	R	V		V	V					
Phaenicophilidae														
Black-crowned Palm-tanager	E						R	R						
Grey-crowned Palm-tanager	E						R	R						
White-winged Warbler	E						R	R						
Green-tailed Warbler	E						R	R						
Spindalidae														
Bahamas Green-backed Spindalis	E	E												
Bahamas Black-backed Spindalis	E	R	R											
Cuban Spindalis	E			E										
Grand Cayman Spindalis	E				E									
Hispaniolan Spindalis	E						R	R						
Puerto Rican Spindalis	E								E					
Jamaican Spindalis	E					E								
Nesospingidae														
Puerto Rican Tanager	E								E					
Calyptophilidae														
Western Chat-tanager	E						R	R						
Eastern Chat-tanager	E						R	R						
Cardinalidae														
Rose-breasted Grosbeak	R	R	R	R	R	R	R	R	R	R		H		V
Black-headed Grosbeak	V			V										
Dickcissel	R	R		R	R	V			V					
Indigo Bunting	R	R	R	R	R	R	H	R	R	R		V		V
Blue Grosbeak	R	R	R	R	R	R		R	V	V				V
Lazuli Bunting	V			V										

Common Name	StE	K&N	A&B	Mon	Gua	Dom	Mar	StL	V&G	Gre	Bar	Ave	Swa	Pro	San
Olive-capped Warbler															
Pine Warbler					V		V								
Audubon's Warbler															
Myrtle Warbler	V	V	V		R	V	V		V		R		R	V	R
Yellow-throated Warbler				V	V						V		R	R	R
Bahama Warbler															
Vitelline Warbler													R		
Prairie Warbler	R	R	V	R	R				V	V			R	H	R
Adelaide's Warbler															
Barbuda Warbler			E												
St Lucia Warbler								V	E						
Black-throated Grey Warbler															
Townsend's Warbler															
Golden-cheeked Warbler															
Black-throated Green Warbler		V	V		R	V	V			V	V		V	R
Canada Warbler					H			V			V			H	V
Western Wilson's Warbler															
Eastern Wilson's Warbler													V		
Phaenicophilidae															
Black-crowned Palm-tanager															
Grey-crowned Palm-tanager															
White-winged Warbler															
Green-tailed Warbler															
Spindalidae															
Bahamas Green-backed Spindalis															
Bahamas Black-backed Spindalis															
Cuban Spindalis															
Grand Cayman Spindalis															
Hispaniolan Spindalis															
Puerto Rican Spindalis															
Jamaican Spindalis															
Nesospingidae															
Puerto Rican Tanager															
Calyptophilidae															
Western Chat-tanager															
Eastern Chat-tanager															
Cardinalidae															
Rose-breasted Grosbeak	H	H	V		R	V	V	V	V		V	V	R	R	R
Black-headed Grosbeak															
Dickcissel					V						V	V	R	V	V
Indigo Bunting		V	V	V	R	V					V		R	V	R
Blue Grosbeak		V			V						V		R	V	R
Lazuli Bunting															

Common name	Regional status	Bah	T&C	Cub	Cay	Jam	Hai	DR	PR	VI	Ang	StM	StB	Sab
Eastern Painted Bunting	R	R	V	R	V	V		V	V					
Scarlet Tanager	R	R	R	R	R	R	H	V	R	V	V	V	V	V
Summer Tanager	R	R	R	R	R	R		V	V	V				V
Western Tanager	V	V		H				V						
Northern Cardinal	I	I			?									
Thraupidae														
Swallow Tanager	V?				V?									
Red-legged Honeycreeper	R			R		?								
Lesser Antillean Saltator	E													
Bahama Bananaquit	NE	R	R	R										
Greater Antillean Bananaquit	E			R	R	R	R							
Common Bananaquit	R								R	R	R	R	R	R
Yellow-faced Grassquit	R			R	R	R	R	R	R					
Orangequit	E					E								
Puerto Rican Bullfinch	E								E					
Greater Antillean Bullfinch	E	R	R			R	R	R						
Cuban Bullfinch	E			E										
Grand Cayman Bullfinch	E				E									
Yellow-shouldered Grassquit	E					E								
Cuban Grassquit	E	I		E										
Barbados Bullfinch	E													
Lesser Antillean Bullfinch	E									R	R	R	R	R
St Lucia Black Finch	E													
Black-faced Grassquit	R	R	R	R		R	R	R	R	R	R	R	R	R
Blue-black Grassquit	R													
Lined Seedeater	V													
Yellow-bellied Seedeater	R													
Chestnut-bellied Seed-finch	I													
Saffron Finch	V+I			V		I			I					
Grassland Yellow-finch	R													
St Vincent Tanager	E													
Grenada Tanager	E													

Common Name	StE	K&N	A&B	Mon	Gua	Dom	Mar	StL	V&G	Gre	Bar	Ave	Swa	Pro	San
Eastern Painted Bunting															V
Scarlet Tanager	V	V	R	V	R	V	V	V	V	V	R	V	R	V	R
Summer Tanager		V			R				V	V	V		R	R	R
Western Tanager															
Northern Cardinal													I		
Thraupidae															
Swallow Tanager															
Red-legged Honeycreeper															
Lesser Antillean Saltator		V			R	R	R	R							
Bahama Bananaquit															
Greater Antillean Bananaquit															
Common Bananaquit	R	R	R	R	R	R	R	R	R	R	R			R	R
Yellow-faced Grassquit															
Orangequit															
Puerto Rican Bullfinch		Ex													
Greater Antillean Bullfinch															
Cuban Bullfinch															
Grand Cayman Bullfinch															
Yellow-shouldered Grassquit															
Cuban Grassquit															
Barbados Bullfinch											E				
Lesser Antillean Bullfinch	R	R	R	R	R	R	R	R	R	R	V				
St Lucia Black Finch								E							
Black-faced Grassquit	R	R	R	R	R	R	R	R	R	R	R			R	R
Blue-black Grassquit										R					
Lined Seedeater					V										
Yellow-bellied Seedeater		V							R	R					
Chestnut-bellied Seed-finch							I								
Saffron Finch															
Grassland Yellow-finch			R		R			R	R	R	R				
St Vincent Tanager									E						
Grenada Tanager										E					

APPENDIX 2. SPECIES/TAXA NOT INCLUDED

Here we principally include (1) species known only as vagrants to the region from introduced populations elsewhere (e.g. Mute Swan), (2) those potential vagrants for which insufficient details exist to treat them even as hypothetical for the present (e.g. Common Goldeneye), and (3) a handful of species more recently introduced to the region that have failed to become established to date (Nanday Parakeet). However, we have not attempted to assemble a complete list of all those species introduced or apparently so in the region, or for which occasional records of escapees exist. For information on such birds we recommend consulting some of the many sources cited in the References and Further Reading, such as Blake (1975), Bond (1956), Bradley (2000), Buckley *et al.* (2009), Lever (2005) Navarro Pacheco (2018) and Raffaele (1989), among others. A handful of additional cases we consider likely to interest the reader are also presented, among them a spate of recent records of purported natural vagrants in Cuba, none of which seem likely to wander to our region 'under their own steam'. On the other hand, Rodríguez Castaneda *et al.* (2017) and Navarro Pacheco (2018) listed a large number of exotic species from the same country, for which proof of self-sustaining populations is lacking in most cases and seems extremely unlikely in the main. These are generally not listed here. Finally, we consider that a purported new species of Parulidae in the Sierra Maestra of south-east Cuba, reported by Dinets & Kolenov (2017, in a rather little-known Brazilian journal with virtually no track record in publishing ornithology!), was far more likely to have involved Connecticut Warbler.

Mute Swan *Cygnus olor* (see p. 30).—First-winter, Abaco, Bahamas, Jan 2005 (*NAB* 59: 342), which 'the committee did not regard as a vagrant but rather just 'origin unknown'. (B. Hallett *in litt.* 2018). Since then, there appear to have been a handful of further records from Abaco and elsewhere in the Bahamas, e.g. *NAB* 63: 518. An adult was photographed on St Kitts on an unknown date (photographs verified).

Common Goldeneye *Bucephala clangula* (see p. 34).—Bond (1940) considered it to be a 'rare winter straggler to the West Indies, whence recorded from Cuba and Barbados'. We are unaware of any Cuban record, and Bond himself later rejected the Barbados report, which was based on Cory (1889); see Buckley *et al.* (2009). Bond (1962) mentioned the species for Barbuda, but there seems to be no justification for this. Frequently listed for the Bahamas, based on a bird off Eleuthera (Riley 1905), but even this report is too vague for acceptance.

Egyptian Goose *Alopochen aegyptiaca* (see p. 36).—Several records from Grand Bahama and New Providence since 2010 (*NAB* 69: 508, eBird).

Bare-eyed Pigeon *Patagioenas corensis*.—Regular sightings on St Martin since 2011 almost certainly refer to escaped birds brought from Curacao or Aruba (Leeward Antilles); no further reports since the passage of Hurricane Irma in Aug–Sept 2017 (B. van Es *in litt.* 2018).

White-throated Swift *Aeronautes saxatalis*.—Descriptions by experienced observers of 'flocks of Chimney and White-throated Swifts' in May 1974 (*Gosse Bird Club Broadsh*. 24: 20, Levy in prep.) and three of the latter with Barn Swallows in Dec 2008 (*Gosse Bird Club Broadsh*. 89: 22, Levy in prep.) are compelling, but require further evidence.

Uniform Crake *Amaurolimnas concolor*.—Raffaele *et al.* (1998) noted that in the West Indies, where only recorded in Jamaica, the species was 'extirpated sometime after 1881', while Haynes-Sutton *et al.* (2009) state that there are no records post-1911, but dedicated searches have not been attempted. For conservation status we follow BirdLife International (2018; with further detail in Szabo *et al.* 2012) in treating the endemic subspecies *concolor* as officially Extinct.

Bermuda Petrel *Pterodroma cahow*.—Fossil bones from Crooked I, Bahamas (Wetmore 1938, Olson & Hiltgartner 1982, Buden 1987).

Barolo Shearwater *Puffinus* (*lherminieri*) *baroli*.—Included on the Puerto Rican checklist (SOPI 2016) based on a Jan 1977 sighting involving three birds (Lee 1989). The description provided is far from sufficient to establish that this taxon has occurred in the region. Future claims will also need to discount other extralimital taxa, particularly Boyd's Shearwater *P.* (*l.*) *boydi*.

American Woodcock *Scolopax minor*.—Four reports from Jamaica (C. Levy *in litt.* 2018, e.g. Scott 1892, *Gosse Bird Club Broadsh*. 5: 22, 42: 11) are insufficiently well described for admission to our list, but vagrants would not be unexpected, given occasional records from Dry Tortugas, off S Florida (van Gils *et al.* 2019); one shot in France in Oct 2006 (Ferrand *et al.* 2008) demonstrates that this species is even capable of crossing the Atlantic.

Common Sandpiper *Actitis hypoleucos*.—A detailed report by experienced birders of a Common Sandpiper in St George's Harbour, Grenada in Apr 2001 (*NAB* 58: 450) is considered doubtful by one of the named observers, who was not involved in the sighting (E. Massiah *in litt.* 2018).

Yellow-legged Gull *Larus michahellis*.—Multiple reports over the last two decades, claiming both nominate and *L. m. atlantis*, some of which, in the case of Puerto Rico (SOPI 2016) and Barbados (https://ebird.org/view/checklist/S40551997), are or have been accepted by national authorities. On the basis of submitted photographic evidence, we are unable to confirm any of these reports and the consensus of opinion among experts is weighted in favour of alternative identifications. Nevertheless, given that *L. m. atlantis* and perhaps *michahellis* have strayed to North America (Howell *et al.* 2014), future occurrence in the West Indies is a strong possibility. Separation of Yellow-legged Gull from hybrid American Herring × Lesser Black-backed Gull is a formidable challenge.

Common Kingfisher *Alcedo atthis*.—Rodríguez *et al.* (2005) reported one collected in mangroves in C Cuba, Apr 2003. The specimen was not deposited in a properly managed collection and was subsequently stated to have been destroyed. An unassisted crossing of the Atlantic, especially from W Europe, appears improbable, and even though Russian birds might regularly migrate up to 3000 km from their breeding grounds (Woodall 2001) the possibility of a misplaced journey south along the E Pacific and from there to the West Indies, is equally remote.

Lanner Falcon *Falco biarmicus* / **Prairie Falcon** *Falco mexicanus*.—A juvenile falcon observed on St Kitts on two dates in Nov 2006 was claimed first as a Lanner and subsequently as a Prairie Falcon. The record has found its way onto various checklists as one or other species. The description does not exclude the much more likely juvenile *tundrius* Peregrine Falcon, the plumage of which might suggest a more exotic species.

Cockatiel *Nymphicus hollandicus*.—Puerto Rican population not self-sustaining (Falcon & Tremblay 2018).

Grey Parrot *Psittacus erithacus*.—Not truly established on Puerto Rico or Martinique.

Senegal Parrot *Poicephalus senegalus*.—First reported from Puerto Rico in Sept 1992, with Monk Parakeets (Salguero-Faría & Roig-Bachs 2000). Included for Martinique and Guadeloupe by Levesque *et al.* (2005), and retained for Guadeloupe by

Levesque & Delcroix (2015), but not listed for Martinique by Belfan & Conde (2016). Also reported as an escapee on Jamaica, 1997 (eBird). We do not consider the species to be currently established anywhere in the region.

Yellow-crowned Amazon *Amazona ochrocephala* (see p. 186) / **Yellow-headed Amazon** *A. oratrix* (see p. 186)/ **Yellow-naped Amazon** *A. auropalliata* (see p. 184).—All three species within the Yellow-headed Amazon complex have been introduced into the region at one time or another. Specific identity is difficult to determine in some cases and, to exacerbate the problem, individuals of mixed parentage could also occur. Very local in Puerto Rico since 1970s (*A. oratrix* and *A. ochrocephala*) but apparently still not established; became established on Grand Cayman in early 1990s (photograph indicates *A. auropalliata*); and introduced c. 1990s on Barbados (species unknown). None of these populations are considered self-sustaining.

Nanday Parakeet *Aratinga nenday*.—Never established on Puerto Rico.

Red-and-green Macaw *Ara chloropterus* (see p. 190).—Initially reported in Puerto Rico by local people in 1980s (Salguero-Faría & Roig-Bachs 2000), with most records (usually pairs or small groups with commoner Blue-and-yellow Macaw) centred on San Juan and Guaynabo. The population does not appear self-sustaining without further releases. In Cuba, Navarro Pacheco (2018) reported that the species is 'recently established with local breeding populations' and has hybridized with perhaps both Blue-and-yellow and Scarlet Macaws *A. macao* there. Navarro Pacheco claimed that all three species are established in Cuba, but does not state where, since when or provide information concerning numbers or breeding. Without proof, we consider the case for a viable introduced population unproven.

Blue-crowned Parakeet *Psittacara acuticaudatus*.—Escapees recorded on Puerto Rico in 1989–1994.

Budgerigar *Melopsittacus undulatus*.—Never established on Puerto Rico or Barbados.

Couch's Kingbird *Tyrannus couchii* (see p. 194).—A specimen (FMNH 43096, no longer extant), speculated to have been taken on Providencia by Robert Henderson (Yojanan Lobo-y-Henriques 2014, Donegan *et al.* 2014), was not mentioned in Cory's (1887) list of birds obtained and is not accepted as part of the Colombian avifauna (Avendaño *et al.* 2017, Donegan *et al.* 2018).

Eurasian Blackcap *Sylvia atricapilla*.—A female claimed to have been trapped near La Habana, Cuba, Oct 2012, reportedly survived 19 months in captivity (Rodríguez *et al.* 2017). Notwithstanding the existence of photographs, one of them published, given an association (same principal observer) with two other extremely unlikely firsts for the West Indies (Common Kingfisher and White-winged Snowfinch) and the surprising failure to deposit the bird in a museum once it died, we consider natural vagrancy to be unproven. Has been recorded in SE Greenland.

Common Myna *Acridotheres tristis*.—One subsequently identified from verbal descriptions (B. Purdy *in litt.*) spent several days in a garden on Grand Bahama in Feb 2010; it was conceivably a natural vagrant from the recently established Florida population, but more likely ship-assisted. Another was photographed in La Habana, Cuba, in Apr 2019 (eBird).

Common Hill Myna *Gracula religiosa*.—Probably introduced into Puerto Rico in the late 1960s (Lever 2005) where it was described as an uncommon and local breeding resident (Raffaele 1989). It has not bred for decades.

Russet-backed Thrush *Catharus ustulatus*.—Usually considered conspecific with Swainson's Thrush *C. swainsoni* (but split by del Hoyo & Collar 2016), recognition of two species creates a formidable identification challenge. This species' known migration routes take it west of our region and, having examined specimen data, the literature and available photographs, we have been unable to find any records for West Indies. However, it remains a plausible vagrant, to be looked for especially on the western edge of the basin.

Purple Grenadier *Granatina ianthinogaster*.—Regular records of escaped cagebirds on Martinique (Belfan & Conde 2016), but not established.

Red-cheeked Cordon-bleu *Uraeginthus bengalus*.—Regular records of escaped cagebirds on Martinique (Belfan & Conde 2016), but not established.

White-winged Snowfinch *Montifringilla nivalis*.—Rodríguez Castañeda *et al.* (2017) reported one caught in W Cuba, Feb 2014, which survived in an aviary until Apr 2014. Photos undeniably portray this species, although apparently not the geographically closest race. The moult sequence reported for the bird while in captivity is at variance of what is known of the species in the wild (Shirihai & Svensson 2018). We refute the authors' conclusion that it 'probably arrived in Cuba through vagrancy'. Outside its core range, this species has only ever been reported to move comparatively short distances.

Yellow-fronted Canary *Crithagra mozambica*.—Never truly established on Puerto Rico, plus a single record on Guadeloupe.

Black-backed Goldfinch *Spinus* (*psaltria*) *psaltria*.—Introduced to Cuba from Mexico, with records between at least 1930s and 1970s, but probably now extirpated. Also present in Cuba in at least the first half of the 1800s (perhaps a native population).

Black-throated Sparrow *Amphispiza bilineata*.—Listed without comment for Andros (Bahamas) by Raffaele *et al.* (1998), but without documentation we omit it. While this species might seem to be an improbable vagrant, it has reached NW Florida (Menk & Stevenson 1978).

Audubon's Oriole *Icterus graduacauda*.—Mentioned as a vagrant in Puerto Rico by Raffaele *et al.* (1998), we are unable to find the source, nor is it retained on the national checklist (SOPI 2016). This largely sedentary Mexican species would be unlikely to find its way naturally to Puerto Rico.

Altamira Oriole *Icterus gularis*.—A specimen collected in NE Cuba, Nov 1954, has been considered part of the region's naturally occurring avifauna (Navarro Pacheco 2018). Given that the species is not known to migrate, and no wandering has been reported (Jaramillo & Burke 1999, Fraga 2011), we consider it unlikely to have occurred naturally.

Yellow-tailed Oriole *Icterus mesomelas*.—In a list of species represented by 'escaped cage-birds, or have been introduced in, or have been erroneously accredited to, the West Indies', Bond (1956) mentioned that he had examined a specimen from Cuba. Reported to have formerly comprised part of Charles Ramsden's collection (Navarro Pacheco 2018), Kirkconnell *et al.* (in press), however, failed to locate the specimen in any museum, Santiago de Cuba (at which most of Ramsden's material has been destroyed; Wiley *et al.* 2008) and USNM included. Resident well to the south and west of the Greater Antilles, this oriole is clearly unlikely to have reached Cuba in a natural state. Listing for Isla de Aves (Padrón López *et al.* 2015) is in error (M. Lentino pers. comm.).

Red-crested Cardinal *Paroaria coronata*.—Two old Puerto Rico records (1973, 1976), as well as a pair constructing a nest in Aug 1987, constitute the earliest documentation, but a local population persisted until c. 2001; also one on St Croix in Dec 1982 (Raffaele 1989), but the species has no self-sustaining population in the West Indies.

REFERENCES AND FURTHER READING

Aceituno, A. & Medina, D. (2009). Estudio preliminar de la avifauna de Islas del Cisne, Honduras. Pp. 105–108 in Rich, T.D., Arizmendi, C., Demarest, D.W. & Thompson, C. (eds.) *Proc. Fourth Int. Partners in Flight Conf.: Tundra to Tropics.* Partners in Flight, McAllen, TX.

Akresh, M.E. & King, D.I. (2015). Observations of new bird species for San Salvador Island, the Bahamas. *Carib. Natur.* **26**: 1–10.

Aldridge, B.M. (1987). Sampling migratory birds and other observations on Providenciales Islands B.W.I. *N. Amer. Bird Bander* **12**: 13–18.

American Ornithologists' Union (AOU) (1998). *Check-list of North American Birds.* Seventh edn. American Ornithologists' Union, Washington DC.

Anderson, D.L, Bonta, M. & Thorn, P. (1998). New and noteworthy bird records from Honduras. *Bull. Brit. Orn. Club* **118**: 178–183.

Anon. (2015). Pictorial highlights. *N. Amer. Birds* **68**: 293–297.

Arnoux, E. (2012). Variabilités phénotypique et génétique chez la Grive à pieds jaunes, *Turdus lherminieri*, à différentes échelles. Ph.D. thesis. Université de Bourgogne.

Austin, J.J., Bretagnolle, V. & Pasquet, E. (2004). A global molecular phylogeny of the small *Puffinus* shearwaters and implications for systematics of the Little-Audubon's Shearwater complex. *Auk* **121**: 847–864.

Avendaño, J.E., Isabel Bohórquez, C.I., Rosselli, L., Arzuza-Buelvas, D., Estela, F.A., Cuervo, A.M., Stiles, F.G. & Renjifo, L.M. (2017). Lista de chequeo de las aves de Colombia: Una síntesis del estado del conocimiento desde Hilty & Brown (1986). *Orn. Colombiana* **16**: eA01–82.

Banks, R.C. (1978). Nomenclature of the Black-bellied Whistling-duck. *Auk* **95**: 348–352.

Banks, R. C. (1988). An old record of the Pearly-breasted Cuckoo in North America and a nomenclatural critique. *Bull. Brit. Orn. Club* **108**: 87–91.

Banks, R.C. & Hole, R. (1992). Birds collected in Jamaica by William T. March, 1861–1866. *El Pitirre* **5(3)**: 15.

Barnés, V. (1936). New records for the Puerto Rican avifauna. *Auk* **53**: 350–351.

Barnés, V. (1947). Additions to the Puerto Rican avifauna with notes on little-known species. *Auk* **64**: 400–406.

Barré, N., Feldmann, P., Tayalay, G., Roc, P., Anselme, M. & Smith, P.W. (1996). Status of the Eurasian Collared-dove (*Streptopelia decaocto*) in the French Antilles. *El Pitirre* **9(3)**: 2–4.

Bayly, N. (2018). Confirmación fotográfica de la presencia de *Setophaga tigrina* y *Setophaga palmarum* en Colombia. *Bol. Soc. Antioqueña Orn.* **27**: 4–6.

Beatty, H.A. (1943). Records and notes from St. Croix, Virgin Islands. *Auk* **60**: 110–111.

Belfan, D. & Conde, B. (2016). *Liste des Oiseaux de la Martinique.* Association Le Carouge, Fort de France.

Bennett, E., Bolton, M. & Hilton, G. (2009). Temporal segregation of breeding by storm petrels *Oceanodroma castro* (*sensu lato*) on St Helena, South Atlantic. *Bull. Brit. Orn. Club* **129**: 92–97.

Biro, M. & Ludlow, M.M. (2015). First record of Spectacled Thrush (*Turdus nudigenis*) in Nevis, West Indies. *Nevis Orn. Soc. Occas. Pap.* **16**: 1–3.

Bisbal F. (1988). Los vertebrados terrestres de las Dependencias Federales de Venezuela. *Interciencia* **33**: 103–111.

Blake, C.H. (1975). Introductions, transplants, and invaders. *Amer. Birds* **29**: 923–926.

Boal, C.W., Sibley, F.C., Estabrook, T.S. & Lazell, J. (2006). Insular and migrant species, longevity records, and new species records on Guana Island, British Virgin Islands. *Wilson J. Orn.* **118**: 218–224.

Bochenski, Z.M. (1994). The comparative osteology of grebes (Aves: Podicipediformes) and its systematic implications. *Acta Zool. Cracov* **37**: 191–346.

Boeken, M. (2018). New avifaunal records and checklist for the island of Saba, Caribbean Netherlands. *J. Carib. Orn.* **31**: 57–64.

Bond, J. (1930). The resident West Indian warblers of the genus *Dendroica. Proc. Acad. Nat. Sci. Philadelphia* **82**: 329–337.

Bond, J. (1936). *Birds of the West Indies.* Academy of Natural Sciences, Philadelphia.

Bond, J. (1940). *Check-list of Birds of the West Indies.* First edn. Academy of Natural Sciences, Philadelphia.

Bond, J. (1942). Additional notes on West Indian birds. *Proc. Acad. Nat. Sci. Philadelphia* **94**: 89–106.

Bond, J. (1950). Results of the Catherwood-Caplin West Indies Expedition, 1948. Part II. Birds of Cayo Largo (Cuba), San Andrés and Providencia. *Proc. Acad. Nat. Sci. Philadelphia* **102**: 43–68.

Bond, J. (1952). Second supplement to the *Check-list of Birds of the West Indies* (1950). Academy of Natural Sciences, Philadelphia.

Bond, J. (1956). *Check-list of Birds of the West Indies.* Fourth edn. Academy of Natural Sciences, Philadelphia.

Bond, J. (1957). Second supplement to the *Checklist of Birds of the West Indies* (1956). Academy of Natural Sciences, Philadelphia.

Bond, J. (1959). Fourth supplement to the *Checklist of Birds of the West Indies* (1956). Academy of Natural Sciences, Philadelphia.

Bond, J. (1961). Sixth supplement to the *Checklist of Birds of the West Indies* (1956). Academy of Natural Sciences, Philadelphia.

Bond, J. (1962). Seventh supplement to the *Checklist of Birds of the West Indies* (1956). Academy of Natural Sciences, Philadelphia.

Bond, J. (1963). Eighth supplement to the *Checklist of Birds of the West Indies* (1956). Academy of Natural Sciences, Philadelphia.

Bond, J. (1964). Ninth supplement to the *Checklist of Birds of the West Indies* (1956). Academy of Natural Sciences, Philadelphia.

Bond, J. (1965). Tenth supplement to the *Checklist of Birds of the West Indies* (1956). Academy of Natural Sciences, Philadelphia.

Bond, J. (1967). Twelfth supplement to the *Checklist of Birds of the West Indies* (1956). Academy of Natural Sciences, Philadelphia.

Bond, J. (1968). Thirteenth supplement to the *Check-list of Birds of the West Indies* (1956). Academy of Natural Sciences, Philadelphia.

Bond, J. (1972). Seventeenth supplement to the *Check-list of Birds of the West Indies* (1956). Academy of Natural Sciences, Philadelphia.

Bond, J. (1973). Eighteenth supplement to the *Check-list of Birds of the West Indies* (1956). Academy of Natural Sciences, Philadelphia.

Bond, J. (1976). Twentieth supplement to the *Check-list of Birds of the West Indies* (1956). Academy of Natural Sciences, Philadelphia.

Bond, J. (1980). Twenty-third supplement to the *Check-list of Birds of the West Indies* (1956). Academy of Natural Sciences, Philadelphia.

Bond, J. (1982). Twenty-fourth supplement to the *Check-list of Birds of the West Indies* (1956). Academy of Natural Sciences, Philadelphia.

Bond, J. (1984). Twenty-fifth supplement to the *Checklist of Birds of the West Indies* (1956). Academy of Natural Sciences, Philadelphia.

Bond, J. (1986). Twenty-seventh supplement to the *Checklist of Birds of the West Indies* (1956). Academy of Natural Sciences, Philadelphia.

Bradley, P.E. (2000). *The Birds of the Cayman Islands: An Annotated Checklist*. BOU Checklist Series 19. British Ornithologists' Union, Tring, UK.

Bradley, P.E. & Rey-Millet, Y-V. (2013). *A Photographic Guide to the Birds of the Cayman Islands*. Christopher Helm, London.

Bradshaw, C.G., Kirwan, G.M. & Williams, R.S.R. (1997). First record of Swainson's Hawk *Buteo swainsoni* in the West Indies. *Bull. Brit. Orn. Club* **117**: 315–316.

Brown, A.C. (2012). Extirpation of the Snowy Plover (*Charadrius alexandrinus*) on St. Martin, West Indies. *J. Carib. Orn.* **25**: 31–34.

Brown, A.C. & Collier, N. (2005a). New and rare bird records from St. Martin, West Indies. *Cotinga* **23**: 52–58.

Brown, A.C. & Collier, N. (2005b). First records of White-eyed Vireo, Blue-winged Warbler, and Blue-winged Warbler × Golden-winged Warbler hybrid for St. Martin. *J. Carib. Orn.* **18**: 69–71.

Brown, A.C. & Collier, N. (2007). New bird records from Anguilla and St. Martin. *J. Carib. Orn.* **20**: 50–52.

Brudenell-Bruce, P.G.C. (1975). *The Birds of New Providence and the Bahama Islands*. Collins, London.

de Bruijne, J.W.A. (1970). Black-browed Albatross (*Diomedea melanophris*) in the Caribbean. *Ardea* **58**: 264.

Buckley, P.A. & Buckley, F.G. (2004). Rapid speciation by a Lesser Antillean endemic, Barbados Bullfinch *Loxigilla barbadensis*. *Bull. Brit. Orn. Club* **124**: 108–123.

Buckley, P.A., Massiah, E.B., Hutt, M.B., Buckley, F.G. & Hutt, H.F. (2009). *The Birds of Barbados: An Annotated Checklist*. BOU Checklist 24. British Ornithologists' Union & British Ornithologists' Club, Peterborough.

Buden, D.W. (1987). *The Birds of the Southern Bahamas: An Annotated Check-list*. BOU Checklist 8. British Ornithologists' Union, London.

Buden, D.W. (1991). Bird band recoveries in the Bahama Islands. *Carib. J. Sci.* **27**: 63–70.

Buden, D.W. (1992). The birds of the Exumas, Bahamas Islands. *Wilson Bull.* **104**: 674–698.

Buden, D.W. & Olson, S.L. (1989). The avifauna of the cayerias of southern Cuba, with the ornithological results of the Paul Bartsch Expedition of 1930. *Smiths. Contrib. Zool.* **477**: 1–34.

Buden, D.W. & Sprunt, A. (1993). Additional observations on the birds of the Exumas, Bahama Islands. *Wilson Bull.* **105**: 514–518.

Burke, W. (1994). Alpine Swift (*Tachymarptis melba*) photographed on St. Lucia, Lesser Antilles - third record for the Western Hemisphere. *El Pitirre* **7(3)**: 3.

Castellón Maure, A.I., Delcroix, F., Navarro Pacheco, N. & Varela Montero, R. (2017). Primer registro del Pitirre de Cassin (*Tyrannus vociferans*) para las Indias Occidentales y nuevos reportes del Bobito Americano (*Sayornis phoebe*) y del American Pipit (*Anthus rubescens*) para Cuba. *Annotated Checklist of the Birds of Cuba* **1**: 35–36.

Chabrolle, A. & Levesque, A. (2015). Spotted Crake (*Porzana porzana*) at Guadeloupe: second record for the New World. *N. Amer. Birds* **68**: 171–173.

Chandler, R. (2009). *Shorebirds of the Northern Hemisphere*. Christopher Helm, London.

Chesser, R.T., Burns, K.J., Cicero, C., Dunn, J.L., Kratter, A.W., Lovette, I.J., Rasmussen, P.C., Remsen, J.V., Stotz, D.F., Winger, B.M. & Winker, K. (2018). Fifty-ninth supplement to the American Ornithologists' Union *Check-list of North American Birds*. *Auk* **135**: 798–813.

Childress, R.B. & Hughes, B. (2001). The status of the West Indian Whistling-duck (*Dendrocygna arborea*) in St. Kitts-Nevis, January–February 2000. *El Pitirre* **14**: 107–112.

Clark, A.H. (1905). Birds of the southern Lesser Antilles. *Proc. Boston Soc. Nat. Hist.* **32**: 203–312.

Coffey, J. & Ollivierre, A. (2019). *Birds of the Transboundary Grenadines*. Birds of the Grenadines.

Collar, N.J. (2001). Family Trogonidae (trogons). Pp. 80–127 in del Hoyo, J., Elliott, A. & Sargatal, J. (eds.) *Handbook of the Birds of the World*. Vol. 6. Lynx Edicions, Barcelona.

Collazo, J.A., Saliva, J.E. & Pierce, J. (2000). Conservation of the Brown Pelican in the West Indies. Pp. 39–45 in Schreiber, E.A. & Lee, D.S. (eds.) *Status and Conservation of West Indian Seabirds*. Society for Caribbean Ornithology, Ruston, LA.

Collinson, J.M., Dufour, P., Hamza, A.A., Lawrie, Y., Elliott, M., Barlow, C. & Crochet, P.-A. (2017). When morphology is not reflected by molecular phylogeny: the case of three 'orange-billed terns' *Thalasseus maximus*, *Thalasseus bergii* and *Thalasseus bengalensis* (Charadriiformes: Laridae). *Biol. J. Linn. Soc.* **121**: 439–445.

Colón, H.E. (1982). A record of the Black Duck *Anas rubripes* in Puerto Rico. *Carib. J. Sci.* **17**: 5–6.

Connor, H.A. & Loftin, R.W. (1985). The birds of Eleuthera Islands, Bahamas. *Fla. Field Natur.* **13**: 77–93.

Cooke, M.T. (1945). Transoceanic recoveries of banded birds. *Bird-Banding* **16**: 123–129.

Cooper, N.W., Ewert, D.N., Wunderle, J.M., Helmer, E.H. & Marra, P.P. (2019). Revising the wintering distribution and habitat use of the Kirtland's warbler using playback surveys, citizen scientists, and geolocators. *Endangered Species Res.* **38**: 79–89.

Cory, C.B. (1887). A list of the birds taken by Mr. Robert Henderson, in the islands of Old Providence and St Andrews, Caribbean Sea, during the winter of 1886-87. *Auk* **4**: 180–181.

Cory, C.B. (1889). *The Birds of the West Indies, including all species known to occur in the Bahama Islands, the Greater Antilles, the Caymans, and the Lesser Antilles, excepting the islands of Tobago and Trinidad*. Estes & Lauriat, Boston.

Cory, C.B. (1892). A list of birds taken on Maraguana, Watling's Island, and Inagua, Bahamas, during July, August, September, and October, 1891. *Auk* **9**: 48–49.

DaCosta, J.M., Miller, M.J., Mortensen, J.L., Reed, J.M., Curry, R.L. & Sorenson, M.D. (2019). Phylogenomics clarifies biogeographic and evolutionary history, and conservation status of West Indian tremblers and thrashers (Aves: Mimidae). *Mol. Phylogenet. Evol.* **136**: 196–205.

Delmore, K.E., Fox, J.W. & Irwin, D.E. (2012). Dramatic intraspecific differences in migratory routes, stopover sites and wintering areas, revealed using light-level geolocators. *Proc. Roy. Soc. Lond. (Ser. B Biol. Sci.)* **279**: 4582–4589.

Devas, R.P. (1970). *Birds of Grenada, St. Vincent and the Grenadines*. Second edn. Carenage Press, St. George's, Grenada.

Devillers, P. (1977). The skuas of the North American Pacific coast. *Auk* **94**: 417–429.

Dhondt, A.A. & Dhondt, K.V. (2008). Two bird species new for Hispaniola. *J. Carib. Orn.* **21**: 46–47.

Dinets, V. & Kolenov, S. (2017). An undescribed New World warbler (Aves, Parulidae) in the mountains of Cuba? *Neotrop. Biol. & Conserv.* **12**: 235–237.

Donegan, T.M. & Huertas, B. (2011). The subspecies of Brown-throated Parakeet *Aratinga pertinax* on San Andres island. *Conserv. Colombiana* **15**: 35–37.

Donegan, T.M. & Huertas, B. (2015). Noteworthy bird records on San Andrés Island, Colombia. *Conserv. Colombiana* **22**: 8–12.

Donegan, T. & Huertas, B. (2018). Notes on some migratory birds rare, new or poorly known on Isla Providencia, Colombia. *Conserv. Colombiana* **25**: 56–63.

Donegan, T., Quevedo, A., Verhelst, J.C., Cortés, O., Pacheco, J.A. & Salaman, P. (2014). Revision of the status of bird species occurring or reported in Colombia 2014. *Conserv. Colombiana* **21**: 3–11.

Donegan, T., Quevedo, A., Verhelst, J.C., Cortés-Herrera, O., Ellery, T. & Salaman, P. (2015). Revision of the status of bird species occurring or reported in Colombia 2015, with discussion of BirdLife International's new taxonomy. *Conserv. Colombiana* **23**: 3–48.

Donegan, T., Ellery, T., Pacheco, J.A., Verhelst, J.C. & Salaman, P. (2018). Revision of the status of bird species occurring or reported in Colombia 2018. *Conserv. Colombiana* **25**: 4–47.

Downer, A. (1968). White-crowned Sparrow. *Gosse Bird Club Broadsh.* **10**: 18.

Dugand, A. (1947). Aves marinas de las costas e islas colombianas. *Caldasia* **4**: 379–398.

Ebels, E.B. (2002). Transatlantic vagrancy of Palearctic species to the Caribbean region. *Dutch Birding* **24**: 202–209.

Eitniear, J.C. & Colón-López, S. (2005). Recent observations of Masked Duck (*Nomonyx dominica* [sic]) in the Caribbean. *Carib. J. Sci.* **41**: 861–864.

Eitniear, J.C. & Morel, M. (2012). A large concentration of Masked Ducks (*Nomonyx dominicus*) in Puerto Rico. *J. Carib. Orn.* **25**: 92–94.

Escola, F., Calchi, R., Hernández, C. & Casler, C. (2015). Primer registro de la Golondrina de las Cavernas *Petrochelidon fulva* para Venezuela y Sur América. *Cotinga* **33**: 118–119.

Evans, P.G.H. (1990). *Birds of the Eastern Caribbean*. Macmillan, London.

Evans, P.G.H. & Jones, A. (1997). *Dominica. A Guide to Birdwatching*. Dominican Ministry of Tourism.

Falcon, W. & Tremblay, R.L. (2018). From the cage to the wild: introductions of Psittaciformes to Puerto Rico. *PeerJ* **6**: p.e5669.

Feilden, H.W. (1889). On the birds of Barbados. *Ibis* **31**: 477–503.

Feldmann, P. & Pavis, C. (1995). Alpine Swift (*Tachymarptis melba*) observed in Guadalupe, Lesser Antilles: a fourth record for the Western Hemisphere. *El Pitirre* **8(2)**: 2.

Feldmann, P., Benito-Espinal, E. & Keith, A.R. (1999). New bird records from Guadeloupe and Martinique, West Indies. *J. Field Orn.* **70**: 80–94.

Feo, T.J., Musser, J.M., Berv, J. & Clark, C.J. (2015). Divergence in morphology, calls, song, mechanical sounds, and genetics supports species status for the Inaguan hummingbird (Trochilidae: *Calliphlox "evelynae" lyrura*). *Auk* **132**: 248–264.

Fernández-Ordóñez, J.C., Narciso, S. & Mata, T. (2016). Primeros registros de la Tórtola Aliblanca *Zenaida asiatica* en Venezuela. *Rev. Venez. Orn.* **6**: 58–61.

Ferrand, Y., Grossmann, F., Guillaud-Rollin, Y. & Jiguet, F. (2008). Première mention de la Bécasse d'Amérique *Scolopax minor* pour la France et le paléarctique occidental. *Ornithos* **15**: 128–131.

Fisher, A.K. & Wetmore, A. (1931). Report on birds collected by the Pinchot Expedition of 1929 to the Caribbean and Pacific. *Proc. US Natl. Mus.* **79**: 1–66.

Fleischer, R.C., Kirchman, J.J., Dumbacher, J.P., Bevier, L., Dove, C., Rotzel, N.C., Edwards, S.V., Lammertink, M., Miglia, K.J. & Moore, W.S. (2006). Mid-Pleistocene divergence of Cuban and North American Ivory-billed Woodpeckers. *Biol. Lett.* **2**: 466–469.

Fleming, J.H. (1901). European Lapwing in the Bahamas. *Auk* **18**: 272.

Fraga, R.M. (2011). Family Icteridae (New World blackbirds). Pp. 684–807 *in* del Hoyo, J., Elliott, A. & Christie, D.A. (eds.) *Handbook of the Birds of the World*. Vol. 16. Lynx Edicions, Barcelona.

Francis, J. (2012). First record of White-winged Dove (*Zenaida asiatica*) on Nevis. *J. Carib. Orn.* **25**: 39.

Frost, M.D. & Massiah, E.B. (2003). Observations of rare and unusual birds on Grenada. *J. Carib. Orn.* **16**: 63–65.

Frost, M.D. & Burke, R.W. (2005). Two observations of Alpine Swift (*Apus melba*) on Barbados. *J. Carib. Orn.* **18**: 79–80.

García-Quintas, A. & Marichal, E. (2016). First record of Great Shearwater (*Ardenna gravis*) in Cuba. *Fla. Field Natur.* **44**: 175–177.

Garrido, O.H. (1985). Cuban endangered birds. Pp. 992–999 in Buckley, P.A., Foster, M.S., Morton, E.S., Ridgely, R.S. & Buckley, F.G. (eds.) *Neotropical Ornithology*. Orn. Monogr. 36. American Ornithologists' Union, Washington DC.

Garrido, O.H. (1997). *Sicalis flaveola*—nueva especie para la avifauna cubana. *El Pitirre* **10**: 55.

Garrido, O.H. (2000a). A new subspecies of Oriente Warbler *Teretistris fornsi* from Pico Turquino, Cuba, with ecological comments on the genus. *Cotinga* **14**: 88–93.

Garrido, O.H. (2000b). ¿Es la Bijirita del Pinar *Dendroica pityophila* (Aves: Parulidae) especie monotípica? *El Pitirre* **13**: 8–11.

Garrido, O.H. (2001). Was Red-legged Honeycreeper *Cyanerpes cyaneus* in Cuba introduced from Mexico? *Cotinga* **15**: 58.

Garrido, O.H. (2007). Subespecie nueva de *Asio dominguensis* para Cuba, con comentarios sobre *Asio flammeus* (Aves: Strigidae). *Solenodon* **6**: 70–78.

Garrido, O.H. & Kirkconnell, A. (2000). *Field Guide to the Birds of Cuba*. Cornell University Press, Ithaca, NY.

Garrido, O.H. & Kirkconnell, A. (2008). The first record of Black-headed Grosbeak *Pheucticus melanocephalus* in the West Indies. *Cotinga* **30**: 72.

Garrido, O.H. & Schwartz, A. (1969). Anfibios, reptiles y aves de Cayo Cantiles. *Poeyana* Ser. A. **67**: 1–44.

Garrido, O.H. & Wiley, J.W. (2003). The taxonomic status of the Puerto Rican Bullfinch (*Loxigilla portoricensis*) (Emberizidae) in Puerto Rico and St. Kitts. *Orn. Neotropical* **14**: 91–98.

Garrido, O.H., Reynard, G.B. & Kirkconnell, A. (1997a). Is the Palm Crow, *Corvus palmarum* (Aves: Corvidae), a monotypic species? *Orn. Neotropical* **8**: 15–21.

Garrido, O.H., Parkes, K.C., Reynard, G.B., Kirkconnell, A. & Sutton, R. (1997b). Taxonomy of the Stripe-headed Tanager, genus *Spindalis* (Aves: Thraupidae) of the West Indies. *Wilson Bull.* **109**: 561–594.

Garrido, O.H., Peterson, A.T. & Komar, O. (1999). Geographic variation and taxonomy of the Cave Swallow (*Petrochelidon fulva*) complex, with the description of a new subspecies from Puerto Rico. *Bull. Brit. Orn. Club* **119**: 80–91.

Garrido, O.H., Kirwan, G.M. & Capper, D.R. (2002). Species limits within Grey-headed Quail-dove *Geotrygon caniceps* and implications for the conservation of a globally threatened species. *Bird Conserv. Int.* **12**: 169–187.

Garrido, O.H., Wiley, J.W. & Kirkconnell, A. (2005). The genus *Icterus* in the West Indies. *Orn. Neotropical* **16**: 449–470.

Garrido, O.H., Wiley, J.W. & Reynard, G.B. (2009). Taxonomy of the Loggerhead Kingbird (*Tyrannus caudifasciatus*) complex (Aves: Tyrannidae). *Wilson J. Orn.* **121**: 703–713.

Garrido, O.H., Wiley, J.W., Kirkconnell, A., Bradley, P., Günther-Calhoun, A. & Rodríguez, D. (2014). Revision of the endemic West Indian genus *Melopyrrha* from Cuba and the Cayman Islands. *Bull. Brit. Orn. Club* **134**: 134–144.

Garrido, O.H., Kirkconnell, A. & Wiley, J.W. (2016). First record of Surf Scoter (*Melanitta perspicillata*) for Cuba and notes on an eighteenth century record for Jamaica. *Fla. Field Natur.* **44**: 19–22.

Gemmill, D. (2015). *Birds of Vieques Island, Puerto Rico: Status, Abundance, and Conservation*. J. Carib. Orn. Spec. Iss. Birds Caribbean, Charlottesville, VA.

Gill, F. & Wright, M. (2006). *Birds of the World: Recommended English Names*. Christopher Helm, London.

van Gils, J., Wiersma, P. & Kirwan, G.M. (2019). American Woodcock (*Scolopax minor*). In: del Hoyo, J., Elliott, A., Sargatal, J., Christie, D.A. & de Juana, E. (eds) *Handbook of the Birds of the World Alive*. Lynx Edicions, Barcelona (retrieved from https://www.hbw.com/node/53866 on 25 April 2019).

Gochfeld, M., Hill, D.O. & Tudor, G. (1973). A second population of the recently described Elfin Woods Warbler and other bird records from the West Indies. *Carib. J. Sci.* **13**: 231–235.

Gochfeld, M., Burger, J., Saliva, J. & Gochfeld, D. (1988). Herald Petrel new to the West Indies. *Amer. Birds* **42**: 1254–1258.

Graves, G.R. (2014). Historical decline and probable extinction of the Jamaican Golden Swallow *Tachycineta euchrysea euchrysea*. *Bird Conserv. Int.* **24**: 239–251.

Gricks, N.P. (1994a). Vagrant White Stork *Ciconia ciconia* (Aves: Ciconiidae) found in Antigua: a first record for the West Indies. *El Pitirre* **7(1)**: 2.

Gricks, N.P. (1994b). Additional records for Antigua-Barbuda. *El Pitirre* **7(3)**: 12.

Groome, J.R. (1970). *A Natural History of the Island of Grenada, West Indies*. Caribbean Printers Ltd., Arima, Trinidad.

van Halewyn, R. & Norton, R.L. (1984). The status and conservation of seabirds in the Caribbean. Pp. 169–222 in Croxall, J.P., Evans, P.G.H. & Schreiber, R.W. (eds) *Status and Conservation of the World's Seabirds*. International Council for Bird Preservation, Cambridge, UK.

Hamel, P.B. & Gauthreaux, S.A. (1982). The field identification of Bachman's Warbler (*Vermivora bachmanii* Audubon). *Amer. Birds* **36**: 235–240.

Hayes, F.E. (2001). First sight records of Swainson's Hawk (*Buteo swainsoni*) for Trinidad and Chacachacare Island, with comments on its status and trans-Caribbean migration. *El Pitirre* **14**: 63–65.

Hayes, F.E. (2004). Variability and interbreeding of Sandwich Terns and Cayenne Terns in the Virgin Islands, with comments on their systematic relationship. *N. Amer. Birds* **57**: 566–572.

Hayes, F.E. (2006). Variation and hybridization in the Green Heron (*Butorides virescens*) and Striated Heron (*B. striata*) in Trinidad and Tobago, with comments on species limits. *J. Carib. Orn.* **19**: 12–20.

Hayes, F.E. & Hayes, B.D. (2006). First record of Striated Heron (*Butorides striata*) for the Greater Antilles at St. John, United States Virgin Islands. *N. Amer. Birds* **59**: 464–465.

Hayes, F.E. & Kenefick, M. (2002). First record of Black-tailed Godwit *Limosa limosa* for South America. *Cotinga* **17**: 20–22.

Hayes, F.E., Paice, M.R., Blunden, T., Smith, P.W., Smith, S.A., White, G. & Frost, M.D. (2006). Status of the American Oystercatcher (*Haematopus palliatus*) in St. Vincent and the Grenadines. *J. Carib. Orn.* **19**: 48–51.

Hayes, F.E., White, G.L., Frost, M.D., Sanasie, B., Kilpatrick, H. & Massiah, E.B. (2002). First records of Kelp Gull *Larus dominicanus* for Trinidad and Barbados. *Cotinga* **18**: 85–88.

Hayes, T. & Thorstrom, R. (2014). First record of a Mississippi Kite (*Ictinia mississippiensis*) in the Dominican Republic. *J. Carib. Orn.* **27**: 25–26.

Hayes, W.K., Bracey, E.D., Price, M.R., Robinette, V., Gren, E. & Stahala, C. (2010). Population status of Chuck-will's-widow (*Caprimulgus carolinensis*) in the Bahamas. *Wilson J. Orn.* **122**: 381–384.

Haynes-Sutton, A., Downer, A. & Sutton, R. (2009). *A Photographic Guide to the Birds of Jamaica*. Christopher Helm, London.

Hellmayr, C.E. & Conover, B. (1948). *Catalogue of the Birds of the Americas and Adjacent Islands*. Publications of the Field Museum of Natural History (Zoological Series) 13 pt. **1(3)**, Chicago.

Hilty, S.L. (2011). Family Thraupidae (tanagers). Pp. 46–329 in del Hoyo, J., Elliott, A. & Christie, D.A. (eds.) *The Handbook of the Birds of the World*. Vol. 16. Lynx Edicions, Barcelona.

Hilty, S.L. & Brown W.L. (1986). *A Guide to the Birds of Colombia*. Princeton University Press, Princeton, NJ & London.

Hoffman, W., Woolfenden, G.E. & Smith, P.W. (1999). Antillean Short-eared Owls invade southern Florida. *Wilson Bull.* **111**: 303–313.

Holland, C.S. & Williams, J.M. (1978). Observations on the birds of Antigua. *Amer. Birds* **32**: 1095–1105.

Hood-Daniel, N. (1988). Birds at Spur Tree, August 1988. *Gosse Bird Club Broadsh.* **51**: 10.

Hough, J. (2000). Identification of Cliff Swallow and Cave Swallow. *Birding World* **13**: 368–374.

Howell, S.N.G. (2012). *Petrels, Albatrosses, and Storm-Petrels of North America: A Photographic Guide*. Princeton University Press, Princeton, NJ.

Howell, S.N.G. & Dunn, J. (2007). *Gulls of the Americas*. Houghton Mifflin, Boston, MA.

Howell, S.N.G. & Patteson, J.B. (2008). Variation in the Black-capped Petrel – one species or more? *Alula* **14**: 70–83.

Howell, S.N.G., Lewington, I. & Russell, W. (2014). *Rare Birds of North America*. Princeton University Press, Princeton, NJ & Oxford.

del Hoyo, J. & Collar, N.J. (2014). *HBW and BirdLife International Illustrated Checklist of the Birds of the World*. Vol. 1. Lynx Edicions, Barcelona.

del Hoyo, J. & Collar, N.J. (2016). *HBW and BirdLife International Illustrated Checklist of the Birds of the World*. Vol. 2. Lynx Edicions, Barcelona.

Hudson, R. (1968). The Great Skua in the Caribbean. *Bird Study* **15**: 33–34.

Humphries, M.B., Gonzalez, M.A. & Rickefs, R.E. (2019). Phylogeography and historical demography of Carib Grackle (*Quiscalus lugubris*). *J. Carib. Orn.* **32**: 11–16.

Iliff, M.J. & Sullivan, B.L. (2004). Little Stint (*Calidris minuta*) in North America and the Hawaiian Islands: a review of status and distribution. *N. Amer. Birds* **58**: 316–323.

Ingels, J., Claessens, O., Luglia, T., Ingremeau, P. & Kenefick, M. (2010). White Wagtail *Motacilla alba*, a vagrant to Barbados, Trinidad and French Guiana. *Bull. Brit. Orn. Club* **130**: 224–226.

Jaramillo, A. & Burke, P. 1999. *New World Blackbirds*. Christopher Helm, London.

Jönsson, K.A., Fabre, P.-H. & Irestedt, M. (2012). Brains, tools, innovation and biogeography in crows and ravens. *BMC Evol. Biol.* **12**: 72.

Joseph, L., Wilke, T., Bermingham, E., Alpers, D. & Ricklefs, R. (2004). Towards a phylogenetic framework for the evolution of shakes, rattles and rolls in *Myiarchus* tyrant-flycatchers (Aves: Passeriformes: Tyrannidae). *Mol. Phylogenet. Evol.* **31**: 139–152.

Kale, H.W., Hundley, M.H. & Tucker, J.A. (1969). Tower-killed specimens and observations of migrant birds from Grand Bahama Island. *Wilson Bull.* **81**: 258–263.

Keith, A.R. (1997). *The Birds of St. Lucia, West Indies: An Annotated Check-list*. BOU Check-list 15. British Ornithologists' Union, Tring.

Keith, A.R., Wiley, J.W., Latta, S.C. & Ottenwalder, J.A. (2003). *The Birds of Hispaniola. Haiti and the Dominican Republic: An Annotated Checklist*. BOU Check-list 21. British Ornithologists' Union, Tring.

Kenefick, M., Restall, R. & Hayes, F.E. (2011). *Birds of Trinidad & Tobago*. Second edn. Christopher Helm, London.

Kepler, C.B. (1971). First Puerto Rican record of the Antillean Palm Swift. *Wilson Bull.* **83**: 309–310.

Kerr, K.C.R. & Dove, C.J. (2013). Delimiting shades of gray: phylogeography of the Northern Fulmar, *Fulmarus glacialis*. *Ecol. & Evol.* **3**: 1915–1930.

Kirkconnell, A. (2019). Obituary: James (Jim) W. Wiley, 20 January 1943–19 September 2018. *Bull. Brit. Orn. Club* **139**: 3–4.

Kirkconnell, A. & Kirwan, G.M. (2008). Aves de Cayo Paredón Grande, Archipiélago de Sabana-Camagüey, Cuba. *J. Carib. Orn.* **21**: 26–36.

Kirkconnell, A., Kirwan, G.M., Garrido, O.H., Mitchell, A. & Wiley, J.W. (in press). *The Birds of Cuba: An Annotated Checklist*. BOU Check-list 26. British Ornithologists' Club, Tring.

Kirkconnell Posada, A., Kirkconnell, A. & Kirwan, G.M. (2018). First record of White-faced Ibis *Plegadis chichi* in the West Indies. *Bull. Brit. Orn. Club* **138**: 272–274.

Kirwan, G.M. & Kirkconnell, A. (2002). The avifauna of Pálpite, Ciénaga de Zapata, Cuba, and the importance of the area for globally threatened and endemic birds. *El Pitirre* **15**: 101–109.

Kirwan, G.M., Williams, R.S.R. & Bradshaw, C.G. (1999). Interesting avifaunal records from the Dominican Republic. *Cotinga* **11**: 27–29.

Kirwan, G.M., Flieg, G.M., Hume, R. & LaBar, S. (2001). Interesting distributional and temporal records from Cuba, winter 2000–2001. *El Pitirre* **14**: 43–46.

Kirwan, G.M., Kirkconnell, A. & Flieg, G.M. (2010). *A Birdwatchers' Guide to Cuba, Jamaica, Hispaniola, Puerto Rico and the Caymans*. Prion Ltd., Cley.

König, C., Weick, F. & Becking, J.-H. (2008). *Owls of the World*. Second edn. Christopher Helm, London.

Kopp, M., Hans-Ulrich, P., Osama, M., Lisovski, S., Ritz, M.S., Phillips, R.A. & Hahn, S. (2011). South Polar Skuas from a single breeding population overwinter in different oceans though show similar migration patterns. *Marine Ecol. Prog. Ser.* **435**: 263–267.

Kushlan, J.A. & Prosper, J.W. (2009). Little Egret (*Egretta garzetta*) nesting on Antigua: a second nesting site in the Western Hemisphere. *J. Carib. Orn.* **22**: 108–111.

Labrada, O. & Blanco, P. (2011). Permanencia invernal y primer registro de nidificación de la Avoceta (*Recurvirostra americana*) en Cuba. *J. Carib. Orn.* **24**: 71–73.

Lack, D. & Lack, A. (1973). Birds on Grenada. *Ibis* **115**: 53–59.

Lack, D., Lack, E., Lack, P. & Lack, A. (1972). Transients in Jamaica, 1970–1971. *Gosse Bird Club Broadsh.* **18**: 1–5.

Lack, D., Lack, E., Lack, P. & Lack, A. (1973). Birds on St Vincent. *Ibis* **115**: 46–52.

Lanyon, W.E. (1967). Revision and probable evolution of the *Myiarchus* flycatchers of the West Indies. *Bull. Amer. Mus. Nat. Hist.* **136**: 331–370.

Larsen, N. & Levesque, A. (2008). Range expansion of White-winged Dove (*Zenaida asiatica*) in the Lesser Antilles. *J. Carib. Orn.* **21**: 61–65.

Latta, S.C. & Parkes, K.C. (2001). A possible *Dendroica kirtlandii* hybrid from Hispaniola. *Wilson Bull.* **113**: 378–383.

Latta, S.C. & Rodríguez, P.G. (2018). Notable bird records from Hispaniola and associated islands, including four new species. *J. Carib. Orn.* **31**: 34–37.

Latta, S., Rimmer, C., Keith, A., Wiley, J., Raffaele, H., McFarland, K. & Fernandez, E. (2006). *Birds of the Dominican Republic and Haiti*. Princeton University Press, Princeton, NJ.

Ławicki, Ł. & van den Berg, A.B. (2016). WP reports. *Dutch Birding* **38**: 102–116.

Leblond, G. (2007). Observation d'une Cicogne blanche *Ciconia ciconia* en Martinique (Petites Antilles). *Alauda* **75**: 244–245.

Lee, D. S. (1988). The Little Shearwater (*Puffinus assimilis*) in the western North Atlantic. *Amer. Birds* **42**: 213–220.

Lemoine, V. (2005). Little Ringed Plover (*Charadrius dubius*) in Martinique: first for the West Indies. *N. Amer. Birds* **59**: 669.

Lenoble, A. (2015). The Violet Macaw (*Anodorhynchus purpurascens* Rothschild, 1905) did not exist. *J. Carib. Orn.* **28**: 17–21.

Lever, C. (2005). *Naturalised Birds of the World*. T. & A.D. Poyser, London.

Levesque, A. (2013). La Tourterelle à Ailes Blanches (*Zenaida asiatica*), nouvelle espèce nicheuse en Guadeloupe. *J. Carib. Orn.* **26**: 55–56.

Levesque, A. (2016). Collared Pratincoles (*Glareola pratincola*) at Guadeloupe. *N. Amer. Birds* **69**: 178–185.

Levesque, A. & Delcroix, F. (2015). Liste des oiseaux de la Guadeloupe (8ème édition). Grande-Terre, Basse-Terre, Marie-Galante, les Saintes, la Désirade, Îlets de la Petite-Terre. *Rapport AMAZONA* **37**.

Levesque, A. & Jaffard, M.-E. (2002). Fifteen new bird species in Guadeloupe (F.W.I.). *El Pitirre* **15**: 5–6.

Levesque, A. & Malglaive, L. (2004). First documented record of Marsh Harrier for the West Indies and the New World. *N. Amer. Birds* **57**: 564–565.

Levesque, A. & Saint-Auret, A. (2007). New or vagrant bird species from Guadeloupe (F. W. I.) in autumn 2003. *J. Carib. Orn.* **20**: 61–64.

Levesque, A. & Yésou, P. (2005). Occurrence and abundance of tubenoses (Procellariiformes) at Guadeloupe, Lesser Antilles, 2001–2004. *N. Amer. Birds* **59**: 674–679.

Levesque, A. & Yésou, P. (2018). Black-capped Petrel (*Pterodroma hasitata*) occurrence near Guadeloupe, Lesser Antilles, 2001–2008. *J. Carib. Orn.* **31**: 20–22.

Levesque, A., Duzont, F. & Ramsahaï, A. (2005). Précisions sur cinq espèces d'oiseaux dont la nidification a été découverte en Guadeloupe (Antilles Françaises) depuis 1997. *J. Carib. Orn.* **18**: 45–47.

Levesque, A., Duzont, F. & Hecker, N. (2012). New bird species recorded in Guadeloupe, French West Indies, 2003–2011. *N. Amer. Birds* **66**: 214–223.

Levesque, A., Chabrolle, A., Delcroix, F. & Delcroix, É. (2019). First record of the Sedge Wren (*Cistothorus platensis*) for the Bahamas and the Caribbean. *J. Carib. Orn.* **32**: 31–33.

Levy, C. (2008). History of ornithology in the Caribbean. *Orn. Neotropical* **19(Suppl.)**: 415–426.

Levy, C. (2015). Great-tailed Grackle (*Quiscalus mexicanus*) spreading in Jamaica. *J. Carib. Orn.* **28**: 15–16.

Levy, C. (in prep.). *Checklist of the Birds of Jamaica*.

Llanes Sosa, A., Pérez Mena, E., González, H., Pérez Hernández, A. & Rodríguez Casariego, D. (2016). Nuevos registros de aves para la península de Guanahacabibes, que incluyen el primer registro de *Cardellina pusilla pileolata* para Cuba. *Poeyana* **502**: 63–71.

Lovette, I.J., Bermingham, E., Seutin, G. & Ricklefs, R.E. (1998). Evolutionary differentiation in three endemic West Indian warblers. *Auk* **115**: 890–903.

Lowe, P.R. (1909). Notes on some birds collected during a cruise in the Caribbean Sea. *Ibis* **51**: 304–347.

Lowe, P.R. (1911). *A Naturalist on Desert Islands*. H.F. & G. Witherby, London.

Luksenburg, J.A. & Sangster, G. (2013). New seabird records from Aruba, southern Caribbean, including three pelagic species new for the island. *Marine Orn.* **41**: 183–186.

Mackin, W.A. (2016). Current and former populations of Audubon's Shearwater (*Puffinus lherminieri*) in the Caribbean region. *Condor* **118**: 655–673.

Madden, H., Hensen, R., Piontek, S., Walton, S., Verdaat, H., Geelhoed, S.C.V., Stapel, J. & Debrot, A.O. (2015). New bird records for the island of St. Eustatius, Dutch Caribbean, with notes on other significant sightings. *J. Carib. Orn.* **28**: 28–34.

Manthey, J.D., Geiger, M. & Moyle, R.G. (2017). Relationships of morphological groups in the northern flicker superspecies complex (*Colaptes auratus* & *C. chrysoides*). *Syst. & Biodiver.* **15**: 183–191.

Martínez, O., Cotayo, L., Kirkconnell, A. & Wiley, J.W. (2016). First record of Lapland Longspur *Calcarius lapponicus* in the Caribbean. *Bull. Brit. Orn. Club* **136**: 295–299.

Massiah, E.B. & Frost, M.D. (2003). Is Channel-billed Toucan (*Ramphastos vitellinus*) established on Grenada? *J. Carib. Orn.* **16**: 68–95.

Mathys, B.A. (2011). First record of Aplomado Falcon (*Falco femoralis*) for the West Indies. *Wilson J. Orn.* **123**: 179–180.

McCandless, J.B. (1961). Bird life in south-western Puerto Rico. I. Fall migration. *Carib. J. Sci.* **1**: 3–12.

McFarland, K.P., Rimmer, C.C., Goetz, J.E., Aubry, Y., Wunderle, J.M., Sutton, A., Townsend, J.M., Sosa, A.L. & Kirkconnell, A. (2013). A winter distribution model for Bicknell's Thrush (*Catharus bicknelli*), a conservation tool for a threatened migratory songbird. *PLoS ONE* **8(1)**: e53986.

McGuire, H.L. (2002). Taxonomic status of the great white heron (*Ardea herodias occidentalis*): an analysis of behavioral, genetic, and morphometric evidence. Final Report. Florida Fish & Wildlife Conservation Commission, Tallahassee.

McGuire, H.L., Taylor, S.S. & Sheldon, F.S. (2019). Evaluating the taxonomic status of the Great White Heron (*Ardea herodias occidentalis*) using morphological, behavioral and genetic evidence. *Auk* **136**: https://doi.org/10.1093/auk/uky010.

McNish, T. (2003). *Lista de Chequeo de la Fauna Terrestre del Archipiélago de San Andrés, Providencia y Santa Catalina, Colombia*. M&B Producciones y Servicios Limitada, Bogotá.

McNish, T. (2011). *La Fauna del Archipiélago de San Andrés, Providencia y Santa Catalina, Colombia, Sudamérica*. Privately published, Colombia.

McNair, D.B., Yntema, L.D., Lombard, C.D., Cramer-Burke, C. & Sladen, F.W. (2006). Records of rare and uncommon birds from recent surveys on St. Croix, United States Virgin Islands. *N. Amer. Birds* **59**: 536–551.

Meier, A.J., Noble, R.E. & Raffaele, H.A. (1989). The birds of Desecheo Island, Puerto Rico, including a new record for Puerto Rican territory. *Carib. J. Sci.* **25**: 24–29.

Mejía, D.A., Paulino, M.M., Wallace, K. & Latta, S.C. (2009). Great-tailed Grackle (*Quiscalus mexicanus*): a new species for Hispaniola. *J. Carib. Orn.* **22**: 112–114.

Menk, G.E. & Stevenson, H.M. (1978). A Florida specimen of the Black-throated Sparrow. *Fla. Field Natur.* **6**: 21.

Merkord, C.L., Rodríguez, R. & Faaborg, J. (2006). Second and third records of Western Marsh-Harrier (*Circus aeruginosus*) for the Western Hemisphere in Puerto Rico. *J. Carib. Orn.* **19**: 42–44.

Miller, J.R., Miller, J. & Findholt, S.L. (2018). Distribution, population status, nesting phenology, and habitat of the West Indian Woodpecker (*Melanerpes superciliaris nyeanus*) on San Salvador Island, Bahamas. *J. Carib. Orn.* **31**: 26–33.

Miller, M.P., Haig, S.M., Gratto-Trevor, C.L. & Mullins, T.D. (2010). Subspecies status and population genetic structure in Piping Plover (*Charadrius melodus*). *Auk* **127**: 57–71.

Mitchell, A. (2009). First record of Dark-eyed Junco (*Junco hyemalis*) for Cuba and some other interesting records. *J. Carib. Orn.* **22**: 98–99.

Mlodinow, S.G. (2002). Bullock's Oriole (*Icterus bullockii*) on Grand Bahama: a second record for the West Indies, with notes on other vagrants from western and central North America. *El Pitirre* **15**: 67–70.

Monroe, B.L. (1968). *A Distributional Survey of the Birds of Honduras*. Orn. Monogr. 7. American Ornithologists' Union, Washington, DC.

Moreno, J.A. (1997). Review of the subspecific status and origin of introduced finches in Puerto Rico. *Carib. J. Sci.* **33**: 233–238.

Morrison, R.I.G. (1980). First specimen of the Little Stint (*Calidris minuta*) for North America. *Auk* **97**: 627–628.

Mukhida, F., Cestero, J. & Cestero, B. (2011). First record of the Long-tailed Duck (*Clangula hyemalis*) for the Caribbean in Anguilla. *J. Carib. Orn.* **24**: 32–33.

Murphy, W.L. (2001). Noteworthy observations of pelagic seabirds wintering at sea in the southern Caribbean. *Sea Swallow* **50**: 18–25.

Navarro Pacheco, N. (2018). *Annotated Checklist of the Birds of Cuba*. Second edn. Ed. Nuevos Mundos, La Habana.

Newton, A. & Newton, E. (1859). Observations on the birds of St. Croix, West Indies, made between February 20th and August 6th, 1857, by Alfred Newton, and between March 4th and September 28th, 1858 by Edward Newton. *Ibis* **3(1)**: 138–150.

Norton, R.L. & Clarke, N.V. (1989). Additions to the birds of the Turks and Caicos Islands. *Fla. Field Natur.* **17**: 32–39.

Norton, R.L., Chipley, R.M. & Lazell, J.D. (1989). A contribution to the ornithology of the British Virgin Islands. *Carib. J. Sci.* **25**: 115–118.

Oberle, M.W. (2000). *Puerto Rico's Birds in Photographs*. Second edn. Ed. Humanitas, San Juan, Puerto Rico.

Oberle, M.W. (2008). *Cantos de Aves del Caribe/Caribbean Bird Song*. CDs. Macaulay Library, Cornell Lab of Ornithology, Ithaca, NY.

Oberle, M. & Haney, C. (2002). Gaviota de Sabine y otras gaviotas raras en San Juan. *El Bien-te-veo* **5(2)**: 3–4.

Olson, S.L. (2015). History, morphology, and fossil record of the extinct Puerto Rican Parakeet *Psittacara maugei* Souancé. *Wilson J. Orn.* **127**: 1–12.

Olson, S.L. & Hilgartner, W.B. (1982). Fossil and subfossil birds from the Bahamas. Pp. 22–56 in Olson, S.L. (ed.) *Fossil Vertebrates from the Bahamas*. Smiths. Contrib. Paleobiol. **48**.

Ortiz, R., Rimmer, C.C., Askanas, H. & Mota, I. (2012). Lincoln's Sparrow (*Melospiza lincolnii*): new record for the Dominican Republic. *J. Carib. Orn.* **25**: 89–91.

Oswald, J.A., Harvey, M.G., Remsen, R.C., Foxworth, D.U., Cardiff, S.W., Dittmann, D.L., Megna, L.C., Carling, M.D. & Brumfield, R.T. (2016). Willet be one species or two? A genomic view of the evolutionary history of *Tringa semipalmata*. *Auk* **133**: 593–614.

Overton, L.C. & Rhoads, D.D. (2004). Molecular phylogenetic relationships based on mitochondrial and nuclear gene sequences for the todies (*Todus*, Todidae) of the Caribbean. *Mol. Phylogenet. Evol.* **32**: 524–538.

Pacheco Garzón, A. (2012). Estudio y conservación de las aves de la Isla de San Andrés. *Conserv. Colombiana* **16**: 1–54.

Padrón López, Y., Lentino, M., Rey, C., Ortiz, E., Viera, Y. & Almendrales, A. (2015). Nuevas especies de aves para el Refugio de Fauna Silvestre Isla de Aves. *Rev. Venez. Orn.* **5**: 52–56.

Paice, M.R. & Speirs, R. (2010). The avifauna of Mustique Island (St Vincent and the Grenadines). *J. Carib. Orn.* **23**: 61–84.

Parada Isada, A., Martínez Llanes, O. & Degnan, L.J. (2014). First report of Common Merganser (*Mergus merganser*) for Cuba and the Greater Antilles. *Fla. Field Natur.* **42**: 148–150.

Paterson, A. (1968). New species records for the Bahamas. *Bull. Brit. Orn. Club* **88**: 109–110.

Paulino, M.M., Mejía, D.A. & Latta, S.C. (2013). First record of Great-tailed Grackle (*Quiscalus mexicanus*) breeding in the West Indies. *J. Carib. Orn.* **26**: 63–65.

Paulson, D.R. (1966). New records of birds from the Bahama Islands. *Not. Nat. (Phil.)* **394**: 1–15.

Paulson, D.R., Orians, G.H. & Leck, C.F. (1969). Notes on birds of Isla San Andrés. *Auk* **86**: 755–758.

Paynter, R.A. (1956). Birds of the Swan Islands. *Wilson Bull.* **68**: 103–110.

Pérez-Rivera, R.A. (1987). Additional records and notes on migratory water birds in Puerto Rico, West Indies. *Carib. J. Sci.* **23**: 368–372.

Pérez-Rivera, R.A. (1998). *Cacatua alba*—neuvo informe para Puerto Rico. *El Pitirre* **11**: 37.

Perlut, N.G., Klak, T.C. & Rakhimberdiev, E. (2017). Geolocator data reveal the migration route and wintering location of a Caribbean Martin (*Progne dominicensis*). *Wilson J. Orn.* **129**: 605–610.

Peters, J.L. (1940). *Check-list of Birds of the World.* Vol. 4. Museum of Comparative Zoology, Harvard University Press, Cambridge, MA.

Phelps, W.H. & Phelps, W.H. Jr. (1957). Las aves de Isla de Aves, Venezuela. *Bol. Soc. Venez. Cienc. Nat.* **88**: 63–72.

Pinchon, P.R. (1976). *Faunes des Antilles Françaises. Les Oiseaux.* Second edn. M. Ozanne & Cie, Fort-de-France.

Pinchon, P.R. & Vaurie, C. (1961). The Kestrel (*Falco tinnunculus*) in the New World. *Auk* **78**: 92–93.

Post, P.W. & Wiley, J.W. (1977). The Shiny Cowbird in the West Indies. *Condor* **79**: 119–121.

Prins, T.G., Reuter, J.H., Debrot, A.O., Wattel, J. & Nijman, V. (2009). Checklist of the birds of Aruba, Curaçao and Bonaire, south Caribbean. *Ardea* **97**: 137–268.

Proctor, C.J., Inman, S.E., Zeiger, J.M. & Graves, G.R. (2017). Last search for the Jamaican Golden Swallow (*Tachycineta e. euchrysea*). *J. Carib. Orn.* **30**: 69–74.

Prŷs-Jones, R. & Evans, P.G.H. (1985). A synopsis of the status and ecology of the birds of Dominica. Unpubl. MS.

Raffaele, H.A. (1981). New records of bird species for Puerto Rico and one for the West Indies. *Amer. Birds* **35**: 142–143.

Raffaele, H.A. (1989). *A Guide to the Birds of Puerto Rico and the Virgin Islands.* Revised edn. Princeton University Press, Princeton, NJ & London.

Raffaele, H., Wiley, J.W., Garrido, O.H., Keith, A.R. & Raffaele, J. (1998). *Birds of the West Indies.* Christopher Helm, London.

Restall, R., Rodner, C. & Lentino, M. (2006). *Birds of Northern South America: An Identification Guide.* Christopher Helm, London.

Reynard, G.B. (2008). *Cantos de las Aves de la República Dominicana.* CD. Macaulay Library, Cornell Lab of Ornithology, Ithaca, NY.

Reynard, G.B. & Garrido, O.H. (2005). *Cantos de las Aves de Cuba.* CD. Macaulay Library, Cornell Lab of Ornithology, Ithaca, NY.

Reynard, G.B. & Sutton, R.L. (2000). *Bird Songs in Jamaica.* CDs. Cornell Lab of Ornithology, Ithaca, NY.

Rheindt, F.E., Christidis, L. & Norman, J.A. (2009). Genetic introgression, incomplete lineage sorting and faulty taxonomy create multiple polyphyly in a montane clade of *Elaenia* flycatchers. *Zool. Scripta* **38**: 143–153.

Ridgway, R. (1884). On a collection of birds made by Messrs. J.E. Benedict and W. Nye, of the United States Fish Commission steamer "Albatross". *Proc. US Natl. Mus.* **6**: 172–180.

Ridgway, R. (1888). Catalogue of a collection of birds made by Mr. Chas. H. Townsend, on islands in the Caribbean Sea and in Honduras. *Proc. US Natl. Mus.* **10**: 572–597.

Riley, J.H. (1905). List of birds collected or observed during the Bahama Expedition of the Geographic Society of Baltimore. *Auk* **22**: 349–360.

Rimmer, C.C. & McFarland, K.P. (1998). Two new avian records for Hispaniola: Swainson's Warbler and Song Sparrow. *El Pitirre* **11**: 15–17.

Robb, M., Mullarney, K. & The Sound Approach (2008). *Petrels Night and Day.* The Sound Approach, Poole, Dorset.

Robbins, M.B. & Parker, T.A. (1997). Voice and taxonomy of *Caprimulgus* (*rufus*) *otiosus* (Caprimulgidae), with a reevaluation of *Caprimulgus rufus* subspecies. Pp. 601–607 in Remsen, J.V. (ed.) *Studies in Neotropical Ornithology Honoring Ted Parker.* Orn. Monogr. 48. American Ornithologists' Union, Washington DC.

Robinson, W. (1905). An addition to the avifauna of Cuba. *Auk* **22**: 315.

Roché, J.C., Bénito-Espinal, É. & Hautcastel, P. (2000). *Oiseaux des Antilles.* CDs. Sittelle, Les Sagnes.

Rodríguez, Y., Garrido, O.H., Wiley, J.W. & Kirkconnell, A. (2005). The Common Kingfisher (*Alcedo atthis*): an exceptional first record for the West Indies and the Western Hemisphere. *Orn. Neotropical* **16**: 141.

Rodríguez, Y., Navarro, N. & Fernández Ordoñez, J.C. (2017). First record of Eurasian Blackcap (*Sylvia atricapilla*) for Cuba and the West Indies. In: Navarro, N. & Reyes, E. (eds.) *Annotated Checklist of the Birds of Cuba.* First edn. Ed. Nuevos Mundos, St Augustine, FL.

Rodríguez Castaneda, Y. & Wiley, J.W. (2015). Probable first breeding record of Indigo Bunting (*Passerina cyanea*; family Cardinalidae) in the West Indies. *J. Carib. Orn.* **28**: 22–24.

Rodríguez Castaneda, Y., Wiley, J.W. & Garrido, O.H. (2017). Additional records of Lazuli Bunting (*Passerina amoena*) and first records of several wild-caught exotic birds for Cuba. *J. Carib. Orn.* **30**: 134–142.

Rodríguez Santana, F. (2010). Reports of Cooper's Hawks (*Accipiter cooperi*), Swainson's Hawks (*Buteo swainsoni*), and Short-tailed Hawks (*Buteo brachyurus*) in Cuba. *J. Raptor Res.* **44**: 146–150.

Ruegg, K. (2007). Divergence between subspecies groups of Swainson's Thrush (*Catharus ustulatus ustulatus* and *C. u. swainsoni*). *Orn. Monogr.* **63**: 67–77.

Russell, S.M., Barlow, J.C. & Lamm, D.W. (1979). Status of some birds on Isla San Andres and Isla Providencia, Colombia. *Condor* **81**: 98–100.

Russello, M.A., Stahala, C., Lalonde, D., Schmidt, K.L. & Amato, G. (2010). Cryptic diversity and conservation units in the Bahama parrot. *Conserv. Genet.* **11**: 1809–1821.

Rypdal, E. (2018). Repeatability in isotopic signatures is linked to consistent individual migration strategies and individual specialization in a long-distance migratory seabird. M.Sc. thesis. Norwegian University of Science & Technology, Oslo.

Salaman, P., Bayly, N., Burridge, R., Grantham, M., Gurney, M., Quevedo, A., Urueña, L.E. & Donegan, T. (2008). Sixteen bird species new for Colombia. *Conserv. Colombiana* 5: 80–85.

Salguero-Faría, J.A. & Roig-Bachs, C. (2000). Senegal Parrot, Blue-crowned Parakeet, Olive-throated Parakeet, and Green-winged Macaw: new psittacine records for Puerto Rico. *El Pitirre* **13**: 45–46.

Sandoval, L. & Arendt, W.J. (2011). Two new species for Nicaragua and other notes on the avifauna of the Atlantic Region and Paso del Istmo Biological Corridor. *Cotinga* **33**: 53–60.

Sangster, G., Collinson, J.M., Crochet, P.A., Knox, A.G., Parkin, D.T., Svensson, L. & Votier, S.C. (2011). Taxonomic recommendations for British birds: seventh report. *Ibis* **153**: 883–892.

Sari, E.H.R. & Parker, P.G. (2012). Understanding the colonization history of the Galápagos flycatcher (*Myiarchus magnirostris*). *Mol. Phylogenet. Evol.* **63**: 244–254.

Satgé, Y.G., Rupp, E. & Jodice, P.G.R. 2019. A preliminary report of ongoing research of the ecology of Black-capped Petrel (*Pterodroma hasitata*) in Sierra de Bahoruco, Dominican Republic – I: GPS tracking of breeding adults. Unpubl. report, South Carolina Cooperative Fish & Wildlife Research Unit, Clemson University, Clemson, South Carolina.

Schwartz, A. & Klinikowski, R.F. (1965). Additional observations on West Indian birds. *Notulae Naturae* **376**: 1–16.

Scott, W.E.D. (1892). Observations on the birds of Jamaica, West Indies. *Auk* **9**: 9–15.

Scott, W.E.D. (1893). Observations on the birds of Jamaica, West Indies. *Auk* **10**: 339–342.

Sessa-Hawkins, M. & Hermes, C. (2019). After the storm. *World Birdwatch* **40(4)**: 24–26.

Sharpe, R.B. (1897). [On *Lanius ludovicianus* in the Bahamas]. *Bull. Brit. Orn. Club* **7**: 17.

Shelley, G.E. (1891). Cuculidae. Pp. 209–434 in Sclater, P.L. & Shelley, G.E. (eds.) *Catalogue of the Birds in the British Museum*. Vol. 19. British Museum (Natural History), London.

Shirihai, H. & Svensson, L. (2018). *Handbook of Western Palearctic birds*. Vol. 2. Bloomsbury, London.

Sladen, F.W. (1989). First record of a Bar-tailed Godwit, *Limosa lapponica* (Aves, Scolopacidae), in the West Indies. *Carib. J. Sci.* **25**: 91.

Smith, P.W. (1987). The Eurasian Collared-dove arrives in the Americas. *Amer. Birds* **41**: 1371–1379.

Smith, P.W. (1995). The Eurasian Collared-Dove reaches the Lesser Antilles. *El Pitirre* **8(3)**: 3.

Smith, P.W. & Smith, S.A. (1999a). A sight record of Ringed Kingfisher (*Megaceryle torquata*) for Grenada. *El Pitirre* **12**: 49–50.

Smith, P.W. & Smith, S.A. (1999b). The breeding of Wilson's (*Charadrius wilsonia*) and Collared (*Charadrius collaris*) Plovers in the southern Lesser Antilles. *El Pitirre* **12**: 50–51.

Smith, P.W., Smith, S.A., Ryan, P.G. & Cassidy, R. (1994). First report of Virginia's Warbler from the Bahama Islands, with comments on other records from the West Indies and eastern North America. *El Pitirre* **7(2)**: 2–3.

Smith, R.W. (1971). The Slate-coloured Junco: a new species of bird for Jamaica. *Gosse Bird Club Broadsh.* **17**: 32–34.

SOPI (2016). *Lista Oficial de las Aves de Puerto Rico*. Sociedad Ornitológica Puertorriqueña, Inc., San Juan. https://www.sopipr.org/lista-oficial-de-aves-de-pr

Sykes, P.W., Holzman, S. & Iñigo-Elias, E.E. (2007). Current range of the eastern population of Painted Bunting (*Passerina ciris*). Part II: Winter range. *N. Amer. Birds* **61**: 378–406.

Szabo, J.K., Khwaja, N., Garnett, S.T. & Butchart, S.H. (2012). Global patterns and drivers of avian extinctions at the species and subspecies level. *PLoS ONE* **7(10)**: e47080.

Steadman, D.W., Norton, R.L., Browning, M.R. & Arendt, W.J. (1997). The birds of St. Kitts, Lesser Antilles. *Carib. J. Sci.* **33**: 1–20.

Suárez, W., Kirkconnell, A. & Norman, N. (2005). The Bald Eagle *Haliaeetus leucocephalus* in Cuba. *Cotinga* **23**: 78, 80.

Thorstrom, R. & McQueen, D. (2008). Breeding and status of the Grenada Hook-billed Kite (*Chondrohierax uncinatus mirus*). *Orn. Neotropical* **19**: 221–228.

Toussaint, A., John, L. & Morton, M. (2009). *The Status and Conservation of Saint Lucia's Forest Birds*. National Forest Demarcation and Bio-Physical Resource Inventory Project, Tech. Rep. 12. Finnish Consulting Group International, Helsinki.

Tye, A. & Tye, H. (1991). Bird species on St. Andrew and Old Providence Islands, West Caribbean. *Wilson Bull.* **103**: 493–497.

Uva, V., Päckert, M., Cibois, A., Fumagalli, L. & Roulin, A. (2018). Comprehensive molecular phylogeny of barn owls and relatives (Family: Tytonidae), and their six major Pleistocene radiations. *Mol. Phylogenet. Evol.* **125**: 127–137.

Voous, K.H. (1957). A specimen of the Spotted Crake, *Porzana porzana*, from the Lesser Antilles. *Ardea* **45**: 89–90.

Voous, K.H. (1973). List of recent Holarctic bird species. Non-Passerines. *Ibis* **115**: 612–638.

Voous, K.H. (1983). *Birds of the Netherlands Antilles*. De Walburg Pers, Utrecht & Curaçao.

Wallace, G.E., Wallace, E.A.H., Froehlich, D.R., Walker, B.E., Kirkconnell, A., Socarrás, E., Carlisle, H.A. & Machell, E. (1999). Hermit Thrush and Black-throated Gray Warbler, new for Cuba, and other significant bird records from Cayo Coco and vicinity, Ciego de Ávila province, Cuba, 1995–1997. *Fla. Field Natur.* **27**: 37–51.

Ward-Bolívar, V. & Lasso-Zapata, J. (2012). Primeros registros del Pato Serrucho Pechicastaño (*Mergus serrator*) para las islas de Providencia y San Andrés, Caribe Colombiano. *Orn. Colombiana* **12**: 47–50.

Watts, B.D. (2011). Yellow-crowned Night-Heron (*Nyctanassa violacea*). No. 161. In: Preston, C.R. & Beane, R.D. (eds.) *The Birds of North America*. American Ornithologists' Union, Philadelphia. doi.10.2173/bna/161.

Wege, D.C. & Anadón-Irizarry, V. (eds.) (2008). *Important Bird Areas in the Caribbean: Key Sites for Conservation*. BirdLife International, Cambridge, UK.

Wetmore, A. (1927). Birds of Porto Rico and the Virgin Islands. *NY Acad. Sci. Scientific Survey of Porto Rico and the Virgin Islands* **9**: 245–406.

Wetmore, A. (1938). Bird remains from the West Indies. *Auk* **55**: 51–55.

Wetmore, A. (1958). Extralimital records for the Eastern Kingbird, Tree Swallow, and Blackpoll Warbler. *Auk* **75**: 467–468.

Wetmore, A. & Swales, B.H. (1931). The birds of Haiti and the Dominican Republic. *US Natl. Mus. Bull.* **155**. Smithsonian Institution, Washington, DC.

White, A.W. (2004). Seabirds in the Bahamian Archipelago and adjacent waters: transient, wintering, and rare nesting species. *N. Amer. Birds* **57**: 436–451.

White, A. & Jeremiah, A. (2011). First record of Large-billed Tern (*Phaetusa simplex*) for Grenada and the Lesser Antilles. *J. Carib. Orn.* **24**: 74–76.

White, A., Cummins, R.H. & Boardman, M.R. (2014). A White Tern (*Gygis alba*) in the Bahamas. *N. Amer. Birds* **67**: 384–385.

Wiley, J.W. (2000). A bibliography of ornithology in the West Indies. *Proc. West. Found. Vert. Zool.* **7**: 1–817.

Wiley, J.W. & Kirwan, G.M. (2013). The extinct macaws of the West Indies, with special reference to Cuban Macaw *Ara tricolor*. *Bull. Brit. Orn. Club* **133**: 125–156.

Wiley, J.W., Román, R.A., Rams Beceña, A., Peña Rodríguez, C., Kirkconnell, A., Ortega Piferrer, A. & Acosta Cruz, M. (2008). The bird collections of Cuba. *Bull. Brit. Orn. Club* **128**: 17–27.

Woodall, P.F. (2001). Family Alcedinidae (kingfishers). Pp. 130–249 in del Hoyo, J., Elliott, A. & Sargatal, J. (eds.) *Handbook of the Birds of the World*. Vol. 6. Lynx Edicions, Barcelona.

Yntema, L.D., McNair, D.B., Cramer-Burke, C., Sladen, F.W., Valiulis, J.M., Lombard, C.D., Lance, A.O. & Fromer, S.L. (2017). Records and observations of breeding waterbirds, rare and uncommon birds, and marked individuals on St. Croix, U.S. Virgin Islands. *J. Carib. Orn.* **30**: 88–127.

Yojanan Lobo-y-Henriques, J.C. (2014). Cave Swallow *Petrochelidon fulva* and Couch's Kingbird *Tyrannus couchii*: a discussion of two difficult cases of potential records for Colombia based on museum specimens. *Conserv. Colombiana* **21**: 58–62.

ENGLISH AND SCIENTIFIC INDEX

A
acadicus, Aegolius 154
Acanthis 258
Accipiter 160
ACCIPITRIDAE 158
Acridotheres 376
Actitis 128, 375
acuflavidus, Thalasseus (sandvicensis) 148
acuta, Anas 44
acuticaudatus, Psittacara 376
acutipennis, Chordeiles 58
adelaidae, Setophaga 306
aedon, Troglodytes 230
Aegolius 154
aegyptiaca, Alopochen 36, 375
Aeronautes 375
aeruginosus, Circus 160
aestiva, Setophaga (petechia) 298
aethereus, Phaethon 48
afer, Euplectes 248
affinis, Aythya 40
Agelaius 276
agilis, Amazona 184
agilis, Oporornis 290
Aix 36
ajaja, Platalea 94
ALAUDIDAE 220
alba, Ardea 100
alba, Cacatua 182
alba, Calidris 122
alba, Gygis 132
alba, Motacilla 256
alba, Tyto 152
Albatross
    Atlantic Yellow-nosed 86
    Black-browed 88
albeola, Bucephala 34
albicaudatus, Geranoaetus 164
albicollis, Saltator 322
albicollis, Zonotrichia 264
albifrons, Amazona 184
albifrons, Anser 32
albiventer, Tachycineta 224
albus, Eudocimus 94
ALCEDINIDAE 172
Alcedo 375
ALCIDAE 152
alcyon, Megaceryle 172
Alle 152
alle, Alle 152
Allenia 234
alnorum, Empidonax 204
Alopochen 36, 375
alpestris, Eremophila 220
alpina, Calidris 122
altiloquus, Vireo 212
Amandava 250

amandava, Amandava 250
Amaurolimnas 375
Amazon
    Black-billed 184
    Cuban 184
    Green-cheeked 186
    Hispaniolan 184
    Imperial 186
    Orange-winged 186
    Puerto Rican 186
    Red-crowned 186
    Red-necked 186
    St Lucia 186
    St Vincent 188
    White-fronted 184
    Yellow-billed 184
    Yellow-crowned 186, 376
    Yellow-headed 186, 376
    Yellow-naped 184, 376
Amazona 184, 376
amazonica, Amazona 186
americana, Aythya 38
americana, Chloroceryle 172
americana, Fulica 84
americana, Mareca 42
americana, Mycteria 92
americana, Recurvirostra 110
americana, Setophaga 296
americana, Siphonorhis 58
americana, Spiza 318
americanus, Coccyzus 74
americanus, Numenius 118
Ammodramus 262
amoena, Passerina 318
Amphispiza 376
anaethetus, Onychoprion 140
Anas 42
ANATIDAE 28
angelae, Setophaga 294
angolensis, Sporophila 332
angustirostris, Todus 170
Anhinga 108
Anhinga 108
anhinga, Anhinga 108
ANHINGIDAE 108
ani, Crotophaga 74
Ani
    Greater 74
    Smooth-billed 74
Anous 132
anoxanthus, Loxipasser 328
Anser 32
anthracinus, Buteogallus 164
Anthracothorax 66
Anthus 254
Antigone 84
antillarum, Myiarchus 200

antillarum, Sternula 142
Antrostomus 60
apiaster, Merops 168
APODIDAE 62
approximans, Vireo (pallens) 214
Apus 66
apus, Apus 66
Ara 190, 376
ARAMIDAE 84
Aramus 84
ararauna, Ara 190
Aratinga 376
arausiaca, Amazona 186
arborea, Dendrocygna 30
Archilochus 74
Ardea 98
ARDEIDAE 96
Ardenna 90
Ardeola 98
ardosiaceus, Turdus 244
Arenaria 120
arminjoniana, Pterodroma 88
arquata, Numenius 118
asiatica, Zenaida 54
Asio 154
assimilis, Agelaius 276
astrild, Estrilda 250
ater, Molothrus 278
Athene 154
atratus, Coragyps 158
atricapilla, Lonchura 252
atricapilla, Sylvia 376
atricilla, Larus 136
atroviolacea, Ptiloxena 278
atthis, Alcedo 375
audoboni, Setophaga 304
Auk, Little 152
aura, Cathartes 158
aurantius, Turdus 244
auratus, Colaptes 174
auriculata, Zenaida 54
aurita, Zenaida 54
auritus, Nannopterum 108
aurocapilla, Seiurus 284
auropalliata, Amazona 184, 376
aurulentus, Anthracothorax 68
autumnalis, Dendrocygna 28
Avadavat, Red 250
Avocet, American 110
Aythya 38

B
bachmanii, Vermivora 286
bahamensis, Anas 44
bahamensis, Coccyzus 78
bahamensis, Coereba (flaveola) 324
bairdii, Calidris 122

bairdii, Campephilus (principalis) 174
Bananaquit 324
Bananaquit
    Bahama 324
    Common 324
    Greater Antillean 324
barbadensis, Loxigilla 328
barbirostris, Myiarchus 198
Barn-owl
    American 152
    Ashy-faced 152
    Common 152
    Lesser Antilles 152
baroli, Puffinus (lherminieri) 375
bartholemica, Coereba (flaveola) 324
Bartramia 116
bassanus, Morus 106
Becard, Jamaican 192
Bee-eater, European 168
bellii, Vireo 214
bengalus, Uraeginthus 376
bernicla, Branta 32
biarmicus, Falco 375
bicknelli, Catharus 240
bicolor, Cyanophaia 70
bicolor, Dendrocygna 30
bicolor, Ducula 56
bicolor, Melanospiza 330
bicolor, Tachycineta 224
bilineata, Amphispiza 376
Bishop
    Northern Red 248
    Orange 248
    Yellow-crowned 248
bishopi, Catharopeza 292
bistriatus, Burhinus 108
Bittern
    American 96
    Common Little 96
    Least 96
    Little 96
Blackbird
    Brewer's 280
    Cuban 278
    Jamaican 276
    Red-shouldered 276
    Red-winged 276
    Rusty 280
    Tawny-shouldered 276
    Yellow-headed 268
    Yellow-hooded 284
    Yellow-shouldered 276
Blackcap, Eurasian 376
blancoi, Contopus (latirostris) 208
Bluebird, Eastern 236
Bobolink 268
Bobwhite
    Black-breasted 26
    Crested 26
    Eastern 26
    Northern 26
Bombycilla 246
BOMBYCILLIDAE 246
bonana, Icterus 274

bonariensis, Molothrus 278
Booby
    Brown 106
    Masked 106
    Red-footed 106
borealis, Calonectris 90
borealis, Numenius 116
Botaurus 96
brachyura, Chaetura 64
brachyurus, Buteo 166
brachyurus, Ramphocinclus 232
Brant 32
Brant, Black 32
Branta 32
brasilianus, Nannopterum 108
Brent 32
Brent, Light-bellied 32
brewsteri, Siphonorhis 60
Brotogeris 182
brunneicapillus, Contopus (latirostris) 208
Bubulcus 98
Bucephala 34, 375
Budgerigar 376
Bufflehead 34
Bullfinch
    Barbados 328
    Cuban 326
    Grand Cayman 326
    Greater Antillean 326
    Lesser Antillean 328
    Puerto Rican 326
bullockiorum, Icterus 272
Bulweria 92
bulwerii, Bulweria 92
Bunting
    Eastern Painted 320
    Indigo 318
    Lazuli 318
    Painted 320
    Snow 260
BURHINIDAE 108
Burhinus 108
Buteo 164
Buteogallus 164
Butorides 98

C

Cacatua 182
CACATUIDAE 182
caerulea, Egretta 102
caerulea, Passerina 318
caerulea, Polioptila 228
caerulescens, Anser 32
caerulescens, Setophaga 302
cahow, Pterodroma 375
Cairina 36
CALCARIIDAE 260
Calcarius 260
calendula, Regulus 246
Calidris 120
Calonectris 90
CALYPTOPHILIDAE 316
Calyptophilus 316
Campephilus 174

campestris, Euneornis 326
canadensis, Antigone 84
canadensis, Branta 32
canadensis, Cardellina 310
Canary, Yellow-fronted 376
caniceps, Geotrygon 52
canicularis, Eupsittula 188
canora, Phonipara 328
canorus, Cuculus 78
canutus, Calidris 120
Canvasback 38
capensis, Zonotrichia 264
CAPRIMULGIDAE 58
Caracara 180
Caracara
    Crested 180
    Northern Crested 180
Cardellina 310
Cardinal
    Northern 322
    Red-crested 376
CARDINALIDAE 316
Cardinalis 322
cardinalis, Cardinalis 322
Carib
    Green-throated 68
    Purple-throated 68
caribaea, Patagioenas 50
caribaeus, Contopus 206
caribaeus, Vireo 216
caribbaea, Pterodroma 88
carolina, Porzana 82
carolinensis, Anas (crecca) 46
carolinensis, Antrostomus 60
carolinensis, Dumetella 232
carolinus, Euphagus 280
carolinus, Melanerpes 178
caspia, Hydroprogne 142
castanea, Setophaga 298
castro, Hydrobates 86
Catbird, Grey 232
Catharacta 150
Catharopeza 292
Cathartes 158
CATHARTIDAE 158
Catharus 238, 376
caudifasciatus, Tyrannus 196
cayennensis, Hydropsalis 58
cayennensis, Vanellus (chilensis) 114
caymanensis, Melanerpes (superciliaris) 178
cedrorum, Bombycilla 246
celata, Leiothlypis 288
cerulea, Setophaga 296
cerverai, Cyanolimnas 82
cerverai, Ferminia 228
Chachalaca
    Rufous-tipped 26
    Rufous-vented 26
Chaetura 64
CHARADRIIDAE 110
Charadrius 112
Chat, Yellow-breasted 268
Chat-tanager

Eastern 316
Western 316
cheriway, Caracara 180
cherriei, Elaenia 192
chihi, Plegadis 94
chilensis, Vanellus 114
Chlidonias 144
Chloroceryle 172
chlopterus, Ara 190, 376
chloropterus, Psittacara 190
chlororhynchos, Thalassarche 86
Chlorostilbon 70
chlorurus, Pipilo 262
Chondestes 262
Chondrohierax 158
Chordeiles 58
chrysia, Geotrygon 52
chrysocaulosus, Colaptes (auratus) 174
Chrysolampis 66
Chrysomus 284
chrysoparia, Setophaga 308
chrysoptera, Vermivora 286
Chuck-will's-widow 60
Ciconia 92
ciconia, Ciconia 92
CICONIIDAE 92
Cinclocerthia 234
cinerea, Ardea 98
cinereiventris, Chaetura 64
cinerescens, Elaenia (martinica) 192
cinereus, Xenus 128
Circus 160
ciris, Passerina 320
cirrocephalus, Larus 136
Cistothorus 230
citrea, Protonotaria 288
citrina, Setophaga 294
Clangula 34
clangula, Bucephala 34, 375
clypeata, Spatula 40
Coccyzus 74
Cockatiel 375
Cockatoo
  Umbrella 182
  White 182
  White-crested 182
cocoi, Ardea 100
Coereba 324
Colaptes 174
colchicus, Phasianus 28
Colinus 26
Collared-dove
  African 48
  Eurasian 48
collaria, Amazona 184
collaris, Aythya 38
collaris, Charadrius 114
colubris, Archilochus 74
Columba 48
columbarius, Falco 180
columbianus, Cygnus 30
COLUMBIDAE 48
Columbina 56
concolor, Amaurolimnas 375

Contopus 206
Conure
  Brown-throated 188
  Cuban 190
  Hispaniolan 190
  Orange-fronted 188
  Red-masked 190
cooperii, Accipiter 162
cooperi, Myiarchus (tyrannulus) 202
Coot, American 84
Coragyps 158
Cordon-bleu, Red-cheeked 376
corensis, Patagioenas 375
Cormorant
  Double-crested 108
  Neotropical 108
Corncrake 80
coronata, Paroaria 376
coronata, Setophaga 304
CORVIDAE 218
Corvus 218
cotta, Myiopagis 194
Coturnicops 80
couchii, Tyrannus 194, 376
Cowbird
  Brown-headed 278
  Giant 278
  Shiny 278
CRACIDAE 26
Crake
  Spotted 82
  Uniform 375
  Yellow-breasted 82
Crane, Sandhill 84
crassirostris, Vireo 216
crecca, Anas 44
crepitans, Rallus 80
Crex 80
crex, Crex 80
crinitus, Myiarchus 202
cristatus, Colinus 26
cristatus, Orthorhyncus 70
cristatus, Pavo 28
Crithagra 376
Crossbill, Hispaniolan 258
Crotophaga 74
Crow
  Cuban 218
  Cuban Palm 220
  Fish 218
  Hispaniolan Palm 220
  House 220
  Jamaican 218
  Palm 220
  White-necked 218
cryptoleuca, Progne 226
cubanensis, Antrostomus 62
cubensis, Tyrannus 198
Cuckoo
  Bay-breasted 76
  Black-billed 76
  Chestnut-bellied 76
  Common 78
  Dark-billed 76

  Mangrove 76
  Pearly-breasted 74
  Yellow-billed 74
CUCULIDAE 74
cucullata, Spermestes 250
cucullata, Tangara 332
cucullatus, Icterus 272
cucullatus, Lophodytes 34
cucullatus, Ploceus 248
cucullatus, Spinus 260
Cuculus 78
cunicularia, Athene 154
Curlew
  Eskimo 116
  Eurasian 118
  Long-billed 118
cyanea, Passerina 318
cyaneoviridis, Tachycineta 224
Cyanerpes 322
cyaneus, Cyanerpes 322
cyanocephala, Starnoenas 50
cyanocephalus, Euphagus 280
Cyanolimnas 82
Cyanophaia 70
cyanoptera, Spatula 40
cyanoptera, Vermivora 286
Cygnus 30, 375
Cypseloides 62

D

dactylatra, Sula 106
decaocto, Streptopelia 48
deglandi, Melanitta 34
delawarensis, Larus 138
delicata, Gallinago 126
delicata, Setophaga 308
Delichon 220
Dendrocygna 28
Dickcissel 318
diomedea, Calonectris 90
DIOMEDEIDAE 86
discolor, Setophaga 306
discors, Spatula 42
Diver, Great Northern 84
Dolichonyx 268
domesticus, Passer 254
dominica, Pluvialis 110
dominica, Setophaga 306
dominicanus, Larus 138
dominicensis, Icterus 274
dominicensis, Progne 226
dominicensis, Spindalis 314
dominicensis, Spinus 260
dominicensis, Tyrannus 196
dominicus, Anthracothorax 68
dominicus, Dulus 246
dominicus, Nomonyx 30
dominicus, Tachybaptus 46
dougallii, Sterna 146
Dove
  Barbary 48
  Caribbean 54
  Eared 54
  Grenada 54

387

Jamaican 54
Mourning 56
Rock 48
White-bellied 54
White-headed 50
White-winged 54
Zenaida 54
Dovekie 152
Dowitcher
  Long-billed 126
  Short-billed 126
dubius, Charadrius 112
Duck
  American Black 44
  Carolina 36
  Long-tailed 34
  Masked 30
  Mottled 44
  Muscovy 36
  Ring-necked 38
  Ruddy 30
  Tufted 38
  Wood 36
Ducula 56
DULIDAE 246
Dulus 246
Dumetella 232
Dunlin 122

E
Eagle, Bald 162
edwardsii, Calonectris 90
Egret
  American Great 100
  Cattle 98
  Great 100
  Great White 100
  Little 104
  Reddish 102
  Snowy 102
  Western Cattle 98
Egretta 102
egretta, Ardea (alba) 100
ekmani, Antrostomus 62
Elaenia 192
Elaenia
  Caribbean 192
  Chinchorro 192
  Greater Antillean 194
  Hispaniolan 192
  Jamaican 194
  Large Jamaican 194
  Small Jamaican 194
  Yellow-bellied 192
Elanoides 160
elegans, Rallus 80
elisabeth, Myadestes 238
Emerald
  Cuban 70
  Hispaniolan 70
  Puerto Rican 70
Empidonax 204
Eremophila 220
erithacus, Psittacus 375

erythrogaster, Hirundo (rustica) 222
erythrogenys, Psittacara 190
erythropthalmus, Coccyzus 76
erythropus, Tringa 130
erythrorhynchos, Pelecanus 104
Estrilda 248
ESTRILDIDAE 248
euchrysea, Tachycineta 224
Eudocimus 94
Eulampis 68
euleri, Coccyzus 74
euleri, Lathrotriccus 202
Euneornis 326
Euodice 250
euops, Psittacara 190
Euphagus 280
Euphonia 256
Euphonia
  Hispaniolan 256
  Jamaican 256
  Lesser Antillean 256
  Puerto Rican 256
Euplectes 248
Eupsittula 188
evelynae, Nesophlox 72
exilis, Ixobrychus 96

F
falcinellus, Plegadis 94
Falco 180, 375
Falcon
  Aplomado 182
  Bat 182
  Lanner 375
  Peregrine 182
  Prairie 375
FALCONIDAE 180
fallax, Elaenia 194
faxoni, Catharus (guttatus) 240
fedoa, Limosa 118
femoralis, Falco 182
ferina, Aythya 38
Ferminia 228
fernandinae, Colaptes 174
fernandinae, Teretistris 268
ferruginea, Calidris 122
Finch
  House 258
  Saffron 332
  St Lucia Black 330
Flamingo
  American 46
  Caribbean 46
flammea, Acanthis 258
flammeus, Asio 156
flaveola, Coereba 324
flaveola, Sicalis 332
flavescens, Setophaga 306
flavifrons, Euphonia 256
flavifrons, Vireo 212
flavipes, Tringa 130
flaviventer, Hapalocrex 82
flaviventris, Empidonax 204
flaviventris, Lathrotriccus (euleri) 202

flavogaster, Elaenia 192
flavoviridis, Vireo 210
Flicker
  Cuban 174
  Fernandina's 174
  Northern 174
  Yellow-shafted 174
Florisuga 66
Flycatcher
  Acadian 204
  Alder 204
  Brown-crested 202
  Common Vermilion 202
  Euler's 202
  Fork-tailed 198
  Great Crested 202
  Grenada 202
  La Sagra's 200
  Lawrence's 202
  Least 206
  Lesser Antillean 200
  Puerto Rican 200
  Rufous-tailed 200
  Sad 198
  Scissor-tailed 198
  Stolid 200
  Vermilion 202
  Willow 204
  Yellow-bellied 204
forficatus, Elanoides 160
forficatus, Tyrannus 198
formosa, Geothlypis 290
fornsi, Teretistris 268
Forpus 188
forsteri, Sterna 146
fortirostris, Quiscalus (lugubris) 280
franciscanus, Euplectes 248
Fregata 104
FREGATIDAE 104
Frigatebird, Magnificent 104
FRINGILLIDAE 256
frugivorus, Calyptophilus 316
fulgens, Dendrocygna (autumnalis) 28
Fulica 84
fulicarius, Phalaropus 128
fuligula, Aythya 38
Fulmar, Northern 88
Fulmarus 88
fulva, Petrochelidon 222
fulva, Pluvialis 110
fulvigula, Anas 44
fumigatus, Turdus 242
furcata, Tyto (alba) 152
fusca, Allenia 234
fusca, Setophaga 298
fuscatus, Margarops 234
fuscatus, Onychoprion 140
fuscescens, Catharus 240
fuscicollis, Calidris 124
fuscus, Larus 138

G
gabbii, Tyrannus (caudifasciatus) 196
Gadwall 42

galbula, Icterus 270
galeata, Gallinula 84
galeatus, Numida (meleagris) 26
Gallinago 126
Gallinula 84
Gallinule
 Common 84
 Purple 82
Gallus 28
gallus, Gallus 28
Gannet, Northern 106
Garganey 40
garzetta, Egretta 104
Gavia 84
GAVIIDAE 84
Gelochelidon 142
genei, Larus 134
genibarbis, Myadestes 238
georgiana, Melospiza 266
Geothlypis 290
Geotrygon 50
Geranoaetus 164
gilvus, Mimus 236
gilvus, Vireo 210
glacialis, Fulmarus 88
Glareola 132
glareola, Tringa 130
GLAREOLIDAE 132
Glaucidium 154
Glaucis 66
glaucoides, Larus 140
glaucops, Tyto 152
Gnatcatcher
 Blue-grey 228
 Cuban 228
Godwit
 Bar-tailed 118
 Black-tailed 118
 Hudsonian 118
 Marbled 118
 Western Black-tailed 118
Goldeneye, Common 34, 375
Goldfinch
 American 258
 Black-backed 376
Goosander 36
Goose
 Black-bellied Brent 32
 Brent 32
 Canada 32
 Egyptian 36, 375
 Greater White-fronted 32
 Orinoco 36
 Pale-bellied Brent 32
 Ross's 32
 Snow 32
Grackle
 Barbados 280
 Boat-tailed 282
 Carib 280
 Common 280
 Great-tailed 282
 Greater Antillean 282
 Purple 280

Gracula 376
graduacauda, Icterus 376
graellsii, Larus (fuscus) 138
gramineus, Pooecetes 264
grammacus, Chondestes 262
grammicus, Pseudoscops 156
Granatina 376
Grassquit
 Black-faced 330
 Blue-black 330
 Cuban 328
 Yellow-faced 324
 Yellow-shouldered 328
gravis, Ardenna 90
Grebe
 Black-necked 46
 Eared 46
 Least 46
 Pied-billed 46
 Red-necked 46
Greenshank, Common 130
Grenadier, Purple 376
grisea, Ardenna 90
grisegena, Podiceps 46
griseus, Limnodromus 126
griseus, Vireo 216
Grosbeak
 Black-headed 318
 Blue 318
 Rose-breasted 316
Ground-dove, Common 56
Ground-tanager, Green-tailed 312
GRUIDAE 84
guarauna, Aramus 84
guildingii, Amazona 188
Guineafowl
 Helmeted 26
 West African 26
gularis, Egretta 104
gularis, Icterus 376
Gull
 American Herring 138
 Arctic Herring 138
 Black-headed 136
 Bonaparte's 134
 Common Black-headed 136
 Franklin's 136
 Glaucous 140
 Great Black-backed 140
 Grey-headed 136
 Grey-hooded 136
 Iceland 140
 Kelp 138
 Laughing 136
 Lesser Black-backed 138
 Little 134
 Ring-billed 138
 Sabine's 134
 Slender-billed 134
 Yellow-legged 375
gundlachi, Accipiter 162
gundlachii, Buteogallus 164
gundlachii, Chordeiles 58
gundlachii, Mimus 236

gundlachii, Vireo 216
guttatus, Catharus 240
gutturalis, Cinclocerthia 234
Gygis 132

H
haemastica, Limosa 118
HAEMATOPODIDAE 108
Haematopus 108
Haemorhous 258
Haliaeetus 162
haliaetus, Pandion 158
Hapalocrex 82
Harrier, Northern 160
hasitata, Pterodroma 88
Hawk
 Broad-winged 166
 Common Black 164
 Cooper's 162
 Cuban Black 164
 Gundlach's 162
 Red-shouldered 166
 Red-tailed 168
 Ridgway's 164
 Sharp-shinned 160
 Short-tailed 166
 Swainson's 166
 White-tailed 164
helenae, Mellisuga 72
Helmitheros 284
henslowii, Passerculus 266
herminieri, Melanerpes 176
Hermit, Rufous-breasted 66
herodias, Ardea 100
Heron
 Cocoi 100
 Great Blue 100
 Great White 100
 Green 98
 Green-backed 98
 Grey 98
 Little Blue 102
 Louisiana 102
 Purple 100
 Squacco 98
 Striated 98
 Tricolored 102
 Whistling 100
 White-necked 100
hiaticula, Charadrius 112
Himantopus 110
himantopus, Calidris 120
himantopus, Himantopus 110
hippocrepis, Sturnella (magna) 270
hirsutus, Glaucis 66
HIRUNDINIDAE 220
Hirundo 222
hirundo, Sterna 146
hispaniolensis, Contopus 208
hollandicus, Nymphicus 375
holosericeus, Eulampis 68
Honeycreeper, Red-legged 322
hrota, Branta (bernicla) 32
hudsonicus, Numenius (phaeopus) 116

hudsonius, Circus 160
humeralis, Agelaius 276
Hummingbird
   Antillean Crested 70
   Bahama 72
   Bee 72
   Blue-headed 70
   Lyre-tailed 72
   Ruby-throated 74
   Ruby-topaz 66
   Rufous 74
   Vervain 72
hybrida, Chlidonias 144
Hydrobates 86
HYDROBATIDAE 86
Hydrocoloeus 134
Hydroprogne 142
Hydropsalis 58
hyemalis, Clangula 34
hyemalis, Junco 264
Hylocichla 238
hyperboreus, Larus 140
hypochrysea, Setophaga (palmarum) 302
hypoleucos, Actitis 375

I
ianthinogaster, Granatina 376
ibis, Bubulcus 98
Ibis
   American White 94
   Glossy 94
   Scarlet 94
   White 94
   White-faced 94
Icteria 268
ICTERIDAE 268
icterocephalus, Chrysomus 284
Icterus 270, 376
icterus, Icterus 272
Ictinia 162
immer, Gavia 84
Imperial-pigeon, Pied 56
imperialis, Amazona 186
inexpectata, Torreornis 262
inornata, Patagioenas 50
insularis, Sitta 228
insularis, Tyto (glaucops) 152
interpres, Arenaria 120
Ixobrychus 96

J
Jabiru 92
Jabiru 92
Jacana 116
jacana, Jacana 116
Jacana
   Chestnut-backed 116
   Northern 116
   Wattled 116
JACANIDAE 116
jacarina, Volatinia 330
Jacobin, White-necked 66
Jaeger
   Arctic 150

   Long-tailed 148
   Parasitic 150
   Pomarine 150
jamaica, Euphonia 256
jamaicensis, Buteo 168
jamaicensis, Corvus 218
jamaicensis, Laterallus 80
jamaicensis, Leptotila 54
jamaicensis, Nyctibius 56
jamaicensis, Oxyura 30
jamaicensis, Turdus 242
jubata, Neochen 36
jugularis, Eulampis 68
Junco 264
Junco
   Dark-eyed 264
   Slate-colored 264
Junglefowl, Red 28

K
Kestrel
   American 180
   Common 180
   Eurasian 180
Killdeer 112
Kingbird
   Cassin's 194
   Couch's 194, 376
   Eastern 196
   Giant 198
   Grey 196
   Hispaniolan 196
   Loggerhead 196
   Puerto Rican 196
   Tropical 194
   Western 194
Kingfisher
   Belted 172
   Common 375
   Green 172
   Ringed 172
Kinglet, Ruby-crowned 246
kirtlandii, Setophaga 294
Kiskadee, Great 194
Kite
   Black 162
   Cuban 160
   Cuban Hook-billed 160
   Everglade 164
   Grenada Hook-billed 158
   Hook-billed 158
   Mississippi 162
   Snail 164
   Swallow-tailed 160
Kittiwake
   Atlantic 134
   Black-legged 134
Knot, Red 120
krameri, Psittacula 190

L
LANIIDAE 218
Lanius 218
lapponica, Limosa 118

lapponicus, Calcarius 260
Lapwing
   Cayenne 114
   Northern 114
   Southern 114
LARIDAE 132
Lark
   American Horned 220
   Horned 220
Larus 134, 375
Laterallus 80
Lathrotriccus 202
latimeri, Vireo 214
latirostris, Contopus 208
laudabilis, Icterus 274
lawrencii, Margarobyas 156
Leiothlypis 288
lembeyei, Polioptila 228
lentiginosus, Botaurus 96
Leptotila 54
lepturus, Phaethon 48
leucocephala, Amazona 184
leucocephala, Patagioenas 50
leucocephalus, Haliaeetus 162
leucogaster, Sula 106
leucognaphalus, Corvus 218
leucometopia, Geotrygon 52
Leuconotopicus 178
Leucopeza 290
leucophrys, Zonotrichia 264
leucopterus, Chlidonias 144
leucopteryx, Icterus 270
leucorhous, Hydrobates 86
leucorodia, Platalea 94
lherminieri, Puffinus 92
lherminieri, Turdus 242
limicola, Rallus 80
Limnodromus 126
Limnothlypis 288
Limosa 118
limosa, Limosa 118
Limpkin 84
lincolnii, Melospiza 266
lineatus, Buteo 166
lineola, Sporophila 330
livia, Columba 48
Lizard-cuckoo
   Bahama 78
   Cuban 76
   Gonave 78
   Hispaniolan 78
   Jamaican 78
   Puerto Rican 78
lobatus, Phalaropus 128
Lonchura 252
longicauda, Bartramia 116
longicaudus, Stercorarius 148
longirostris, Coccyzus 78
Longspur, Lapland 260
Loon, Common 84
Lophodytes 34
Loxia 258
Loxigilla 328
Loxipasser 328

ludoviciana, Piranga 320
ludovicianus, Lanius 218
ludovicianus, Pheucticus 316
lugubris, Quiscalus 280
luteola, Sicalis 332
Lymnocryptes 126
Lyretail, Inaguan 72
lyrura, Nesophlox 72

M
Macaw
　Blue-and-gold 190
　Blue-and-yellow 190
　Red-and-green 190, 376
maccormicki, Catharacta 152
macroura, Vidua 254
macroura, Zenaida 56
macularius, Actitis 128
maculatus, Pardirallus 82
magister, Myiarchus (tyrannulus) 202
magister, Vireo 210
magna, Sturnella 270
magnificens, Fregata 104
magnolia, Setophaga 296
maja, Lonchura 252
major, Crotophaga 74
major, Quiscalus 282
malabarica, Euodice 250
malacca, Lonchura 252
Mallard 42
mango, Anthracothorax 68
Mango
　Green 68
　Green-breasted 66
　Hispaniolan 68
　Jamaican 68
　Puerto Rican 68
Mannikin
　Bronze 250
　Chestnut 252
　Nutmeg 252
Mareca 42
Margarobyas 156
Margarops 234
marila, Aythya 40
marinus, Larus 140
Marsh-harrier
　Eurasian 160
　Western 160
Martin
　Caribbean 226
　Collared Sand 222
　Common House 220
　Cuban 226
　Northern House 220
　Purple 226
　Sand 222
martinica, Chaetura 64
martinica, Elaenia 192
martinicensis, Troglodytes (aedon) 230
martinicus, Porphyrio 82
maugaeus, Chlorostilbon 70
mauri, Calidris 124
maximus, Thalasseus 148

maynardi, Leuconotopicus (villosus) 178
Meadowlark
　Cuban 270
　Eastern 270
Megaceryle 172
megaplaga, Loxia 258
Megascops 156
melacoryphus, Coccyzus 76
melancholicus, Tyrannus 194
Melanerpes 176
Melanitta 34
melanocephalus, Pheucticus 318
melanoleuca, Tringa 130
melanophris, Thalassarche 88
melanopsis, Icterus 274
Melanospiza 330
melanotos, Calidris 124
melba, Tachymarptis 66
meleagris, Numida 26
Mellisuga 72
mellivora, Florisuga 66
melodia, Melospiza 266
melodus, Charadrius 114
Melopsittacus 376
Melopyrrha 326
Melospiza 266
melpoda, Estrilda 248
Merganser
　Common 36
　Hooded 34
　Red-breasted 36
merganser, Mergus 36
Mergus 36
Merlin 180
merlini, Coccyzus 76
MEROPIDAE 168
Merops 168
mesomelas, Icterus 376
mexicanus, Falco 375
mexicanus, Haemorhous 258
mexicanus, Himantopus (himantopus) 110
mexicanus, Quiscalus 282
mexicanus, Todus 170
michahellis, Larus 375
Microligea 312
micromegas, Nesoctites 172
migrans, Milvus 162
migratorius, Turdus 244
Milvus 162
MIMIDAE 232
Mimus 236
minima, Mellisuga 72
minimus, Catharus 240
minimus, Empidonax 206
minimus, Lymnocryptes 126
minor, Chordeiles 58
minor, Coccyzus 76
minor, Scolopax 375
minuta, Calidris 122
minutilla, Calidris 124
minutus, Anous 132
minutus, Corvus (palmarum) 220
minutus, Hydrocoloeus 134
minutus, Ixobrychus 96

mississippiensis, Ictinia 162
Mniotilta 286
Mockingbird
　Bahama 236
　Northern 236
　Tropical 236
modestus, Vireo 214
Molothrus 278
monachus, Myiopsitta 182
montana, Geotrygon 52
montana, Xenoligea 312
Montifringilla 376
Moorhen, Laughing 84
Morus 106
moschata, Cairina 36
mosquitus, Chrysolampis 66
Motacilla 256
motacilla, Parkesia 284
MOTACILLIDAE 254
mozambica, Crithagra 376
multicolor, Todus 170
Munia
　Black-headed 252
　Checkered 252
　Chestnut 252
　Scaly-breasted 252
　Tricoloured 252
　White-headed 252
MUSCICAPIDAE 246
musculus, Troglodytes (aedon) 230
musica, Euphonia 256
mustelina, Hylocichla 238
Myadestes 238
Mycteria 92
mycteria, Jabiru 92
Myiarchus 198
Myiopagis 194
Myiopsitta 182
Myna
　Common 376
　Common Hill 376
mystacea, Geotrygon 52

N
nana, Eupsittula 188
Nannopterum 108
nanus, Vireo 214
nasicus, Corvus 218
nebularia, Tringa 130
nenday, Aratinga 376
Neochen 36
Nesoctites 172
Nesophlox 72
Nesopsar 276
NESOSPINGIDAE 316
Nesospingus 316
niger, Chlidonias 144
niger, Cypseloides 62
niger, Pachyramphus 192
niger, Quiscalus 282
niger, Rynchops 132
nigerrimus, Nesopsar 276
Night-heron
　Black-crowned 96

Yellow-crowned 96
Nighthawk
  Antillean 58
  Common 58
  Lesser 58
Nightjar
  Cuban 62
  Hispaniolan 62
  Puerto Rican 60
  Rufous 60
  St Lucia 60
  White-tailed 58
nigra, Melopyrrha 326
nigrescens, Setophaga 308
nigricans, Branta (bernicla) 32
nigricephala, Spindalis 314
nigricollis, Podiceps 46
nigricollis, Sporophila 330
nigrogularis, Icterus 270
nilotica, Gelochelidon 142
nivalis, Montifringilla 376
nivalis, Plectrophenax 260
nivosus, Charadrius 114
noctis, Loxigilla 328
noctitherus, Antrostomus 60
Noddy
  Black 132
  Brown 132
  White-capped 132
Nomonyx 30
northropi, Icterus 272
noveboracensis, Coturnicops 80
noveboracensis, Parkesia 284
nudigenis, Turdus 242
nudipes, Megascops 156
nugator, Myiarchus 202
Numenius 116
Numida 26
NUMIDIDAE 26
Nuthatch, Bahama 228
Nyctanassa 96
NYCTIBIIDAE 56
Nyctibius 56
Nycticorax 96
nycticorax, Nycticorax 96
Nymphicus 375

O
oberi, Icterus 274
oberi, Myiarchus 200
obscurus, Pyrocephalus (rubinus) 202
occidentalis, Pelecanus 104
oceanicus, Oceanites 86
Oceanites 86
OCEANITIDAE 86
ochrocephala, Amazona 186, 376
ochropus, Tringa 128
ODONTOPHORIDAE 26
Oenanthe 246
oenanthe, Oenanthe 246
Oldsquaw 34
olivacea, Piranga 320
olivaceus, Tiaris 324
olivaceus, Vireo 210

olor, Cygnus 30, 375
Onychoprion 140
Oporornis 290
Orangequit 326
oratrix, Amazona 186, 376
Oriole
  Altamira 376
  Audubon's 376
  Bahama 272
  Baltimore 270
  Bullock's 272
  Cuban 274
  Eastern Hooded 272
  Hispaniolan 274
  Hooded 272
  Jamaican 270
  Martinique 274
  Montserrat 274
  Orchard 272
  Puerto Rican 274
  St Lucia 274
  Yellow 270
  Yellow-tailed 376
Ortalis 26
Orthorhyncus 70
oryzivora, Lonchura 252
oryzivorus, Dolichonyx 268
oryzivorus, Molothrus 278
osburni, Vireo 212
Osprey 158
Osprey, Western 158
ossifragus, Corvus 218
otiosus, Antrostomus (rufus) 60
otus, Asio 154
Ovenbird 284
Owl
  Ashy-faced 152
  Bare-legged 156
  Barn 152
  Burrowing 154
  Common Short-eared 156
  Jamaican 156
  Long-eared 154
  Northern Long-eared 154
  Northern Saw-whet 154
  Short-eared 156
  Stygian 154
Oxyura 30
Oystercatcher, American 108

P
Pachyramphus 192
pallens, Vireo 214
palliatus, Haematopus 108
pallida, Spizella 262
pallidus, Contopus 206
Palm-swift, Antillean 64
Palm-tanager
  Black-crowned 310
  Grey-crowned 310
palmarum, Corvus 220
palmarum, Phaenicophilus 310
palmarum, Setophaga 302
Palmchat 246

palustris, Cistothorus 232
palustris, Microligea 312
Pandion 158
PANDIONIDAE 158
paradisaea, Sterna 146
Parakeet
  Blue-crowned 376
  Brown-throated 188
  Canary-winged 182
  Cuban 190
  Hispaniolan 190
  Jamaican 188
  Monk 182
  Nanday 376
  Olive-throated 188
  Orange-fronted 188
  Red-masked 190
  Ring-necked 190
  Rose-ringed 190
  White-winged 182
parasiticus, Stercorarius 150
Pardirallus 82
Parkesia 284
Paroaria 376
Parrot
  Black-billed 184
  Cuban 184
  Grey 375
  Hispaniolan 184
  Imperial 186
  Orange-winged 186
  Puerto Rican 186
  Red-crowned 186
  Red-necked 186
  Rose-throated 184
  Senegal 375
  St Lucia 186
  St Vincent 188
  White-fronted 184
  Yellow-billed 184
Parrotlet, Green-rumped 188
Partridge-dove, Blue-headed 50
Parula, Northern 296
PARULIDAE 284
Passer 254
Passerculus 266
PASSERELLIDAE 260
PASSERIDAE 254
Passerina 318
passerina, Columbina 56
passerina, Spizella 260
passerinus, Forpus 188
Patagioenas 50, 375
Pauraque
  Jamaican 58
  Least 60
Pavo 28
Peafowl
  Common 28
  Indian 28
pectoralis, Colinus (virginianus) 26
pelagica, Chaetura 64
PELECANIDAE 104
Pelecanus 104

Pelican
   American White 104
   Brown 104
penelope, Mareca 42
pensylvanica, Setophaga 300
percussus, Xiphidiopicus 176
peregrina, Leiothlypis 288
Peregrine 182
peregrinus, Falco 182
personus, Turdus (fumigatus) 242
perspicillata, Melanitta 34
pertinax, Eupsittula 188
petechia, Setophaga 298
petersi, Coccyzus (longirostris) 78
Petrel
   Bermuda 375
   Black-capped 88
   Bulwer's 92
   Capped 88
   Jamaican 88
   Trindade 88
Petrochelidon 222
Pewee
   Cuban 206
   Hispaniolan 208
   Jamaican 206
   Lesser Antillean 208
   Puerto Rican 208
   St Lucia 208
PHAENICOPHILIDAE 310
Phaenicophilus 310
phaeopus, Numenius 116
Phaethon 48
PHAETHONTIDAE 48
Phaetusa 142
PHALACROCORACIDAE 108
Phalarope
   Grey 128
   Red 128
   Red-necked 128
   Wilson's 126
Phalaropus 128
pharetra, Setophaga 294
PHASIANIDAE 28
Phasianus 28
Pheasant
   Common 28
   Grey-rumped 28
   Ring-necked 28
Pheucticus 316
philadelphia, Geothlypis 290
philadelphia, Larus 134
philadelphicus, Vireo 208
Phoebe, Eastern 204
phoebe, Sayornis 204
phoeniceus, Agelaius 276
phoenicobia, Tachornis 64
PHOENICOPTERIDAE 46
Phoenicopterus 46
Phonipara 328
PICIDAE 172
Piculet, Antillean 172
Pigeon
   Bare-eyed 375

Feral 48
Jamaican Band-tailed 50
Plain 50
Red-necked 50
Ring-tailed 50
Rock 48
Scaly-naped 50
White-crowned 50
pileolata, Cardellina (pusilla) 310
Pintail
   Bahama 44
   Northern 44
   White-cheeked 44
pinus, Setophaga 304
pinus, Spinus 258
Pipilo 262
Pipit
   American 254
   Buff-bellied 254
   Sprague's 254
pipixcan, Larus 136
Piranga 320
Pitangus 194
pityophila, Setophaga 302
Platalea 94
platypterus, Buteo 166
platyrhynchos, Anas 42
Plectrophenax 260
Plegadis 94
PLOCEIDAE 248
Ploceus 248
Plover
   American Golden 110
   Black-bellied 110
   Collared 114
   Common Ringed 112
   Grey 110
   Little Ringed 112
   Pacific Golden 110
   Piping 114
   Semipalmated 112
   Snowy 114
   Thick-billed 112
   Wilson's 112
plumbea, Setophaga 292
plumbeus, Turdus 244
Pluvialis 110
pluvialis, Coccyzus 76
Pochard, Common 38
Podiceps 46
podiceps, Podilymbus 46
PODICIPEDIDAE 46
Podilymbus 46
Poicephalus 375
poliocephalus, Phaenicophilus 310
Polioptila 228
POLIOPTILIDAE 228
polyglottos, Mimus 236
polytmus, Trochilus 70
pomarinus, Stercorarius 150
Pooecetes 264
Poorwill
   Jamaican 58
   Least 60

Porphyrio 82
portoricensis, Icterus 274
portoricensis, Melanerpes 176
portoricensis, Melopyrrha 326
portoricensis, Spindalis 314
Porzana 82
porzana, Porzana 82
Potoo, Northern 56
pratincola, Glareola 132
Pratincole, Collared 132
pretrei, Spindalis (zena) 312
prevostii, Anthracothorax 66
principalis, Campephilus 174
Priotelus 168
PROCELLARIIDAE 88
Progne 226
Protonotaria 288
psaltria, Spinus (psaltria) 376
Pseudoscops 156
Psittacara 190, 376
PSITTACIDAE 182
Psittacula 190
Psittacus 375
Pterodroma 88, 375
Ptiloxena 278
Puffinus 92, 375
puffinus, Puffinus 92
pugnax, Calidris 120
puncticulata, Lonchura 252
purpurea, Ardea 100
pusilla, Calidris 124
pusilla, Cardellina 310
Pygmy-owl, Cuban 154
Pyrocephalus 202
pyrrhonota, Petrochelidon 222

Q

Quail-dove
   Blue-headed 50
   Bridled 52
   Crested 50
   Grey-fronted 52
   Grey-headed 52
   Hispaniolan 52
   Jamaican 50
   Key West 52
   Ruddy 52
   White-fronted 52
querquedula, Spatula 40
Quiscalus 280
quiscula, Quiscalus 280

R

radiolatus, Melanerpes 176
Rail
   Black 80
   Clapper 80
   King 80
   Southern Spotted 82
   Spotted 82
   Virginia 80
   Yellow 80
   Zapata 82
RALLIDAE 80

ralloides, Ardeola 98
Rallus 80
RAMPHASTIDAE 172
Ramphastos 172
Ramphocinclus 232
Recurvirostra 110
RECURVIROSTRIDAE 110
Redhead 38
Redpoll 258
Redpoll, Common 258
Redshank, Spotted 130
Redstart, American 294
Reef-egret, Western 104
Reef-heron, Western 104
Reeve 120
REGULIDAE 246
Regulus 246
religiosa, Gracula 376
richardsoni, Melanospiza 330
ricordii, Chlorostilbon 70
ridgwayi, Buteo 164
ridibundus, Larus 136
Riparia 222
riparia, Riparia 222
Rissa 134
Robin
    American 244
    Bare-eyed 242
roseigaster, Temnotrogon 168
roseogrisea, Streptopelia 48
rossii, Anser 32
rostrata, Geothlypis 292
Rostrhamus 164
ruber, Eudocimus 94
ruber, Phoenicopterus 46
rubescens, Anthus 254
rubinus, Pyrocephalus 202
rubra, Piranga 320
rubripes, Anas 44
rubripes, Turdus 244
rufescens, Egretta 102
Ruff 120
ruficapilla, Leiothlypis 288
ruficauda, Cinclocerthia 234
ruficauda, Ortalis 26
rufigularis, Coccyzus 76
rufigularis, Falco 182
rufum, Toxostoma 236
rufus, Antrostomus 60
rufus, Selasphorus 74
rustica, Hirundo 222
ruticilla, Setophaga 294
Rynchops 132

S
sabini, Xema 134
sagrae, Myiarchus 200
Saltator 322
Saltator, Lesser Antillean 322
salvini, Spindalis (zena) 314
sanctaeluciae, Ramphocinclus (brachyurus) 232
Sanderling 122

Sandpiper
    Baird's 122
    Buff-breasted 124
    Common 375
    Curlew 122
    Green 128
    Least 124
    Pectoral 124
    Semipalmated 124
    Solitary 128
    Spotted 128
    Stilt 120
    Terek 128
    Upland 116
    Western 124
    White-rumped 124
    Wood 130
sandvicensis, Thalasseus 148
sandwichensis, Passerculus 266
Sapsucker, Yellow-bellied 174
savana, Tyrannus 198
savannarum, Ammodramus 262
saxatalis, Aeronautes 375
Sayornis 204
Scaup
    Greater 40
    Lesser 40
scitulus, Trochilus 72
sclateri, Chaetura (cinereiventris) 64
sclateri, Euphonia 256
scolopaceus, Limnodromus 126
SCOLOPACIDAE 116
Scolopax 375
Scoter
    Surf 34
    White-winged 34
Screech-owl
    Bare-legged 156
    Cuban 156
    Puerto Rican 156
Seed-finch, Chestnut-bellied 332
Seedeater
    Lined 330
    Yellow-bellied 330
Seiurus 284
Selasphorus 74
semipalmata, Tringa 130
semipalmatus, Charadrius 112
semperi, Leucopeza 290
senegalus, Poicephalus 375
serrator, Mergus 36
serripennis, Stelgidopteryx 226
Setophaga 292
Sheartail, Inagua 72
Shearwater
    Antillean 92
    Audubon's 92
    Barolo 375
    Cape Verde 90
    Cory's 90
    Great 90
    Manx 92
    Scopoli's 90
    Sooty 90

Shelduck, Common 36
Shoveler, Northern 40
Shrike, Loggerhead 218
Sialia 236
sialis, Sialia 236
sibilatrix, Syrigma 100
Sicalis 332
siju, Glaucidium 154
Silverbill, Indian 250
simplex, Phaetusa 142
Siphonorhis 58
Siskin
    Antillean 260
    Pine 258
    Red 260
Sitta 228
SITTIDAE 228
Skimmer, Black 132
skua, Catharacta 150
Skua
    Arctic 150
    Great 150
    Long-tailed 148
    Pomarine 150
    South Polar 152
smithsonianus, Larus 138
Snipe
    Jack 126
    Wilson's 126
Snowfinch, White-winged 376
sociabilis, Rostrhamus 164
Solitaire
    Cuban 238
    Rufous-throated 238
solitaria, Tringa 128
solitarius, Vireo 212
Sora 82
sordidulus, Contopus 206
Sparrow
    Black-throated 376
    Chipping 260
    Clay-colored 262
    Cuban 262
    Eastern Song 266
    Grasshopper 262
    Henslow's 266
    House 254
    Java 252
    Lark 262
    Lincoln's 266
    Rufous-collared 264
    Savannah 266
    Song 266
    Swamp 266
    Vesper 264
    White-crowned 264
    White-throated 264
    Zapata 262
sparverius, Falco 180
Spatula 40
speculiferus, Nesospingus 316
Spermestes 250
Sphyrapicus 174
SPINDALIDAE 312

Spindalis 312
Spindalis
  Bahamas Black-backed 312
  Bahamas Green-backed 312
  Cuban 312
  Grand Cayman 314
  Hispaniolan 314
  Jamaican 314
  Puerto Rican 314
  Western 312
spinosa, Jacana 116
Spinus 258, 376
Spiza 318
Spizella 260
splendens, Corvus 220
sponsa, Aix 36
Spoonbill
  Eurasian 94
  Roseate 94
Sporophila 330
spragueii, Anthus 254
spurius, Icterus 272
squamosa, Patagioenas 50
squatarola, Pluvialis 110
Starling
  Common 232
  European 232
Starnoenas 50
Steganopus 126
Stelgidopteryx 226
stellaris, Cistothorus 230
STERCORARIIDAE 148
Stercorarius 148
Sterna 146
Sternula 142
Stilt
  Black-necked 110
  Black-winged 110
Stint, Little 122
stolidus, Anous 132
stolidus, Myiarchus 200
Stork
  American Wood 92
  White 92
  Wood 92
Storm-petrel
  Band-rumped 86
  Leach's 86
  Madeiran 86
  Wilson's 86
Streamertail 70, 72
Streamertail
  Black-billed 72
  Red-billed 70
strepera, Mareca 42
Streptopelia 48
Streptoprocne 62
striata, Butorides 98
striata, Setophaga 300
striatus, Accipiter 160
striatus, Melanerpes 176
STRIGIDAE 154
Sturnella 270
STURNIDAE 232

Sturnus 232
stygius, Asio 154
subis, Progne 226
subita, Setophaga 308
subruficollis, Calidris 124
subulatus, Todus 170
Sula 106
sula, Sula 106
SULIDAE 106
sulphuratus, Pitangus 194
superciliaris, Melanerpes 178
surinamensis, Chlidonias (niger) 144
swainsoni, Buteo 166
swainsoni, Catharus 238
swainsonii, Chlorostilbon 70
swainsonii, Limnothlypis 288
swalesi, Turdus 242
Swallow
  American Barn 222
  American Cliff 222
  Bahama 224
  Bank 222
  Barn 222
  Caribbean Cave 222
  Cave 222
  Cliff 222
  Golden 224
  Northern Rough-winged 226
  Tree 224
  Violet-green 224
  White-winged 224
Swan
  Mute 30, 375
  Tundra 30
  Whistling 30
Swift
  Alpine 66
  American Black 62
  Ashy-rumped 64
  Black 62
  Chimney 64
  Common 66
  Grey-rumped 64
  Lesser Antillean 64
  Short-tailed 64
  White-collared 62
  White-throated 375
Sylvia 376
Syrigma 100

T

Tachornis 64
Tachybaptus 46
Tachycineta 224
Tachymarptis 66
Tadorna 36
tadorna, Tadorna 36
Tanager
  Grenada 332
  Hispaniolan Highland 312
  Puerto Rican 316
  Scarlet 320
  St Vincent 332
  Summer 320

  Swallow 322
  Western 320
  Western Stripe-headed 312
Tangara 332
taylori, Melopyrrha 326
taylori, Tyrannus (caudifasciatus) 196
Teal
  Blue-winged 42
  Cinnamon 40
  Common 44
  Eurasian 44
  Green-winged 46
Temnotrogon 168
temnurus, Priotelus 168
TERETISTRIDAE 268
Teretistris 268
Tern
  American Black 144
  Arctic 146
  Black 144
  Bridled 140
  Cabot's 148
  Caspian 142
  Common 146
  Common Gull-billed 142
  Common White 132
  Eurasian Black 144
  Forster's 146
  Gull-billed 142
  Large-billed 142
  Least 142
  Roseate 146
  Royal 148
  Sandwich 148
  Sooty 140
  Whiskered 144
  White-winged 144
  White-winged Black 144
Tersina 322
tertius, Calyptophilus 316
Thalassarche 86
Thalasseus 148
thalassina, Tachycineta 224
Thick-knee, Double-striped 108
Thrasher
  Brown 236
  Pearly-eyed 234
  Scaly-breasted 234
  St Lucia 232
  White-breasted 232
THRAUPIDAE 322
THRESKIORNITHIDAE 94
Thrush
  Bicknell's 240
  Cocoa 242
  Eastern Hermit 240
  Eastern Red-legged 244
  Forest 242
  Grey-cheeked 240
  Hermit 240
  La Selle 242
  Lesser Antillean 242
  Northern Red-legged 244
  Olive-backed 238

Russet-backed 376
Spectacled 242
Swainson's 238
Western Red-legged 244
White-chinned 244
White-eyed 242
Wood 238
thula, Egretta 102
Tiaris 324
tigrina, Setophaga 296
tinnunculus, Falco 180
TITYRIDAE 192
TODIDAE 170
Todus 170
todus, Todus 170
Tody
   Broad-billed 170
   Cuban 170
   Jamaican 170
   Narrow-billed 170
   Puerto Rican 170
torquata, Megaceryle 172
torquatus, Phasianus (colchicus) 28
Torreornis 262
Toucan, Channel-billed 172
Towhee, Green-tailed 262
townsendi, Setophaga 308
townsendi, Spindalis (zena) 312
Toxostoma 236
traillii, Empidonax 204
Trembler
   Brown 234
   Grey 234
trichas, Geothlypis 292
tricolor, Egretta 102
tricolor, Steganopus 126
tridactyla, Rissa 134
Tringa 128
tristis, Acridotheres 376
tristis, Spinus 258
TROCHILIDAE 66
Trochilus 70
Troglodytes 230
troglodytes, Estrilda 250
TROGLODYTIDAE 228
Trogon
   Cuban 168
   Hispaniolan 168
TROGONIDAE 168
Tropicbird
   Red-billed 48
   White-tailed 48
Troupial 272
Troupial, Venezuelan 272
TURDIDAE 236
Turdus 242
Turnstone, Ruddy 120
TYRANNIDAE 192
tyrannulus, Myiarchus 202
Tyrannus 194, 376
tyrannus, Tyrannus 196
Tyto 152
TYTONIDAE 152

U
uncinatus, Chondrohierax 158
undulatus, Melopsittacus 376
Uraeginthus 376
urbicum, Delichon 220
ustulatus, Catharus 376

V
validus, Myiarchus 200
valisineria, Aythya 38
Vanellus 114
vanellus, Vanellus 114
varia, Mniotilta 286
varius, Sphyrapicus 174
Veery 240
ventralis, Amazona 184
Vermivora 286
vermivorum, Helmitheros 284
versicolor, Amazona 186
versicolor, Geotrygon 50
versicolor, Tangara 332
versicolurus, Brotogeris 182
verticalis, Tyrannus 194
vetula, Coccyzus 78
Vidua 254
viduata, Dendrocygna 28
VIDUIDAE 254
vieilloti, Coccyzus 78
villosus, Leuconotopicus 178
violacea, Melopyrrha 326
violacea, Nyctanassa 96
virens, Contopus 206
virens, Icteria 268
virens, Setophaga 308
Vireo 208
Vireo
   Bell's 214
   Black-whiskered 212
   Blue Mountain 212
   Blue-headed 212
   Cuban 216
   Eastern Warbling 210
   Flat-billed 214
   Jamaican 214
   Mangrove 214
   Philadelphia 208
   Providence 214
   Puerto Rican 214
   Red-eyed 210
   San Andres 216
   Thick-billed 216
   Warbling 210
   White-eyed 216
   Yellow-green 210
   Yellow-throated 212
   Yucatan 210
VIREONIDAE 208
virescens, Butorides (striata) 98
virescens, Empidonax 204
virginiae, Leiothlypis 290
virginianus, Colinus 26
virginianus, Colinus 26
viridigenalis, Amazona 186
viridis, Anthracothorax 68

viridis, Tersina 322
vitellina, Setophaga 306
vitellinus, Ramphastos 172
vittata, Amazona 186
vociferans, Tyrannus 194
vociferus, Antrostomus 60
vociferus, Charadrius 112
Volatinia 330
vulgaris, Sturnus 232
Vulture
   American Black 158
   Black 158
   Turkey 158

W
Wagtail
   White 256
   White-faced 256
Warbler
   American Yellow 298
   Adelaide's 306
   Arrowhead 294
   Audubon's 304
   Bachman's 286
   Bahama 306
   Barbuda 308
   Bay-breasted 298
   Black-and-white 286
   Black-throated Blue 302
   Black-throated Green 308
   Black-throated Grey 308
   Blackburnian 298
   Blackpoll 300
   Blue-winged 286
   Brewster's 286
   Canada 310
   Cape May 296
   Cerulean 296
   Chestnut-sided 300
   Connecticut 290
   Eastern Palm 302
   Eastern Wilson's 310
   Elfin Woods 294
   Golden 300
   Golden-cheeked 308
   Golden-winged 286
   Green-tailed 312
   Hooded 294
   Kentucky 290
   Kirtland's 294
   Lawrence's 286
   Magnolia 296
   Mourning 290
   Myrtle 304
   Nashville 288
   Northern Yellow 298
   Olive-capped 302
   Orange-crowned 288
   Oriente 268
   Palm 302
   Pine 304
   Plumbeous 292
   Prairie 306
   Prothonotary 288

Semper's 290
St Lucia 308
Swainson's 288
Tennessee 288
Townsend's 308
Virginia's 290
Vitelline 306
Western Palm 302
Western Wilson's 310
Whistling 292
White-winged 312
Wilson's 310
Worm-eating 284
Yellow-headed 268
Yellow-rumped 304
Yellow-throated 306
Warbler-tanager, White-winged 312
Waterthrush
    Louisiana 284
    Northern 284
Waxbill
    Black-rumped 250
    Common 250
    Orange-cheeked 248
Waxwing, Cedar 246
Weaver, Village 248
wellsi, *Leptotila* 54
Wheatear, Northern 246
Whimbrel 116
Whimbrel
    American 116
    Eurasian 116
Whip-poor-will
    Eastern 60
    Puerto Rican 60

Whistling-duck
    Black-bellied 28
    Fulvous 30
    Northern Black-bellied 28
    Southern Black-bellied 28
    West Indian 30
    White-faced 28
Whydah, Pin-tailed 254
Wigeon
    American 42
    Eurasian 42
Willet 130
wilsonia, *Charadrius* 112
wilsonii, *Chondrohierax* 160
Wood-pewee
    Eastern 206
    Western 206
Woodcock, American 375
Woodpecker
    Bahamas Hairy 178
    Cayman 178
    Cuban Green 176
    Cuban Ivory-billed 174
    Guadeloupe 176
    Hairy 178
    Hispaniolan 176
    Ivory-billed 174
    Jamaican 176
    Puerto Rican 176
    Red-bellied 178
    West Indian 178
Woodstar
    Bahama 72
    Inagua 72
    Lyre-shaped 72

Wren
    Antillean House 230
    Eastern Marsh 232
    House 230
    Marsh 232
    Northern House 230
    Sedge 230
    Southern House 230
    Zapata 228

X

Xanthocephalus 268
xanthocephalus, *Xanthocephalus* 268
xanthomus, *Agelaius* 276
Xema 134
Xenoligea 312
Xenus 128
Xiphidiopicus 176

Y

Yellow-finch, Grassland 332
Yellowlegs
    Greater 130
    Lesser 130
Yellowthroat
    Bahama 292
    Common 292
    Eastern 292

Z

zena, *Spindalis* 312
Zenaida 54
zonaris, *Streptoprocne* 62
Zonotrichia 264

QUICK INDEX

Albatrosses 86–88
Amazons 184–188
Anhingas 108
Anis 74
Auks 152
Avadavats 250
Avocets 110
Bananaquit 324
Barn-owls 152
Becards 192
Bee-eaters 168
Bishops 248
Bitterns 96
Blackbirds 268–284
Bluebirds 236
Bobolink 268
Bobwhites 26
Boobies 106
Bullfinches 326–328
Buntings 260, 318–320
Caracaras 180
Cardinals 322
Caribs 68
Chachalacas 26
Chat-tanagers 316
Chuck-will's-widow 60
Cockatoos 182
Collared-doves 48
Conures 188–190
Coots 84
Cormorants 108
Cowbirds 278
Crakes 82
Cranes 84
Crossbills 258
Crows 218–220
Cuckoos 74–78
Curlews 116–118
Dickcissel 318
Dowitchers 126
Ducks and Geese 28–46
Eagles 162
Egrets 98–104
Elaenias 192–194
Emeralds 70
Euphonias 256
Falcons 182
Finches 258
Flamingos 46
Flickers 174
Flycatchers 204–206
Frigatebirds 104
Gallinules 82–84
Gnatcatchers 228
Godwits 118
Goldfinches 258
Grackles 280–282
Grassquits 324, 328–330
Grebes 46

Grosbeaks 316–318
Guineafowl 26
Gulls 134–140
Harriers 160
Hawks 160–168
Hermits 66
Herons 98–102
Honeycreepers 322
Hummingbirds 66–74
Ibises 94
Jabiru 92
Jacanas 116
Jacobins 66
Jaegers 148–150
Juncos 264
Kestrels 180
Kingbirds 194–198
Kingfishers 172
Kinglets 246
Kiskadees 194
Kites 158–164
Kittiwakes 134
Lapwings 114
Larks 220
Limpkin 84
Lizard-cuckoos 76–78
Loons 84
Macaws 190
Mangos 66–68
Mannikins 250–252
Martins 220–222, 226
Meadowlarks 270
Mergansers 34–36
Mockingbirds 236
Munias 252
Night-herons 96
Nighthawks 58
Nightjars 58–62
Noddies 132
Nuthatches 228
Orangequit 326
Orioles 270–274
Ospreys 158
Ovenbird 284
Owls 152–156
Oystercatchers 108
Palm-swifts 64
Palm-tanagers 310
Palmchat 246
Parakeets 182, 188–190
Parrots 184–188
Parrotlets 188
Peafowl 28
Pelicans 104
Petrels 88, 92
Pewees 206–208
Phalaropes 126–128
Pheasants 28
Phoebes 204

Piculets 172
Pigeons and Doves 48–56
Pintails 44
Pipits 254
Plovers 110–114
Poorwills 58–60
Potoos 56
Pratincoles 132
Pygmy-owls 154
Quail-doves 50–52
Rails 80–82
Saltators 322
Sandpipers 116–130
Scaups 40
Scoters 34
Screech-owls 156
Shearwaters 90–92
Shrikes 218
Siskins 258–260
Skimmers 132
Skuas 148–152
Snipes 126
Solitaires 238
Sparrows 252–254, 260–266
Spindalises 312–314
Spoonbills 94
Stilts 110
Storks 92
Storm-petrels 86
Streamertails 70–72
Swallows 222–226
Swans 30
Swifts 62–66
Tanagers 316, 320–322, 332
Teal 40–44
Terns 132, 140–148
Thick-knees 108
Thrashers 232–236
Thrushes 238–244
Todies 170
Toucans 172
Tremblers 234
Trogons 168
Tropicbirds 48
Vireos 208–216
Vultures 158
Warblers 284–310
Waterthrushes 284
Waxbills 248–250
Waxwings 246
Whimbrels 116
Whistling-ducks 28–30
Wigeon 42
Woodpeckers 174–178
Woodstars 72
Wrens 228–232
Yellow-finches 332
Yellowlegs 130
Yellowthroats 292

400

USA

BAHAMA ISLANDS

1
2
3
4

7
6
8
Cuba
9
10

Mexico

GREAT

11

Jamaica
13 12

Belize

Honduras

Nicaragua

29

Costa Rica

Panama

Colombia